MW00444368

THE
BIRDS
OF
NEVADA

Mountain Bluebird, *Sialia currucoides*, Nevada's State Bird. Photo by Albert A. Alcorn

THE
BIRDS
OF
NEVADA

by

J. R. Alcorn

Fairview West Publishing
Fallon, Nevada 1988

FAIRVIEW WEST PUBLISHING
510 South Maine St., Fallon Nevada 89406

J. R. Alcorn
Copyright © 1988
All Rights Reserved

Printed by Braun-Brumfield, Inc.
Ann Arbor, Michigan, U.S.A.

ISBN: 0-9620221-0-1
Library of Congress No.: 88—080743

Contents

Contents

Contents

Contents

Contents

PREFACE

For Ray Alcorn, THE BIRDS OF NEVADA is the culmination of his life's work. He has spent nearly 60 of his 76 years gathering information on Nevada birds. His abilities in the field are unsurpassed and his knowledge of Nevada ornithology is extensive. He has traveled throughout the state and his notes, numbering over 10,000 pages, are the basis for this book.

While his personal notes alone warrant a book, Ray wanted to contribute more to this project. So, over the years, he spoke with bird enthusiasts and encouraged them to keep records. He interviewed biologists at Nevada refuges and recorded their comments. He researched thousands of pages of published material, going back to the 1800's when ornithological reports concerning Nevada were a part of expeditions exploring the new western territory. He interviewed "old-timers," pioneers to the state and natives such as famous Paiute Indian woman Wuzzie George. Finally, he began writing the manuscript.

The result of his years of effort is THE BIRDS OF NEVADA, not just an account of Nevada birds, but a history of Nevada birds.

As I have worked with Ray on this book over the past eight years, I've gained an understanding of THE BIRDS OF NEVADA and of the man who wrote it. His work is a testimony of patience and persistence, described best as a labor of love—a love for science, for history and for Nevada.

Over the years, many things have been said about Ray Alcorn's remarkable contributions to Nevada ornithology. THE BIRDS OF NEVADA, however, speaks loudest of all.

Colleen Ames

ACKNOWLEDGEMENTS

The cooperation and help I have received from so many has made this book possible. To those who have been in the field or behind the typewriter and to all who offered advice and encouragement along the way, I am grateful.

Special thanks go to those who assisted me in tabulating censuses, searching literature, editing and typing the manuscript. They include: Doris Guire Alcorn, Colleen Ames, Trina Ames, Mary Baldwin, Charlie Bliss, Theresa Carey, Yvette Carey, Illa Cress, Kelli Davenport, Frank Gonzales, Ron Healy, Joyce McCarthy, Kathy Napier, Marcia Parrish and Valerie Pacheco.

Thanks go to the following people who contributed information on Nevada birds: Jim Curran, Gary Herron, Larry Neal, Cathy Osugi, Jerry Page, Bob Quiroz, Norm Saake, Lois Saxton, San Juan Stiver and Steve Thompson.

I am especially indebted to Art Biale who made his bird records of over 25 years for the Eureka area available to me; to Larry LaRochelle for information on the birds at Ruby Lake National Wildlife Refuge; to Larry Napier and Pete Schwabenland for information on the birds at Stillwater Wildlife Management Area and Lahontan Valley; to the late Chuck Lawson for information on the birds of southern Nevada; to Kathy Napier who assisted in a search and tabulation of the literature and examined records at the United States National Museum in Washington, D.C.; to Nick Cady, editor of the UNR Press, for his editorial guidance and encouragement; and to Vernon Mills, a competent bird observer and friend of over 50 years, who furnished many records for the Fallon area and spent countless hours in the field with me.

I am very thankful for the support of my family in this project. My late wife, Doris Guire Alcorn, typed the first draft of the manuscript in 1970. She and I were happily married for over 40 years and she was my constant companion and helper in numerous biological projects before her death from cancer in 1976.

My son, Albert A. Alcorn, worked for the Fish and Game Department and his records provide valuable information for many areas in Nevada. His photography skills, both in the field and in the darkroom, have been a great deal of help. His editorial assistance and support throughout this project have meant a great deal.

Last but not least, I am grateful for the assistance of Miss Colleen Ames over the past eight years. She helped prepare and edit the manuscript and she designed the layout of the book. Her efficient research of the literature, her computer skills and her other talents lightened the work in many ways.

INTRODUCTION

The purpose of this book is to give the reader 1) a list of all reported birds in Nevada, 2) their common and scientific names, 3) their worldwide distribution, 4) the status of each species in Nevada, 5) pertinent records of occurrence, and 6) other information such as food habits and nesting.

1. LIST OF THE BIRDS

THE BIRDS OF NEVADA lists 456 species of birds: resident, migratory, introduced, hypothetical and extinct. Years ago, the acceptable way to add a bird to a state checklist was to shoot it, prepare a study skin and have it identified at a university where other study skins were available for comparison. Now, many birds are added to checklists by photographs or by accurate descriptions of competent observers. Some species in this book would not meet the requirements of a bird records committee for inclusion in a state list. However, I have chosen to add these species because they provide important information on Nevada birds.

2. COMMON AND SCIENTIFIC NAMES

The order of presentation of species is based upon checklist order given by the **American Ornithologist's Union Checklist of North American Birds** (6th ed., 1983). Previous names are listed if changes have occurred.

3. WORLDWIDE DISTRIBUTION

The worldwide distribution of each bird is given briefly using information summarized from the **American Ornithologist's Union Checklist of North American Birds** (6th ed., 1983).

4. STATUS IN NEVADA

The status of each bird in THE BIRDS OF NEVADA is based upon published records, the author's notes of over fifty years, refuge checklists and information from observers throughout the state.

5. RECORDS OF OCCURRENCE

The information contained in the "Records of Occurrence" section has been gathered from many sources, including published and unpublished records.

The records for each species are reported in a west-to-east, north-to-south order, beginning in Washoe County.

The exception to this is Linsdale's 1936 report, which appears at the beginning of the "Records of Occurrence" section. His work, BIRDS OF NEVADA, was a comprehensive study on birds in the state and gives species records and status prior to 1936.

Using this information, together with the other records presented, it is possible to determine the history and trends of a species in Nevada.

6. NESTING, FOOD HABITS AND OTHER INFORMATION

These sections primarily contain the author's observations and comments. Other people's observations and some published material are also included.

THE
BIRDS
OF
NEVADA

Order **TINAMIFORMES**

Family **TINAMIDAE:** Tinamous

Crested Tinamou   *Eudromia elegans*

Distribution. Occurs in Argentina and Chile.

Status in Nevada. Introduced in the late 60's and early 70's in several localities in southern Nevada. There are no recent records.

Records of Occurrence. According to the Nevada Department of Wildlife (MS, 1974) these birds were released in southern Nevada as follows: "149 were released six miles southwest of Searchlight, Clark County in 1969; 10 were released in the McCullough Mountains, Clark County in 1970; 50 were released at Christmas Tree Pass on March 2, 1971; 35 were released in the East Mormon Mountains, Lincoln County on November 20, 1972."

"No sightings have been received on this species since the 1972 release. It is presently assumed that those releases resulted in failure to establish this bird. The program has been discontinued."

Order **GAVIIFORMES**

Family **GAVIIDAE:** Loons

Red–throated Loon   *Gavia stellata*

Distribution. From Aleutian Islands eastward to Newfoundland; south to Tuburon Island, Mexico.

Status in Nevada. Rare.

Records of Occurrence. Hugh Kingery (1978) reported Red–throated Loons at Walker Lake October 19.

Kingery (1984) reported, "Nevada reported its fifth Red–throated Loon, observed for 10 minutes October 29 at Pahranagat NWR."

Allan D. Cruickshank (1970) reported two of these loons on Lake Mead December 21, 1969. He (1972) reported that four seen December 18, 1971 had been observed regularly since October 1 of that year.

Charles Lawson (TT, 1976) reported, "I have seen one of these birds within 15 feet at Lake Mead."

Kingery (1982) reported Red–throated Loons at Las Vegas November 11–30, 1981 and he (1983) reported, "Las Vegas' Red–throated Loon stayed November 11–December 4."

Arctic Loon   *Gavia arctica*

Distribution. In the Arctic of both hemispheres. In North America south to northern Mexico.

Status in Nevada. Uncommon in winter months. A few winter on the larger lakes in southern Nevada.

Records of Occurrence. V. T. Wilson and R. H. Norr (1951) reported B. H. Cater found a freshly killed bird near the Ruby Lake National Wildlife Refuge (NWR), Elko County, October 21, 1950. Kingery (1984) reported one Arctic Loon at this same location October 24, 1983.

Kingery (1978) reported an Arctic Loon in breeding plumage briefly visited Pyramid Lake July 7. He (1980) reported one at the same location July 9.

D. B. Marshall (1951) picked up the remains of an immature Arctic Loon on the shore of Soda Lake, Churchill County on April 9, 1950.

Kingery (1978) reported one at Topaz Lake October 15–16, 1977 and another January 1, 1978.

Kingery (1987) reported an Arctic Loon at Logandale on April 11 was the first spring record for the year for southern Nevada.

Lawson (TT, 1976) commented, "The Arctic Loon is an uncommon wintering bird in southern Nevada." He (1977) reported one at Tule Springs Park, Clark County, October 19, 1976.

Kingery (1977) reported, "As with last year, Nevada had five Arctic Loon records, three at Lake Mohave, December 2; and two there February 14."

Kingery (1982) reported six at Las Vegas in October and November, 1981. He (1984) reported, "This winter's only Arctic Loon dove and fed January 21 in the most southern point in the Region [Mountain West]–non arctic Davis Dam, Nevada." He (1985) reported, "Las Vegas had more Arctic Loons (3–4) in December than Common Loons . . . at Davis Dam, 150 miles to the south."

Common Loon   *Gavia immer*

Distribution. From the Aleutian Islands eastward to Newfoundland; south to Mexico.

Status in Nevada. Regular but not a numerous migrant throughout the state. In some years a few may winter on Walker Lake. It is a common wintering species in southern Nevada. Statewide, the Common Loon is usually found on the larger lakes, but has been seen on smaller ponds and rivers and one was observed in an irrigation canal near Fallon.

Records of Occurrence. Sheldon NWR (1982) reports the Common Loon as a transient visitor.

At Wildhorse Reservoir one was seen May 14, 1972. Other solitary birds have been seen at this same reservoir in the spring of most years (Alcorn MS). At Mason Creek near Elko one was seen May 15, 1972 (Alcorn MS).

Larry LaRochelle (MS, 1967) reported banding one Common Loon at Ruby Lake NWR June 2, 1962. Loons are listed on the refuge checklist as an accidental visitor in spring or fall.

Ruby Lake personnel reported an unusual sighting of a Common Loon May 13, 1970 and another in winter plumage May 11 and 16, 1972.

Kingery (1981) reported loons stayed at Ruby Lakes into December of 1980. He (1984) reported one at Wells on June 16 and (1987) one at Jiggs July 6.

The author has seen loons in other parts of Nevada including Ryepatch Reservoir, Lahontan Reservoir, the Reno area, Little Soda Lake, Indian Lakes, Pyramid Lake, Walker Lake, Pahranagat Lakes and Lake Mead. All were autumn or spring records and, except for Walker Lake and Lake Mead, normally involved one or two birds.

Robert T. Orr and James Moffitt (1971) reported at Lake Tahoe the Common Loon is, "A fairly common migrant, apparently neither wintering nor breeding."

Henshaw (1877) reported individuals on Washoe Lake in November, 1876.

Common Loons were recorded eight times in the Reno area during November and May of 1941–42 (Alcorn MS).

Fred A. Ryser regularly observed solitary birds on Virginia Lake, the Truckee River and other areas in the Reno vicinity. He (1965) reported a solitary Common Loon on numerous occasions on Virginia Lake, the Truckee River and a pond in northwest Reno in January, February and March of 1965. Ryser (1969) also reported a loon near Reno December 21.

Cruickshank (1971) reported one at Mason Valley December 27, 1970.

Doris Alcorn (MS) observed one at the east end of Lahontan Reservoir May 31, 1975.

Chester Markeley and LeRoy Giles (MS, 1955) observed and heard calling a Common Loon on Papoose Lake, northeast of Fallon April 22, 1955. They were able to approach close enough to see the neck band distinctly.

Peter Schwabenland (MS, 1966) reported at Stillwater NWR in three years time (1963–66) he saw only one Common Loon. Larry Napier (MS, 1974) reported that at this same refuge from 1966 to 1974, he noted a few in spring and fall, but not every year. He considers them rare.

Vernon Mills (PC) and Deanna Lemburg (PC) report solitary Common Loons at Little Soda Lake, west–northwest of Fallon. Loons visit this deep water lake yearly, possibly attracted to the Lahontan Redshiner fish (*Richarsonius*) that are abundant there.

As this author traveled the highway along the west side of Walker Lake September 29, 1965, about 100 loons were seen scattered over a wide area (Alcorn MS).

Norm Saake (MS) frequently saw these birds at Walker Lake in April, May and November of 1970. He saw about 40 on April 9, and over 50 on November 17, 1970. The most loons Saake saw at Walker Lake in a day during the 1972–73 winter was 25, mostly solitary birds.

Jim Curran (PC) saw two or three Common Loons each month at Walker Lake from October, 1972 to March, 1973. He speculated that some wintered on this lake. On April 20, 1975 he reported seeing one loon make 12 dives, each lasting 52 to 57 seconds in length.

Jean M. Linsdale (1936) reported Dr. E. Raymond Hall saw dozens of dead loons along the shore of Walker Lake May 13, 1932. Hall was unable to determine why these birds had died.

Nevada Fish and Game personnel shot about 40 loons at Walker Lake in the fall of 1964 to protect newly planted trout. On April 20, 1975, the Nevada Department of Fish and Game and the U.S. Fish and Wildlife Service were planting trout in this lake. The State planted seven to 12 inch trout and the federal government planted five inch trout, many of which were sick or exhausted from the long trip from Idaho. Curran (PC) reported loons were there in force, feeding on the trout. One observer estimated 400 loons present; Curran reported 200–300.

Art Biale (MS) reported the Common Loon in Eureka as rare. However, he did see one at Illapah Reservoir October 11, 1970. In 1984, Biale (PC) reported no recent records.

Kingery (1985) reported a Common Loon, "Found itself in trouble and aided by man: one at Eureka, Nevada October 20 was brought to a pond unable to fly, but left 24 hours later." He (1987) reported a Common Loon at Eureka June 18.

Albert A. Alcorn (MS) reported seeing a solitary Common Loon on a small pond near Baker May 12, 1967. He reported others at Pahranagat Valley on May 14, 1965 and Lake Mead in March of 1964.

Patricia R. Snider (1969) reported at least one loon wintered on Lake Mead (from January 25), and two seen March 20, 1969. She (1971) reported a single Common Loon near Las Vegas November 8, 1970.

Gale Monson (1972) reported a loon seen at Tule Springs Park near Las Vegas in May of 1972.

Lawson (TT, 1976) reported, "In southern Nevada this is a very common wintering species. We normally see it beginning in early October and it lasts right through until at least mid–April."

Comment. The occurrence of the Common Loon in most parts of Nevada is very irregular. At Walker Lake there may be from one to several hundred individuals one week and none the next.

Most loons have been seen during fall and spring migrations and have been observed most often during April and May. Sometimes the planting of trout into lakes has coincided with an abundance of loons. This creates a problem because these birds easily catch and eat the unwary fish. Efforts to shoot the loons have resulted in many birds being killed and may explain the reason for the dead loons found by Hall at Walker Lake in 1932.

Ray Corlett (MS, 1964) reported about 40 loons were shot at Walker Lake by Nevada Fish and Game personnel in 1964. These loons were shot to portect newly planted trout. Until recent years, it was common practice for sportsmen and wildlife personnel to shoot fish-eating birds in an effort to protect newly planted fish. Other measures are now taken to alleviate this problem.

Saake (MS, 1984) commented, "We have changed the fish planting dates so we do not plant during migration. Also we now use planting barges."

These two changes seem to have mitigated this problem.

Yellow–billed Loon   *Gavia adamsii*

Distribution. From arctic North America and eastern Siberia. Rarely to Vancouver Island, Minnesota and Long Island.

Status in Nevada. Accidental.

Records of Occurrence. J. R. Remsen, Jr., and Laurence C. Binford (1975) reported one Yellow–billed Loon near Incline Village, Lake Tahoe on January 6, 1973. This is the only Nevada record.

Order **PODICIPEDIFORMES**

Family **PODICIPEDIDAE:** Grebes

Pied–billed Grebe *Podilymbus podiceps*

Distribution. In most of North America from British Columbia to New Brunswick; south through Central and South America to Argentina.

Status in Nevada. Occurs statewide, common but not abundant.

Records of Occurrence. Sheldon NWR (1982) reported the Pied–billed Grebe as a nester.

Linsdale (1936) reported Hall (MS) found broods of three to eight young Pied–billed Grebes about two-thirds grown near Halleck, Elko County, from July 31 to August 8, 1925. "They were seen along the East Fork of the Humboldt River and in sloughs connected with the river."

I. N. Gabrielson (1949) reported, "Two found at Ruby Lake NWR, Elko County, June 1, 1948."

LaRochelle (MS, 1967) reported at the Ruby Lake NWR, "A common though not abundant bird to be found as long as open water remains. Several stayed throughout the winter of 1962–63. No known nesting but they must surely nest here. I have yet to see flightless young."

Kingery (1984) reported, "Ruby Lake NWR attracted 400–600 Pied–billed Grebes in May." He later (1984) reported, "800 nesting Pied–billed Grebes."

In the Reno area from September through March, 1941–42, one or two birds were seen on 41 different days (Alcorn MS).

Ryser (1985) reports this bird appears to be as abundant in the winter in the Reno area as it is in summer, "If not more so."

Alcorn (1946) reported Mills photographed a nest containing young in the Fallon area on young July 14, 1942.

Napier (MS, 1974) saw an adult Pied–billed Grebe with four small young in the Stillwater area August 14, 1967. He estimated 40 young were produced at Stillwater Wildlife Management Area (WMA) in 1972 and 180 produced in 1973. The increase in population was attributed to more accurate censusing methods. Young or nests were seen by Napier as follows: August 14, 1967 (4 young with an adult female); June 20, 1968 (one nest with 4 eggs); July 25, 1969 (one adult on a nest); July 25, 1969 (4 young); and May 5, 1970 (2 young).

Kingery (1986) reported, "Pied–billed Grebes nested early at Stillwater: biologists discovered a complete clutch of eggs April 25."

Saake (MS) reported Pied–billed Grebes nesting in Mason Valley.

Biale (MS) recorded this bird in the Eureka area 30 times in a 10–year period (1964–74). It was seen all months except January, June, July and August.

In Pahranagat Valley Pied–billed Grebes were recorded all seasons but were most numerous in spring and summer. An adult with six young was seen near Hiko July 11, 1963 (Alcorn MS).

The Audubon Christmas Bird Count (CBC) repeatedly reveals Pied–billed Grebes in southern Nevada.

Lawson (TT, 1976) commented, "This grebe is a permanent resident here in southern Nevada. It not only breeds at Lake Mead, but also at places where you can find water such as

Tule Springs, Corn Creek, Pahranagat Lakes, and so forth.'' Lawson also reported Pied–billed Grebes sometimes produce two broods a year. "It is not uncommon to see a very young Pied–billed Grebe in early September."

Horned Grebe   *Podiceps auritus*

Distribution. From Alaska eastward to Nova Scotia; south to southern California and Florida.

Status in Nevada. An uncommon migrant in spring and fall with small numbers wintering on the larger lakes in southern Nevada.

Records of Occurrence. Sheldon NWR (1982) reported the Horned Grebe as a migrant during spring and fall.

Robert Ridgway (1877) reported the Horned Grebe as quite numerous in August and September at Franklin Lake, Elko County.

Linsdale (1951) reported, "Willet (MS) reported Horned Grebes at Anaho Island, Pyramid Lake in early November, 1914. Richardson (MS) saw two at Pyramid Lake April 19, 1942."

Virgil K. Johnson obtained a specimen at Pyramid Lake January 22, 1961.

Orr and Moffit (1971) reported a careful watch for Horned Grebes at Lake Tahoe was unrewarded in the fall of 1924, 1925 and 1926. They collected two, the only ones encountered, on the California side of the lake in March, 1927.

Cruickshank (1971) reported one seen in Mason Valley December 27, 1970 during the CBC.

Mills (PC) saw one at the southwest end of Lahontan Reservoir April 13, 1975.

Ryser (1973) reported, "An unusual, persistent presence of Horned Grebes in the Reno and Stillwater areas in the early spring of 1967."

Napier (MS, 1974) reported seeing Horned Grebes at Stillwater WMA March 17, 1967 (2 seen) and May 6, 1967 (5 seen). These were the only ones reported by Stillwater biologists from 1949 to 1967.

Horned Grebes are unreported by Alcorn in Lahontan Valley and by Biale in the Eureka area.

A. A. Alcorn (MS) reported two Horned Grebes in the Hiko area April 14, 1965 and three October 15, 1965.

Lawson (TT, 1976) reported, "A rare to uncommon wintering bird in southern Nevada." He (1977) saw one in breeding plumage at Las Vegas Wash May 10, 1969.

Kingery (1979) reports one seen each year on the Henderson CBC.

Red–necked Grebe   *Podiceps grisegena*

Distribution. From northwest Alaska to northeastern Siberia; south to southern California and Japan.

Status in Nevada. A rare transient.

Records of Occurrence. Howard (1939) reported a prehistoric record for the Red–necked Grebe (two bones) from Lovelock Cave, Humboldt Mountains.

Kingery (1984) reported, "A rare Red–necked Grebe stopped at Reno, Nevada, April 8."

Orr and Moffitt (1971) listed this bird as a rare fall migrant for Lake Tahoe. They reported J. E. Pomin noted only a few in the fall at Lake Tahoe over a 40–year period.

Cruickshank (1971) reported one Red–necked Grebe during the Mason Valley CBC December 27, 1970.

Lawson (TT, 1976) reported, "Some people from the National Park Service saw one on Lake Mead December 13, 1973."

Lawson (MS) also reported C. Richards photographed one in winter plumage at Lake Mead Marina on December 7, 1980. It was still present January 3, 1981.

Kingery (1982) reported one at Las Vegas from October 26 to November 4, 1981. He (1984) reported, "Red–necked Grebes visited Las Vegas December 9–11 and Davis Dam January 28."

Eared Grebe   *Podiceps nigricollis*

Distribution. Found in Eurasia, Africa and western North America.

Status in Nevada. Resident statewide. More abundant in migrations. Summer nesting colonies recorded in most sections of the state. Found in winter (usually in southern Nevada) only on unfrozen lakes, ponds and streams.

Records of Occurrence. Sheldon NWR (1982) reported the Eared Grebe as a nesting summer resident.

In Elko County, these grebes have been seen in summer months on most ponds and reservoirs. They move out of the area as the waters freeze over in winter.

About 100 were seen on the west side of Wildhorse Reservoir in Elko County on August 17, 1971 and several were seen on Favre Lake, (9500 feet elev.) in the Ruby Mountains, Elko County on August 24, 1970 (Alcorn MS). Eared Grebes were found by the author to be common in limited numbers at Ruby Lake in spring, summer and fall.

LaRochelle (MS, 1967) reported at Ruby Lake NWR Eared Grebes are, "Common, beginning towards the end of spring migrations and building up to 1000–2000. Nesting colony of 250 nests in East Swamp during the 1964 season. They generally leave in mid–November."

Kingery (1983) reported, "Eared Grebes had successful nesting: 600 nests, 1200 young at Ruby Lake NWR, back up to the level achieved in 1979."

Kingery (1987) reported, "Eared Grebe nesting success dipped . . . . Ruby Lakes reported the most drastic drop, from 1200 young in 1983 and 1984 to only 100 this year."

Farther south at Ryepatch and Pitt–Taylor Reservoirs, these birds were frequently seen in spring, summer and fall with limited numbers present in most winters (Alcorn MS).

Kingery (1978) reported 500 Eared Grebes wintered on Pyramid Lake during the 1976–77 winter.

Orr and Moffitt (1971) reported Eared Grebes are present nearly year round at Lake Tahoe with small numbers breeding.

Near Minden, two young were seen with an adult July 6, 1942 (Alcorn MS).

In the Fallon area, Eared Grebes may be seen in all months. On Soda Lake, one of the few lakes that does not freeze over in cold winter months, Eared Grebes are common and abundant (Alcorn MS).

Alcorn (1946) reported three nests (4, 4 and 2 eggs) were seen about nine miles north–northeast of Fallon July 3, 1943.

Five nesting colonies were observed (about 100 nests per colony) in the Carson Lake pasture area June 16, 1971. No young were observed, but most nests contained eggs (Alcorn MS).

The Stillwater WMA reported two young with an adult July 28, 1949; two broods (3 and 4 young) August 7, 1950; three young with a pair of adults June 26, 1951; five nests August 4, 1953; one young with two adults July 11, 1955; and one young with an adult August 30, 1955.

Napier (MS, 1974) estimated during the summer of 1972, Eared Grebes were produced in the Stillwater area. He commented, "Spring weather seems to determine when they arrive. They peak in April or May at 2000–10,000. Nesting population probably does not exceed 300 nests on open water. The nests are made of mounds of sego pondweed. Some birds stay until October or November, depending upon how long the warm weather lasts."

Biale (MS) recorded these birds 51 times from May of 1970 to March of 1974. He (PC, 1984) reports they are common in the spring.

Clarence Cottam (1929) reported several thousand grebes died during a snowstorm December 13, 1928 and were found on the snow and on roofs of houses in Caliente, Lincoln County.

A. A. Alcorn (MS) reported two young Eared Grebes on adult's back on a small lake near Hiko June 11, 1965. The young stayed on the adult's back when the adult dove.

From April 6, 1960 to June 27, 1965, A. A. Alcorn recorded these birds on 30 different days near Hiko. A total of 206 were seen, averaging 6.2 birds per group. They were seen once in February and March; 10 times in April; 11 in May; 4 in June; twice in July; and once in September.

R. K. Grater (1939) reported for Clark County, "uncommon resident. Recorded in April, May, June, August and December, 1938 in Hemenway Wash and at St. Thomas. Young were observed on June 14, 1938."

In contrast to Grater's findings, from December of 1959 to June of 1965, this bird was reported 64 times and in all seasons, most often in March, April and May (Alcorn MS).

A State Fish and Game aerial sweep of Lake Mead during the winter of 1966 revealed 40,000–50,000.

Kingery (1978) reported at least 100 wintered on Lake Mead during the 1977–78 winter.

Western Grebe   *Aechmophorus occidentalis*

Distribution. In western North America from southeast Alaska to Mexico; locally in the Pacific states, Great Basin and the Colorado River.

Status in Nevada. Statewide on lakes and ponds, more numerous in western and southern Nevada. In winter found mostly on the larger lakes.

Records of Occurrence. Sheldon NWR (1982) reported the Western Grebe as a summer resident, thought to nest at the refuge.

Only a few records, mostly of individual birds, are available for Elko County. Lone Western Grebes were seen at Wildhorse, Wilson and Willow Creek Reservoirs in spring, summer and fall months (Alcorn MS).

For Elko County, Linsdale (1936) reported Bailey (MS) saw one at Franklin Lake in June, 1898. The bird checklist for Ruby Lake NWR (1974) reported the Western Grebe as rarely observed.

LaRochelle (MS, 1967) reported these grebes were never abundant and rarely seen at Ruby Lake NWR in 1962. He noted two sightings (a pair and a single) on June 26, 1962.

Curran (MS) reported seeing seven on a 30-acre pond about five miles west of Elko on May 23, 1974 as he was censusing birds from a small plane. On this same date farther south, Curran saw 27 Western Grebes scattered along the Humboldt River as he flew the river from upper Ryepatch Reservoir to Winnemucca. The river was in moderate flood stage then.

Kingery (1985) reported, "Western Grebes nested commonly at Ruby Lakes."

Ridgway (1877) observed Western Grebes at Pyramid Lake in abundance in 1867. He reported, "This grebe appeared to be a permanent resident."

Ryser (1966) reported well over 2000 on Pyramid Lake September 23, 1966. Kingery (1978) reported 1000 Western Grebes wintered on Pyramid Lake in 1977–78.

H. W. Henshaw (1877) reported this species as breeding abundantly on Washoe Lake where he found downy young August 31.

Orr and Moffitt (1971) recorded them as an uncommon fall transient at Lake Tahoe.

A Western Grebe, shot and discarded by a duck hunter on December 9, 1957 at Sheckler Reservoir, was examined by the author. Its stomach contained three pebbles and two yellow perch, each about four inches in length.

Linsdale (1936) reported, "On May 26, 1934 about 10 miles north of Stillwater, Churchill County, a flock was seen on open water of a lake and several young ones in white down and about 1/4 grown were watched."

Alcorn (1946) reported one adult and two small young taken in the Fallon area August 25, 1940.

At Stillwater WMA the following observations were recorded by refuge personnel: August 5, 1949, one young on back of adult; May 27, 1955, one adult with one young; July 28, 1967, many young. On April 30, 1971 about 1200 Western Grebes were in this area and an estimated 700 young were produced in 1972. At Lead Lake, Stillwater Marsh, Curran (PC) observed two adult Western Grebes feeding on chub and blackfish May 27, 1974.

Thompson (MS, 1988) reported Western Grebes are easily found in the Fallon area from late February to mid–October. In 1986 they were seen every month except January, July, November and December.

At Walker Lake about 3000 were seen September 29, 1965 (Alcorn MS).

In the Eureka area, the birds are seldom seen. One was observed at Bartines Pond April 7, 1960. Biale (MS) reported seeing single birds on three occasions over a 10–year period: at Eureka on June 1, 1969 and May 30, 1970 and at the J. D. Ranch, Pine Valley on June 6, 1973. Biale (PC, 1986) reported, "These birds are an irregular migrant in the Eureka area."

In the Pahranagat Lakes area individual birds were seen from April to November, 1960–65. They were not seen each month of every year (Alcorn MS).

Kingery (1978) reported, "15,000 Western Grebes wintered on Lake Mead, winter 1977–78." He (1982) reported, "Widespread flocks of Western Grebes included light phase birds reported from Las Vegas."

Lawson (TT, 1976) reported Lake Mead and Lake Mohave have large winter populations with as many as 40,000–50,000 birds. He comments, "Apparently these birds do accumulate during the fall migrations, then some of them move out about mid–December and the population on Lake Mead stabilizes to around 15,000–20,000."

Order **PROCELLARIIFORMES**

Family **HYDROBATIDAE:** Storm–Petrels

Least Storm–Petrel   *Oceanodroma microsoma*

Distribution. Ranges at sea on the west coast from southern California; south to Ecuador.

Status in Nevada. Hypothetical.

Records of Occurrence. Lawson (1977) reported them at Lake Mohave: September 12 and 17, 1976.
   Kingery (1977) reported from 50 to 70 seen resting in the water of Lake Mead near Saddle Island September 14, 1976.

Order **PELICANIFORMES**

Family **SULIDAE:** Boobies and Gannets

Blue–footed Booby   *Sula nebouxii*

Distribution. On islands from the Gulf of California east to southwestern Arizona; south to Columbia and Peru.

Status in Nevada. Accidental straggler.

Records of Occurrence. Monson (1972) reported, "August 27 to November 29, 1971, three immature Blue–footed Boobys were found on the west end of Lake Mead, for a new Nevada record and in the 1971–72 winter season, three stayed at Las Vegas on Lake Mead until December 22 and one until January 23."
   Lawson (TT, 1976) reported, "This bird occurred at Lake Mead in one of the eruption years. They were first seen 27 August, 1971 as reported by Monson. Two individuals were observed and both were immatures. The two were observed for a period of time, then they were joined by a third bird which was apparently a second–year bird."
   Lawson (MS) reported an immature bird seen at Lake Mohave just above Davis Dam, September 7, 1977.

Brown Booby   *Sula leucogaster*

Distribution. In the Indian, Pacific and Atlantic Oceans. In the Pacific Ocean from the Gulf of California to Ecuador.

Status in Nevada. Accidental straggler.

Records of Occurrence. Monson (1972) reported, "One adult and one immature were observed on the west end of Lake Mead August 27 to September 19, 1971." He also reported, "One remained at Las Vegas Bay to December 5."

Family **PELECANIDAE:** Pelicans

American White Pelican   *Pelecanus erythrorhynchos*

Distribution. From British Columbia east to Nova Scotia; south to Nicaragua.

Status in Nevada. A common summer resident in western Nevada, with a large breeding colony on Anaho Island in Pyramid Lake. Transient over entire state in migrations. Leontine Nappe and Don A. Klebenow (1973) reported this pelican's status as, "Decreasing. There has been a gradual decline in the breeding population at Anaho Island in recent years. It is not now endangered, but the long range outlook for the breeding colony at Anaho Island is not good."

Records of Occurrence. Sheldon NWR (1982) reported the American White Pelican is an uncommon spring migrant.

At Wildhorse Reservoir over 20 were seen September 8, 1961 (Alcorn MS).

LaRochelle (MS, 1967) reported at Ruby Lake NWR, this bird was seldom seen. However, four were seen there June 16, 1962 and at Franklin Lake to the north, 50 were seen April 12, 1966.

More recent narrative records from Ruby Lake NWR report, "1970, unusual sighting of 125 white pelicans was made on a large shallow lake at the south end of the refuge October 6, 1970. One bird carried a wing tag; April 17, 1972, a flock of about 40 white pelicans were observed soaring above the South Sump."

Kingery (1986) reported 175 summering at Ruby Lake NWR.

Napier (MS, 1974) reported at Anaho Island NWR, the production of white pelicans, "Varies greatly from year to year. Human intrusion during incubation can cause severe losses by gull predation. Due to infrequent visits to the island, production estimates are very gross. Over 10 years (1964–74) production ranged between 1725 in 1974 and 3400 in 1969. Clutch size—generally two or three is fairly common and a few have four. Production is slightly more than one young per nest."

Kingery (1981) reported, "White Pelicans had great success at Anaho Island NWR, where 3100 made it to flight stage." He (1986) reported, "Anaho Island NWR produced 4171 young American White Pelicans to flight stage."

Mills (PC) reported about 1200 American White Pelicans seen in the Fernley Marsh area April 26, 1971. He expressed concern about their effect on the fish life there.

Kingery (1977) reported, "A massive die–off of pelicans occurred at Fallon, Nevada, with 100+ birds dead from unknown causes." He (1979) reported, "At Pyramid Lake, a major die–off occurred in August, with 200–500 found dead or dying."

Ryser (1985) reports he and Ira LaRivers found an, "Ancient nesting colony . . . on an old beachline on a ridge outside of Hazen, Nevada."

Another ancient nesting colony was reported by Hall (1940) at Rattlesnake Hill near Fallon.

The author located still another old nesting colony on a sandy hill near the end of the Carson River in 1935. The many eggshells found there were evidence of former nesting there.

American White Pelican. Photo by A. A. Alcorn.

Steven P. Thompson (PC, 1986) reported 300 pelican nests on June 17, 1986 at the old nesting colony found by the author in 1935. This was the first nesting at this site in at least 50 years. Thompson (PC, 1988) reported the pelicans returned to this area in 1987, but did not nest there.

During the winter of 1985–86, these birds overwintered in the Stillwater Marsh areas. Thompson (PC) estimated from 1500 to 2000 birds were present then. They were apparently attracted by the thousands of dead and disabled fish, mostly carp, present throughout the winter.

Napier (MS, 1974) reported at Stillwater WMA pelicans, "Generally arrive in March, sometimes in February if the weather is mild. March through June population is small at Stillwater (500–1000). After young are on the wing, many birds come from Anaho Island and stay at Stillwater (3000–8000 peak pop.) until November or December."

Mills (PC) reported seeing about 200 white pelicans flying low around some temporary water that covered part of a dry lake bed in southwestern Lander County April 15, 1973.

Gabrielson (1949) reported, "five present on Walker Lake, Mineral County, October 7, 1932: a single bird seen in northern Eureka County on the Humboldt River, June 1, 1948."

In the Eureka area one was found dead July 27, 1961 (Alcorn MS).

Biale (MS) reported from 1960–74, lone individuals were seen near Eureka August 30, 1960 and April 23, 1961. He (PC, 1984) considers them rare in the Eureka area and has no recent sightings.

From 1960 to 1964 in Moapa Valley, J. R. and A. A. Alcorn (MS) reported these birds 5 times in April; 7 in May; on June 16, 1960 (10 birds); and 4 times in August.

American White Pelicans, shading young from midday sun. Photo courtesy of U.S. Fish and Wildlife Service.

George T. Austin and Glen W. Bradley (1971) for Clark County reported, "Transient in riparian and aquatic areas. Usual occurrences from March 30 to June; and from July 20 to October 5."

Lawson (TT, 1976) commented, "This bird in southern Nevada is simply a migrant. It is not uncommon to see flocks of 300 or more birds."

Lois H. Heilbrun and Gary Rosenburg (1981) reported two on the Henderson CBC December 20, 1980. Kingery (1982) reported, "A single white pelican wintered at Las Vegas."

Food. Alcorn (1943a) reported, "In Churchill County, as opportunity afforded, the writer has obtained pelicans for stomach examination. Most of them were shot by other persons and turned over to me for examination. All were taken from ponds or small lakes in the area. It was noted that the pelicans whose stomachs contained large fish were reluctant to fly when molested. In two instances, pelicans that were alarmed flew from small ponds with difficulty. When they were 50 feet or more from the ground and immediately after a shot was fired, they regurgitated large carp. These carp struck the ground with a "plop"; one was 16 inches in length. A total of 25 adult pelicans obtained for examination reveal that the stomachs of seven were empty or contained remains of unidentified fish. The other 18 yielded information given in the following approximate percentages: chub *(Siphateles obesus)* 1.1%; carp *(Cyprinus carpio)* 17.8%; catfish *(Ameiurus nebulosus* or *Ameiurus melas)* 7.8%; Sacramento perch *(Archoplites interruptus)* 59.2%; yellow perch *(Perca flavescens)* .6%; and largemouth bass *(Huro salmoides)* 13.4%."

"There are more than 15 kinds of fishes in the waters within the area where these pelicans were obtained. Further studies in the Fallon area would conceivably change the percentages and add new kinds of fishes to the known list of food of these birds."

"Hall in 1924 at Anaho Island found that 3% (by numbers) of the fishes eaten by pelicans were what may be considered food fishes for humans. Bond in 1940 at the same place found that 11.4% were food fish. The findings of Hall and Bond at Anaho Island are in sharp contrast to the data obtained in the Fallon vicinity from 1940–42 where 81% of the fishes found in pelican stomachs were food fish."

Brown Pelican   *Pelecanus occidentalis*

Distribution. From British Columbia south to southern Chile; from North Carolina southward along the coasts to Brazil.

Status in Nevada. A straggler to Nevada.

Records of Occurrence. Linsdale (1951) reported a Brown Pelican was observed in Churchill County on May 20, 1934."

Mills and the author checked the report of a dead Brown Pelican five miles east of Fallon and examination revealed it was a white pelican with discolored feathers, possibly oil (Alcorn MS).

Vincent Mowbray (MS, 1986) reported the Brown Pelican as rare in northwestern Nevada.

Ryser (1967) reported small numbers near Alamo, on the Pahranagat Lakes February 15–19, 1967.

Linsdale (1951) reported, "Smiley (1937) found pelicans on Lake Mead on April 15 and 31, 1936. Baldwin (1947) reported Brown Pelicans present in late September, 1946 with white pelicans on the shore of Lake Mead."

Lawson (TT, 1976) reported, "As far as I can determine there are approximately 20 to 21 good, solid records of this bird in southern Nevada with about 38 to 40 individuals being observed in those 20 to 21 records. The most recent record was September 13, 1975 when there was an immature individual at Lake Mead. As far as I know there are no records of adults. Ed Scovill of Boulder City has one photograph taken on the upper portion of Lake Mohave with 16 individuals in that one photograph. So, it is rare to uncommon, depending upon the seasonal eruptions that occur from time to time as the birds move off their nesting grounds and begin to disperse."

Lawson (1977) listed Brown Pelicans at Lake Mead August 31, 1975 and at Las Vegas Wash September 13, 1975.

Family **PHALACROCORACIDAE:** Cormorants

Double–crested Cormorant   *Phalacrocorax auritus*

Distribution. From southwestern Alaska to Newfoundland; south to Baja California, Florida and the British Honduras.

Status in Nevada. Summer resident with breeding colonies at various localities in western and southern Nevada. Permanent resident in southern Nevada. Transient over most of the state in migrations. This bird is most often seen on lakes and major rivers in western and southern Nevada.

Records of Occurrence. Sheldon NWR (1982) reported the Double–crested Cormorant as an uncommon migrant spring and summer.

The Double–crested Cormorant is seldom seen at Ruby Lake NWR, in the Eureka area or in the Ely area. However, about 15 were seen in the South Sump area of Ruby Lake NWR on October 2, 1971 (Alcorn MS).

Napier, (MS, 1974) reported at Anaho Island NWR, "A cormorant nesting colony off shoreline. Human disturbance affects these birds more than pelicans. Boaters coming close to colony flushed cormorants and gulls predate on eggs. Since 1950 production was highest at 3000, but in the last 10 years (1964–74) highest was 2000 (1967) and lowest was 350 (1974)."

The cormorant has been observed at Pyramid Lake by Hall, Bond, Alcorn and Marshall. From their reports, cormorant population has increased considerably in recent years. The Double–crested Cormorant nests on Anaho Island and on the Needle Rock Islands (northern Pyramid Lake). On July 29, 1960, the author visited the four small Needle Islands and wrote, "About 1500 cormorants flew from the southwest–most island as we approached by boat from the south. The east side of this island had about 200 cormorant nests on it and the nests contained dead, dried young cormorant bodies. Mr. John Kieger thinks that about 200 cormorants were seen by him nesting here earlier in the summer. Although he saw 200 old nests on another island, Kieger said there were no nesting birds on it this summer."

Napier (MS, 1974) reported, "Cormorants are common at the Fernley WMA during the summer months; up to 100 birds."

A small nesting colony has persisted for many years in the cottonwood trees on the northwest shore of Lahontan Reservoir (Alcorn MS).

Saake (MS, 1981) reported cormorants have started nesting at Sheckler Reservoir (west of Fallon) with 20–30 nests in the summer of 1981.

Thompson (PC, 1986) reports about 700 Double–crested Cormorants stayed at the Carson Sink through December of 1985, where dead and dying fish were abundant. These birds left the area for part of January to return in February of 1986.

Napier (MS, 1974) reported at Stillwater WMA, "Use at Stillwater is limited to spring and fall. Generally March to April and August to November. They seldom number more than 10."

Biale (MS) reported Double–crested Cormorants in the Eureka area only three times: April 26, 1963, July 4, 1970 and May 9, 1984 (Newark Valley).

J. R. and A. A. Alcorn (MS) reported seeing from one to 40 birds on 38 different days in the Pahranagat Valley from 1960 to 1965. Ten or more were seen February 26, 1962; three on February 7, 1963; six on February 16, 1965; 7 times in March; 10 in April; 9 in May; 3 in June, July and August; and one was seen October 18, 1963. Nesting in this area was observed when J. R. Alcorn saw 14 adult cormorant perched on nests and about 15 others on the nearby Pahranagat Lake April 9, 1961.

A. A. and J. R. Alcorn (MS) saw over 20 cormorant nests in the tree tops at the north end of Upper Pahranagat Lake May 4, 1962. Adult cormorants were setting on some of these nests. From the ground this author could see three nests with young cormorants in them. One nest contained young about half grown.

A. A. Alcorn (MS) saw about 35 cormorants, some on nests, at Upper Pahranagat Lake (north shore) March 29, 1965.

Double-crested Cormorant young in nesting colony. Photo by Laura E. Mills.

Gordon W. Gullion et al., (1959) for southern Nevada reported, "A common permanent resident on the lakes in the Pahranagat Valley and on the Colorado River and its reservoirs. In spite of efforts to destroy it, a nesting colony has persisted for several years in the cottonwood grove at the north end of Pahranagat Lake. Nestlings were present in this colony May 23, 1951."

J. R. and A. A. Alcorn (MS) observed these birds in the Moapa Valley in all seasons from 1960 to 1965.

Lawson (MS, 1981) reported the Double–crested Cormorant as a common permanent resident in Clark County.

Food. In an effort to determine food habits of the cormorant, 120 cormorants were taken by various people from Indian Lakes, Carson River, Rattlesnake Reservoir, Little Soda Lake and canals in the Fallon area. The author examined their stomachs and found they contained: 76 carp (28%), 60 yellow perch (22%), 37 trout (14%), 28 largemouth bass (11%), 25 bullheads (9%), 24 Sacramento perch (9%), 17 chub (6%) and two crawfish (1%).

Olivaceous Cormorant   *Phalacrocorax olivaceus*

Distribution. From northern Mexico, Louisiana; south to Tierra del Fuego in South America.

Status in Nevada. Hypothetical.

Records of Occurrence. Kingery (1979) reported Nevada's first record for the Olivaceous Cormorant was during the Las Vegas CBC December 16, 1978. The second record was from Lake Mead March 12, 1978.

Family **FREGATIDAE:** Frigatebirds

Magnificent Frigatebird   *Fregata magnificens*

Distribution. At sea along the Pacific coast of California; south to northern Peru. Wanders to the coasts of California and southern United States. After storms can be found accidentally in the interior of North America.

Status in Nevada. Accidental.

Records of Occurrence. The late Mrs. Anna Bailey Mills told this author in the 1930s that she saw a frigatebird flying high overhead at their ranch three miles west–southwest of Fallon, Nevada.

The late Mr. Wendell Wheat (PC) reported he had seen two frigatebirds flying eastward at his place on the Carson River near Carson City July 13, 1967.

G. C. Baldwin (1947) reported one in the vicinity of Eldorado Canyon, 35 miles below Boulder Dam September 2, 1946.

Lawson (1973) reported Art Johnson (National Park Service) observed a female frigatebird soaring above Davis Dam, Lake Mohave, September 3, 1971.

Lawson (TT, 1976) reported, "Most of the records of this bird are from the Davis Dam area." He (1977) reported seeing one at the dam September 17, 1976.

Kingery (1984) reported, "A letter to the newspaper provided an adequate description of a Magnificent Frigatebird flying over a Las Vegas street November 18."

Order **CICONIIFORMES**

Family **ARDEIDAE:** Bitterns and Herons

American Bittern   *Botaurus lentiginosus*

Distribution. From southeastern Alaska to Newfoundland; south to Panama.

Status in Nevada. Statewide resident and winter visitant in marshy areas.

Records of Occurrence. Sheldon NWR (1982) reported the American Bittern as occasionally seen spring, summer and fall with the possibility of some nesting in the area.

Linsdale (1936) reported these birds from eight of the 17 Nevada counties. Although they occur over a wide area, the American Bittern is not an abundant bird in Nevada.

This author saw one American Bittern in Elko May 28, 1965.

LaRochelle (MS, 1967) reported these birds were recorded each month of the year at Ruby Lake NWR. However, most were present from early spring until mid–winter. In this same area, J. R. and A. A. Alcorn (MS) recorded them in July, August, September and October of 1965.

Kingery (1980) reported one at Ruby Lake in December, 1979 and he (1983) reported, "Ruby Lakes reported 200 pair of nesting American Bitterns."

This author has seen solitary birds in the Minden area in April, June and November and one was observed near Fort Churchill on the Carson River near Lahontan Reservoir by Mills September 11, 1960.

Alcorn (1946) reported specimens taken in the Fallon area August 28, 1940 and November 13, 1939. He reported them more numerous in summer than winter. Others were seen by him in this area April 25, 1962, April 5, 1966, October 18, 1963 and November 13, 1964 (several).

In more recent years American Bitterns have occasionally been heard in summer months at the nearby Stillwater WMA. Schwabenland (MS, 1966) reported seeing a few every year in summer and in December.

Napier (MS, 1974) reported for Stillwater WMA, "Only occasional ones have been seen from 1967 to 1974. Only seen in summer, may nest, but population is very small."

At Mahala Slough, west of Fallon on May 29, 1941, one was observed from a distance of 20 feet. In the bittern's bill was a live gopher snake about 15 to 18 inches long. The bittern was holding the snake by the neck near its head, and at intervals it would shake the snake, work it back in forth in its bill and repeatedly bite and shake the snake. The snake soon appeared to be dead and the bittern swallowed it head first, then flew to alight in the grass near a small pond (Alcorn MS).

Along the Reese River west of Austin solitary birds were seen in May of 1960 and 1961 (Alcorn MS).

In the Eureka area one was observed September 13, 1961 (Alcorn MS) and Biale (MS) saw one in this same area May 27, 1972. He (PC, 1984) considers them uncommon in the Eureka area.

Three were seen near Hiko in the Pahranagat Valley August 8, 1963 (Alcorn MS). Kingery (1977) reported them at Pahranagat into December.

Snider (1971) reported American Bitterns at Las Vegas May 21 and August 29, 1970.

Oliver K. Scott (1963) reported single birds at Ash Meadows January 2 and December 21, 1962.

Kingery (1978) reported one wintered at Tule Springs near Las Vegas (1977–78). He (1985) reported, "An American Bittern above Davis Dam, Nevada September 13, stood in a patch of creosote bush."

Grater (1939) first reported bitterns at St. Thomas, Clark County in April and May, 1938.

Lawson (TT, 1976) commented, "I would say that it is probably an uncommon migrant and summer resident in Clark County. It can be heard calling during the summer and I suspect that, of course, it does breed in Las Vegas Wash."

Least Bittern   *Ixobrychus exilis*

Distribution. From Oregon to New Brunswick; south through Mexico and Central America to South America.

Status in Nevada. Summer visitant and transient.

Records of Occurrence. Ruby Lake NWR personnel reported one adult on May 27, 1984.

Linsdale (1936) reported one noted August 5, 1925 by Hall near Halleck, Elko County.

Ralph Ellis (1935) reported collecting a male from the west side of Ruby Lake, three miles north of the Elko County line on July 18, 1927.

Ridgway (1877) reported one in May in willows along the lower Truckee River.

Linsdale (1936) reported two observed by Stager (MS) in marshes south of Carson City, Ormsby County on June 13, 1933.

Alcorn (1946) reported a specimen obtained four miles west of Fallon in September of 1935.

Larry Neal reported a juvenile Least Bittern found seven miles southeast of Fallon on July 20, 1986.

This author noted one seen at the Carson Sink October 14, 1945. "I was standing in a clump of tules and one Least Bittern alighted about 11 feet from me. I observed the bird for several minutes before it flew to another clump of tules."

Kingery (1983) reported, "One at Kirch WMA near Lund."

Linsdale (1936) reported, "On June 8, 1928, Hall (MS) flushed one from a wet meadow at Fish Lake, Esmeralda County. The bird flew to marsh vegetation in a nearby waterhole where it could not be found again."

Lawson (MS, 1976) saw one in Pahranagat Valley and Snider (1970) reported a single bird near Las Vegas May 16.

Grater (1939) reported, "Transient visitant. Records for August and September, 1938 at Hemenway Wash and St. Thomas. First records for southern Nevada."

Lawson (TT, 1976) commented for southern Nevada, "It is probably more common than we realize. Due to its secretive nature and its small size it is easy to overlook in a marsh. We do have a number of records here in southern Nevada and I would say that it is a rare to uncommon migrant and summer resident."

Kingery (1987) reported, "A Least Bittern at Overton April 26 gave southern Nevada its first record in 10 years."

Great Blue Heron   *Ardea herodias*

Distribution. From Alaska to Nova Scotia; south to southern Mexico and the West Indies.

Status in Nevada. A common summer resident over most of the state; uncommon in winter in northernmost Nevada. Two races have been reported in Nevada: *A.h. treganzai* and *A.h. hyperonca.*

Records of Occurrence. Sheldon NWR (1982) reported the Great Blue Heron as an uncommon, permanent resident, occasionally seen in the winter.

LaRochelle (MS, 1967) for Ruby Lake NWR reported Great Blue Herons as common throughout the year. He reported, "Up to 75 birds in the area, but do not know of any nesting."

Kingery (1984) reported, "Ruby Lake held steady producing 125 young (as last year)." He (1985) reported, "Great Blue Herons nested normal in numbers at Ruby Lakes and established a new nest site at nearby Franklin Lake with two to three young in 10 to 15 nests built on small islands covered with greasewood."

The author's records indicate most Great Blue Herons leave the Ruby Lakes area in winter months. Kingery (1987) however, reported 100 Great Blue Herons wintered at Ruby Lake NWR during the winter of 1986.

Great Blue Herons also nest on Anaho Island at Pyramid Lake. The nests are usually built in a dense growth of alkali weed *(Bassia)* or in the tops of greasewood *(Sarcobatus)* bushes. The author, accompanied by Mills, G. H. Hansen, and Frank Richardson went to this island June 14, 1942. On the east side, not more than 50 feet from the waters edge we saw a few blue heron nests. One nest contained four eggs, another three very young blue herons. We also saw young blue herons, nearly full grown, off the nest and nearly able to fly. This author judged that about 10 pair of blue herons were nesting at this location.

About 15 blue heron nests containing young were seen at this same location July 21, 1942. The nests, located in heavy alkali weed growth, were made of this same material. They were scattered along near the shore for about 200 yards. One nest contained four small young. For the most part, the number of young in each nest was from two to four. Some young birds appeared to be almost fully grown. In addition to the above, there were about 15 empty nests. On the northeast side of Anaho Island on May 26, 1944, 14 nests contained from two to four young herons. Eight nests containing eggs were seen at the same location as the 14 listed above. Eggs in these eight nests were counted as follows: 3, 4, 3, 4, 3, 4, 3, 1.

About 50 adult blue herons were seen flying at once over this nesting area May 29, 1959. The nests, made mostly of *Bassia*, were frequently built on the top of a large Sarcobatus. Nineteen young, one–half to two–thirds grown were counted. There were nine nests each containing from one to four young. Nine nests examined April 21, 1961 were made mostly of alkali weed. Three nests contained five eggs each and six nests contained four eggs each (Alcorn MS).

In the Reno, Minden and Yerington area, these birds were observed in all months. A small nesting colony has been observed yearly in the cottonwood tree tops near the northeast shore of Lahontan Reservoir, Lyon County. Two trees in this area were examined May 23, 1959. One of these cottonwood trees, with green leaves, contained 13 cormorant nests and eight blue heron nests. The second tree contained 12 heron and 11 cormorant nests. Some of the heron nests contained three–fourths grown young. Young cormorants could be heard in some of the nests (Alcorn MS).

In the Fallon area from January, 1960 to December, 1967, these herons were frequently seen in all months (Alcorn MS). The author interviewed Wuzzie George, an elderly Paiute Indian woman, on March 26, 1969. She reported the local Indians ate Great Blue Herons and their eggs.

The author and Mills visited an area two miles north–northeast of Stillwater on June 2, 1944 to examine the treetop nesting colony of Great Blue Herons. About 175 nests were seen, most of which contained young herons. These herons continue to nest in the cottonwood trees in this area today (1987).

Napier (MS, 1974) reported, "Year round resident, although most leave in winter. Dead birds have been found in winter apparently from exposure and lack of food. They begin coming to nest sights late in February. Nesting is prolonged during the summer. Young birds can be found in the nests from late April to early May to mid–June. Stillwater's nesting population does not exceed 200."

Kingery (1987) reported 550 Great Blue Heron nests in Lahontan Valley.

Biale (PC, 1984) does not consider this bird common or abundant in the Eureka area. This

Great Blue Herons display territorial behavior (above). A Great Blue Heron tree-top nesting colony at Stillwater (below). Photos by A. A. Alcorn.

is probably due to the lack of marshes in the vicinity. Combining the author's records with Biale's (1964–74) these birds were seen in all months a total of 53 times.

In the Pahranagat Valley from August 11, 1960 to June 28, 1965, these herons were recorded at least once each month, with most records available for April, May, October and November (Alcorn MS).

At Upper Pahranagat Lake on May 4, 1962 adult herons were feeding young in some of 12 treetop nests. One adult was seen setting on a nest in this same area May 23, 1962 (Alcorn MS).

In the Moapa Valley these herons were observed in all months, but no nesting was observed from 1959 to 1965 (Alcorn MS).

Lawson (TT, 1976) comments, "this bird is a common, permanent resident in southern Nevada."

Great Egret   *Casmerodius albus*

Distribution. Almost worldwide. In the New World, from southern Oregon to Maine; south to the Straits of Magellan.

Status in Nevada. A summer resident and transient.

Records of Occurrence. Sheldon NWR (1982) reported the Great Egret as a spring and fall migrant.

Linsdale (1951) listed this bird as a summer visitant over most of the state.

LaRochelle (MS, 1967) reported the Great Egret as common from late spring to early winter at the Ruby Lake NWR. Kingery (1984) reported 250 to 300 young at this same location.

Saake (MS) reported a rookery at the southwest corner of the upper Humboldt Sink May 25, 1970 during an aerial bird census.

In the Mason Valley (Yerington) area, these birds were regularly observed from April to late September. Two nests were seen on the Carson River, five miles east of Fort Churchill May 30, 1960. The nests were near the top of a tall, dead cottonwood tree. An adult was setting on each nest. Adult birds were setting on the same nests June 12, 1960 (Alcorn MS).

From April, 1960 to December, 1966, these egrets were observed in the Fallon area 7 times in January; 6 in February; 3 in March; 10 in April; 8 in May; 3 in June; 5 in July; 3 in August; twice in September; 3 in October; once in November; and 6 times in December (Alcorn MS).

Napier (MS, 1974) reported these birds were recorded at Stillwater WMA during the winter months of December and January, 1966–68. From 1971 to 1974, however, the last sightings were around October and none seen until March. These winters were severe, evidently forcing the birds out.

Saake (MS, 1981) reported the biggest rookery in Nevada of this fairly uncommon bird is in west nutgrass of the Stillwater Marsh and at nearby Carson Lake.

Kingery (1984) reported, "Stillwater produced 200 Great Egrets." He (1985) reported 80 young produced.

Great Egret. Photo by Norm Saake.

Kingery (1987) reported, "Stillwater refuge hosted 160 nesting pair of Great Egrets, 185 pair in 1986. They also wandered from their Nevada nest sites, with reports from Denio June 11, Eureka June 17 and Kirch July 20."

Biale (MS, 1986) considers the Great Egret uncommon in the Eureka area. He reported three records: April 24, 1960, September 17, 1972 and April 26, 1986. The author observed another in this area May 17, 1967.

Kingery (1975) reported, "One Great Egret stopped at Diamond Valley [Eureka County]."

A. A. Alcorn (MS) in Pahranagat Valley reported two on April 13, 1965; one May 11 and 17, 1960; nine June 16, 1960; and three August 11, 1960.

J. R. and A. A. Alcorn (MS) observed from one to five in the Moapa Valley from December, 1959 to April, 1965. Two were seen January 18, 1960 and others were seen 6 times in March; twice in April; once in May and July; twice in August; and two were seen December 15, 1959.

Kingery (1977) reported a Great Egret at Logandale January 20, 1977.

Kingery (1984) reported, "The count of 16 at Las Vegas December 10 was unusual." He (1985) also reported, "Davis Dam attracted one Great Egret January 11."

Lawson (TT, 1976) for southern Nevada reported, "It is more common as a migrant than as a resident. We seldom ever see one in the summertime and we seldom see more than two or three birds during migration."

Snowy Egret   *Egretta thula*

Distribution. From northern California to Maine; south locally to South America.

Status in Nevada. Summer resident locally in the northern and western part of the state. Common fall and spring migrant in southern Nevada.

Records of Occurrence. Linsdale (1936, 1951) reported, "a summer resident in small numbers in valleys throughout the state."

Sheldon NWR (1982) reported the Snowy Egret as a spring and fall migrant.

Gabrielson (1949) reported two on Virgin Creek, Humboldt County, May 31, 1948.

At present (1988) this bird is abundant and common, especially along the Humboldt River and in western Nevada. These birds were repeatedly seen along the Humboldt River from Halleck to Carlin from April into September, 1961–68. Over 100 were observed roosting in the willows along this river near Elko in August (Alcorn MS).

LaRochelle (MS, 1967) reported Snowy Egrets as common from late spring until late fall with up to 65 present at Ruby Lake NWR.

Kingery (1983) reported, "Snowy Egret nesting dropped at Ruby Lakes from 150 pair in 1980 to 75 this year."

Kingery (1980) reported a helicopter survey in north–central Nevada on May 28 counted 420 Snowy Egrets.

In the Mason Valley (Yerington) area, these egrets were seen from April to October (Alcorn MS).

Alcorn (1946) reported for Lahontan Valley, "Abundant from April through September of each year. Records of stragglers are available for October, November and December. A nesting colony was seen about three miles north of Stillwater in July, 1940 and nests containing eggs or young were seen in the same locality July 5, 1941."

In the Fallon area from 1960–64, inclusive, these birds were observed once in February; 7 in April; 43 in May; 10 in June; 11 in July; 3 in August; 8 in September; and 4 times in October (Alcorn MS).

Napier (MS, 1974) reported, "These birds cannot take much cold weather. They are first seen late in March and stay until late September or October. They no longer nest at Stillwater, but colonies exist at the Canvasback Gun Club and Carson Lake. Birds feed at Stillwater. Summer population is 100–300."

Kingery (1984) reported, "Stillwater produced 510 Snowies." He (1985) reported 140 produced at Stillwater.

Thompson (PC, 1988) reported fewer Snowy Egrets at Stillwater WMA in 1986 and 1987.

Snowy Egrets have frequently been seen along the Reese River from west of Austin upstream to the Reese River Ranger Station.

Biale (MS) reported the Snowy Egret in the Eureka area 12 times from 1964–73. This author's records for this same area are of five on May 16, 1960; six on May 16, 1961; 14 on August 10, 1961; and three on June 4, 1962. Biale (PC, 1984) reports seeing them at least once every year in the Eureka area.

Five Snowy Egrets were seen at Comins Lake south of Ely June 5, 1962 (Alcorn MS).

A lone Snowy Egret was standing on the highway near the Diablo Highway Maintenance

Station in Railroad Valley, Nye County May 5, 1962. This is many miles from the nearest water and the bird was probably migrating and stopped to rest (Alcorn MS).

Kingery (1987) reported a 50% decrease (15 birds total) in migrating and breeding Snowy Egrets at Kirch WMA.

In the Pahranagat Valley these birds were observed each month from April into September (Alcorn MS).

In the Moapa Valley from 1960 to 1965, the earliest record of this bird was February 21. They were normally seen from March through August (Alcorn MS).

Gullion et al., (1959) reported, "Recorded from wet areas in most valleys and on the shores of Lakes Mead and Mohave. Early spring arrivals appear about April 28 (1951). Two unusual records: one bird on the Boulder Dry Lake (1700 feet elev., Eldorado Valley, Clark County) May 3, 1951; two birds on Sarcobatus Flat (4000 feet elev., Nye County), at least 26 miles from the nearest pools of water at Springdale (3800 feet elev., Nye County) on May 28, 1953."

Austin and Bradley (1971) for Clark County reported, "Transient in riparian and aquatic areas with common occurrences from March 30 to October 4. Unusual occurrences on March 2, November 2 and 21 and December 5 and 8."

Kingery (1976) reported a Snowy Egret wintered at Lake Mead.

Lawson (TT, 1976) commented, "I don't know of any nesting populations in southern Nevada and most of them occur in late April and May and again in late August to early October. I have seen upwards to 35 of them during spring migrations in Vegas Wash."

Little Blue Heron   *Egretta caerulea*

Distribution. From southern California to Nova Scotia; south along the coasts of Mexico and Central America and to Peru and Uruguay.

Status in Nevada. An uncommon straggler to southern Nevada.

Records of Occurrence. Austin and Bradley (1968) reported, "Baldwin, (1944) lists one record for Clark County from Lake Mead on November 13, 1943. Austin, who is familiar with this species from field work in the southeast has observed immature birds at Tule Springs on May 9, (with two Snowy Egrets) and September 2, 1964, and at Henderson Slough on August 20, 1964. All three birds showed dark primaries and the black–tipped, bluish bill."

Lawson (MS, 1981) reported an immature Little Blue Heron found dead in Nevada, near Davis Dam the previous year. Remains are in an Arizona collection.

Kingery (1983) reported, "Little Blue Herons wandered this summer. One at Mesquite, on June 17, seen only in flight, provided Nevada's sixth record."

Lawson (TT, 1976) comments, "I would say that this is an occasional to rare transient or straggler in southern Nevada. There are a number of records in Audubon Field Notes, but I don't think there are as many records as indicated."

Tricolored Heron   *Egretta tricolor*

Distribution. From central Baja California to the Atlantic coast and north to Maine; south through Mexico and Central America to Brazil. Irregularly to central North America.

Status in Nevada. Hypothetical.

Records of Occurrence. Grater (1939c) reported one along the shore of Lake Mead, near the site of Saint Thomas, Clark County from November 21, 1938 to December 8, 1938.

Kingery (1986) reported, "A Tricolored Heron flew out of a carex marsh at Kirch WMA on June 10, for Nevada's second record."

Cattle Egret   *Bubulcus ibis*

Distribution. In the Western hemisphere from central California east to Colorado and north into Canada and northeastern United States; south to Gulf States and Florida through Middle America to South America.

Status in Nevada. Uncommon resident in selected localities. In recent years it has established itself as a resident in the various parts of the state with many summer records including near Halleck and Ruby Lakes in the summer of 1979 and Fallon, 1980. Gary Herron (PC, 1984) reports these birds are expanding their range with an increase in numbers in Elko County.

Records of Occurrence. Kingery (1973) reported, "Ruby Lake NWR, Nevada had its first record of Cattle Egret, a bird found, May 7." He (1978) reported one at Ruby Lake NWR May 5 and (1980) one August 3–7. In the same year he reported a total of 15 at Elko and Ruby Lake NWR.

Kingery (1984) reported, "Cattle Egrets remained in the Region [Mountain West] into late fall with two at Reno November 4–30+." Ryser (1985) noted they were reported from Washoe Valley, Nevada in August, 1975.

Three were seen in Mason Valley about three miles north of Yerington December 21, 1983 (Alcorn MS).

Saake (MS) reported, "I saw two Cattle Egrets on Carson Lake June 11, 1979. They were along the west end of Madsen Dike."

Kingery (1980) reported one pair nested a Carson Lake, furnishing apparently the first Nevada nesting record.

Saake (MS) reported one nest in the West Nutgrass Unit in Lahontan Valley in 1980 and Mills (PC) suspected nesting at Sheckler Reservoir in 1980.

Gerald Alberson (PC, 1987) reported three Cattle Egrets feeding near a swather as he was cutting alfalfa three miles south of Fallon on June 1, 1987. "They followed along, sometimes only a few feet from the swather," and were catching what he thought to be crickets.

Kingery (1980) reported five to 10 in Lahontan Valley during the 1979 summer.

Thompson (PC, 1988) reported the Cattle Egret population increasing in the Lahontan Valley area. He observed a nest in 1987.

Kingery (1980) reported 16 in mid–October at Las Vegas, and throughout the summer near Halleck. He (1977) reported one at Lake Mohave December 5–13, 1976.

Richard C. Castetter and Herbert O. Hill (1979) reported one August 10–19, 1977 at the Nevada Test Site.

Lawson (1977) reported one at Logandale October 2–3, 1975 (photo); and one discovered by Leon Hill at the Pahranagat NWR in December of 1969. They have been seen almost every winter since that date.

Kingery (1978) reported one or two Cattle Egrets at Overton during the 1977–78 winter. He (1983) reported 15 Cattle Egrets at Overton April 17.''

Lawson (TT, 1976) considers the Cattle Egret a rare transient visitor in Clark County, possibly a rare wintering bird.

Green–backed Heron   *Butorides striatus*

Distribution. Formerly the Green Heron. From Vancouver Island to New Brunswick; south through Mexico to South America.

Status in Nevada. An uncommon summer resident in southern Nevada; transient in other parts of the state.

Records of Occurrence. Kingery (1986) reported, ''Ruby had a Green–backed Heron April 28 through May 1.''

Linsdale (1951) reported, ''Summer resident in the Colorado River Valley and recorded in late summer in the western part of the state. Ned Johnson saw one five miles west of Reno May 1, 1948. Alcorn (1940) reported a specimen obtained in September, 1934, four miles west of Fallon and two on August 24, 1939 from the Carson River, Douglas County. R. M. Bond (1940) in May, 1939, saw one at Hiko Lake and one in Rose Valley, Lincoln County. Cottam (1947) saw two on June 30, 1947 at the mouth of Muddy Creek south of Overton.''

Ryser (1972) reported one heron was seen in the Truckee Meadows October 1 and two were seen in Reno in mid–October.

Ralph Lettenmaier (PC) reported one struck by an auto was picked up and release in a Reno park in September, 1968.

Although Green–backed Herons have occasionally been seen in western Nevada during the summer, there are no nesting records (Alcorn MS).

Mills (PC) observed one on the Truckee Canal, five miles east of Fernley June 8, 1958.

Napier (MS, 1974) reported that during his eight years (1967–74) at Stillwater WMA this bird was seen on three occasions.

Dave Paullin and Curran (PC) observed one of these herons at Big Indian Lake, north–northeast of Fallon September 10, 1970.

Saake (MS) reported one in the west nutgrass, Stillwater WMA the first week of June, 1981.

Biale (MS) saw solitary Green–backed Herons in the Eureka area April 25, 1970 and December 3, 1971. He (PC, 1984) has not seen any since.

One was observed near Hiko May 4, 1965 (Alcorn MS).

Kingery (1986) reported, ''A Green–backed Heron November 11 + at Kirch WMA, Nevada fed around the refuge headquarters on fishes endemic to the desert springs.''

One was observed on the Virgin River near Bunkerville July 2, 1961 and another was sighted by A. A. Alcorn near the same locality April 29, 1965 (Alcorn MS).

In the Moapa Valley these birds were seen 10 times from August, 1961 to June 24, 1964. All were solitary birds, except for two seen August 27, 1963. Birds were seen in the months of April, May, June, July, August and October (Alcorn MS).

Lawson (TT, 1976) commented, ''This species may be more common here in southern Nevada than most of us realize because it is simply more difficult to see.'' He considers it as an uncommon to common summer resident and migrant in southern Nevada. Lawson reported six in sight at one time at Lake Mead in the fall.

Kingery (1980) said, "Las Vegas reported an immature Green–backed Heron seen July 24 to 30 possibly from a local nesting sight."

Heilbrun and Rosenburg (1982) reported two Green–backed Herons on the Henderson CBC December 19, 1981.

Kingery (1983) reported, "At Las Vegas, two Green–backed Herons wintered at a Fish Hatchery; probably small numbers winter there regularly."

Black–crowned Night Heron   *Nycticorax nycticorax*

Distribution. From central Washington and east–central Alberta east to Nova Scotia; south to Texas.

Status in Nevada. Summer resident over entire state and a winter resident in the western and southern part.

Records of Occurrence. Linsdale (1936) reported this heron from 12 of 17 Nevada counties. Recent records are from all sections of the state.

Sheldon NWR (1982) reported the Black–crowned Night Heron as seen occasionally in spring, summer and fall and an occasional nester.

LaRochelle (MS, 1967) reported these birds common at the Ruby Lake NWR from early spring until mid–winter. "Up to 125 use the area with no known nesting."

The 1969 Ruby Lake NWR narrative report noted two large colonies observed, each containing approximately 40 birds.

Kingery (1980) reported, "From Ruby Lake came a disturbing report. The 30 pair nesting on that refuge suffered from thin eggshells, possibly picked up from pesticides in Mexico." He (1984) reported, "100 produced at Ruby Lakes, down from 200." Kingery (1986) reported, "Ruby Lakes had only 10 night heron nests."

In the Reno area these herons are seen throughout the year (Alcorn MS).

Ryser (1985) reported, "from early 1930s until recently, the largest winter rookery here was located in west Sparks. For years, the number of herons in this rookery approached 100 birds."

Saake (MS) reported two nesting at Washoe Lake April 15, 1970.

Alcorn (1946) reported this heron was seen in all seasons in the Fallon area, but most abundantly and widespread in August. A nesting colony, three miles north of Stillwater was seen July 5, 1941. All nests were constructed of tules and were built about a foot above the water in a growth of tules. Six nests contained from one to three young all less than half–grown. About 50 adults were seen in the area and there were probably more nests. The adults made no attempt to defend the nesting area.

Schwabenland (MS, 1966) reported for Stillwater WMA these birds were common all year with two nesting colonies of 300–400 herons.

Napier (MS, 1974) reported at Stillwater WMA, "These herons are present all year and probably still nest at Stillwater in summer. Population during mid–winter is 25–50 and during the summer over 500."

Kingery (1986) reported, "Stillwater reported 5000 Black–crowned Night Herons, mostly immatures, during September."

This author observed one of these herons standing on the edge of the water with a bullfrog in its beak May 12, 1942. Examination of the frog revealed it was 12 1/2 inches in length from its nose to the tip of its hind legs. The nearly dead frog was badly scratched on its body and a hind leg was broken (Alcorn MS).

The stomach of another heron taken in the Fallon area August 8, 1946 contained 3, one to two-inch black bullhead (catfish), and a dragonfly nymph (Alcorn MS).

They were observed in summer along the Reese River southwest of Austin, and were recorded at Big Creek and Kingston Canyon in May, June and July (Alcorn MS).

Biale (MS) recorded these herons in the Eureka area only on five occasions in a 10–year period (1964–74) from May to August.

Biale (PC, 1986) reports, "These are not a common bird in the Eureka area."

Gullion et al., (1959) reported for southern Nevada, "Although a fairly common permanent resident in brushy riparian sites along the Colorado River, in Pahrump Valley (2800 feet elev., Nye County) and along the Meadow Valley Wash and in Pahranagat Valley, one unusual record was of a single individual seen at about 2500 feet elev. in the creosote bush—ber–sage vegetation type on the Quo Vadis slop north of Boulder Dry Lake. This bird was 13 airline miles from the nearest riparian or marsh habitat."

In the Overton area, Moapa Valley from December, 1959 to December, 1964, from one to 20 of these herons were seen 4 times in January; 3 in February; 5 in March; once in April; 4 in May; once in July and August; and twice in December (Alcorn MS).

Austin and Bradley (1971) reported for Clark County, "Permanent resident in riparian and aquatic areas. Particularly common from March 30 to May 15 and from July 8 to October 15."

Lawson (TT, 1976) commented, "I don't know of any breeding colonies here in southern Nevada and most of these birds we do see are simply transients, both spring and fall."

Family **THRESKIORNITHIDAE:** Ibises and Spoonbills

White–faced Ibis   *Plegadis chihi*

Distribution. From California north to Oregon; east to North Dakota; south to South America.

Status in Nevada. Summer resident in the northern half of the state. Recorded statewide in valleys in marshy areas.

In that these birds have been sighted so often, no effort will be made to list all sightings.

Linsdale (1936) reported these birds from six of 17 Nevada counties. Present records cover 11 counties. Habitat destruction and pesticide poisoning has caused population declines in recent years.

Records of Occurrence. According to Robert J. Oakleaf (1974b) about 70–100 White–faced Ibis nests were observed at Franklin Marsh in Ruby Valley, Elko County, 1973. He reported 120 nesting adults in this same marsh in 1974.

LaRochelle (MS, 1967) reported at Ruby Lake NWR up to 65 ibis arrive in mid and late spring, leave in mid–fall, with up to 10 pair nesting yearly.

Kingery (1977) reported, "Ruby Lake had 300 young in a swamp coveted by power boaters." He later (1978) reported, "In Nevada 800 pair nested in seven scattered sites. At Ruby Lake NWR they raise 400 young. At Carson Lake 400 pair had poor success owing to poor water conditions and trampling by cattle." He (1979) reported 500 young White–faced Ibis' at Ruby Lake NWR. He stated, "Ruby Lake posted big increases [in 1984] and Stillwater fledged a pitiful 25 White–faced Ibises."

Kingery (1986) reported, "Ruby Lakes had 260 pair of White–faced Ibises nesting."

Oakleaf (1974a) reported increased water in the Humboldt Sink, Pershing County in recent years provided new nesting sites for these birds. The first nesting in this sink occurred in 1973 when about 140 pair nested. "In 1974 the nesting population approximately doubled."

Curran (MS) reported about 600 ibis at the rookery in Humboldt Sink during the summer of 1974.

Kingery (1983) reported, "White–faced Ibises established a small new colony near Battle Mountain with 20 pair."

Nappe and Klebenow (1973) reported the nesting colony of ibis in Lahontan Valley, Churchill County as one of three major nesting sites in the western United States.

Alcorn (1946) reported a nesting colony in Stillwater Marsh, three miles north of Stillwater in 1940 and 1941. On May 12, 1944 one nest contained four eggs and other nests were in various stages of completion. The nests were constructed of and in a growth of tules. Most were about two feet above the surface of the three foot deep water.

Schwabenland (MS, 1966) reported two nesting colonies in this marsh with up to 1500 adults in 1965–66.

Napier (MS, 1974) reported no nesting of ibis in the Stillwater Marsh from 1967–74 although about 25 adults showed up in the spring. Later in the summer up to several hundred arrived to feed in the marsh. Napier also reported in the nearby Carson Lake area, no censusing for nesting ibis was done prior to 1970 at which time 600 nests were reported.

Cathy T. Osugi (1973) reported, "The White–faced Ibis' nesting colony at Carson Lake is one of the three largest colonies in the western United States. Estimates of the number of nests in the Carson Lake colony are as follows: 1970–600 nests; 1971–250 nests; 1972–1300 nests. An estimated 3900 birds were produced in 1972.

Napier (MS) reported in 1973 an estimated 3312 nests were reported for the Carson Lake area. In 1974 there was no nesting in the area. The diversion of water and loss of habitat in the Stillwater and Carson Lake areas from 1973–77 caused the loss.

Kingery (1978) reported 400 pair returned in 1978 to Carson Lake to resume nesting. He (1981) reported, "Nesting reports included 6600 reaching flight stage at Fallon (Stillwater WMA) and Walker Lake."

Saake (MS, 1982) reported 300 pair at Carson Lake and 400 pair at Stillwater.

White–faced Ibis occasionally winter in the Fallon area. Mills (PC) repeatedly observed a small group of eight or more about five miles southwest of Fallon in December of 1958 and January of 1959. This author and Mills visited the area January 22, 1959 and the ibis were standing on a drain ditch bank about 20 feet from open water. In December of 1965 and January of 1966, about 40 were observed almost daily west of Fallon. They were seen along a drain ditch containing unfrozen water (Alcorn MS).

Schwabenland (MS, 1966) reported ibis wintering in the Stillwater area in 1965–66.

In the Eureka area, these birds are seldom seen. Two flocks containing 10 and 17 birds were seen May 16, 1960 (Alcorn MS).

Biale (MS) reported from 1964–74 he saw them three times in May and on June 23, 1971.

Gabrielson (1949) reported a number of ibis seen August 15, 1933 in Railroad Valley, Nye County. They were feeding with other water birds in the overflow from an artesian well.

In the Pahranagat Valley J. R. and A. A. Alcorn (MS) reported from two to 15 birds on seven occasions in May and August from May, 1960 to May, 1964.

J. R. and A. A. Alcorn (MS) reported these birds were seen in Moapa Valley in April, May, August and September, 1960–64.

Austin and Bradley (1971) reported, "Transient in riparian and aquatic areas. It occurred April 20, June 1, August 4 and September 14."

Lawson (TT, 1976) commented, "I would say that this species in southern Nevada is an uncommon to common spring and fall migrant."

Food. Alcorn (1946) reported in the Fallon area, "In August flocks containing 50 to 2000 birds were commonly seen feeding in the alfalfa fields during irrigation. These fields are irrigated by flooding and these birds were then able to obtain earthworms and other food items."

"At the J. S. Mills ranch, three miles west–southwest of Fallon, Churchill County on August 10, 1941, at 8:30 a.m., Mr. V. L. Mills showed my father, Mr. W. H. Alcorn, and me a place on the Mills' ranch where the ibis were doing damage to the alfalfa. The field had been irrigated (flooded) within the last two days and a small amount of water was noted to cover some parts of the fields. There were about 2000 White–faced Glossy Ibis standing in the field. The alfalfa, that was almost ready to cut (full grown) was flattened out on the ground. The birds appeared to be feeding on the field and due to their numbers they had knocked the alfalfa down and thereby made it difficult or possibly impossible to cut for hay."

"Previous to this month most of the ibis were concentrated on or near the nesting ground and attracted little attention from ranchers at this end of the valley. Now, ibis in large and small numbers are seen in all parts of the valley and many ranchers are complaining that the birds are causing trouble with the alfalfa."

"On the J. S. Mills ranch on August 12, 1941, V. L. Mills killed seven ibis and this author examined their stomachs. Three were empty and of the other four stomachs, one contained seven angle worms, two contained four angle worms and one contained one angle worm and a small dark beetle *(Coleoptera)*."

"V. L. Mills on August 19, 1941 shot three White–faced Glossy Ibis and gave them to me for analysis of their stomachs. All stomachs appeared full and identification was made on a percentage basis as follows: one contained 1% grasshoppers, 10% small black beetles and 89% angle worms; one contained 1% small black beetles and 99% angle worms; and one contained 5% grass, 5% small black beetles and 90% angle worms."

Roseate Spoonbill   *Ajaia ajaja*

Distribution. Rarely to southern California and southwestern Arizona; across Texas to the east coast (North Carolina) to south America.

Status in Nevada. Rare straggler.

Records of Occurrence. Castetter and Hill (1979) reported the first sighting was of an immature at the sewage ponds at the Nevada Nuclear Test Site, Nye County June 5 and 6, 1977.

Lawson (1979) reported Richard Voss was driving north on U.S. 95 about 48 kilometers north of Las Vegas, Clark County, near an area known as Gunsight Pass July 1, 1976. He discovered four immature birds standing in the middle of the highway. He stopped and was observing them when a passing truck put the birds to flight and struck two of them. One specimen is in the Nevada State Museum and the other is at the University of Nevada, Las Vegas.

George Austin (1977) found two at Bowman Reservoir, Logandale, Clark County August 22 and another at the Overton WMA on the same day. All were immature.

Family CICONIIDAE: Storks

Wood Stork   *Mycteria americana*

Distribution. Found casually in the interior of North America. More commonly found along the Gulf coast to South America.

Status in Nevada. Irregular transient.

Records of Occurrence. Linsdale (1951) reported, "Summer visitant, mainly to the southern portion of the state. A. B. Howell (MS) saw one along the Humboldt River, west of Beowawe in July of 1930. At the Mills Ranch on the Carson River near Fallon, two individuals were seen wading in a slough July 25, 1930. Five of the birds have been seen at one time earlier in the summer by members of the Mills family (Hall, MS)."

Alcorn (1946) reported Wood Storks from the Fallon area in June and July, 1935 and 1936.

Linsdale (1936) reported, "Russell picked up one of two skeletons found on a fence near Hiko, Pahranagat Valley, Lincoln County June 6, 1931 and Hall (MS) examined a mounted specimen May 25, 1932 in the collection of Joe and Dean Thiriot which had been taken in the spring of 1929 in the Pahranagat Valley."

A. A. Alcorn (MS) saw 42 Wood Storks in the Pahranagat Valley 10 miles southeast of Alamo October 17, 1959. In this same valley two miles south of Hiko he observed approximately 40 Wood Storks flying in a northerly direction February 21, 1965. A flock had been seen on the Thiriot ranch about that same time.

W. W. Cooke (1913) recorded a specimen from Nevada, taken in July, 1871, but no definite locality is given.

Gabrielson (1949) reported several Wood Storks present September 30, 1939 near Overton on Muddy Creek, Clark County.

Austin and Bradley (1971) for Clark County reported, "Transient in riparian and aquatic areas, with usual occurrences in May, July, August 27 and October."

Lawson (TT, 1976) commented, "There are numerous records for this species in Clark County and at Pahranagat NWR, Lincoln County. I have seen it in Las Vegas Wash and there are records from Corn Creek and Tule Springs, so I would say that this bird is an occasional to rare summer visitant."

Marabou Stork   *Leptoptilos crumeniferus*

Distribution. Southern Asia and tropical Africa. Escaped captive birds have been reported from various parts of the world.

Status in Nevada. Rare, irregular occurrence.

Records of Occurrence. Osugi (MS) reported, "On September 9, 1974, I received a phone call from Saake. He said a strange bird was observed at the Canvasback Gun Club by club members."

On this same date, Osugi went to the gun club with biologist Napier. They located the bird standing on a dike next to a pond. Later, Saake and Curran arrived. Napier and Saake photographed the bird. The bird slowly walked away from Napier and Saake and Curran approached from the opposite direction to within 100 yards, when the bird flew. Total observation time was one hour. It was identified by Saake as a Marabou Stork. This bird was repeatedly seen until September 14, and at one time an observer approached to within 50 feet before it flew.

Order **PHOENICOPTERIFORMES**

Family **PHOENICOPTERIDAE:** Flamingos

Greater Flamingo   *Phoenicopterus ruber*

Distribution. Resident of Yucatan, Cuba, the Bahamas and Brazil to the mouth of the Amazon. Wanders to Florida and along the east coast to Nova Scotia.

Status in Nevada. Rare, of accidental occurrence. From one to three have been reported from western and southern Nevada. Those seen are probably escapees. All available records are given.

Records of Occurrence. Herron (PC, 1983) reported three flamingos were seen on the lake at the MGM Grand in Reno from May 11 to 13, 1983. One was a brightly colored bird.

Saake (MS) saw one at Fernley WMA May 4, 1970 and another at this same place in 1971.

One was observed at Foxtail Lake, Stillwater Marsh, December 3, 1965 (Alcorn MS). Another, possibly the same bird seen in 1965, was seen by Napier on October 21, 1966 at this same place.

Joe Jarvis (PC) observed one at the north end of Walker Lake January 9, 1967.

Austin and Bradley (1968) reported, "On October 20, 1962 on the shore of Lake Mead near Overton, an individual was observed for 30 minutes. Although in all probability it was an escapee, its color was bright."

Order **ANSERIFORMES**

Family **ANATIDAE:** Swans, Geese and Ducks

Fulvous Whistling Duck   *Dendrocygna bicolor*

Distribution. From California to Louisiana; south into South America.

Status in Nevada. Irregular uncommon visitor.

Records of Occurrence. Linsdale (1936) states, "The only definite record for the state was furnished by Henshaw (1877) who examined one of three specimens shot by H. G. Parker, from several large flocks which visited Washoe Lake early in 1877. The species had never before been seen at the lake. Salvadori (1895) listed a skin in the Salvin Godman collection taken in winter at Washoe Lake, Nevada by H. G. Parker.''

The *American Ornithologist's Union Checklist of North American Birds* (5th ed.) reported this duck breeds at Washoe Lake. However, the author has no knowledge of their breeding in this area.

Vic Oglesby (PC) reported seeing from two to 11 of these ducks in the last two weeks of April and the first week of May, 1960 at Mason Valley WMA, north of Yerington.

Alcorn (1941) reported, ''One specimen was taken by Mr. William Fisher from a flock of about 20 birds along the Carson River, 14 miles west of Fallon November 16, 1940.''

One was shot at the Harmon Reservoir about five miles east of Fallon October 19, 1951 (Alcorn MS).

Austin and Bradley (1965) reported, ''Bob Poole collected one at Dike Lake in Pahranagat Valley, Lincoln County December 2, 1962. This appears to be the first record of this species in southern Nevada.''

San Juan Stiver (PC, 1986) reported one seen at Sunnyside WMA in 1977 and another at Overton in 1974.

Black–bellied Whistling Duck   *Dendrocygna autumnalis*

Distribution. From southern Texas and Arizona; south into South America. Occasionally to California.

Status in Nevada. A rare straggler.

Records of Occurrence. Saake (MS) reported a hunter shot one of these birds north of Yerington at Mason Valley WMA October 12, 1974. No other records are available for Nevada.

Tundra Swan   *Cygnus columbianus*

Distribution. Formerly the Whistling Swan. From Alaska to Baffin Island; south across the United States to the gulf region.

Status in Nevada. Fall, winter and spring visitant in Nevada.

Records of Occurrence. Sheldon NWR (1982) reports the Tundra Swan as an occasional migrant in spring, fall and winter.

LaRochelle (MS, 1967) reported these swans at Ruby Lake NWR are an uncommon winter transient, seldom staying more than a few days. Some 1700 passed through March 17, 1965.

From narrative reports at Ruby Lake NWR, it was reported only 19 were observed in 1969. ''In 1970, fall migration was mid–November. Sixty were observed on the refuge. In mid–December, 1971, 22 birds were recorded, in 1972, the fall migration brought the first flock of 66 swans on October 30. This was 36 days earlier than the first observations in 1971. On December 12, 1972 a peak population of 160 occurred. A large migration passed through the Ruby Valley on the afternoon and night of December 5, 1972. Several flocks of about 100 swans were observed and heard flying over the valley at a high altitude.''

Tundra Swan. Photo by A. A. Alcorn.

Kingery (1979) reported 12 swans wintered at Ruby Lake, and he (1981) reported, "Wintering Whistling Swans included 30 at Ruby Lake." He (1986) reported, "Ruby Lake NWR, Nevada, reported the most Tundra Swans—1000 flying south October 28 to November 5."

In western Nevada, these swans have repeatedly been seen in late fall, winter and spring. They have been recorded from the Reno, Lake Tahoe, Washoe Lake, Yerington, Lahontan Reservoir, Fallon and Walker Lake areas. Most of these birds are in the Fallon area where they feed in the various small ponds, lakes and reservoirs and in the marsh areas near Stillwater and Carson Lake (Alcorn MS).

Saake (MS) says, "These swans follow the sego pondweed along the lower Humboldt and Stillwater area. In mid–winter, in years when these waters freeze over, the swans migrate out and return in spring."

Mills (PC) reported an unusually large flock of about 200 flying west over his place three miles west of Fallon January 5, 1942.

Schwabenland (MS, 1966) reported at Stillwater WMA up to 8000 swans may be counted there, with a record of 12,700 swans censused January 5, 1957.

Napier (MS, 1974) reported for Stillwater WMA, "Usually appear late in October and build to late December when freezing weather forces many out. Some stay all winter on open holes in ice. Numbers again build to a peak late in January or early February. They depart early April. In the past eight years (1967–74) the peak has been smaller (3000–5000)."

Saake (MS, 1983) reported 500 permits issued yearly between 1969 and 1982 for hunting swans in Lahontan Valley. In 1983, 650 permits were issued. He (PC, 1988) reported in 1986 and 1987, although 650 permits were available, all of them were not issued. He feels the condition of the wetlands has affected the number of people hunting and is the reason for the drop in the number of permits issued.

Some sentiment against shooting these birds, especially in the first year of the hunt, resulted in a number of hunters obtaining permits when they had no intention of shooting a swan. In recent years, sentiment against hunting these swans has decreased and harvesting numbers have increased.

According to Napier (MS, 1974) the largest swans taken were two males weighing 20 lbs 2 ozs. The smallest swan weighed 8 lbs 9 ozs. The widest wing span was of an immature female (87 1/2 inches). The longest bird, tip of bill to tip of tail was an adult male (52 1/2 inches). Additional information shows an average of 157 swans harvested each year (1970–1986) during the swan season.

These swans were repeatedly seen in small marsh areas on the Reese River west of Austin from November through March (Alcorn MS).

Biale (MS) reported for the Eureka area from 1964–74 birds were seen January 3, 1971; 4 times in February; 9 in March; October 29, 1972; and 4 times in November and December.

Kingery (1987) reported 214 Tundra Swans at Kirch WMA November 10, "Five times more than last year's peak."

From December, 1959 to December, 1964, J. R. and A. A. Alcorn (MS) observed from one to 59 of these birds in the Pahranagat Valley during winter months of November through February.

Kingery (1977) reported 140 at Pahranagat Valley November 20, 1976.

In the Overton area from 1960–64, from one to five Tundra Swans were seen 13 times from November through February (Alcorn MS).

Lawson (TT, 1976) commented, "The Whistling Swan in Clark County is an occasional to rare bird. We do see one or two, I guess, every year. However, most of them occur in Lincoln County at Pahranagat NWR where it is an uncommon to common winter resident."

Trumpeter Swan   *Cygnus buccinator*

Distribution. From Alaska to Saskatchewan; south into the western United States. Also along the Gulf Coast and Atlantic coast to North Carolina. Colonies were introduced and established at wildlife refuges in Nevada and southwestern South Dakota.

Status in Nevada. Resident in Elko County and visitant probably as far south as Stillwater Marsh, Churchill County.

This swan is the largest of all American waterfowl. It weighs from 20–30 pounds and has a seven to eight foot wingspread. It was extensively hunted and by 1900, was nearly extinct in the United States. Protected by federal and state laws, its recovery has been so successful that it was removed from the list of rare wildlife in December of 1968.

Records of Occurrence. The occurrence of this bird in Nevada was reported by Henshaw (1877) who observed these birds on Washoe Lake and in the sink of the Carson River during the fall.

There is some question about Henshaw's report as indicated by W. J. Hoffman (1881) who stated Ridgway found swans numerous in the Pyramid Lake area, but did not obtain specimens and therefore, it is not certain whether they were Trumpeter or Whistler.

Linsdale (1951) reported, "A letter from Baine H. Cater on June 20, 1949 contains information that Trumpeter Swans originally captured in October, 1947 at Red Rock Lakes, Montana, had been held captive at Ruby Lake NWR with intention of releasing them in Nevada."

LaRochelle (MS, 1967) reported Trumpeter Swans were, "Transplanted to Ruby from Red Rock Lakes refuge in Montana in 1955 when 20 were brought in. Now there are 14 birds of which there are five pair. Production of 13 cygnets in 1965 is the highest recorded. Six cygnets reached flight stage. Birds winter in collection ditch and spring pools when the marsh is frozen over."

Two adults with five young were observed at this refuge June 20, 1966. Two specimens in the United States National Museum (USNM) are from this refuge. One was obtained September 20, 1966 and one April 2, 1969.

Narrative reports from 1969 at Ruby Lake NWR stated, "Winter population peaked at 29. Eight pair were left on the refuge after the population dispersed. Three birds were produced, but only two cygnets, one in each of two broods, survived to flight stage. Fall migration brought a peak of 33, nine of these were cygnets."

The 1970 report stated, "Winter population stabilized at 17 birds. Three breeding pair were left. Only one brood of four cygnets was produced. Three survived to flight stage. Fall migration was 32 birds."

Kingery (1987) reported, "Ruby Lakes had an average population of 25, with 49 at the peak January 15. Six pair raised five cygnets, three more than last year."

Curran (PC) reported a pair of Trumpeter Swans nesting on the Marys River, eight miles north of Deeth on May 21, 1974. He reported, "The nest was a very large pile, with an adult on nest."

Saake, (MS, 1975) reported, "The Trumpeter Swans use Franklin Lake more than Ruby." He reported seeing a pair every spring over the past three years along the Humboldt River, north of Elko. Often a pair and a single were seen.

Kingery (1979) reported 20 Trumpeter Swans wintered at Ruby Lake, with 19 young produced. He (1981) reported, "Trumpeter Swans had their best year ever at Ruby Lake NWR with 15 cygnets included in the December 1, 1980 count of 50."

Kingery (1983) reported only two broods, a total of nine cygnets were produced. The following year (1984) he reported, "Three pair among Ruby Lake's 21 Trumpeter Swans nested, producing five cygnets."

Later, Kingery (1986) reported, "Ruby Lake's Trumpeter Swans maintain a static population of 25; the group produced three cygnets in 1985."

Kingery (1981) reported the cygnets fledged by Trumpeter Swans during the past years at Ruby Lake had found one new probable nesting site, the Newark Valley near Eureka.

Biale (PC, 1984) reported one bird found, "At the J. D. Ranch in Pine Valley in the late spring or early summer of 1963. It had a broken wing. The bird was banded, so I sent it to Patuxent."

Kingery (1984) reported, "Seven on February 5 at Carson City."

Lawson (TT, 1976) commented, "To my knowledge we have no record of this species in southern Nevada."

Mute Swan   *Cygnus olor*

Distribution. Introduced and established in the northern United States and parts of Canada. Also found in Siberia, Africa, India and Korea.

Status in Nevada. A few domesticated birds in parks.

Records of Occurrence. Cruickshank (1971) reported two seen in the Truckee Meadows (Reno area) December 26, 1970 during the CBC.

Greater White–fronted Goose   *Anser albifrons*

Distribution. From northern Alaska to western Greenland; south to Mexico, primarily west of the Mississippi.

Status in Nevada. An uncommon fall, winter and spring visitant. Two races occur: *A. a. frontalis* and *A. a. gambelli*.

Records of Occurrence. Sheldon NWR (1982) reported the Greater White–fronted Goose as an occasional fall migrant.

H. H. Dill (1946) reported this goose present at Ruby Lakes from November 11 to 15, 1943 in numbers from 14 to 214. He reported it as a fairly common fall migrant in the Ruby Lakes area in the early 30s.

LaRochelle (MS, 1967) reported only four sightings of four single birds over a 3–year period.

Up to five birds were seen in the Reno area in the winters of 1941–43, and about 100 were seen in this same area April 8, 1960 (Alcorn MS).

Ryser (1964) reported Greater White–fronted Geese along Kleppe Lane in Reno during April of 1964.

Ryser (1985) reported, "Up until recent years this goose was a regular winter visitant in western Nevada. Concentrations of 50–300+ would appear and loiter for a few weeks in January or February in the wet fields adjacent to Kleppe Lane on the eastern edge of Sparks. With the decline of ranching the marshes and wet fields disappeared, industrial developments such as warehousing took over, and the geese stopped coming here."

Linsdale (1951) reported, "Richardson (MS) saw about 10 geese three miles east of Fernley in May, 1946."

Thompson (MS, 1988) reported a few early spring records for the Fernley area.

Alcorn (1940) obtained a specimen near Soda Lake April 3, 1938 and he (1946) reported this goose as transient but not numerous in the Lahontan Valley and reported records for March, May, September and December.

In the fall of 1959 these geese were reported from various areas of the Lahontan Valley. Alex Oser reported about 100 in the Stillwater area while hunting in December. About 25 were seen at Carson Lake December 27. Twenty–three were seen west of Fallon January 17, 1971 (Alcorn MS).

Napier (MS, 1974) reported about 20 at Carson Lake on February 8, 1971. Mills reported 17 east of Fallon March 14, 1971.

Schwabenland (MS, 1966) reported at Stillwater WMA up to 10 birds were seen each fall, but they didn't stay long. At the same refuge Napier (MS, 1974) reported about 25 present in late fall of most years.

Saake (MS, 1981) reported the Greater White–fronted Goose as the earliest migrant with a few seen at Washoe Lake; Smith, Mason and Lahontan Valleys; and Ash Meadows in September.

Biale (MS) reported these birds in the Eureka area April 3, 1960 and 17 birds on October 31 and November 1, 1971. Biale (PC, 1984) reported he has not seen any since then.

Kingery (1986) reported, "Five at Kirch December 2 to January 6." Cottam (1936) saw six of these geese eight miles south of Alamo January 1, 1925.

A. A. Alcorn (MS) examined one shot by a hunter in the Pahranagat Valley in December, 1959. He also saw two near Hiko January 9, 1965. In Moapa Valley he saw two flocks of 12 and six birds September 22, 1961. He banded one February 5, saw another February 9, 1963 and saw four on March 18, 1964.

George Austin (1970) reported, "There are several records for northern Nevada (Alcorn 1946) but only two for southern Nevada (Cottam 1936, Grater 1939). There are additional records of single birds at Corn Creek (field station of Desert NWR) on 8 June 1964, late September 1962 and 28 December 1964 (Hansen, pers. comm.), Tule Springs Park (13 mi. N.W. of Las Vegas) from 16 February to 2 April 1967 and Overton WMA on 4 October, 1964."

Lawson (TT, 1976) commented, "This goose is a rare to uncommon winter visitor in southern Nevada. We do see them at Lake Mead, Tule Springs, and even on the International Golf Course behind the Hilton Hotel. A flock of 17 wintered there one year."

Kingery (1987) reported six Greater White–fronted Geese at Las Vegas March 5.

Graylag Goose   *Anser anser*

Distribution. Northern Europe, northern Asia, northeastern India and China; domesticated in the United States.

Status in Nevada. Rare.

Records of Occurrence. Kingery (1978) reported one Graylag Goose seen at Las Vegas flying freely February 26 through May 13. Vince Mowbray (PC) regarded this bird as an escapee. These geese are a common species in wildlife breeder's collections and escapees may occasionally be seen.

Snow Goose   *Chen caerulescens*

Distribution. The Lesser and Greater Snow Goose and the Blue Goose have been consolidated into one species with the common name "Snow Goose." Across the arctic coasts of North America to Greenland; south to southern Mexico.

Status in Nevada. This goose is a common fall and spring visitant, especially at the Stillwater and Carson Lake areas of Churchill County. A few winter in the southern part of the state. The blue color phase birds are occasionally seen.

Records of Occurrence. Sheldon NWR (1982) reported the Snow Goose as an occasional fall migrant.

LaRochelle (MS, 1967) at Ruby Lake NWR reported only one sighting of a single bird with a group of Canada Geese during March of 1965.

The narrative report for Ruby Lake NWR noted one Snow Goose present from November 7–13, 1971. In 1972, the report stated one Snow Goose spent most of November on the refuge in the company of Canada Geese.

Gabrielson (1949) reported, ''10 present on Washoe Lake, Washoe County, November 16, 1934.''

Ray Emerson (PC) reported in the Stillwater area in the fall of 1940 the first Snow Geese arrived October 12.

Oglesby (PC) estimated 30,000 Snow Geese in Lahontan Valley December 5, 1959 during an aerial wildlife census of the valley.

Napier (MS, 1974) reported, ''They usually arrive early in October and peak in November. Winter forces birds out of the valley. Spring migration is quite small compared to fall. The last birds move through in March or sometimes early April. Stillwater receives limited Snow Goose use (peak 1000 +) compared to Carson Lake. Pastures and flooded grasslands there provide more browse.''

David B. Marshall and J. R. Alcorn (1952) reported one of the ''blue'' form at Carson Lake in December, 1950. Other sightings of these geese were reported from there each winter for over 30 years.

Schwabenland (MS, 1966) reported at Stillwater WMA the ''blue'' form was, ''seen now and then.'' He observed five with a flock of 1500 Snow Geese in December, 1964. A group of four were seen at the Carson Lake pasture December 3 and 4, 1964.

Napier (MS, 1974) reported he saw one of the ''blue'' form at Stillwater WMA November 9, 1966 with about 500 Snow Geese. He saw five other ''blue geese'' in a flock of 3000 Snow Geese November 13, 1966. One was a lone adult and the other four were a family group.

Saake (MS, 1983) reported each year a few ''Blue Geese'' were seen. He noted they are more frequently seen during spring migration and the greatest number seen in western Nevada in one year was seven.

Kingery (1986) reported, ''15,000 at Stillwater February 28.''

Large numbers of Snow Geese are sometimes seen at the north end of Walker Lake (Alcorn MS).

One was seen at Bartines Pond in the Eureka area June 4, 1962 (Alcorn MS).

Biale (MS, 1983) reported up to six seen in the Eureka area in the fall every two or three years. He noted they are usually seen with a flock of Canada Geese.

In the Pahranagat Valley groups of three to eight Snow Geese were occasionally seen in November, December, January and March, but not in every year (Alcorn MS).

In Moapa Valley, two were seen in November and December of 1959 and January 3, 1960 (probably the same birds); two on October 31, 1960; three on January 10, 1961; one March 9–14, 1962; one January 22 and March 20 of 1963; 16 on February 21 and three on March 18 of 1964 (Alcorn MS).

Lawson (TT 1976) commented, ''The Snow Goose is an uncommon to common winter visitor here in southern Nevada. We do get threes and fours of them in Overton and at Tule Springs, at Lake Mead and Lake Mohave. I have seen them below Davis Dam also.''

Ross' Goose   *Chen rossii*

Distribution. From Arctic Canada to Saskatchewan; south in the western United States, Texas, the Gulf Coast and Florida.

Status in Nevada. Winter visitant.

Records of Occurrence. Carol Evans, wildlife biologist, reported (MS, 1984) one Ross' Goose at Ruby Lake NWR March 17, 1984. It was with a group of 10–15 Canada Geese.

Lettenmaier (PC) saw a Ross' Goose on Virginia Lake in January of 1969.

Kingery (1982) reported one at Reno April 15. Gary Rosenburg (1982) reported one in Truckee Meadows during the CBC December 19, 1981.

Marshall and Alcorn (1952) reported, "This species is not listed by Linsdale. Hunters report seeing occasional geese of this species in Churchill County since 1933 and it is not uncommon for one or more to be killed each fall. We examined two killed at Carson Lake on November 4, 1951. One of these birds was sent to the Charles R. Conner Museum at Pullman, Washington and the other is to be mounted for display at the Sagebrush Cafe in Fallon. A larger number of these birds than usual turned up in the Fallon area in 1951, when a minimum of five were shot by hunters."

At Stillwater WMA, two of these geese were treated for botulism November 2, 1952 (Alcorn MS).

An immature female shot in the Fallon area December 16, 1962 had been banded at Smiley, Saskatchewan October 17, 1962 (Alcorn MS).

Eugene Duffney shot one three miles east–northeast of Stillwater in November of 1964 (Alcorn MS).

Napier (MS, 1974) reported a few present at Stillwater WMA with Snow Geese. Several are shot by hunters in most years.

Saake (PC, 1983) reported, "These geese constitute 5% of the white geese that are seen in the state." He noted most are seen in western Nevada.

Kingery (1984) reported four at Lund March 12–26. He (1982) reported three at Alamo March 21.

Austin (1970) reported, "Two birds were at Overton WMA on 20 November, 1968 when one was shot. The other remained in company with seven Snow Geese until 27 November, 1968. This is the first southern Nevada record." He (1976) reported one at Lorenzi Park (Las Vegas) November 23, 1970 and another April 5, 1971.

Lawson (TT, 1976) commented, "I think there are five to 10 records of this bird here in Clark County, but at best it is only an occasional to rare winter visitor."

Heilbrun and Rosenburg (1981) reported one on the Desert Game Range December 27, 1980 during the CBC.

Kingery (1984) reported, "A surprising count of 17 Ross' Geese at Las Vegas December 26; one stayed to winter and a second possibly wintered at Glendale, Nevada, 60 miles east of Las Vegas."

Emperor Goose   *Chen canagica*

Distribution. From Aleutians south to California.

Status in Nevada. Accidental.

Records of Occurrence. Overton WMA manager, Harold Peer, (PC) examined a hunter–killed Emperor Goose at the Overton WMA December 11, 1960.

Brant   *Branta bernicla*

Distribution. From arctic North America along the west coast to Baja California and the east coast to North Carolina.

Status in Nevada. An uncommon straggler.

Records of Occurrence. Linsdale (1951) reported, "Two early records are for Pyramid and Washoe Lakes."

This author saw one at Anaho Island, Pyramid Lake March 24, 1961 (Alcorn MS).

Linsdale (1951) reported V. K. Johnson and V. Woodbury identified one on Virginia Lake, Reno in the fall of 1947.

Scott (1967) reported one at Reno, May 5, 1967.

Oglesby (PC) reported a hunter shot one at Old River Reservoir near Fallon October 5, 1957. This bird had been banded on the Lower Kashnuk River July 16, 1950 (Alcorn MS).

Napier (MS, 1974) reported one shot by a hunter at Stillwater WMA was brought through the checking station on October 25, 1969.

Mills (PC) reported one at Rattlesnake Reservoir, near Fallon April 29 and 30, 1970.

Mills (PC) reported one May 12, 1970. Saake, Curran, and Savage reported 14 at Carson Lake, south of Fallon on October 28, 1970.

Biale (MS) reported one in the Eureka area April 24, 1960. The author obtained one 20 miles west of Eureka May 16, 1960. This was probably the same bird seen by Biale in April.

Grater (1939) reported Brants in the Virgin Basin in March and August, 1938.

Baldwin (1944) saw one on Lake Mead January 14, 1944; and about 15 at Lake Mead March 30 and 31.

Canada Goose   *Branta canadensis*

Distribution. From arctic Alaska eastward to Greenland; south across most of the United States to Mexico.

Status in Nevada. Permanent resident in western Nevada, a summer resident in northern part of the state and winter resident in the southern part. Transient over entire state. Four races have been reported in Nevada: *B. c. moffitti, B. c. occidentalis, B. c. leucopareia* and *B. c. minima*.

Records of Occurrence. Canada Geese have been reported from all areas of the state, occurring mainly in aquatic areas of the various valleys.

Sheldon NWR (1982) reported the Canada Goose as a permanent resident, nesting observed.

LaRochelle (MS, 1967) reported at Ruby Lake NWR these geese arrive from February to March and build up to 300 with nesting and production of 300–450 young. They leave the area in early January.

The narrative reports for this refuge reveal in 1969 and 1970 an estimated 250 goslings were raised to flight stage. In 1971, an estimated 215 were raised and in 1972, a total of 110 breeding pair raised 270 goslings.

The author interviewed Wuzzie George, an elderly Indian woman, on March 26, 1969. George reported catching Canada Geese with nets in the Stillwater Marsh area.

Saake (MS, 1983) reported Canada Geese nest intermittently along the Humboldt River with concentrations south and west of Winnemucca and on both sides of Elko. Many nest at Franklin Lake.

Oglesby (MS, 1960) reported the largest nesting concentration of Canada Geese in the state at Washoe Lake. He counted 150 nests there in 1956. In many parts of western Nevada, and especially in the Reno area, these geese are easily observed on a daily basis throughout the year.

Saake (MS) reported five Aleutian Canada Geese on the west shore of Washoe Lake in mid–November, 1980.

In the Fallon area, all four races may be observed in one day and are frequently seen in the hunter's possession. Mills saw a flock of 54 Cackler Geese in this area February 22, 1969 (Alcorn MS).

Mills reported a brood of baby Canada Geese in the Fallon area April 7, 1975; two broods April 19, 1970; and a brood of five young, about a week old, April 24, 1945.

Napier (MS) reported five eggs in a nest May 25, 1969 at Stillwater WMA. This probably represents a second effort at nesting because most geese in this area hatch in April.

Napier (MS, 1974) for Stillwater WMA reported, "A common nester, breeding population is nearly resident. Most move no farther than Lovelock, Yerington or Washoe Lake."

"In July, locals congregate at Stillwater and population increases to 1000. Numbers then drop off to 200–400 until migrants arrive. Peak at 2000+ in December. Many of these are Lesser Canada Geese. A few Cackling Geese are occasionally seen."

In the Eureka area, an adult was seen on a nest containing six eggs May 16, 1960 (Alcorn MS). Biale reported seeing Canada Geese in the Eureka area from 1964–74 a total of 5 times in January; 8 in February; 14 in March; 9 in April (nest April 12, 1970); 4 in May; once in June; none in July; once in August; twice in September; 4 in October; 23 in November; and 42 times in December.

Oglesby (MS, 1960) reported Canada Geese nest in Diamond Valley, Eureka County and in Newark Valley and Comins Lake, White Pine County.

In the Pahranagat Valley from 1959–65, these geese were recorded 6 times in January; twice in February; 8 times in March; once in April; none from May through October; 3 in November; and 5 times in December. A hunter shot one in Pahranagat Valley of the race *minima* October 24, 1964 (Alcorn MS).

Gullion et al., (1959) reported, "A winter visitor, *B. c. moffitti* moves into this area as early as October 23 (1951, Mohave Valley) and remains in some abundance in the Mohave, Virgin, Moapa, (1300 feet elev., Clark County) and Pahranagat Valleys until as late as March 31, (1949 and 1951). Two birds taken by hunters on November 4, 1951 near Bunkerville (1650 feet elev., Virgin Valley, Clark County) had been banded at Blackfoot Reservoir, Caribou County, Idaho as adults. One was banded in June, 1947 and the other in June, 1951."

"*B. c. leucopareia,* the 'Lesser Canada Goose' was identified among flocks of *B. c. moffitti* seen in the Virgin Valley on November 4 and 23, 1951. A single *B. c. minima,* the 'Cackling Goose' was seen in a flock of the larger races along the Virgin River west of Bunkerville on November 4, 1951. Other birds of this race were seen at Overton (1300 feet elev., Moapa Valley) on February 28, 1952."

A. A. Alcorn (MS) reported for Moapa Valley about 1000 of these birds were in that valley January 22, 1963. He saw three Cackler *(minima)* Geese in this valley October 9, 1961. J. R.

and A. A. Alcorn (MS) recorded these geese in the Moapa Valley from 1959 to 1965 16 times in January; 13 in February (48 banded February 5, 1963); 16 in March; 5 in April; 4 in May; twice in June; July 24, 1964; August 6, 1964; September 17, 1961; none in October; 5 in November; and 6 times in December (Alcorn MS).

Austin and Bradley (1971) reported Canada Geese are, "Winter resident in riparian and aquatic areas of Clark County. Common occurrence from September 19 to May 2. Unseasonal occurrence on August 9."

Lawson (MS, 1981) reported seven Aleutian Geese at Overton WMA from November 8 to 22, 1979.

Ruddy Shelduck   *Tadorna ferruginea*

Distribution. Southeast Europe and Central Asia.

Status in Nevada. Accidental. Possible escapees from a game farm or zoo.

Records of Occurrence. Saake (MS) reported one shot August 19, 1976 by a hunter at the Alkali Lake WMA 13 miles north of Wellington, Lyon County. The bird was mounted and is on display at the Mason Valley WMA.

Approximately 10 were seen by Saake at this same lake August 19 and 25, 1980 and again, September 8, 1980. Saake (MS) reported, "They always seemed to stay as a distinct group and did not mix with other waterfowl in the area."

Saake (MS) later reported, "On May 27, 1981, I observed four Ruddy Shelducks on the extreme north shore of Pyramid Lake. There were about 2000 molting Canada Geese in the general area, and the shelducks were with one of the largest flocks. While most of the geese were flightless, the four shelducks were capable of flight."

Common Shelduck   *Tadorna tadorna*

Distribution. Breeds from British Isles to western Siberia; winters south to Egypt, China and Japan.

Status in Nevada. Accidental. Probably escapees with some recent reproduction in the wild.

Records of Occurrence. Kingery (1977) reported, "The seven at Fallon in September rank as likely escapees; three at the same place last spring, also were classified as escapees."

Kingery (1978) reported shelducks at Walker Lake for two years: "Of the seven last fall, one remained this fall—a male seen September 28."

Kingery (1980) reported, "Western Nevada has recorded for four years straight, a flock of Common Shelducks, usually three to four, but one aerial flight this year counted seven at Stillwater WMA."

Wood Duck   *Aix sponsa*

Distribution. From British Columbia to Nova Scotia; south to Florida and central Mexico.

Status in Nevada. Summer resident and transient in western Nevada; winter resident in southern Nevada.

At present time, this duck is a fairly common, but inconspicuous, summer resident in limited numbers in western Nevada. Because of its secretive nature it may have been overlooked in earlier years.

Records of Occurrence. Linsdale (1936) reported, "Formerly rather uncommon in western part of the state. One pair reported seen in July, 'among the cottonwoods of the Truckee', in Washoe County (Ridgway, 1877). Henshaw (1877) listed this species as rather uncommon near Carson City, Ormsby County."

Sheldon NWR (1982) reported the Wood Duck as a rare transient visitor in spring and fall.

Robert Quiroz (PC) saw a pair of Wood Ducks fly from Columbia Creek, Elko County May 26, 1968. The date suggests nesting in that area.

LaRochelle (MS, 1967) reported for Ruby Lake NWR the Wood Duck is a fall and winter visitor of up to 75 birds. He commented, "I banded the first of these in 1965 (7 banded). One was returned from southern California's Lake Isabella."

Kingery (1987) reported, "One to two appeared at Ruby Lake NWR, Nevada August 18 and September 30."

In Reno a pair was observed February 4, 1941 and in 1942, pairs were observed twice in March and April and December 1 (Alcorn MS).

Ryser (1967) reported a pair nesting across the Truckee River from the Thomas Ranch June 11, 1967.

Mills (PC) saw three broods of Wood Ducks on the Truckee River June 28, 1959. One brood consisted of nine young. He was unable to get a count of the other broods. All were half–grown.

In an effort to furnish nesting sites for these ducks, this author, with the help of Oglesby, put up five nest boxes in the same vicinity Mills had seen young in in 1959. They were put up April 6, 1961.

Mills saw a brood of over nine about one–third grown July 14, 1963. They ducked into the willows when they saw him and were difficult to count.

Wilson and Norr (1949) reported, "Marriage observed a Wood Duck nest containing nine eggs in an old magpie nest near Carson City, Nevada, May 22."

These ducks were repeatedly seen in the spring and summer months by Mills and Alcorn, along the Carson River from Fort Churchill to the upper end of Lahontan Reservoir. Two broods were observed in this area July 4, 1961 (Alcorn MS).

In the Fallon area from 1959 to 1969, Wood Ducks were observed once in January; twice in February; 6 times in March; 9 in April; 13 in May; 7 in June; once in July; 6 in August; 7 in September; 3 in October; and 4 times in November (Alcorn MS).

Broods of young have been seen in the Fallon area each year on the Carson River and nearby canals since 1964. Kenneth Kent (PC) reported on his ranch 14 miles west of Fallon, the Wood Ducks often flew out of the alfalfa fields before the first cutting of alfalfa in June. A visit to his ranch April 26, 1971 revealed 34 Wood Ducks, mostly pairs, that flew from one of his dry alfalfa fields. Others were repeatedly seen to fly from this field in May and June.

While it was not determined what these ducks were doing in this field, the cottonwood tree–lined Carson River flows along the north edge of these fields and nest boxes have been maintained at this place for the past 25 years. Examination of these boxes May 13, 1968 revealed the boxes contained one, two, and 10 eggs; also a female flew from a box containing 15 eggs. On June 8, 1968, there were 18 eggs covered with down in the box containing only two eggs May 13. On June 8, a female was seen on the river at this place. As she flapped along on the water in a crippled fashion, six young judged to be two weeks old, were seen (Alcorn MS).

Saake (MS, 1984) reported nesting along the Truckee, Carson and Walker Rivers and feels that these ducks are limited by the lack of nesting sites.

Schwabenland (MS, 1966) reported at Stillwater WMA this duck was observed once or twice over a 3–year period.

Napier (MS, 1974) reported several Wood Ducks are shot each year at Stillwater Marsh.

Biale (MS) reported seeing one female Wood Duck killed by a hunter in Diamond Valley (Eureka area) November 21, 1964.

Kingery (1978) reported one Wood Duck at Diamond Valley October 8–15.

Biale (PC, 1986) reported one male in Antelope Valley October 14, 1984.

Castetter and Hill (1979) reported at the Nevada Test Site one pair of Wood Ducks were seen October 4–8, 1975.

In the Moapa Valley at Overton WMA, two Wood Ducks were seen October 31, 1960.

Austin (1970) reported, "The remains of an individual found in October (Gullion, 1952) is the only previous southern Nevada record. The species has recently wintered in small numbers from December 8 (1967, Corn Creek) to March 28 (1968, Lorenzi Park, Las Vegas). A male was found dead at Corn Creek in the spring of 1966."

Kingery (1985) reported, "The first Wood Duck in three years at Las Vegas was pursued by a Cooper's Hawk when last seen."

Lawson (TT, 1976) commented, "This duck is a rare to uncommon winter resident here in Clark County from October until March."

Green–winged Teal   *Anas crecca*

Distribution. The American group *carolinensis* is found from Alaska to Newfoundland; south through Mexico to Honduras. The *crecca* group is known as the Common or European Teal in Europe.

Status in Nevada. Summer resident in limited numbers in north and winter resident in south. Most abundant during spring and fall migrations. One record of the *crecca* group was reported from west of Fallon.

Records of Occurrence. Sheldon NWR (1982) reported the Green–winged Teal as a summer resident, recorded all year.

W. P. Taylor (1912) reported a nest containing five eggs on the Quinn River, Humboldt County and F. S. Hanford (1903) reported this duck as a common breeder at Washoe Lake, Washoe County.

The presence of a few pair in summer months at Wildhorse Reservoir may indicate some nesting there (Alcorn MS).

LaRochelle (MS, 1967) at Ruby Lake NWR reported, "Though some 250–400 are present in the spring, no pairs have been seen nor young sighted. They are present year–round except in severest winters. Their numbers seldom exceed 1000."

Saake (MS, 1981) reported them mostly migrant with very limited nesting in Nevada.

In the Reno area these ducks were abundant in winter months from October through March, with the largest numbers (up to 100 birds) observed in November and December (Alcorn MS).

Alcorn (1946) reported them present in the Fallon area in summer in limited numbers; a few young were captured and raised until fully grown.

A teal collected by Charles York west of Fallon in November, 1969 was of the *crecca* group (Alcorn MS).

Schwabenland (MS, 1966) reported at Stillwater WMA, "This duck is very abundant in spring and fall with up to 90,000 in the fall of 1964. Up to 60,000 in spring. A little nesting occurs here, occasionally a brood is sighted and a few pair are seen in summer."

Napier (MS, 1974) for Stillwater WMA reported, "Generally present in abundant numbers during spring and fall migration. A few stick it out during winter. I believe a few nest in the valley. Pairs are recorded on the breeding surveys. I cannot recall seeing a brood, however. On April 4, 1972, Osugi and I saw a male European Teal with a flock of Green–winged Teal at Mahala Slough marsh area."

A few Green–winged Teal may nest along Reese River west of Austin. A male was seen there June 4, 1962 and another June 1, 1964. In this area limited numbers are recorded throughout the year (Alcorn MS).

Biale (MS) recorded these birds 81 times in the Eureka area between April, 1964 and December, 1967. They were seen in all months and a brood of young were observed July 2, 1967.

From January, 1970 to December, 1973 he saw these ducks 9 times in January; 11 in February; 17 in March; 12 in April; 13 in May; twice in June; none in July or August; once in September; 3 in October; 7 in November and 11 times in December.

In eastern Nevada, one seen at Comins Lake June 5, 1962 was thought to be part of a nesting pair. A pair on a small pond near Baker were thought to be nesting May 31, 1967 (Alcorn MS).

Kingery (1987) reported 1805 Green–winged Teal at Kirch WMA February 9.

J. R. and A. A. Alcorn (MS) reported for Pahranagat Valley from February, 1960 to March 1965, they saw this duck twice in January and February; 12 times in March; 8 in April; 9 in May; June 14, 1965; none in July, August or September; twice in October; 5 in November; and once in December. The largest number seen was during winter months when up to 3800 were censused in the valley (Alcorn MS).

In Moapa Valley from January, 1960 to April, 1965, this duck was seen 6 times in January; 8 in February; 12 in March; 7 in April; 3 in May and June; none in July; twice in August and September; 4 in October; twice in November; and once in December (Alcorn MS).

Lawson (TT, 1976) commented, "For Clark County, this teal is one of our more abundant migrant and wintering species."

American Black Duck *Anas rubripes*

Distribution. From northern Saskatchewan to Newfoundland; south to the Gulf states and Florida. Has spread westward in recent years.

Status in Nevada. A rare straggler to Nevada.

Records of Occurrence. A Black Duck, banded by the New York Conservation Department, was recovered by W. D. Lewis at six miles west–southwest of Fallon November 4, 1963. The duck, an adult female was banded at Wilson Hill, New York on September 20, 1962.

Ryser (1965) reported one American Black Duck at Virginia Lake, Reno, April 1, 1965.

Cruickshank (1965) reported eight at Overton December 29, 1964 during the CBC.

Mallard   *Anas platyrhynchos*

Distribution. From Alaska and Canada; south casually to Panama.

Status in Nevada. A common resident in most valleys at marshes, rivers and lakes. Most move out of northern Nevada during cold winter months when lakes and streams are frozen and out of southern Nevada during the hot summer months.

Linsdale (1936) reported Mallards from 10 of 17 Nevada counties. Recent records cover all 17 counties. Because of the large number of sightings and locality records, no effort is made to list all of the occurrences.

Records of Occurrence. Sheldon NWR (1982) reported the Mallard as a common permanent resident.

LaRochelle (MS, 1967) reported at Ruby Lake NWR, "Our most numerous duck. To be found year round, up to 7000. Commonest nesting duck with 150 to 275 pair bringing off from 7.2 to 7.8 young per pair. Generally band 500–800 each year."

Kingery (1981) reported 3000 Mallards at Ruby Lake in 1980.

Schwabenland (MS, 1966) reported this duck as abundant all year at Stillwater WMA with the greatest numbers in fall, when 12,000–15,000 arrive. They nest in this area with, "500 to 600 birds raised each year."

Napier (MS, 1974) reported, "Common nesters in the valley and about fourth in abundance next to the Cinnamon Teal, Redhead and Gadwall. One of the earliest ducks to nest here, the first brood usually hatches in early April (late March if weather is mild). Present all year, but population during migration is relatively low. A few birds stay along canals and ditches during the winter."

Biale (MS) reported seeing these ducks in the Eureka area on 128 different days and in all months from 1970 to 1973.

In the Hiko vicinity from January, 1960 to March, 1965 these birds were recorded in all months (Alcorn MS).

A. A. Alcorn (MS) frequently recorded these birds in Moapa Valley in January and February with up to 200 observed some days. A search of suitable area failed to reveal any of these ducks in this valley on May 30, 1963. They were not seen in summer months except for 11 seen on July 24, 1963 and 14 exactly a year later on July 24, 1964 (Alcorn MS).

Austin and Bradley (1971) reported for Clark County, "Winter resident in riparian and aquatic areas, with common occurrence from August 9 to June 7. Unseasonal occurrence on June 30 and July 7."

Lawson (TT, 1976) commented, "We don't have them in the summertime in southern Nevada."

Northern Pintail   *Anas acuta*

Distribution. From northern North America; south to Columbia. Also present in Europe, Russia, Siberia and Asia.

Status in Nevada. Summer resident in all except the southern part of the state. Winter resident in southern Nevada.

Records of Occurrence. Sheldon NWR (1982) reported the Northern Pintail as a permanent resident.

LaRochelle (MS, 1967) reported 19 pair nesting at Ruby Lake NWR in 1965.

Kingery (1985) reported, "Ruby Lake NWR had 6800 Northern Pintails . . . way above normal."

Kingery (1979) reported 2000 pintails flew over Carson City February 27.

In the Fallon area from January, 1960 to January, 1974 these ducks were recorded each month. However, they were seen on fewer occasions and in fewer numbers in December and January. They were most abundant during spring and early fall migrations (Alcorn MS).

Schwabenland (MS, 1966) reported these ducks as very abundant at the Stillwater WMA with up to 70,000 birds in fall and up to 60,000 in spring. Nesting resulted in 900 young raised during the summer of 1965.

For this same area, Napier (MS, 1974) reported this is one of the most abundant ducks in spring and fall migrations. Spring peak occurs late February or early March. Most leave the area by December. He reported the pintail is a minor nester in the Stillwater area compared to other species and noted, "An early nester, similar to the Mallard."

Biale (MS) saw these birds in the Eureka area between January of 1970 and January of 1974 a total of 5 times in January; 8 in February; 17 in March; 10 in April; 16 in May; 6 in June; 7 in July and August; 13 in September; twice in October; none in November; and 5 times in December. Although none were seen in November during this time, they were often seen during this month in other years.

Biale (MS) reported a nest with eight eggs under a rabbitbrush about 15 yards from water May 30, 1964.

J. R. and A. A. Alcorn (MS) reported this duck in Pahranagat Valley from January, 1960 to March, 1965 a total of 4 times in January; 3 in February; 13 in March; 8 in April; 13 in May; 4 in June; twice in August (about 500 seen August 11, 1960); October 18, 1963 (about 1000 seen); 5 times in November; and 100 seen December 23, 1964.

At the north end of Upper Pahranagat Lake on May 4, 1962, a female pintail flew from a nest containing six eggs. It was located about 40 yards from the water's edge in a grassy situation. Another nest was seen May 23, 1962 in this same area (Alcorn MS).

J. R. and A. A. Alcorn (MS) saw this duck in the Moapa Valley, Clark County from January, 1960 to February, 1965 a total of 9 times in January; 7 in February; 9 in March; 5 in April, May and June; on July 24, 1964; 3 in August; and twice in September, October, November and December.

Austin and Bradley (1971) reported, "A winter resident in riparian and aquatic areas of Clark County. Common occurrence from August 5 to May 7. Unseasonal occurrence on June 7."

Lawson (TT, 1976) commented, "I don't have any summer records for this species here (southern Nevada), but it is a common migrant and winter resident."

Kingery (1983) reported, "Pintails produced one pair with seven young at Las Vegas, an unusual breeding record."

Blue–winged Teal   *Anas discors*

Distribution. From east–central Alaska to Nova Scotia; south through Central America to Ecuador and Brazil.

Status in Nevada. Uncommon summer resident and transient. The female of this duck is often confused with the female of the abundant Cinnamon Teal. Because the two species are usually indistinguishable in the field, they are frequently misidentified by hunters.

Records of Occurrence. Linsdale (1936, 1951) reported this duck as a summer resident and rather common about the wet meadows during the nesting season.

From 1930 to 1980, the author's experience has been the opposite of that reported by Linsdale, as this teal was more commonly found in eastern rather than western Nevada. However, the search for this duck in the earlier part of this period was only in the Fallon area.

Sheldon NWR (1982) reported the Blue–winged Teal as an occasional summer resident with nesting known in the area.

In Paradise Valley, Humboldt County, a pair was observed April 25, 1969 (Alcorn MS).

LaRochelle (MS, 1967) reported the first known nesting of this teal at Ruby Lake NWR was observed in 1965, when he found a pair with a nest of eight eggs, from which seven young hatched.

Kingery (1983) reported, "15 pair of Blue–winged Teal at Ruby Lakes produced 40 young."

Kingery (1977, 1978) reported 22 and 10 Blue–winged Teal in the Reno area.

A male was seen near Minden May 8, 1942 (Alcorn MS).

Alcorn (1946) reported, "Only one record in the Fallon area is available for this teal. On April 1, 1939, Mills saw and obtained a solitary male at Mahala Slough."

Later the author wrote, "Mr. Vernon Mills reported June 14, 1959 that within the past 10 days he has seen Blue–winged Teal at several places. The first was two pair at the Harmon Pasture, four miles east of Fallon. He then saw a pair four miles north of Fallon. I went with Mills to the area four miles north of Fallon June 14, 1959 and we saw one, crescent–faced, male Blue–winged Teal. Mills said that last year was the first year he had seen Blue–winged Teal in the Fallon area during the breeding season (May to June). He thought that several pair may have nested here in 1958."

From 1959 to 1970 these birds have been seen by various people in the Fallon, Stillwater and Carson Lake area once in March; 6 times in April; 7 in May; 10 in June; 5 in July; and December 16, 1963 (Alcorn MS).

Schwabenland (MS, 1966) reported at Stillwater WMA occasional pairs were seen in spring and summer months. Possibly some nesting occurs.

Napier (MS, 1974) reported, "Occasional pairs of this duck in the Lahontan Valley during the breeding season. Probably more around Carson Lake than other places in the valley."

Alcorn and Biale (MS) reported seeing these ducks in the Eureka area between April of 1964

A male Blue-winged Teal. Photo by A. A. Alcorn.

and July of 1972 a total of once in March; 5 times in April; 19 in May; June 19, 1966; and July 3, 1972. At an earlier date, one was seen October 2, 1960 in this area.

Biale (PC, 1984) observed a pair at Bartine's Ranch April 14, 1984. He reports they are uncommon in the Eureka area.

In eastern Nevada a male was seen at Comins Lake June 5, 1962 (Alcorn MS).

In the Pahranagat Valley from two to four of these ducks were seen from 1960–65 twice in April; 6 times in May; June 3, 1960; and July 4, 1961 (Alcorn MS).

Gullion et al., (1959) reported for southern Nevada, ''Noted in the Pahranagat Valley on May 8 and 14, 1952.''

J. R. and A. A. Alcorn (MS) saw a few birds in the Moapa Valley from 1960 to 1965 in April and May.

Austin and Bradley (1971) reported for Clark County, ''Transient in riparian and aquatic areas. Common occurrences on April 7 to May 15 and from August 20 to October 17. Unseasonal occurrence on July 26, November 27 and December 22.''

Kingery (1978) reported, ''Blue–winged Teal, regarded as rare in Nevada, occurred at Reno, Las Vegas and Eureka and five pair produced 20 young at Ruby Lake.''

Lawson (TT, 1976) commented, ''Rare to uncommon migrant and summer resident. We do have breeding records at Pahranagat and at Vegas Wash.''

Cinnamon Teal   *Anas cyanoptera*

Distribution. From southern British Columbia to central North Dakota; southward to Columbia and South America.

Status in Nevada. Common summer resident and transient. The Cinnamon Teal is one of the most common nesting ducks in Nevada.

Records of Occurrence. Sheldon NWR (1982) reported the Cinnamon Teal as an uncommon summer resident and occasional winter resident. Nesting occurs in the area.

LaRochelle (MS, 1967) reported for Ruby Lake NWR, "At times our most numerous teal, with up to 700 on the area. Leaves early fall and seldom found later. Has varied from 45 to 110 pair with an average brood of 6.6."

Kingery (1987) reported 9000 Cinnamon Teal at Ruby Lakes NWR, "way above normal."

In the Fallon area from January 1, 1961 to January 1, 1966 these ducks were seen 4 times in January; 5 in February; 30 in March; 32 in April; 44 in May; 41 in June; 19 in July; 4 in August; once in September; 10 in October; and none in November and December, although several were shot by hunters in both of these months (Alcorn MS).

Schwabenland (MS, 1966) reported this duck as abundant in fall and spring at Stillwater WMA with up to 15,000 sighted in the fall of 1965. "Nesting second only to Redhead with 2000 to 2500 young produced each year."

At this same refuge, Napier (MS, 1974) reported, "Records for each month of the year. Very few for December and January. They begin increasing in March. Most abundant nesting duck on Stillwater. Since 1967 the Cinnamon Teal has outnumbered the Redhead in nesting. Cinnamon Teal does not show a definite hatching peak. Appears birds begin nesting when they arrive, so hatchings are spaced from early May into June. Cinnamon Teal are leaving by late September, for few show up in the bag during the opening weekend of the hunting season."

Biale (MS) reported seeing these teal in the Eureka area from January of 1964 to December of 1967 twice in February; 6 times in March; 17 in April; 53 in May; 9 in June; and July 4, 1965. From January, 1970 to December, 1973, Biale reported them 3 times in February; 13 in March; 12 in April; 17 in May; 10 in June; and five times in July.

The author's records for the Eureka area reveal a flock of about 10 on September 22, 1960; a flock of 30 on August 10, 1961; nine on August 15, 1961; and about 20 on August 29, 1961 (Alcorn MS).

In Pahranagat Valley from 1960–1965, these birds were observed 10 times in March; 13 in April and May; 3 in June; once in July, August, September; and about 50 birds were seen October 18, 1963 (Alcorn MS).

In the Moapa Valley from 1960 to 1964, inclusive, these ducks were seen twice in January and February; 12 times in March and April; 10 in May and June; once in July; 5 in August and September; twice in October; and on December 3, 1963. A female with young was seen at the Overton WMA June 6, 1962 (Alcorn MS).

Austin (1970) reported, "I have records for every month, but mostly from early March to late May, and mid–August to mid–October. It bred on the Overton WMA in 1968 (Hutchison), supplementing the breeding record of Cottam (1947)."

Austin and Bradley (1971) reported for Clark County, "Transient in riparian and aquatic areas. Present all year with greatest abundance from February 27 to June 7 and from August 4 to October 13."

Lawson (TT, 1976) commented, "Probably our most abundant teal. It is a common to abundant migrant and summer resident. Apparently it does move out of southern Nevada for the most part of the winter."

Kingery (1983) reported, "A female with four young on May 27 evidenced the first breeding Cinnamon Teal at Las Vegas in several years."

Northern Shoveler   *Anas clypeata*

Distribution. From Alaska to Prince Edward Island; south through Mexico to Costa Rica.

Status in Nevada. Summer resident in northern part of the state, resident in western Nevada and winter resident in the southern part of the state. These ducks are most abundant during migration.

Records of Occurrence. Sheldon NWR (1982) reported the Northern Shoveler as an uncommon summer resident.

LaRochelle (MS, 1967) has no record of this duck nesting at the Ruby Lake NWR. Although not an abundant bird, he reported up to 600 had been recorded.

In western Nevada, these ducks were observed in the vicinity of Reno, Washoe Lake, Minden, Yerington and upper Lahontan Reservoir in March, April, May, November and December (Alcorn MS).

This duck was abundant along the east shore of Winnemucca Lake, where hundreds were seen in groups of up to 15 in October, 1930. Since that time, this lake has dried up (Alcorn MS).

In the Fallon area these ducks were observed from 1960 to 1967 a total of 4 times in January; once in February; 17 in March; 10 in April; 3 in May; once in June; 3 in July; none in August or September; 7 in October; 4 in November; and 3 times in December (Alcorn MS).

Schwabenland (MS, 1966) reported, "Abundant during fall and spring at Stillwater WMA with up to 35,000 counted. Nesting on the increase with 150 young raised in 1965."

Napier (MS, 1974) reported, "Very abundant during migrations March to April and September to October. It is not one of the major nesting species in Stillwater Marsh."

Saake (MS, 1981) reported the shoveler is primarily a migrant, with limited nesting.

Kingery (1987) reported 210,000 Northern Shovelers at Stillwater during fall migration in 1986.

This author and Biale (MS) for the Eureka area reported seeing these ducks from 1964 to 1974 a total of 9 times in March; 8 in April; 7 in May; 3 in June; once in August; and 5 times in September. Biale (PC, 1984) reported he knows of no nesting in this area.

J. R. and A. A. Alcorn (MS) for Pahranagat Valley from 1960 to 1965 saw these birds once in January and February; 8 times in March; 10 in April and May; once in September; and twice in November. The greatest number seen was April 13, 1965 when about 450 were observed.

Gullion et al., (1959) reported these ducks were recorded in the Pahranagat and Virgin Valleys from as early as August 9 (1951) to as late as March 31 (1951).

Alcorn (MS) saw one in Pahrump Valley August 22, 1961.

In Moapa Valley shovelers were observed from December, 1959 to June, 1964 a total of 4 times in January; 7 in February; 12 in March; 6 in April and May; 3 in June; twice in August and November; and 3 times in December (Alcorn MS).

Lawson (TT, 1976) commented, "This species is uncommon to common during migrations and in winter."

Gadwall   *Anas strepera*

Distribution. From Alaska to New Brunswick; rarely to Nova Scotia and south to Mexico. Also present in Eurasia.

Status in Nevada. Resident in northern and western part of the state. Winter resident in southern Nevada.

Records of Occurrence. Sheldon NWR (1982) reported the Gadwall as a common permanent resident with nesting in the area.

LaRochelle (MS, 1967) at Ruby Lake NWR reported, "Never conspicuous, but nearly always present. Numbers about 300–500 in spring with up to 25 pair nesting. The average brood is 6.5 at Ruby for the last four years (1963–66). Fall numbers up to 1900."

Kingery (1986) reported, "Ruby Lakes area had a record number of Gadwalls: 22,300 September 17 and 12,000 October 13."

At Idlewild and Virginia Lake Parks, Reno, these ducks were repeatedly observed from October through April (Alcorn MS).

Schwabenland (MS, 1966) reported, "Abundant at Stillwater WMA with up to 25,000 present in the fall. A common nester with over 1000 birds raised each year."

Napier (MS, 1974) reporting for the same area stated, "They are present all year, but most leave late December to early January. Abundant during summer and fall, and are one of the most abundant duck nesters in the valley."

Saake (MS, 1981) reported heavy use at Walker Lake, especially in September and October.

The author's and Biale's combined record for Eureka from 1960 to March, 1974 reveal this bird was seen on January 27, 1974; twice in March; 15 times in April; 28 in May; 8 in June; none in July; twice in August; September 22, 1960 (8 birds seen); October 2, 1960 (10 + seen); none in November; and December 18, 1966. Biale (PC, 1984) says, "It is a common duck and may breed here, most common in migration season."

In eastern Nevada these ducks were seen in May and June at Comins Lake and a small pond near Baker (Alcorn MS).

J. R. and A. A. Alcorn (MS) reported in the Hiko, Pahranagat Valley vicinity from April, 1960 to June, 1965, these birds were recorded 5 times in March; 8 in April; 10 in May; twice in June and October; and once in November. The greatest number seen was approximately 275 on April 13, 1965.

J. R. and A. A. Alcorn (MS) reported for Moapa Valley between December of 1959 and November of 1964, these birds were seen twice in January; 4 times in February; 8 in March; 5 in April and May; twice in June and October; once in November; and twice in December.

Gullion et al., (1959) reported for southern Nevada, "Uncommon winter visitor, recorded from ponds and streams throughout the area from as early as August 7 (1952) to as late as May 26 (1951). Recorded localities included Meadow, Pahranagat and Pahrump Valleys; Ash Meadows; and Mead and Mohave Lakes."

Austin and Bradley (1971) reported, "Winter resident in riparian and aquatic areas of Clark County with common occurrences from August 10 to May 29."

Lawson (TT, 1976) commented, "It is easy to see large numbers of these at Pahranagat Lake, also good numbers of Gadwalls winter in the Fort Mohave area of Clark County and south of Davis Dam."

Eurasian Wigeon   *Anas penelope*

Distribution. From Alaska to Newfoundland; south to Texas. Also in Iceland, the British Isles and across Siberia.

Status in Nevada. A rare transient in Nevada.

Records of Occurrence. Hugh E. Kingery and Charles S. Lawson (1985) reported a Eurasian Wigeon at Elko on March 5, 1985.

Ned K. Johnson and Frank Richardson (1952) reported Johnson saw a male on several occasions at Idlewild Park, Reno in the fall of 1944. The bird was associated with semi–domesticated Mallards and allowed a very close approach.

Saake (MS, 1981) reported one shot at the Mason Valley WMA in the fall of 1979.

Napier (MS, 1974) saw one male Eurasian Wigeon with a flock of American Wigeon at the Stillwater Marsh February 25, 1968.

Curtis McGowan Jr. (PC, 1985) reported shooting a Eurasian Wigeon at Carson Lake Pasture November 6, 1985.

Kingery (1983) reported, "Three pair of Eurasian Wigeon reportedly visited Railroad Valley Marsh, 80 miles east of Tonopah, Nevada.

Lawson (TT, 1976) commented, "We have no record of this species here in Clark County."

American Wigeon   *Anas americana*

Distribution. From Alaska to southern Nova Scotia; south along the Pacific, Atlantic and Gulf coast to Costa Rica.

Status in Nevada. An uncommon summer resident in the northern part of the state and a winter resident in western and southern Nevada. Common and abundant during migrations and an occasional breeder.

Records of Occurrence. Sheldon NWR (1982) reported the American Wigeon as an uncommon resident, with possible nesting in the area.

LaRochelle (MS, 1967) reported, "Present at Ruby Lake NWR except in mid–winter. Up to 2000 birds during fall. No known nesting. A pair observed at this refuge June 20, 1966 suggests that a few may be nesting in the area."

Ruby Lake NWR (1981) reported the American Wigeon as common spring, fall and winter; uncommon in summer with nesting occurring.

At Idlewild and Virginia Parks in Reno, from 10 to 500 American Wigeon were repeatedly seen in winter months. In 1942 they were seen 9 times in January; 17 in February; 15 in March; 10 in April; twice in May; none seen in June, July or August; 6 in September; 9 in October; twice in November; and 5 times in December (Alcorn MS).

In the Fallon area these ducks are present in limited numbers in all months with larger numbers seen in fall and spring migrations (Alcorn MS).

Mills (PC) reported a pair June 13, 1959 and another pair July 12, 1959 and thought they might be nesting.

Schwabenland (MS, 1966) reported at Stillwater WMA these ducks are abundant in the fall months with up to 16,000 present. He did not observe any nesting in the area.

Napier (MS) reported a brood seen at Stillwater Marsh August 19, 1968.

Napier (MS, 1974) reporting for Stillwater WMA stated, "A few are around all year. The largest numbers occur during migration. In the fall a large influx of birds occurs for several weeks then move on. A few birds stay to nest. Pairs at Stillwater probably do not exceed 10.

A hen wigeon was flushed from a nest June 25, 1968 at Stillwater. The nest, situated in 21–inch high salt grass contained 10 wigeon eggs and one Redhead egg."

Saake (MS, 1981) reported very limited nesting in western Nevada. In 1983 he reported American Wigeon were more common in northeastern Nevada.

Thompson (PC, 1988) reported he has not observed any nesting over the past three years in the Lahontan Valley area.

For the Eureka area, Biale reported this duck was seen from 1964 through 1973 a total of 4 times in January; twice in February; 7 in March and April; 5 in May; 3 in June; none in July or August; twice in September and October; 5 in November; and 14 times in December. Their presence June 22 and 23, 1966 suggests possible nesting in this area.

A male was seen at Comins Lake, Ely area on June 5, 1962 (Alcorn MS).

In Pahranagat Valley from 1960 to 1965, these birds were observed 3 times in January and February; 7 in March; twice in April; 3 in May; twice in June; none in July, August or September; and 3 times in November and December (Alcorn MS).

In Moapa Valley from 1960 to 1965, these ducks were seen 5 times in January; 7 in February and March; 4 in April; 5 in May; twice in June and July; none in August; once in September; 3 in October; none in November; and a hunter had one in his bag December 16, 1963 (Alcorn MS).

Lawson (TT, 1976) commented, "This wigeon is a common migrant and winter species here in Clark County."

Canvasback   *Aythya valisineria*

Distribution. From Alaska east to southern Ontario and along the Pacific coast; south to Mexico and the Gulf States.

Status in Nevada. Summer resident and transient and winter resident in southern Nevada.

Records of Occurrence. Linsdale (1936) reported, "Transient, common on lakes and ponds. Reports of this duck in Nevada in summer may be due to confusion of the species with Redhead."

At this time (1988) this duck regularly nests in several areas in Nevada and is a winter resident in the southern part.

Sheldon NWR (1982) reported the Canvasback as an uncommon permanent resident and nester.

In Elko County, several groups of 10 to 20 birds were observed at Wildhorse Reservoir April 12, 1971. At Ruby Lake NWR they were seen from April through September, 1965–66 (Alcorn MS).

LaRochelle (MS, 1967) reported at Ruby Lake NWR 800 to 1200 of these ducks arrive in mid–spring and 50 to 140 pair produce an average of 6.9 per brood.

Kingery (1982) reported 3300 Canvasback at Ruby Lake NWR in August, 1981.

In western Nevada, Canvasback have been seen in Reno, Washoe Lake, Minden and Yerington areas in fall, winter and spring months (Alcorn MS).

Schwabenland (MS, 1966) reported for Stillwater WMA, "Abundant in fall months, with up to 25,000 counted. Limited nesting in this area produced up to 50 young in 1965."

At this same refuge, Napier (MS, 1974) reported, "Most abundant during fall migrations in late October and early November. Small breeding populations–probably does not exceed 10 pair."

Kingery (1977) reported 24,230 Canvasback visited Stillwater WMA, a number equivalent to 29% of the Pacific Flyway population counted in January, 1976.

Biale (MS) reported this duck in the Eureka area 16 times from 1964 to 1974. They were seen 3 times in March; April 12, 1970; 6 in May; twice in June; July 2 and 3, 1972; and twice in October.

Biale (PC, 1984) considers them fairly common on the ponds in springtime and uncommon in fall.

In eastern Nevada, six were seen on a small pond near Baker April 20 and one October 27, 1967 (Alcorn MS).

J. R. and A. A. Alcorn (MS) from 1960 to 1964 in Pahranagat Valley saw these ducks twice in January; 4 times in February; 7 in March; twice in April; June 2, 1963; October 18, 1963; 6 in November; and December 23, 1964, when over 500 were seen. The greatest number seen in this valley was about 1225 Canvasback on February 16, 1965.

Gullion et al., (1959) reported for southern Nevada, "Three records: Overton, February 13, 1951, one bird; Pahranagat Valley, March 31, 1951, about 15 birds; and Lake Mohave, January 7, 1952, one bird."

J. R. and A. A. Alcorn (MS) for Moapa Valley from 1959 to 1964 saw this duck January 3, 1963; March 20, 1963; twice in May; June 1, 1963; June 3, 1964; July 24, 1964; twice in October; 3 times in November; and in December, 1959.

Austin (1970) reported, "This is a fairly common winter resident from early October to early April. Wauer (pers. comm.) found it present at Overton WMA during the summer of 1966 when a female with eight young was seen June 9. This is the first breeding record for southern Nevada, although it breeds in the northern part of the state (Marshall and Alcorn, 1952)."

Austin and Bradley (1971) reported for Clark County, "Winter resident in aquatic areas. Common from October 4 to May 11. Unseasonal appearances on June 9 and July 20."

Lawson (TT, 1976) commented, "It is a common to uncommon resident and migrant in Clark County."

Redhead   *Aythya americana*

Distribution. From south–central Alaska to Ontario; south to Guatemala.

Status in Nevada. Summer resident in northern part of the state, resident in western Nevada and winter resident in southern Nevada.

Records of Occurrence. Linsdale (1936) stated, "Summer resident and transient; common about many medium–sized lakes and ponds, especially in western Nevada. Early to arrive in spring and late to leave in fall. On June 9 and 10, 1932 and May 16, 1933, the Redhead was the most numerous species of duck to be seen on Little Washoe Lake."

Sheldon NWR (1982) reported the Redhead as a common to occasional resident with nesting in the area.

LaRochelle (MS, 1967) reported at Ruby Lake NWR, "At times both our most numerous duck and nester. Up to 450 nesting pair with an average 6.2 young per pair. Up to 18,000 in 1950s."

In Idlewild and Virginia Parks, Reno, these ducks were seen in 1941 and 1942 a total of 5 times in January; 9 in February; 3 in March and April; 7 in October and November; and 12 times in December. Also, on May 21, 1962 and June 16, 1961, several were seen in these parks (Alcorn MS).

On the canals and small ponds in the Fallon area from 1961 to 1966, these birds were seen once in January; 3 times in February; 9 in March; 14 in April; 12 in May; 4 in June; 5 in July; twice in August and September; 6 in October; and twice in December (Alcorn MS).

Mills wrote (letter) September 11, 1940 saying, "My brother Claude and I rode to Fallon horseback the other night and there were at least 100 ducks in the canal between the Jones' ditchhouse and town. The water was rather low in the canal and the water weeds were sufficiently exposed to make for easy feeding. All that I got a look at were Redheads." The area mentioned by Mills is 1.3 miles in length and is located about two miles west of Fallon.

Schwabenland (MS, 1966) reported at Stillwater WMA, "Abundant in fall and spring months, up to 30,000 birds. Nesting, top nester in the area, 3200 birds in 1965. Over 7000 young were raised in 1958 and 1959."

Napier (MS, 1974) reporting on this same refuge said, "Abundant in fall and spring. Production has dropped off since mid–1960's due to loss of habitat. It used to be the most abundant nester, but now the Cinnamon Teal outnumbers them during the breeding season."

Paullin (PC) was at Lead Lake in the Stillwater Marsh June 13, 1970. He reported a Redhead with three young swam from shore out into the lake nearby. One baby duck strayed away from the others and a gull swam over to the stray duckling, grabbed it by the nape of the neck, shook and killed it, then swallowed the whole duckling. Four or five Yellow–headed Blackbirds chased the gull as it flew away.

Biale (MS) reported in the Eureka area from 1964 to 1973, inclusive, he saw this duck 3 times in February; 14 in March; 11 in April; 29 in May; 6 in June; and twice in September and October. He (PC, 1984) reports seeing them every year, but he does not consider them abundant in the Eureka area.

In Pahranagat Valley from March, 1960 to February, 1966 these ducks were observed once in January; 3 times in February; 11 in March and April; 14 in May; 4 in June; 3 in July; once in August; none in September; 4 times in October and November; and none in December (Alcorn MS).

Gullion et al., (1959) reported this bird as a, "Winter visitor, particularly in Pahranagat Valley, from at least August 10 (1954) to May 23 (1951). Also recorded in the Virgin Valley on November 5, 1951 and at Lake Mohave January 7, 1952."

J. R. and A. A. Alcorn (MS) observed these ducks in Moapa Valley from December, 1959 to June, 1965, twice in January; 5 times in February; 8 in March; 3 in April; 4 in May; 5 in June; July 24, 1964; September 22, 1961; twice in October; 3 times in November; and December 15, 1959.

At Overton WMA, six pair were seen June 3, 1964. At this same place A. A. Alcorn saw a Prairie Falcon unsuccessfully trying to catch a flying Redhead June 7, 1961. The falcon would fly close to the water, then swoop up in an attempt to catch the duck.

Austin and Bradley (1971) reported for Clark County, "A winter resident in aquatic areas. Common from October 4 to June 9. Unseasonal occurrences August 27, September 10 and July 17 and 20."

Lawson (TT, 1976) commented, "I have seen flocks of 300–500 of them all winter long in the Fort Mohave area below Davis Dam."

A male Ring-necked Duck. Photo by A. A. Alcorn.

Ring–necked Duck   *Aythya collaris*

Distribution. From Alaska to Newfoundland; south to the Canal Zone.

Status in Nevada. Irregular summer resident in selected localities; transient.

Records of Occurrence. Linsdale (1936) reported this duck was a transient in Nevada with records too few to show status.

Sheldon NWR (1982) reported the Ring–necked Duck is of occasional occurrence in all seasons.

At Wildhorse Reservoir 14 were seen April 12, 1971 (Alcorn MS).

A pair was observed June 27, 1962 at a reservoir near Jiggs (Alcorn MS).

LaRochelle (MS, 1967) reported at Ruby Lake NWR, "Common fall and winter resident with up to 200 in the area."

On this refuge the author observed one April 14, 1966. A pair was noted June 20, 1966. Other single birds were seen August 22, September 29 and December 17, 1965 (Alcorn MS).

Kingery (1984) reported, "Ruby Lakes had two nests of Ring–necked Ducks, one successfully. The refuge had 20 nesting pair last year."

Kingery (1980) reported two Ring–necked Ducks at Lake Tahoe December 16, 1979.

At Idlewild and Virginia Parks, Reno, these ducks were seen twice in January; 6 times in February; 5 in March; twice in April; none from May to September; 5 in October; 7 in November; and 11 times in December (Alcorn MS).

Heilbrun and Rosenburg (1982) reported 78 at Truckee Meadows December 19, 1981 during the CBC.

Alcorn (1946) reported this duck in Lahontan Valley as a winter transient, but not abundant. Specimens were taken in January, March, October and November.

Schwabenland (MS, 1966) reported at Stillwater WMA up to 100 birds use the area during the fall months and no nesting has been recorded.

At this same refuge, Napier (MS, 1974) reported a few present during migration. No known nesting.

Biale (MS) reported these birds in the Eureka area on nine occasions: January 23, 1970; March 19, 1967; March 15, 1970; March 15, 1974; April 12, 1970; April 30, 1967; May 5, 1966; May 20, 1973; and October 25, 1964. He (PC, 1984) reports he has not seem them every year and considers them uncommon to the Eureka area. In 1986 (PC) he reported more Ring–necked Ducks than ever before in the Eureka area, especially in April and May.

A. A. Alcorn (MS) saw three Ring–necked Ducks on a pond near Baker October 6, 1967.

In Pahranagat Valley one was seen January 8, 1962; three on January 29, 1963; about 100 on February 16, 1965; approximately 52 on March 1, 1965; about 150 on March 29, 1965; about 25 on April 30, 1965; one May 7, 1965; and two on May 14, 1965 (Alcorn MS).

In Moapa Valley, these ducks were seen December 15, 1959; twice in January; 3 times in February; March 20, 1963; April 23 and June 12, 1964; November 24, 1961; and about 10 on November 26, 1963 (Alcorn MS).

Austin and Bradley (1971) reported for Clark County, "A winter resident in aquatic regions and common between September 15 and May 17."

Greater Scaup   *Aythya marila*

Distribution. From Alaska to Newfoundland; south to Texas.

Status in Nevada. An uncommon but regular winter visitor to western and southern Nevada and transient in Elko County.

Records of Occurrence. Linsdale (1936) reported only Ridgway (1877) listed this duck as a winter visitant to Pyramid Lake.

A specimen (head only) in the USNM was taken by D. E. Lewis at Cave Creek, Ruby Lake NWR November 25, 1966.

Seven were seen by the author in Reno December 11, 1942 (Alcorn MS).

Lois H. Heilbrun and Kenn Kaufman (1979) reported three Greater Scaup in the Truckee Meadows, Reno area, during the CBC.

Mills (PC) saw one in the Fallon area January 2, 1968; one was shot November, 1969 (now a mounted specimen); and another was shot January 10, 1969.

Schwabenland (MS, 1967), Napier (MS, 1974) and Saake (MS, 1971) reported one or two each year come through Stillwater WMA check stations during the hunting season.

Kingery (1984) reported one at Kirch WMA May 21.

Austin and Bradley (1966) for southern Nevada reported, "One was examined from a hunter's bag from the Virgin River near Riverside by Gullion (1952) on November 4, 1951, constitutes the only published record for Clark County. A drake was seen at close range with several Ring–necked Ducks at Tule Springs on March 20, 1964.''

Austin (1970) reported, ''Gullion (1952) and Austin and Bradley (1966) reported the previous southern Nevada records for March and November. I have additional records for Overton WMA: 27 October, 1968; 16 November, 1968; and 25 December, 1968. This indicates a small wintering population.''

Austin and Bradley (1971) reported for Clark County, ''Winter resident in aquatic areas. Common on October 27 to December 25 and on March 30.''

Kingery (1978) reported them at Davis Dam January 14 and four on February 18. He (1982) reported two Greater Scaups at Las Vegas were the first southern Nevada record since 1978. Kingery (1984) reported that one Greater Scaup spent January in Las Vegas and Davis Dam.''

Lawson (TT, 1976) commented, ''Hunter bag checks indicate that they are more abundant than most people believe. I would say that it is a rare to uncommon migrant and winter resident here in Clark County.''

Lesser Scaup   *Aythya affinis*

Distribution. From Alaska to west–central Quebec; south to Columbia.

Status in Nevada. Irregular summer resident in selected localities; transient.

Records of Occurrence. Linsdale (1936) reported, ''Transient and winter visitant throughout the state; definite records—all in May.''

At present (1988) these birds have been observed in many months throughout the state.

Sheldon NWR (1982) reported the Lesser Scaup as an occasional permanent resident, thought to nest in the area.

A group of about 10 Lesser Scaup were seen on Wildhorse Reservoir April 12, 1971 (Alcorn MS).

LaRochelle (MS, 1967) reported at Ruby Lake NWR this duck is most abundant during late fall and winter, with up to 400 present. Nesting of up to 140 pair with broods averaging 4.7.

Saake (MS, 1981) reported the Lesser Scaup is primarily a migrant duck with limited nesting at Wildhorse Reservoir, Ruby Lakes and Mason Valley.

Orr and Moffitt (1971) reported at Lake Tahoe the Lesser Scaup or Bluebill is one of the commonest species of ducks wintering on Lake Tahoe, its number probably being exceeded only by the Goldeneye.

On the ponds in the parks of Reno, small numbers of up to 30 were seen 8 times in January; 14 in February; 11 in March and April; twice in May; once in October; 8 in November; and 15 times in December (Alcorn MS).

In the Fallon area these ducks were seen yearly in fall, winter and spring (through April) with one record of two on May 1, 1961 (Alcorn MS).

Schwabenland (MS, 1966) reported for Stillwater WMA, ''Common during the fall months, up to 2500 in late fall and winter of 1965. No nesting observed.''

Napier (MS, 1974) for the same area reported several hundred were seen during the peak migration with no known nesting.

In the Eureka area this bird is uncommon. This author and Biale (MS) reported them on four occasions between 1960 and 1966, twice in April and twice in May. He (PC, 1986) reported one at the Hay Ranch March 25 and 26, 1986.

J. R. and A. A. Alcorn (MS) for Pahranagat Valley reported from January, 1960 to May, 1965 saw these ducks January 3 and February 4, 1960; 6 times in March; 3 in April; 6 in May; June 2, 1963; and 3 birds on July 4, 1961.

J. R. and A. A. Alcorn (MS) for Moapa Valley observed this bird from January, 1960 to June, 1964 a total of 4 times in January and February; 11 in March; 4 in April; 9 in May; June 16, 1961; June 1, 1963; June 12, 1964; July 29, 1960; and 4 times in November (Alcorn MS).

Austin and Bradley (1971) for Clark County reported, "Winter resident in aquatic areas. Common October 2 to May 4. Unseasonal records on June 1, July 9 and August 25."

Lawson (TT, 1976) commented, "The Lesser Scaup is a common to abundant winter resident and migrant in Clark County."

Harlequin Duck   *Histrionicus histrionicus*

Distribution. From Alaska to Greenland; south to Texas and Florida.

Status in Nevada. Rare.

Records of Occurrence. Linsdale (1936) reported, "Basis for the ascription of this duck to Nevada is the statement by Hanford (1903) that at Washoe County, 'a male was taken at Frankstown near the lake and identified by Mr. Steinmetz'."

Lawson (1977) reported one male seen in the Needles area of Pyramid Lake June 24, 1977 by Dave and Karen Galat. It was photographed by Lawson.

Thompson (PC, 1988) reported one Harlequin Duck on the pyramid in Pyramid Lake on May 12, 1987.

Kingery (1981) reported, "A later report substantiates a male mid–summer Harlequin Duck in somewhat worn breeding plumage, which sat on the shore of Walker Lake July 9." John Buglar reported the same bird seen in late June.

Oldsquaw   *Clangula hyemalis*

Distribution. From Alaska to Greenland; south along both coasts to Texas.

Status in Nevada. An uncommon migrant.

Records of Occurrence. Linsdale (1936) had no record of this bird in Nevada. In that this bird is so seldom reported, all available records are given.

Kingery (1986) reported, "Oldsquaws visited Ruby Lake October 16–19."

Kingery (1984) reported, "An Oldsquaw stopped at Fernley, Nevada, April 11–14."

Marshall and Alcorn (1952) reported, "This species has not previously been listed for the state. According to Mills, a mounted Oldsquaw at the Oats Park School in Fallon that was examined by this author and Mills was killed in the 1920s in the vicinity of Stillwater, Churchill County. A more recent specimen present in the Sagebrush Cafe in Fallon was shot at Carson Lake October 20, 1949."

One was shot by a hunter in the Stillwater WMA November 23, 1952. One of these ducks was captured eight miles west–southwest of Fallon June 9, 1964. A male was shot by Mills four miles northeast of Fallon November 20, 1968 (Alcorn MS).

Saake (MS) shot a male at Carson Lake December 3, 1970 and prepared it as a mounted specimen and Saake (MS, 1981) commented, "These ducks are very uncommon."

Kingery (1977) reported that Fallon's summer Oldsquaw stayed through August 2.

Kingery (1984) reported four at Kirch WMA.

A. A. Alcorn (MS) examined one shot by a hunter near Overton November 13, 1960.

Kingery (1977) reported four at Lake Mohave and one at Lake Mead in March, 1977.

Lawson (TT, 1976) commented, "Two years ago, we found one individual at Fort Mohave which remained through last summer and was joined last winter by four other individuals. In addition, two other birds were seen at Las Vegas Wash."

Lawson (1977) reported one at Colorado River, 10 kilometers below Davis Dam January, 1975; at Davis Dam two were seen November 24, 1976; in southern Nevada eight were seen January 4, 1977; and at Lake Mohave five were seen.

Kingery (1977) reported three to five were at Davis Dam and Lake Mohave all winter.

Black Scoter *Melanitta nigra*

Distribution. From Alaska eastward, scattered across Canada to Newfoundland; south along both coasts.

Status in Nevada. Rare.

Records of Occurrence. Three were seen in Reno October 21, 1976 (Alcorn MS).

Kingery (1977) reported three Black Scoter were seen at Ely October 21–22, and at Sunnyside WMA November 11 a hunter killed a Black Scoter. The bird is now a specimen at the University of Nevada, Reno.

Scott (1968) reported one Black Scoter in southern Nevada.

Lawson (MS, 1981) reported one in the Las Vegas Wash arm of Lake Mead December 30, 1980.

Surf Scoter *Melanitta perspicillata*

Distribution. From Alaska east to parts of Canada; south along the coast to the Gulf of California and to Florida.

Status in Nevada. Occasional migrant to Nevada.

Records of Occurrence. Kingery (1984) reported a Surf Scoter at Ruby Lake in late November, 1983 and another December 6. He (1985) reported one November 23–30.

Kingery (1978) reported two October 15 to November 4 at Topaz Lake.

Alcorn (1946) reported two shot by Ward C. Russell in the Fallon area at Soda Lake October 19, 1940 and another was obtained November 12, 1940.

Napier (MS, 1974) reported seeing two females shot by hunters and checked through the Stillwater Check Station east of Fallon October 19 and one October 22, 1967. He also reported three killed on this area in the 1968 hunting season and two in the 1972 season.

Saake (MS) reported one shot by a hunter in the Fallon area at Carson Lake in 1979 and an adult male at Harmon Reservoir in 1973.

Thompson (PC, 1988) reported a hunter shot an immature Surf Scoter just below Lahontan Reservoir October 27, 1987.

One was shot by a hunter on a small pond at Tippett, White Pine County October 27, 1967 (Alcorn MS).

Austin (1970) reported, "A female was shot at the Overton WMA November 2, 1968. Another was shot during the fall of 1968 at Key Pittman WMA in Pahranagat Valley, Lincoln County. Hayward et al., (1963) gave the previous southern Nevada records of two taken on the Nevada Test Site, Nye County in October, 1961."

Lawson (1977) reported one on the Colorado River below Davis Dam December 5, 1974, first seen November 30 and he (MS, 1980) reported one photographed on Lake Mead December 7, 1980.

Lawson (TT, 1976) commented, "We do have records of this scoter from Lake Mead and Davis Dam area. I would say it should be listed as an occasional to rare fall migrant in Clark County."

Kingery (1981–84) reported Surf Scoters in the Las Vegas and Davis Dam area each winter.

White–winged Scoter   *Melanitta fusca*

Distribution. From Alaska to Newfoundland; south in winter to Baja California and (rarely) Florida.

Status in Nevada. Accidental straggler.

Records of Occurrence. Margaret A. Rubega and David Stejskal (1984) reported one White–winged Scoter on the Ruby Lakes CBC (one was collected). Kingery (1985) reported, "Ruby Lake had eight White–winged Scoter October 28 swimming among decoys November 23–30.

Kingery (1987) reported, "Most unusual was a White–winged Scoter at Reno, November 14 and joined by a second December 13."

Leukering and Rosenburg (1987) reported one White-winged Scoter during the Truckee Meadows CBC December 20, 1986.

Orr and Moffitt (1971) reported the White–winged Scoter as a rare winter straggler to Lake Tahoe.

Alcorn (1946) reported specimens obtained in the Fallon area November 3, 1940, November 12, 1940 and November 2, 1941. Others were shot east of Fallon at Stillwater WMA October 23, 1949 and November 4, 1951.

In this same area, one was collected October 13, 1961 and Mills shot one October 28, 1962 (Alcorn MS).

Napier (MS, 1974) reported from Stillwater WMA, "Since 1967 several have been shot by hunters."

Saake (MS, 1981) reported one or two shot every year in the Fallon area.

Kingery (1978) reported one at Topaz Lake November 4.

Curran (PC) reported one taken at the northwest side of Walker Lake November 15, 1970.

Biale (PC, 1984) reported, "A single bird was found on a road in Spring Valley (about eight miles south of Eureka) on November 14, 1975. The bird was unable to fly, and was released at a pond in Diamond Valley by the Nevada Department of Wildlife.

For southern Nevada, Snider (1969) reported a pair was observed at Henderson Marsh, Lake Mead February 26, 1969.

Monson (1972) reported two at Las Vegas Bay, Lake Mead October 30 and November 3, 1971.

Kingery (1978) reported White–winged Scoters at Las Vegas from November 30 to December 1, 1977. He (1981) reported, "The White–winged Scoter passed through Las Vegas May 4."

Lawson (TT, 1976) commented, "The White–winged Scoter is a rare to occasional species here in southern Nevada as a migrant."

Common Goldeneye *Bucephala clangula*

Distribution. From Alaska to Nova Scotia; south to Florida and Texas.

Status in Nevada. Winter visitor and migrant.

Records of Occurrence. Sheldon NWR (1982) reported the Common Goldeneye as occasionally seen in summer and fall and uncommon in the winter.

LaRochelle (MS, 1967) reported for Ruby Lake NWR, "A conspicuous duck most often seen during winter when some 300–500 use the area. No nesting."

In 1941 and 1942 in Reno, these birds were seen 5 times in November; 7 times in December; and three were observed February 8, 1942. The most seen at any one time was six (Alcorn MS).

Ryser (1968) reported several observed at Virginia Lake, Reno, late in February, 1968. Also, Ryser (1971) reported a male and two females on this same lake January 17, 1971 and listed them as a rare winter visitant in western Nevada.

Orr and Moffitt (1971) reported this duck as a winter visitant at Lake Tahoe, "Common from early December to April." In mid–winter the Common Goldeneye is considered to be the most abundant and widely distributed species of duck on Lake Tahoe.

Schwabenland (MS, 1966) reported at Stillwater WMA, "A few have been seen every year during the fall months."

At this same refuge, Napier (MS, 1974) reported, "Observed between November and early April. Most common during the colder months. Peak numbers are less than 100."

Mills (PC) observed one in the Fallon area June 20, 1966.

Saake (MS) reported these birds were all over western Nevada, especially at Walker Lake November 17, 1970 and in 1981 he reported no nesting, but the largest concentration remained at Walker Lake.

Biale (MS) has seen this bird only one time in the Eureka area—in Newark Valley April 3, 1966.

In Pahranagat Valley one was noted November 25, 1961; two were seen February 16, two on March 17, six on March 29 and one May 14, all in 1965 (Alcorn MS).

Gullion et al., (1959) reported, "On February 13, 1951 one was seen at Overton. On March 27, 1951 a flock of Goldeneyes seen flying over Las Vegas Valley (1800 feet elev., Clark County) was believed to be of this species, and four days later on March 31, about eight of this species were seen on lakes in the Pahranagat Valley. Imhof observed three birds on Lake Mead November 26, 1951."

J. R. and A. A. Alcorn (MS) recorded these ducks in the Moapa Valley area from January,

1960 to May, 1965 twice in January and February; 9 times in March; 3 in April and May; on June 1, 1963; and twice in November.

Austin (1970) reported this species present on the larger lakes in small numbers from November 11, at Henderson (1964) to April 3 (1965) at Corn Creek.

Kingery (1978) reported 975 at Davis Dam February 4.

Lawson (TT, 1976) reported, "This duck is a common to very abundant winter resident and migrant here in southern Nevada from November through April. I have seen flocks of 600–1000 below Davis Dam and in the Fort Mohave area."

Barrow's Goldeneye   *Bucephala islandica*

Distribution. From Alaska to Greenland; south to Arizona and South Carolina.

Status in Nevada. A rare transient.

Records of Occurrence. LaRochelle (MS, 1967) reported at Ruby Lake NWR this bird is a rare transient visitor, with one sighting in 1963.

Kingery (1979) reported five wintered at Ruby Lake and he (1980) reported two at Ruby Lake NWR December 18, 1979.

Heilbrun and Rosenburg (1982) reported six on the CBC at Ruby Lakes December 19, 1981.

Orr and Moffitt (1971) reported for the Lake Tahoe region, "rare summer visitant." They listed records for the California side of the Lake, including a nest of 10 eggs seen by Moffitt July 7, 1912. No records are available for the Nevada side of the lake.

Kingery (1980) reported two at Lake Tahoe December 16, 1979.

Ryser (1971) reported a male and two females on Virginia Lake, Reno, March 21, 1971 and Kingery (1979) reported a male in the Truckee Meadows during the CBC.

Saake trapped a male and two females at the Fernley WMA March 10, 1970.

Schwabenland (MS, 1966) reported for Stillwater WMA, "An occasional bird has been sighted in the Stillwater Marsh."

Cruickshank (1966) listed one at the Desert Game Range December 28, 1965 during the CBC.

Lawson (TT, 1976) notes, "For the past two winters [1974–75] we have found up to 50 individuals of these birds immediately below Davis Dam in the spill area."

Lawson (1977) reported goldeneyes at Davis Dam on December 5, 1974 (17–57 birds) until December 17.

Kingery (1983) reported, "Barrow's Goldeneye did not return to southern Nevada, with the only report being of one January 1."

Bufflehead   *Bucephala albeola*

Distribution. From Alaska to Ontario; south to Florida and Mexico.

Status in Nevada. A common winter visitor.

Records of Occurrence. Sheldon NWR (1982) reported the Bufflehead as occasional in spring and summer and uncommon in fall and winter.

On Wildhorse Reservoir two were seen April 12, 1971 (Alcorn MS).

LaRochelle (MS, 1967) reported at Ruby Lake NWR this was a late fall through early summer bird, but not a nester. Up to 800 are present in winter.

Orr and Moffitt (1971) reported at Lake Tahoe these ducks were observed on the lake from October to May. They write, "The occurrence of Bufflehead on more than one occasion in the middle of May indicates the possibility that a few individuals may breed here."

Alcorn (1946) reported this duck in Lahontan Valley as a winter resident, but not abundant. "Seen in all months from October to June, inclusive. Most common in December, January and February."

Later records reveal its presence in winter months from October through May. The greatest number seen was April 1, 1961 when about 20 were observed (Alcorn MS).

Schwabenland (MS, 1966) reported at Stillwater, Bufflehead were commonly seen during fall and winter months. No nesting was observed.

Napier (MS, 1974) reported for this same area Bufflehead were observed from October through May. "Peak numbers do not exceed 500."

Saake (MS) reported on November 17, 1970 he took an aerial survey to census waterfowl and saw these ducks all over western Nevada, especially at Walker Lake.

This author and Biale (MS) saw this duck in the Eureka area from 1960 to 1972, 5 times in March; 4 in April; and 3 times in May. The greatest number seen was over 50 observed April 7, 1960. Biale (PC, 1984) reported these birds are uncommon in this area, with most birds seen in the spring.

Gabrielson (1949) reported one at Twin Springs, Nye County, November 17, 1934.

In Pahranagat Valley from April, 1960 to May, 1965, two were seen January 29 and two on February 7, 1963; they were seen 9 times in March (up to 175 birds in one day); 7 times in April (up to 400 birds on April 13, 1965); 6 times in May; twice in October; and 3 times in November (Alcorn MS).

Gullion et al., (1959) reported this duck was recorded, "From as early as November 28 (1951) at Lake Mohave to as late as May 14 (1952) at Pahranagat Valley."

J. R. and A. A. Alcorn (MS) reported for Moapa Valley from December, 1959 to April, 1964 these ducks were seen once in January; 4 times in February; 11 in March; 8 in April; 4 in May; October 30, 1961; 4 in November; and 3 times in December.

Kingery (1984) reported, "Several remained for an unusual June stay at Las Vegas."

Lawson (TT, 1976) commented, "It is a common winter resident and migrant in Clark County; with up to 250–400 individuals over–wintering in the Fort Mohave area below Davis Dam. On Lake Mead and Mohave you will see lesser numbers. In Vegas Wash we see them at all the ponds."

Hooded Merganser   *Lophodytes cucullatus*

Distribution. From southeastern Alaska to southern Nova Scotia; south to Mexico and Florida.

Status in Nevada. An uncommon winter visitant with possible nesting in western Nevada.

Records of Occurrence. Linsdale (1936) stated, "Ridgway (1877) wrote of the Hooded Merganser, 'occasionally met with in summer in the wooded valleys of the Truckee and Carson River, but it seemed to be very rare'. Near Carson City, Ormsby County, Henshaw (1877) found this species, 'quite abundant, but occurring in late fall'. A male from west side of Ruby Lake, three miles north of Elko County line, taken December 28, 1927 was reported by Ellis (1935) as in his collection."

Sheldon NWR (1982) reported the Hooded Merganser as a rare, transient visitor in spring and fall.

LaRochelle (MS, 1967) reported these mergansers at Ruby Lake NWR as transient visitors with seldom more than 10 on the area. A. A. Alcorn (MS) saw one November 17, 1965 in the Ruby Lakes area.

Kingery (1979) reported one at Ruby Lake NWR on the CBC and he (1980) reported two at Ruby Lake NWR December 18.

Ryser (1971) reported a pair at Verdi in early March of 1971.

Ryser (1985) commented, "I believe that there still is limited nesting going on in this area, because every spring I encounter several pair on a pond close to the Truckee River at Verdi, Nevada."

Kingery (1977) reported 12 at Verdi February 27. He (1980) also reported eight at South Lake Tahoe December 16.

Alcorn (1946) reported birds seen in the Lahontan Valley November, January and February, and dead birds were examined December, January and February. Recent records for the Fallon area are of up to four birds seen in January, October, November and December.

Schwabenland (MS, 1966) reported an occasional pair at Stillwater WMA in fall months. Napier (MS) reported occasional birds brought through the Stillwater Check Station.

Biale (MS) reported one seen in the Diamond Valley November 26, 1968.

A female was taken at Key Pittman WMA, Lincoln County, November 27, 1966 (Alcorn MS).

In Moapa Valley two were seen March 15, 1962 and two others November 26, 1963 (Alcorn MS).

Austin and Bradley (1971) reported for Clark County, "Transient in riparian and aquatic areas. Common from November 9 to January 16."

Lawson (TT, 1976) commented, "The Hooded Merganser is an occasional to rare winter visitor in southern Nevada. Most of the individuals were seen in Vegas Wash or below Davis Dam."

Common Merganser   *Mergus merganser*

Distribution. From Alaska to Newfoundland and south–central Russia; south to Mexico and Florida.

Status in Nevada. Resident in selected areas and transient statewide.

Records of Occurrence. Sheldon NWR (1982) reported the Common Merganser as uncommon in spring and fall, with occasional occurrences in the winter.

Linsdale (1936) reported Hall (MS) saw an adult with a brood of young on Goose Creek, Elko County July 14, 1935.

On the south fork of the Owyhee River, northwestern Elko County, an adult with one young was seen July 12, 1967. In addition, several adults were seen on this same river on the same date (Alcorn MS).

LaRochelle (MS, 1967) reported for Ruby Lake NWR, "Transient visitor, usually about eight to 15 birds staying for five to seven weeks."

Hall (1926) reported this bird nesting at Pyramid Lake. The author has seen young there since then and wildlife biologists report nesting of Common Merganser still occurs here.

Saake (MS, 1983) reported these birds at the upper Humboldt, and in late November over 1000 at Rye Patch in 1968–69. He also reported nests at Pyramid Lake along the shoreline in weeds and rocks.

In Idlewild and Virginia Lake Parks, Reno in 1942, these birds were seen 7 times in January; 13 in February; 10 in March; 3 in April; once in May; 3 in June; once in October; and 4 times in December. Nine birds was the greatest number seen in any one day (Alcorn MS).

Ryser (1970) reported seeing five on Virginia Lake December 6, 1970.

Orr and Moffitt (1971) reported for Lake Tahoe this bird is a permanent resident breeding commonly about the lake. No seasonal fluctuation in numbers was noted.

Kingery (1980) reported 12 at South Lake Tahoe December 16, 1979.

In western Nevada, these birds have been seen by the author at Pyramid Lake, Derby Dam, on the Truckee River, in Reno, Washoe Lake, on the Carson River near Minden, Wilson Canyon on Walker River, in Mason Valley, Lahontan Reservoir and the main canals and ponds in the Fallon area. Sightings are represented by all months except September (Alcorn MS).

Schwabenland (MS, 1966) reported these birds were common at Stillwater WMA in the fall and winter months with up to 4000 birds being counted. No nesting was observed.

Napier (MS, 1974) reported this bird at Stillwater WMA is generally present in winter from November to April and occasionally in other months. Peak is seldom over 2000 birds.

Kingery (1986) reported, "Stillwater wintered 3000 Common Merganser, unusually high numbers."

Biale (MS) reported seeing them in Diamond Valley, November 26, 1968; and four or five near the Tonkin Ranch in the Eureka area January 11, 1970. He (PC, 1986) reported two at Newark Valley March 16, 1986.

Austin and Bradley (1971) reported for Clark County, "Winter resident in riparian and aquatic areas and common from November 20 to March 18. Unseasonal occurrence was on April 15."

Lawson (TT, 1976) notes, "This merganser is an uncommon to common migrant and winter resident at Lakes Mead and Mohave below Davis Dam in the Overton area and at the Pahranagat Refuge. There are times when we see more Red–breasted Merganser than Common Merganser.

Food Habits. Alcorn (1953) reported that between November 14, 1940 and March 14, 1945, 111 Common Merganser *(Mergus merganser)* stomachs were examined by Karl F. Lagler, Robert R. Miller and the author. These birds came from seven areas in Churchill County, including the Carson River, irrigation canals, Indian Lake, Rattlesnake Reservoir, Harmon Pasture, Hazen Reservoir and Dutch Bill Lake. The 111 stomachs examined contained, 189 carp, 25 Sacramento Perch, 20 yellow perch, 12 suckers, nine bullhead catfish, nine largemouth bass, one red–striped shiner, one bluegill, and one chub. Twenty stomachs were found to be empty or contained fish remains digested beyond identification. Gravel was also found in some of the stomachs.

Red–breasted Merganser   *Mergus serrator*

Distribution. From Alaska to Newfoundland; south to Mexico and all of Florida. Also occurs in Eurasia.

Status in Nevada. Transient statewide and an uncommon winter resident in the southern part.

Records of Occurrence. All of Linsdale's (1936) report is given as follows: "Winter visitant and transient. Found on Pyramid Lake and the Truckee River in December, 1867 (Ridgway, 1877). Henshaw (1877) found it numerous near Carson City, Ormsby County, in fall. A female was noted May 20, 1934 on Soda Lake, Churchill County (Linsdale, MS). On May 14, 1930 at four miles southeast of Millett Post Office, 5500 feet elev, Nye County, a female was watched swimming alone on a pond. An adult male was obtained May 2, 1891 on Vegas Wash, Clark County (Fisher, 1893). Fitch collected a specimen on May 24, 1932 at Crystal Spring, 4000 feet, Pahranagat Valley, Lincoln County. These occurrences in May are all late for localities so far south and may represent non–breeding individuals."

LaRochelle (MS, 1967) reported for Ruby Lake NWR, "Transient visitor of eight to 12 birds, usually in groups of two to five, staying for a short time."

Saake (MS, 1984) reported Terry Retterer observed 10 Red–breasted Merganser January 10, 1984 at Truckee Meadows.

Alcorn (1946) reported specimens in the Lahontan Valley, taken in April and November. Others were seen in the Indian Lakes area near Fallon. One taken November 14, 1940 had a 3–inch, largemouth black bass in its stomach.

Mills (PC) reported six merganser flew from the whirlpool at the outlet below Lahontan Dam January 28, 1960. Also, Mills reported a drake in good plumage was seen three miles east–northeast of Fallon April 12, 1975.

Neither Schwabenland (MS) nor Napier (MS) had records for Stillwater WMA. Saake (MS), however, reported one harvested December 21 or 22, 1974 at Stillwater WMA.

Biale (MS) reported them at Tonkin Reservoir in the Eureka area April 17, 1966; at the J. D. Ranch April 12, 1970; at Illapah Reservoir April 22, 1972; and at Tonkin Reservoir area May 27, 1973.

Kingery (1983) reported a Red–breasted Merganser at Eureka June 21.

Kingery (1984) states, "Kirch WMA reported a female Red–breasted Merganser through June."

Gullion et al., (1959) reported, "Three records: Imhof observed this species on Lake Mead, November 26, 1951, a bird found dead on the U.S. Highway 95 (2500 feet elev.) about 19 miles south of Searchlight, (3600 elev., Clark County) on May 23, 1952; and on June 3, 1952, a road–killed male was examined on the same highway where it crosses the Boulder Dry Lake in Eldorado Valley (1700 elev.)."

J. R. and A. A. Alcorn (MS) report on numerous occasions they thought they might have seen Red–breasted Merganser at Overton WMA, Moapa Valley, but from lack of a male in good plumage or a specimen in hand they were hesitant to list it for the area.

Austin and Bradley (1971) reported for Clark County, "Winter resident in riparian and aquatic areas. Common from November 16 to June 7."

Lawson (TT, 1976) commented, "It is not uncommon to see 65–100 of them in a single day on Lake Mead. We do find good numbers of them all winter long on Lakes Mead and Mohave.''

Lois H. Heilbrun and Douglas Stotz (1980) reported a Red–breasted Merganser at Henderson during the CBC December 15, 1979.

Heilbrun and Rosenburg (1981) reported 49 during the Henderson CBC December 20, 1981.

Ruddy Duck   *Oxyura jamaicensis*

Distribution. From east–central Alaska to Nova Scotia; south to South America.

Status in Nevada. Common resident in all areas except when forced out by icy conditions in mid–winter in northern parts of the state.

Records of Occurrence. Sheldon NWR (1982) reported the Ruddy Duck as a common permanent resident spring, summer and fall, occasional in winter. Nesting is reported in the area.

LaRochelle (MS, 1967) reported these birds as, ''common permanent residents at Ruby Lake NWR with up to 1400 on the area. In 1964, 120 pair produced average broods of 5.3.''

At Idlewild and Virginia Parks, Reno, in 1941–42, up to 25 of these ducks were seen twice in February; 4 times in March; 7 in April; October 26; 9 in November; and 11 in December (Alcorn MS).

In the Fallon area from 1960–65, inclusive, these ducks were observed twice in January; 3 times in February; 16 in March; 10 in April; 11 in May; 3 in June; 5 in July; once in August and September; 4 in October and November; and 3 times in December (Alcorn MS).

Schwabenland (MS, 1966) reported at Stillwater WMA, ''Quite abundant with up to 40,000 having been counted in the fall and 55,000 in spring. Up to 800 have been raised.''

Napier (MS, 1974) reported from Stillwater, ''This duck is common most of the year, except when icy conditions in December, January and February force it out of the area. During the migrations large rafts of Ruddy Ducks appear for a week or two, then move on. Common nester at Stillwater.''

Saake (MS, 1981) reported this duck as a common nester in big marshes in Stillwater, Humboldt and Carson Lake, also in Mason Valley.

Kingery (1986) reported at Stillwater that 1500 Ruddy Ducks were present, ''Unusually high numbers.''

Gabrielson (1949) reported the Ruddy Duck as, ''Common at Walker Lake, Mineral County on October 7 and 8, 1932 when a male was taken, and on November 16, 1934.''

Biale and J. R. Alcorn (MS) reported this duck in the Eureka area from February to October. They were seen 61 times between April, 1964 and October, 1973. Biale (PC, 1984) reported they are a common nester on the few ponds in the area.

J. R. and A. A. Alcorn (MS) observed these ducks in the Pahranagat Valley from February, 1960 to June 28, 1965 twice in January; 3 times in February; 8 in March and April; 13 in May; 4 in June; twice in July; August 8, 1963; September 19, 1960; 3 in October; and 5 times in November. Usually 10 or less were seen, but larger numbers were observed in February, March and April. The largest number seen was approximately 1050 on April 13, 1965.

J. R. Alcorn and A. A. Alcorn (MS) observed these ducks in the Moapa Valley area from December, 1959 to May, 1965 a total of 7 times in January; 5 in February; 12 in March; 4 in

April; 9 in May; 7 in June; July 24, 1964; 3 in September and October; 7 in November; and 5 times in December.

Austin and Bradley (1971) reported for Clark County, "Permanent resident in riparian and aquatic areas, present all year."

Kingery (1979) reported five broods—50 young at gravel pits near Las Vegas.

Lawson (TT, 1976) commented, "This duck is an uncommon to common permanent resident in Clark County."

Order FALCONIFORMES

Family CARTHARTIDAE: American Vultures

Turkey Vulture   *Cathartes aura*

Distribution. From southern British Columbia to western Ontario; across the United States to South America.

Status in Nevada. Common to uncommon summer resident over the entire state. Ryser (1985) reports a decline in numbers in Nevada in recent years.

Records of Occurrence. Sheldon NWR (1982) reported the Turkey Vulture as uncommon spring and fall, common in summer. It is thought to nest in the area.

In the Elko area from 1960–66, these birds were seen twice in March; 7 times in April; 4 in May; 8 in June; 4 in July; 7 in August; and 10 times in September (Alcorn MS).

LaRochelle (MS, 1967) reported for Ruby Lake NWR, "A common summer resident with predictable arrival and departures. First few (two to five) arrive March 17 to 24, and numbers build up to 70–90. Roost behind headquarters in aspens and on cliffs. Leave third week of October."

Kingery (1983) reported the earliest Turkey Vulture at Ruby Lake arrived March 10.

Herron (PC, 1984) reports about 150 birds nest in the Ruby Lakes area and during the spring of 1984, large numbers were seen feeding on deer that had died off during the heavy winter.

Ryser (1985) reports, "Rick Stetter and I found a roost in the Great Basin on a late September afternoon, along Soldier Creek, just off the northwest tip of the Ruby Mountains. At least 30 Turkey Vultures were soaring and gliding over and around a grove of tall cottonwood trees. Other vultures were perched, here and there, in the tops of the trees, and some were landing and others taking off."

Ryser (1967) reported the first flight of these birds seen over Reno March 4, 1967. Kingery (1981) reported, "Late Turkey Vultures included single birds at Reno October 2, 1980."

An unseasonal record for Carson City was of one seen January 27, 1970 (Alcorn MS).

In the area south of Yerington on June 13, 1941, 27 Turkey Vultures were observed by the author. It appeared these vultures had assembled there to eat the many dead ground squirrels *(Citellis t. mollis)*, kangaroo rats and jackrabbits that had died as a result of poisoning operations directed against the Ground Squirrels. I looked for dead vultures, but found none that appeared to be affected by secondary poisoning. One of these vultures was shot and its neck was full of jackrabbit meat (Alcorn MS).

Herron (PC, 1984) reports north of Yerington in the Weeks area, these birds are often seen and he suspects nesting in the area.

Alcorn (1946) reported for Lahontan Valley (Fallon area) these birds are seen each month from March to October, inclusive, with most records for May.

In the Fallon area from January, 1960 to January, 1964, these birds were seen 4 times in March; 7 in April and May; 4 in June; 5 in July; 4 in August; and twice in September. The earliest day was March 19, 1961 and the latest was September 27, 1962. Four miles southwest of Fallon on May 19, 1936, the author noted, "Two blackbirds attacked, in air, a Turkey Vulture. The vulture regurgitated twice in its flight and both times I found ground squirrels (*C. t. mollis*) had been regurgitated. The blackbirds then gave up the chase as the vulture appeared to fly more easily with less food. As the blackbirds gave up the chase a crow was seen to attack the flying vulture and the crow succeeded in plucking out a feather. The vultures are plentiful in this district now. The ranchers are drowning and killing ground squirrels during irrigation and leaving them in the fields. The vultures seem to be attracted by these squirrels as they are feeding on them (Alcorn MS)."

Biale (MS) reported these birds in the Eureka area from January, 1970 to January, 1974 a total of 4 times in March; 23 in April; 27 in May; 13 in June; 15 in July; 14 in August; and 20 times in September. From 1964 to 1974 he did not see them before March 26 (1971) and he never saw them later than September 30 (1971).

Biale (PC, 1984) reports nests observed by others in the Eureka area.

In Pahranagat Valley from 1960–65, inclusive, these vultures were seen 3 times in March; 13 in April; 12 in May; 5 in June; 4 in July; 3 times in August; and on September 19, 1960 (4 were seen) (Alcorn MS).

Johnson (1974) reported the Turkey Vulture is a, "Wide–ranging predator and/or scavenger that normally occurs at very low density and is easily overlooked. In the Grapevine Mountains there is no assurance that any of these were resident in 1939 and 1940; all could have been merely visitants and their exclusion from Miller's list of breeding birds could be justified on that point alone. The probability is high that it was not actually missing during the recent period and that virtually all species resident in 1939–40 were also breeding in 1971 and 1973."

Gullion et al., (1959) for southern Nevada reported "These birds are primarily spring and fall migrants on the desert, with most records falling between March 9 (1951, Overton) and late June, and again in September. Conspicuous migratory flight occurs in April. We have three mid–summer records from the desert: Las Vegas Valley July 14, 1951, four birds; Mohave Valley July 19, 1951, about 15 birds; and Lake Mead July 26–27, 1952, two birds. One bird over the Mohave Valley on January 25, 1952, constitutes the single winter record for this species for Nevada."

In Moapa Valley from 1960 to 1965, inclusive, these birds were observed on February 19, 1962; 9 times in March; 11 in April; 13 in May; 12 in June; 5 in July; 6 in August; 4 times in September; October 1, 1962; and October 7, 1963 (Alcorn MS).

In this valley, A. A. Alcorn (MS) saw 79 of these birds in one flock September 27, 1963 and suggested they were migrating.

Austin and Bradley (1971) for Clark County reported, "Transient in desert scrub, riparian, woodland and montane forest areas. Common from February 20 to October 4, unseasonal record on January 25 and February 1."

Kingery (1978) reported Turkey Vultures appeared at Boulder City February 27.

Lawson (TT, 1976) commented, "It is a common migrant and a rare to uncommon summer resident. Found primarily in the agricultural areas around Las Vegas and in the Moapa–Overton area. There simply is not enough food except in the agricultural areas.''

California Condor   *Gymnogyps californianus*

Distribution. Formerly resident near Pacific coast from Washington (Fort Vancouver, 1827) and Oregon (Drain of the Columbia River); south to northern Baja California. Now all California Condors are in captivity with the last known wild condor captured April 19, 1987.

Status in Nevada. Extinct in Nevada.

Records of Occurrence. According to Richard Dillon (1966) Grizzly Adams saw one of these birds on the Walker River in 1853 as he traveled down the river toward Walker Lake. Adams was reportedly so careless with the fact, there is some question as to whether or not he saw one.

Linsdale (1936) reported, ''Allotment of the California Condor to the recent avifauna of Nevada is based upon the reported finding of a complete ulna in Gypsum Cave, just out of Las Vegas. In announcing the discovery of the specimen, Miller (1931) commented, 'It is perhaps not fossil, but it is doubtless several centuries old since there has been a considerable degree of mineralization of the bone.' ''

Howard (1952) reported bone remains of the California Condor found in Smith Creek Canyon Cave, White Pine County.

Lawson (TT, 1976) commented, ''I actually had someone report one of these, but when I got to the area the bird was no longer there. The description was perfect; however, you are simply going to have to discount that. I really couldn't believe it. It could have been a question of light hitting the bird.''

Herron (PC, 1984) reported a biologist saw a condor soaring near Crystal Lake, Ash Meadows on February 23, 1984.

Kingery (1984) reported, ''The discovery of nesting California Condors at Porterville, California in March gives credence to a report of a soaring condor at Ash Meadows, Nevada, February 23, 150 miles east of Porterville.''

Family **ACCIPITRIDAE:** Kites, Eagles, Hawks and Allies

Osprey   *Pandion haliaetus*

Distribution. From Alaska to Labrador and Newfoundland; south along Pacific and Gulf coasts to South America.

Status in Nevada. Uncommon, a few individuals observed over widely scattered areas. Most commonly seen in Churchill and Clark Counties. Formerly nested at the upper end of Lahontan Reservoir. One recent nest record for Lake Tahoe.

Records of Occurrence. Sheldon NWR (1982) reported the Osprey as a rare transient visitor in spring and fall.

LaRochelle (MS, 1967) reported one sighted at Ruby Lake NWR June 11, 1964 and another lone bird April 11–13, 1966.

Lynn C. Howard (MS, 1971–72) reported for Ruby Lake NWR, "Two sightings of Ospreys were made during the year."

The Osprey is listed by Ruby Lake NWR (1981) as rare spring and fall.

Kingery (1980) reported four Ospreys seen (14 days of observation) at Goshute Mountains hawk watch, September through October.

Kingery (1977) reported Ospreys laid eggs in a nest at Lake Tahoe for the first recorded nest sight in Nevada. He (1978) reported an Osprey at Lake Tahoe successfully fledged young.

Herron (PC, 1984) reported the Osprey at Lake Tahoe laid three eggs and raised three young for 100% nesting success in 1984.

One was seen in the Lovelock area April 14, 1964 (Alcorn MS).

Kingery (1984) reported, "Osprey observations increased by 40% this spring . . . . They included urban visitors taking goldfish from a subdivision pond at Carson City."

Thompson (MS, 1988) reported two separate Osprey sightings in the Fallon area. One was at Stillwater WMA April 23, 1986, the other was a single individual hunting over Little Soda Lake September 15, 1986.

Nappe and Klebenow (1973) reported, "At one time Osprey nested along the Colorado River and in western Nevada. No known nesting sites in the past 10 years."

Mills (PC) saw one near Fernley September 11, 1960 and other lone individuals were seen in April and May at the upper end of Lahontan Reservoir, Lyon County. He reported April 20, 1975 he saw one Osprey at upper Lahontan and another (same day) at Rattlesnake Reservoir. He reported seeing one or more of them every year in the Fallon area, including one April 13, 1985 at Harmon Reservoir and two at Rattlesnake Reservoir May 12, 1985. One had a large carp in its talons.

Alcorn (1946) reported individuals seen in the Lahontan Valley in April, May, August, September and October. A combination of records kept by Mills, Napier and Alcorn for this valley since 1960 reveal records for March, April and May.

The crop and stomach of a male Osprey, shot nine miles northeast of Fallon May 7, 1944 contained the remains of a largemouth black bass (Alcorn MS).

Biale (MS) reported this bird at the Hay Ranch west of Eureka May 13, 1964 and at Tonkin Reservoir and vicinity June 15, 1970 and May 31 and September 4–6, 1971. Biale (PC, 1984) reported one seen at Diamond Valley on September 28, 1980.

A. A. Alcorn (MS) saw one in the Pahranagat Valley April 13, 1965.

In Clark County Austin and Bradley (1971) reported, "Transient in desert scrub and riparian areas. Common from April 18 to May 26 and from July 28 to October. Unseasonal occurrence on November 30 and January."

Lawson (TT, 1976) commented, "It is a rare to uncommon migrant spring and fall in southern Nevada."

Black–shouldered Kite   *Elanus caeruleus*

Distribution. Formerly the White–tailed Kite. From Washington to Baja California and the western U.S. Also in Texas, Florida and the Gulf states; south to Chile and Argentina.

Status in Nevada. An uncommon newcomer to Nevada.

Records of Occurrence. Thomas Lugaski et al., (1972) reported, "Several reports on the occurrence of the White–tailed Kite in the areas of Walker Lake, Mineral County, Nevada, and Fallon, Churchill County, Nevada, have been made to various members of the Biology Department, University of Nevada–Reno and members of the Lahontan Chapter of the National Audubon Society, but all attempts to confirm these sightings were fruitless.''

"Ryser (1972) reported in Mason Valley, Lyon County, Nevada in an area three miles due west of the Fort Churchill Power Station and four miles southwest of Wabuska, Nevada, the authors observed, 'at about 10:00 in the morning, two White–tailed Kites coming towards them from the north. The kites, upon approaching our position, veered to the east and separated, one continuing to fly due east, the other southeasterly. We observed the kite flying due east for about five minutes during which time we noted his overall whitish falcon shape, long, pointed wings, with a distinctive black patch on the fore edge, his extremely long white tail and his familiar 'kiting' attitude while hovering in search of prey.' ''

"Several weeks later on January 24, 1972, Dr. Fred Ryser, Jr., Curator of Birds and Mammals, University of Nevada–Reno, and Charles Lawson, member of the National Audubon Society, reported sighting, on two occasions, White–tailed Kites in the Mason Valley area, thus confirming the sighting of December, 1971.''

Ryser (1972) reported one in the fall near Mount Rose Summit, and he (1973) reported one 23 miles east of Reno March 16, 1973.

Saake (MS, 1981) reported one at Carson Lake in August of the 1970s.

Stephen G. and Melinda A. Pruett–Jones and Richard L. Knight (1980) reported, "Since 1972, there have been several sightings of White–tailed Kites in Nevada. In 1971, two birds were seen on the Mason Valley CBC, and other individuals were recorded at Reno and Pyramid Lake (Kingery, 1972). From 1973 to 1975, there were at least four other sightings, all in the vicinity of Reno (Kingery, 1973–75).''

Mowbray (PC, 1987) reported a Black–shouldered Kite seen September 18, 1982 in the Mormon Farm area, Clark County. He commented that two were seen in that area over a 2–week period.

Kingery (1986) reported, "A Black–shouldered Kite seen at Las Vegas April 25–30 provided southern Nevada's second record.''

Mississippi Kite   *Ictinia mississippiensis*

Distribution. From southwestern and central United States along the Gulf coast and Florida; south to South America (Paraguay).

Status in Nevada. Rare.

Records of Occurrence. Kingery (1979) reported, "An adult and immature Mississippi Kite at Corn Creek, found by tourists, gave Nevada its first record on a day with strong northern winds.''

Kingery (1982) reported, "At Alamo, Nevada on March 21, a Mississippi Kite performed graceful corkscrew aerobatics for five minutes for Nevada's second record.''

Kingery (1986) reported, "A stray Mississippi Kite visited Las Vegas June 6 (a first–summer bird).''

Bald Eagle   *Haliaeetus leucocephalus*

Distribution. From Alaska to Newfoundland; south to Baja California and Florida.

Status in Nevada. Winter visitant. Formerly nested at Pyramid Lake and possibly in southern Nevada.

Two specimens in the Museum of Vertebrate Zoology, University of California at Berkeley are labeled *H. l. leucocephalus,* the southern race of Bald Eagle. One was taken January 18, 1943 and one December 16, 1945; both from the Fallon area.

Records of Occurrence. Saake (MS) saw an adult at New Year Lake, Washoe County, April 1, 1970.

Sheldon NWR (1982) reported the Bald Eagle as occasionally seen in fall and winter.

Linsdale (1936) stated, "Bald Eagles nested on the main island in Pyramid Lake in 1866. The species was considered rare in that neighborhood by Ridgway (1877). Another record for the state is of two adults seen about the middle of March, on a dead mesquite, in Ash Meadows, Nye County (Fisher, 1893)."

Oakleaf (1974a) stated, "Fish and Game sight records of 195 Bald Eagles in recent years show a much wider winter distribution than indicated by literature records. Surveys specifically for Bald Eagles in 1973–74 have also added significantly to distribution records (104 Bald Eagles observed in 1973–74)."

"Sightings indicated two distinct wintering populations occur in the state: a population in the western corner of the state, including the Reno area, Lahontan Valley and the Carson River, and a population scattered the entire north–south length of the eastern quarter of the state."

"Jerry Page counted a combined total of 35 Bald Eagles on two aerial surveys in January and February, 1973."

Oakleaf (1974b) stated, "During the 1974 January survey, 40 Bald Eagles (21 adults, 19 immatures) were counted wintering in Nevada."

Herron (MS, 1984) reported that before 1983, no known nesting of Bald Eagles occurred in Nevada. Up to 114 birds have been reported in one year and he suspects that 150 winter in the state. The distribution of the population is 60% in western Nevada, 35% in eastern Nevada and 5% in southern Nevada.

LaRochelle (MS, 1967) reported for Ruby Lake NWR two or three were seen each winter, often soaring or perched in a tree.

Howard (MS, 1969–72) reported for Ruby Lake NWR three were seen in the winter of 1969; one seen December 1, 1970; one immature on February 6, 1971; an adult on March 19, 1971; and one adult January 11, 1972.

Kingery (1983) reported, "50–60 Bald Eagles were gathered at Ruby Lakes in August."

Nappe and Klebenow (1973) reported, "Robert Quiroz records that up to 10 Bald Eagles are seen each winter in the area south of Wells and also in the vicinity of Spruce Mountains, Elko County."

Accompanied by John Keiger and Sandy Wilbur, the author saw one adult Bald Eagle perched on a large boulder at Anaho Island, Pyramid Lake, Nevada July 29, 1960.

Orr and Moffitt (1971) reported for the Lake Tahoe area, "Infrequent summer visitant, occurring more often in fall and winter, less common in recent years than formerly."

Kingery (1980) reported seven Bald Eagles at South Lake Tahoe December 16, 1979.

Heilbrun and Rosenburg (1981) reported one in the Reno area December 20, 1980 during the CBC.

Curran (PC) reported an unseasonal occurrence of an adult Bald Eagle at Silver Springs, Lyon County on June 23, 1973. He reported it flew across the highway and alighted on some rocks near the dump just east–northeast of the highway intersection.

Alcorn (1946) reported small numbers seen in the Fallon area each winter month from November to February. In the same area, from 1960 to 1970, these birds were observed 9 times in January; 10 in February; 7 in March; and 8 times in December.

Schwabenland (MS, 1966) reported for Stillwater WMA between six and 15 were seen every year during the fall and winter months.

Napier (MS, 1974) reported on this same area, "The birds were commonly seen from December through March. Most are seen in the Indian Lakes area. Peak number is under 10 birds. The number of birds seems to be decreasing in the past eight years."

An example of the population in Lahontan Valley on March 6, 1969 is indicated by the flight of wildlife biologist Jim Good and pilot biologist Ray Glahn, who flew over the area from Fallon to Carson Sink, Pelican Island, Leader Reservoir and then near East Lake and Cottonwood Lake. A total of four Bald Eagles were seen in 1 1/2 hours of flying time (Alcorn MS).

Thompson (MS, 1986) reported 35 Bald Eagles near the mouth of the Carson River January 7, 1986. They were attracted to the area by a large number of dead and dying fish (mostly carp). Thompson attributed the die–off to ice and low oxygen in the water.

Thompson (PC, 1988) reported the highest number of Bald Eagles seen wintering in the Lahontan Valley area was 70 on February 27, 1987. He estimates there may have been as many as 100 present in the area.

Their absence from the Eureka area is evidenced by the fact that Biale (MS) had never seen one in over 30 years of birdwatching in the area. Recently, however, he has reported (PC, 1984) one in Newark Valley March 22, 1981; one February 20, one March 10 and two on March 17 all of 1983. He (MS, 1988) reported one adult and one immature in northwest Diamond Valley December 20, 1987.

One was observed by A. A. Alcorn (MS) near Maynard Lake, Pahranagat Valley, November 14, 1961.

Austin (1968) stated, "This species apparently has not been reported in southern Nevada since 1959. I saw an adult at Dike Lake and an immature just south of Hiko, both of Lincoln County on January 22, 1968. The following day Charles G. Hansen, biologist at Desert NWR, saw two adults near Dike Lake."

Castetter and Hill (1979) reported a Bald Eagle eating jackrabbits at Rock Valley February 16, 1977.

Kingery (1979) reported 42 Bald Eagles in the Nevada winter survey.

Lawson (TT, 1976) commented, "The Bald Eagle is a rare to uncommon winter visitor in southern Nevada. Most of the birds seen were around the Lake Mead area."

Northern Harrier   *Circus cyaneus*

Distribution. Commonly referred to as the Marsh Hawk. From Alaska to as far east as Newfoundland; south to Baja California and Virginia and farther south to southern Columbia.

Status in Nevada. A common statewide resident, but more common in the northern part of the state.

These hawks prefer wet meadows and marshy areas, but are also found in desert scrub and in sagebrush–covered valleys many miles from water.

Records of Occurrence. Sheldon NWR (1982) reported the Northern Harrier as common spring, summer and fall; uncommon in winter. The Northern Harrier also nests in the area.

LaRochelle (MS, 1967) reported at Ruby Lake NWR, "This bird is our most common hawk with some present the year round. Most numerous (35–50) in the fall. Males not usually seen until early spring."

The author's records for Elko County reveal they were not recorded in January or February, but were seen twice in March; 7 times in April; twice in May; 8 in June; once in July; 7 in August; 9 in September; 6 in October; twice in November; and once in December.

Kingery (1983) reported, "40 pair of Marsh Hawks nested at Ruby Lake."

Ridgway (1877) reported, "No marsh of any extent was visited, either in winter or summer, where this hawk could not be seen at almost any time during the day skimming over the tules in search of its prey. The latter consists of small birds of all kinds, the young of waterfowl, lizards and probably small mammals, although the latter were not found in the crop of any of those specimens examined. The stomachs and crops of those killed at Pyramid Lake were filled to their utmost capacity with the remains of small lizards, and nothing else; at the same locality, however, they were often observed to chase small birds, particularly Brewer's and the Black–throated Sparrows, the most numerous species, of which this hawk appears to be a most dreaded enemy, since its appearance creates perfect consternation among all the sparrows in its path, who utter distressed cries, and make confused and desperate efforts to escape by plunging precipitately into the thickest bushes."

At Willow Creek Reservoir, Elko County, a Marsh Hawk was seen to catch a Western Meadowlark in flight on July 11, 1966 (Alcorn MS).

Ryser (1969) reported a considerable number over–wintering in the Truckee Meadows. Practically all of them were in juvenile or female plumage.

Alcorn (1946) reported in the Lahontan Valley this hawk was a common resident with numerous records available. In this area from January, 1960 to December 31, 1967, these birds were seen 84 times in January; 53 in February; 28 in March; 17 in April; 9 in May; 3 in June; twice in July and August; 4 in September; 21 in October; 29 in November; and 48 times in December.

Schwabenland (MS, 1966) reported at Stillwater WMA the largest build–up in population usually occurs in late summer or early fall.

Napier (MS, 1974) for the same area reported the lowest population is, "During the breeding season and in mid–winter. The peak population is in late fall and winter until freeze up, then many move out." He reported nesting in the area. There has been a decline in numbers of these birds in this valley in recent years (Alcorn MS).

Thompson (PC, 1988) reported a continued decline in nesting in the Fallon area; only two nests seen in 1987.

Biale (MS) reported seeing these hawks in the Eureka area from January 1, 1965 to December 31, 1967 a total of 15 times in January; 14 in February; 17 in March; 8 in April; 15 in May; 5 in June; 6 in July; 4 in August; 7 in September; 14 in October; 13 in November; and 21 times in December. In this area from January 1, 1971 to December 31, 1973, Biale saw them 10 times in January and February; 7 in March; 14 in April and May; 6 in June; twice in July; 10 in August; 13 in September; 15 in October; 8 in November; and 16 times in December. He saw a nest July 4, 1965 containing three young ready to leave the nest.

Biale (PC, 1986) reported a nest with five eggs June 13, 1986. It was located at 7500 feet elev., near Ruby Hill Mine area. All five young hatched.

In eastern Nevada, one was seen in the Kern Mountains October 26, 1967; one at Comin's Lake June 5, 1962; and in the Baker area April 2, 1967; 3 times in May; and 4 times in October (Alcorn MS).

In Pahranagat Valley from January, 1960 to December, 1965, inclusive, these hawks were recorded 5 times in January; twice in February; 9 in March; 5 in April; 9 in May; June 14, 1965; August 11, 1960; twice in September; 4 in October and November; and 7 times in December (Alcorn MS).

Gullion et al., (1959) reported, "Marsh Hawks are widespread on desert areas from about September 4 (1953, Gold Butte) to as late as April 8 (1952, Las Vegas Valley). They linger somewhat longer in the more moist desert valleys, remaining as late as May 26 (1951, Meadow Valley) and returning as early as August 6 (1954, Pahrump Valley)."

In Moapa Valley from December, 1959 to December, 1964, inclusive, J. R. and A. A. Alcorn (MS) reported these hawks were seen 18 times in January; 8 in February; 17 in March; 4 in April; 3 in May; June 16, 1960; twice in July; 4 in August; 6 in September; 4 in October; and 8 times in November and December.

In Pahrump Valley in 1960 and 1962, they were recorded in March, August, September and October (Alcorn MS).

Austin and Bradley (1971) for Clark County reported, "Winter resident in riparian areas and transient in desert scrub areas. Common from August 4 to May 8."

Lawson (TT, 1976) commented, "We seldom see one in the summertime. They are very common in the fall and all winter in southern Nevada."

Sharp–shinned Hawk   *Accipiter striatus*

Distribution. From Alaska to Newfoundland; south to South America.

Status in Nevada. Uncommon summer resident in mountain ranges and winter resident in valleys.

Herron (PC, 1984) reported these birds are doing well. "They nest in everything, usually less than 12 feet off the ground and the nest is usually shallow and about 1 1/2 feet wide."

Records of Occurrence. Sheldon NWR (1982) reported the Sharp–shinned Hawk as uncommon spring and fall; occasionally seen in the summer. It possibly nests in the area.

Ryser (1973) reported this bird in forests around the lake in the Pine Forest Mountains in early August, 1973.

A lone bird was seen each day October 22 and 23, 1962 at Stonehouse Creek, Santa Rosa Range (Alcorn MS).

In Elko County, this hawk was seen several times near Jarbidge in September, 1959. Another lone bird was noted near Tuscarora September 28, 1965 (Alcorn MS).

LaRochelle (MS, 1967) reported at Ruby Lake NWR this hawk is not often seen on the refuge and only four sightings were recorded.

One was seen near Lovelock November 22, 1963 (Alcorn MS).

In the Reno area, lone birds were observed in February, March, April, November and December (Alcorn MS).

Ryser (1971) reported these hawks were repeatedly seen in and about Reno March 21, 1971.

Orr and Moffitt (1971) reported for Lake Tahoe, "Common summer and fall visitant; rare in winters."

One was seen near Carson City March 21, 1962, and another lone bird was observed near Virginia City July 20, 1962 (Alcorn MS).

One was seen near Yerington April 17, 1946. Lone individuals were repeatedly seen in this area during winters of 1981–83 (Alcorn MS).

Alcorn (1946) reported in Lahontan Valley this hawk is a common winter visitant, with sight records for all months except June and July.

In the Fallon area from January, 1960 to January, 1967, these birds were recorded 6 times in January; 11 in February; 4 in March; 5 in April; May 11, 1963; June 1, 1964; 4 in October; and 7 times in November and December. During the winter from 1981 to 1983, lone individuals were often seen in this same area (Alcorn MS).

Examination of three Sharp–shinned Hawk stomachs obtained in the Fallon area February 20, 1940 revealed two contained remains of Horned Larks and one contained remains of a White–crowned Sparrow (Alcorn MS).

Lone birds were seen near Austin April 7, 1960 and October 15, 1964. At Kingston Canyon one was seen June 20, 1963. A lone bird was observed at Greenmonster Canyon, Monitor Range, October 2, 1960 (Alcorn MS).

In the Eureka area Biale (MS) reported seeing this bird from 1964 to October 4, 1973; on February 14, 1965; 5 times in April; July 5, 1964; August 30, 1964; 6 times in September; twice in October and November; and December 7, 1967. He (PC, 1984) reports they are not a common hawk, seen mostly in spring and fall in the Eureka area.

Gabrielson (1949) reported, "One in Fish Lake Valley, Esmeralda County, October 8, 1932; and one found dead, October 11, 1932 at Potts Post Office in northern Nye County."

A. A. Alcorn (MS) saw one near Baker April 30, 1967.

Johnson (1973) reported, "One was seen flying over clumps of white fir and serviceberry at 7000 feet elevation in Highland Range, Lincoln County. In view of the date and habitat this locality is probably used for breeding, although proper documentation of this point is still lacking."

Gullion et al., (1959) reported for southern Nevada, "Common on the desert only in fall and spring migration. Fall migration extends from mid–October to late November, and spring migration is primarily in April, although records as early as February 2 (1952, Las Vegas Valley) have been obtained. One bird collected by Gullion in Las Vegas Valley, four miles east of Las Vegas on April 10, 1952 has been identified at the Museum of Vertebrate Zoology as A. s. velox."

In the area from Hiko, Pahranagat Valley, to Lake Mead at the south end of Moapa Valley, lone birds were seen January 10, 1964 and January 4, 1967; February 7, 1963; March 20, 1962; 3 times in April; May 6, 1964; June 14, 1964; 3 times in July; October 1, 1962; 3 times in November; and December 1, 1963 (Alcorn MS).

Austin and Bradley (1971) reported for Clark County, "Winter resident in riparian areas. Transient in desert scrub, woodland and montane forests. In riparian and desert scrub it is common from August 19 to May 4 with unseasonal occurrence in July. In woodland and montane forests it is common from August 30 to November 2."

Lawson (TT, 1976) commented, "It is an uncommon to common permanent resident here in Clark County. They do breed in the Spring Mountains and Sheep Range."

Cooper's Hawk   *Accipiter cooperii*

Distribution. From British Columbia to Nova Scotia (rarely); south to Honduras.

Status in Nevada. Summer resident in mountains and winter visitant to the valleys.

Herron et al., (1985) reported, "From 1974 through 1983, the Department of Wildlife has documented 193 Cooper's Hawk nesting territories that were occupied by adults. From evaluating the available preferred habitat it is estimated that approximately 700 nesting pair are present in Nevada."

Records of Occurrence. Sheldon NWR (1982) reported the Cooper's Hawk as uncommon throughout the year with some nesting in the area.

In Elko County, lone individuals were seen near Jarbidge on September 7 and 8, 1961; one near Wildhorse Reservoir August 19, 1971; and others were seen near Tuscarora September 20 and October 20, 1964 (Alcorn MS).

LaRochelle (MS, 1967) reported at Ruby Lake NWR he saw this hawk on August 19, 1965 and he saw two on April 28, 1970.

One was seen near Lovelock February 9, 1965 (Alcorn MS).

Orr and Moffitt (1971) reported this hawk as fairly common in the Lake Tahoe area in summer and more abundant in autumn with a few individuals wintering.

In western Nevada, one was seen in Reno December 25, 1970; one at Washoe Lake March 6, 1962; one at Virginia City July 20, 1962; and one at Wilson Canyon, Walker River March 3, 1961 (Alcorn MS).

In the Yerington area, one was seen November 29 and December 4, 1962 and lone individuals were repeatedly seen in winter months from 1981–84 at the McGowan bird feeder near Yerington. One caught a White–crowned Sparrow there on January 21, 1982 and another knocked itself out by flying into a house window on January 17, 1983 (Alcorn MS).

At the upper end of Lahontan Reservoir one was seen November 27, 1959 and another December 20, 1961 (Alcorn MS).

Alcorn (1946) reported this bird in Lahontan Valley (Fallon area) was a winter visitant and transient in summer with records in each month of the year except June and July. In this area from January 1, 1960 to January 1, 1968, this hawk was seen 18 times in January; 13 in February; 6 in March; April 25, 1964; August 31, 1967; twice in September; 10 in October; 9 in November; and 11 times in December (Alcorn MS).

Near Fallon on February 28, 1963, a male Cooper's Hawk had captured a starling and as our car approached, it tried to fly away with the starling. The hawk became alarmed and released the starling, which flew away to join a small flock (30 birds) of starlings. On October 24, 1964 at this author's ranch, west of Fallon, a Cooper's Hawk captured one California Quail. This was the second time a Cooper's Hawk had a quail that month (Alcorn MS).

On January 18, 1965 at this same place, one of these hawks was frightened from a partially eaten Red–shafted Flicker (Alcorn MS).

In the general area four miles west of Fallon, these birds were obtained and their stomachs revealed the following: December 24, 1939, crop contained remains of California Quail; March 1, 1940, crop contained remains of Townsend Ground Squirrel; January 17, 1943, crop contained remains of California Quail (Alcorn MS).

Thompson (MS, 1988) reported single Cooper's Hawks seen October 7, 1986 along the East Canal in Fallon and on January 4, 1987 at Carson Lake.

From Fallon to Eureka these birds were seen at Cherry Creek, in the Clan Alpine Mountains on September 29 and October 20, 1960; near Reese River, Nye County on May 19, July 12 and October 6, 1965; near Austin on September 22, 1960, December 12, 1961, November 6, 1964 and June 23, 1965; at Big Creek south of Austin on May 15 and August 19, 1971; at Potts Ranch, Monitor Valley on August 26, 1966; and a nest in an aspen tree at Wilson Creek, Toquima Range contained a large young on July 4, 1969 (Alcorn MS).

Biale (MS) reported from 1964 to February 24, 1974, this hawk was seen in the Eureka area twice in January; 3 times in February; twice in March; 6 in April; 4 in May; June 27, 1965 (nest); 5 in July and August; 15 in September and October; twice in November; and 4 times in December.

In the Eureka area on June 27, 1965 Biale (MS) saw a nest with eggs in a pinon tree. On July 25, 1965 this nest contained a young hawk ready to leave the nest. Since then, he has seen at least six nests in mountain mahogany trees six miles south of Eureka (7800 feet elev.).

A. A. Alcorn (MS) saw lone birds in Spring Valley, White Pine County, May 1, 1967 and near Baker October 6, 1967.

One was seen near Beatty April 25, 1965 (Alcorn MS).

Gullion et al., (1959) reported for southern Nevada, "Absent from the desert in summer; these hawks arrive from the north as early as August 6 (1953, Pahrump Valley) and remain through the winter until as late as May 20 (1951, also Pahrump Valley). During heavy fall migration, from about mid–September to early October, and again in spring migration during April, this species, together with the Sharp–shinned Hawk, became so numerous that Gullion had to suspend quail trapping to prevent heavy game bird losses."

J. R. and A. A. Alcorn (MS) recorded this bird 17 times in southern Nevada from January, 1960 to May, 1965.

One was seen near Hiko April 14, 1965 (Alcorn MS).

In the Virgin Valley, individuals were seen December 15, 1962; January 25, February 22, July 24, and August 29, 1963 (Alcorn MS).

In Moapa Valley, one was seen January 28, 1960; three were seen October 31, 1960; lone individuals were seen November 25, 1961; March 5, 10 and 13, 1962; February 18 and September 14, 1963; and March 18, 1964 (Alcorn MS).

In the Spring Mountains at Trough Springs one was observed and a nest in an aspen tree at Rosebud Spring contained two young July 19, 1963 (Alcorn MS).

Austin and Bradley (1971) for Clark County reported, "Summer resident in montane forests and winter resident in riparian areas. Transient in desert scrub woodland. In woodland and montane forests it is common from April 2 to October 6. In desert scrub and riparian common from August 23 to May 31."

Lawson (TT, 1976) commented, "We have records of them from the Spring Range, records of them from Red Rock Canyon area where there is a nest. They nest in Lee, Cow, Macks and Scout Canyons–they are very well represented in the Mount Charleston area."

Northern Goshawk   *Accipiter gentilis*

Distribution. From Alaska to Newfoundland; south to northern Mexico and Florida.

Status in Nevada. Resident in some mountain ranges and transient in the lower valleys.

Herron (PC, 1984) reports he believes there are 300 pair nesting in the state of Nevada, with 87% of the nests found in aspen trees.

Records of Occurrence. Sheldon NWR (1982) reported the Northern Goshawk as a rare summer resident; also reported in spring, summer and fall.

In Elko County, lone birds have been seen in the vicinity of Jarbidge on September 6, 1961; Jack Creek on October 24, 1971; Tuscarora on September 6, 1964; Elko on April 11, 1966; and Ruby Valley on May 19, 1966 (Alcorn MS).

Orr and Moffitt (1971) reported at Lake Tahoe, "Uncommon summer visitant, becoming more abundant in the fall, but apparently not wintering regularly."

A lone bird was seen at the Lake Tahoe 4–H camp July 21, 1966 (Alcorn MS).

Ryser (1965) reported several successful nestings in the Carson Range in the summer of 1965 and he saw as many as four individuals at one time.

Repeated sightings of this hawk by this author in the Pine Nut Range in early summer of 1980 and 1981 would indicate nesting there (Alcorn MS).

In other areas of the state, lone birds were observed in the Yerington area on January 19, 1961; near Fort Churchill on January 5, 1960, May 3, 1970 and March 14, 1971 (Alcorn MS).

Alcorn (1946) reported individuals obtained in the Lahontan Valley (Fallon area) in February and March and others were seen in September and November.

In the past 20 years, at least one of these birds has been observed in the Fallon area in most years. Most of those seen were lone birds observed in April and May (Alcorn MS).

In the general area from Fallon to Eureka, lone birds were observed at Buffalo Canyon near Eastgate, November 10, 1963; near Ellsworth on March 18, 1966; Becker Canyon on November 6, 1963; near Austin October 25, 1961, November 8, 1964 and October 31, 1967; and at Greenmonster Canyon, Monitor Range on September 26, 1958 (Alcorn MS).

Biale (MS) reported seeing these hawks in the Eureka area February 28, 1971; March 5, 1972; one May 9, 1965 (on a nest); one May 29, 1971; one young in a nest June 10, 1965; two young in a nest June 17, 1966; four young in a nest with two adults nearby June 18, 1972; a nest with an unknown number of young June 24, 1973; an active nest July 4, 1973; one August 31, 1969; 4 times in September; October 25, 1971; twice in November; and one December 7, 1972.

Biale (PC, 1986) reported, "They are uncommon to common in the Eureka area, more common during the nesting season."

Gabrielson (1949) reported, "A single bird seen May 27, 1932 on Lehman Creek in the vicinity of Snake Mountains, White Pine County."

Johnson (1973) reported in southeastern Nevada, "May 22, 1972 in the Clover Mountains, an active nest was found, attended by a female, and photographed (by Blair Csuti), nest 35 feet above ground, ponderosa pine. On the same date, about six miles from the nest site, another female was seen over open pine forest."

"On June 24–25, 1972, two individuals in the fir and aspen forest on Mount Wilson, Lincoln County, a presumed male at 8600 feet and a large bird thought to be a female at 8100 feet. Considerable habitat at upper levels in the Quinn Canyon Mountains appears suitable for this species but it has not yet been found there."

Austin and Bradley (1971) reported for Clark County, "Permanent resident in montane forests occurring all year. Transient in woodland areas."

Oakleaf (1974b) reported, "Work during 1974 indicated that Goshawks are more common than previously thought. Information was collected on 23 Goshawks' nesting sites that have been active within recent years. From preliminary analysis of potential nesting habitat, I estimate that the 23 known nesting territories represent less than 4% of the total population in Nevada.''

Lawson (TT, 1976) commented, "This hawk is a rare permanent resident here in Clark County. There are a few resident breeding pair in the Charleston area."

Common Black–Hawk   *Buteogallus anthracinus*

Distribution. From central Arizona to Texas; south to Venezuela in South America.

Status in Nevada. Rare.

Records of Occurrence. Kingery (1978) reported, "One Black Hawk was at Corn Creek May 27." He (1979) reported, "Nevada added its second and third Black Hawk records with one near Davis Dam September 26 and another at Meadow Valley Wash near Elgin."

Kingery (1980) reported, "One Black Hawk (its fourth for Nevada) was at Las Vegas on September 15." He also reported, "One Black Hawk on April 2. It is the fifth near Fort Mohave."

Kingery (1984) reported, "A Common Black–Hawk visited Davis Dam, Nevada, April 9."

Harris' Hawk   *Parabuteo unicinctus*

Distribution. From southeastern California, Arizona, New Mexico, Texas and (casually) Louisiana and Mississippi; south through Mexico and Central America to central Chile and Argentina.

Status in Nevada. Rare.

Records of Occurrence. Linsdale (1951) reported, "One sight record: on March 30, 1949, W. Pulich saw one at Overton (Monson, 1949)."

Lawson (TT, 1976) commented, "I have two personal sightings, one in 1975 and one prior to that time in the Fort Mohave area, in the old cottonwood trees there. This hawk suffers a great deal from popularity from falconers and there is an awful lot of pressure on the birds. Falconers come over from California to capture them and sell them."

"Another record also from the Sunset Park area here in Las Vegas of a young bird in the fall of 1973. There was some effort by falconers to take that bird."

Lawson (1977) saw one at Mormon Farm southeast of Las Vegas April 18, 1975.

Kingery (1987) reported "Harris' Hawk, apparently wild, was at Overton, Nevada May 2."

Red–shouldered Hawk   *Buteo lineatus*

Distribution. From northern California to southern New Brunswick; south to the Florida Keys and south in Mexico to Veracruz.

Status in Nevada. Uncommon visitant.

Records of Occurrence. Ryser (1972) reported one near Mount Rose Summit in the fall of 1972. He (1985) reported several other sight records from fall to spring.

Herron (PC, 1988) reported Red–shouldered Hawks seen in Reno during the 1987–88 winter. He stated Hugh Judd saw a juvenile November 3, 1987. Judd later reported an adult seen from November 9, 1987 through February, 1988 in the same area, although they maintained separate territories.

Neal (PC, 1988) reported two sightings of the Red–shouldered Hawk in the Fallon area. One was November 10, 1987 and the other January 11, 1988.

Herron (PC, 1984) reported an increase of Red–shouldered Hawks in Nevada. He reported an immature at Hawthorne in July, 1982 and another July 7, 1983.

Kingery (1986) reported, "A Red–shouldered Hawk wintered at Kirch WMA. They may nest in the area since one was collected nearly Ely, Nevada last year."

Snider (1970) reported one seen near Las Vegas December 6, 1969. It stayed until March 1, 1970.

Cruickshank (1970) reported one on the Desert Game Range December 28, 1969 and noted good studies all month of a Red–shouldered Hawk had been possible.

Monson (1973) reported one seen at Mormon Farm, near Las Vegas, Nevada on September 2 and 25, 1972 and from December 2, 1972 to February 17, 1973.

Kingery (1978) reported one appeared intermittently all winter at Corn Creek 1977–78 and he (1978) reported one at Corn Creek September 1–4 and an adult at Pahranagat Refuge was seen for several days around September 23. Also, he (1980) reported, "A Red–shouldered Hawk was at Corn Creek, near Las Vegas, July 30."

Lawson (MS, 1981) reported, "We now have photos from eight feet of an immature July 16, 1981 here in Las Vegas."

Heilbrun and Rosenburg (1981) reported a Red–shouldered Hawk at the Desert Game Range during the CBC December 27, 1980 and they also reported one on the Henderson CBC December 20, 1981.

Herron (PC, 1984) reported these birds seen by various individuals in southern Nevada November 26, 1981, December 29, 1982, January 20, 1983 and March 23, 1983.

Kingery (1984) reported, "Pahranagat Refuge, Nevada hosted two Red–shouldered Hawks July 23." He (1985) noted, "Nevada reported five Red–shouldered Hawks at Las Vegas, Pahranagat Refuge and Kirch WMA."

Kingery (1987) reported, "Nevadans continued to find Red–shouldered Hawks, with single birds at Las Vegas, Mt. Charleston, and Ruby Lakes, plus an adult and immature at Kirch August 12–October 13."

Kingery (1987) reported a Red–shouldered Hawk at Logandale April 11.

Broad–winged Hawk   *Buteo platypterus*

Distribution. From central Alberta to Nova Scotia; south along the coasts into South America.

Status in Nevada. Irregular transient.

Records of Occurrence. Kingery (1981) reported, "The Kingsley Mountains near Wendover, in Nevada produced one Broad–winged Hawk on September 14."

Herron (PC, 1988) reported a total of 33 Broad–winged Hawk sightings in the Goshute Mountains from 1974 to 1984.

Lawson (TT, 1976) reported one at Corn Creek May 7, 1973.

Kingery (1979) stated, "Nevada's second Broad–winged Hawk visited Las Vegas on September 21."

Swainson's Hawk *Buteo swainsoni*

Distribution. From Alaska to Manitoba; south to Argentina. Migrates in great flocks through Mexico and Central America. Casual in eastern North America.

Status in Nevada. Summer resident in the north, central and western valleys and transient. Although there are several reported winter sightings of this bird, this author has yet to see one in winter.

Herron (PC, 1984) reports these birds continue to decline in numbers in the western states—thought to be due to the mortality rates on the winter grounds in South America, where the natives reportedly use them for food."

Herron et al., (1985) reported, "From survey data presently available it is estimated that fewer than 150 pair of Swainson's Hawks nest in Nevada."

Records of Occurrence. Sheldon NWR (1982) reported the Swainson's Hawk as one of occasional occurrence spring and fall and uncommon in summer, with some nesting reported.

One was seen at Badger, northern Washoe County, May 5, 1961 (Alcorn MS).

In Humboldt County, lone individuals were seen at Orovada on April 22, 1942 and in Paradise Valley April 25, 1969 and May 12, 1966 (Alcorn MS).

In Elko County from 1962–71, inclusive, these hawks were seen 4 times in April; 6 in May; 5 in June; 3 in July; 4 in August; September 26, 1962 and an adult was observed July 11, 1965. These hawks were most numerous around high populations of Richardson's Ground Squirrels and they reportedly feed on these animals in large numbers (Alcorn MS).

LaRochelle (MS, 1967) reported at Ruby Lake NWR this hawk is commonly seen during late spring and early fall, but never is numerous. No nesting was observed on this refuge.

These birds are common in the Lovelock area in April and May where they are frequently attracted by the large number of meadow mice *(Microtus montanus)* in the area.

The 1964 CBC reported an immature carefully checked at Unionville, Pershing County, January 1, 1965. The 1960 CBC reported four Swainson's Hawks at Reno, December 25, 1959 as most unusual. No other reports are available for this bird in this area in winter.

Ryser (1973) reported the first one seen in 1973 was in Washoe Valley March 18 and three others were seen in the Truckee Meadows March 23.

One seen near Minden May 19, 1944 was thought to be nesting in that area; others were seen in the Yerington area in April and one seen in late May was thought to be a breeding bird (Alcorn MS).

Alcorn (1946) reported this hawk was a summer resident in the Lahontan Valley (Fallon area). It was frequently seen from April to August, inclusive. "This hawk nests in cottonwood trees in considerable numbers throughout the valley."

The stomach of a female shot two miles northwest of Fallon June 4, 1942 contained the remains of a Townsend's Ground Squirrel (Alcorn MS).

Since 1946 this hawk has diminished in numbers in this valley, probably due to the low numbers of Townsend's Ground Squirrels. From 1960 to 1967, inclusive, this hawk was seen 11 times in April; 8 in May; 5 in June; and 3 times in July and August (Alcorn MS).

Napier (MS, 1974) reported these birds were observed through the summer with several pair nesting in the Stillwater area east of Fallon.

In the Austin area, these hawks were present from April 9 to August 31. On Reese River, Nye County, they were seen several times in May (Alcorn MS).

Biale (MS) reported a nest in the Eureka area May 24, 1966 in a boxelder tree approximately 15 feet above ground contained three eggs. On May 11, 1969 he saw a Swainson's Hawk fly out of a small juniper tree (6200 feet elev.) containing a nest. There were no eggs in the nest located nine feet above ground. The nest was made of dead brush and branches and the lining contained bits of eggshells from the previous year. On July 10, 1970 he saw a nest made of large dry twigs (sage, etc.) lined with fresh juniper, green pinon twigs and mustard weed in a juniper tree 10 feet above ground (6350 feet elev.). The nest contained two young approximately one–week–old and one dirty white egg.

Biale (MS) reported these birds in the Eureka area as early as April 6 (1969) and as late as September 12 (1971). He reported them from 1964 to 1973, inclusive, 9 times in April; 27 in May; 21 in June; 19 in July; 15 in August; and 4 times in September.

In eastern Nevada, a late record of one of these birds was in the Kern Mountains October 26, 1967 and one was seen several times in May, 1967 near Baker (Alcorn MS).

One was seen six miles north–northeast of Beatty June 3, 1959. The date would indicate possible nesting in the area (Alcorn MS).

Johnson and Richardson (1952) reported, "Found breeding in southernmost Nye County: Ash Meadows on June 19, 1951; Pahrump Ranch on June 22, 1951, where a nest with 2–month–old young was observed. These records extend the known breeding range of this species well south in the state, for Smoky Valley in northern Nye County is the southernmost breeding station given by Linsdale."

Cruickshank (1965) reported at the Desert Game Range on December 28, 1964 one was seen during the annual CBC.

Austin and Bradley (1971) reported for Clark County, "Transient in riparian areas. Common from February 5 to June 21 and from August 11 to September 29. Unseasonal on December 10 and 22."

Lawson (TT, 1976) commented, "We most commonly see this species here in southern Nevada as a spring and fall migrant. They simply do not winter in this area."

Zone–tailed Hawk   *Buteo albonotatus*

Distribution. From southern California (rarely) to Texas; south to northern South America.

Status in Nevada. Rare.

Records of Occurrence. Kingery (1980) reported one Zone–tailed Hawk at Pahranagat NWR April 21, 1980.

Lawson (MS, 1981) reported one near Mesquite June 5, 1980. He also reported one seen by Kay O'Connell at the Mormon Farm area near Las Vegas April 19, 1975. It was later observed by Lawson, Mowbray and others.

Red–tailed Hawk   *Buteo jamaicensis*

Distribution. From Alaska to Nova Scotia; south to Panama.

Status in Nevada. Common statewide resident. Two races have been reported from Nevada: *B. j. calurus* and *B. j. harlani.*

Records of Occurrence. So many sightings of this hawk exist in Nevada, that no attempt will be made to list all of them. Herron (PC, 1984) reported, "This hawk is seen everywhere in the state."

Sheldon NWR (1982) reported the Red–tailed Hawk as a permanent nesting resident; common spring, summer and fall; uncommon in winter.

At Lamance Creek, east side of Santa Rosa Mountains, June 10, 1970 an adult was seen circling for altitude and in its claws was a three–foot snake. As it gained altitude, the hawk drifted eastward and probably headed toward its nest (Alcorn MS).

In Elko County these birds from 1964 to 1967 were seen twice in March; 3 times in April; 6 in May; 4 in June; July 12, 1967; 6 in August; 7 in September; 4 times in October; and December 22, 1965 (Alcorn MS).

LaRochelle (MS, 1967) reported sightings at the Ruby Lake NWR for each month of the year.

In the Lovelock area from January, 1960 to December 30, 1967, these hawks were seen 12 times in January; 7 in February; twice in March and April; 2 were seen May 9, 1961 and thought to be nesting; September 18, 1964; 5 in November; and 7 times in December (Alcorn MS).

Ryser (1969) reported a very large overwintering population in the Truckee Meadows. He (1970) reported in February, 1970 a casual inspection from the road would reveal over 30 hawks in the Truckee Meadows.

Orr and Moffitt (1971) reported these birds in the Lake Tahoe region as fairly common in summer and fall, noted less frequently in winter.

Ryser (1968) reported in western Nevada the fall build–up was spectacular.

In the Yerington area these hawks were seen in winter from November through March; with one record May 28, 1962 (Alcorn MS).

Herron (PC, 1984) reported them nesting in large cottonwood trees in the Yerington area.

Alcorn (1946) reported this hawk as a resident in the Lahontan Valley (Fallon area); most abundant in winter, scarce in summer. In this area from January 1, 1960 through December, 1967, this bird was seen 45 times in January; 24 in February; 18 in March; 7 in April; 8 in May; twice in August and September; 10 in October; 16 in November; and 20 times in December. About 14 miles west of Fallon, two adults were frequently seen at a nest near the top of a large cottonwood tree. On May 21, 1968 after a strong wind, a fully feathered young Red–tailed Hawk, not yet able to fly, was seen on the ground near the nest tree.

Napier (MS, 1974) reported this hawk nests in cottonwood trees along the Carson River north and east of Fallon, and it also occurs in the area in winter months.

An adult setting on a nest in a large cottonwood tree at Eastgate, Churchill County was seen June 12, 1971 (Alcorn MS).

Examination of crop and stomach contents of these hawks was undertaken whenever one was found dead with results as follows: Fallon area, February 28, 1945, stomach contained remains of a Townsend Ground Squirrel and one *Microdipodops* mouse; Fallon area, March

Red-tailed Hawk. Photo by A. A. Alcorn.

14, 1945, stomach contained remains of a Townsend's ground squirrel and a Gopher Snake that was about 14 inches long; head of Cherry Creek, Clan Alpine Mountains, August 14, 1945, stomach contained remains of a frog and remains of a Pocket Gopher *(Thomomys)* (Alcorn MS).

Biale (MS) for the Eureka area from 1964 to 1967, inclusive, reported seeing this bird 132 times. It was recorded in all months. From 1970 to October, 1973, inclusive, he reported these birds 4 times in January; 3 in February; 9 in March; 8 in April; 6 in May; 16 in June; 11 in July; 14 in August; 19 in September; 5 in October and November; and 4 times in December. He reported a nest in an aspen tree with two young ready to leave the nest June 27, 1965. A nest 25 feet up in a poplar tree contained two half–grown young June 5, 1972. A nest was also seen May 12, 1973.

Biale (PC, 1986) reported the Red–tailed Hawk as a consistent, common nester in the Eureka area.

In Pahranagat Valley from 1960 to 1965, inclusive, these hawks were seen January 9, 1965; February 7, 1963 and February 16, 1965; 4 times in March; 5 times in April; May 11, 1960; July 1, 1961; twice in November; and 3 times in December (Alcorn MS).

J. R. and A. A. Alcorn (MS) reported these birds in the Moapa Valley from December, 1959 to April, 1964 a total of 5 times in January; 9 in February; 15 in March; twice in April; 3 in May; twice in June; 3 in July; September 7, 1963; twice in October; November 26, 1963; and twice in December.

Austin and Bradley (1971) for Clark County reported, "Permanent resident in woodlands. Summer resident in montane forests. Winter resident in desert scrub and riparian. Woodland and montane forest (common from February 23 to October 12) unseasonal on November 23, and December 9 and 12. Desert scrub and riparian (common from August 20 to May 31) unseasonal on July 10 and 23."

Ryser (1970) reported the race *B. j. harlani* as occurring in Clark County.

Lawson (TT, 1976) reported, "There are numerous nesting records. This hawk, year round, is a common species in southern Nevada."

Ferruginous Hawk *Buteo regalis*

Distribution. From eastern Washington to southern Saskatchewan; south into Mexico.

Status in Nevada. Herron et al., (1985) reported, "In Nevada the Ferruginous Hawk is a common breeding species, nesting primarily in Elko, White Pine, Eureka and northern Lincoln Counties, although isolated nesting pair have been located throughout the state. Ferruginous Hawks have rarely been observed wintering in Nevada."

Herron (PC, 1984) reported the population of this bird in Nevada appears stable, with approximately 220 nesting territories, mostly in eastern Nevada. Nests are usually found in juniper trees.

Records of Occurrence. Oakleaf (1974b) reported, "Only four active nests and 17 sightings during the 1974 breeding season indicate that this species is uncommon in Nevada. Only 14 sightings of Ferruginous Hawks were made during the winter of 1974. These sightings were scattered throughout the entire state with no concentrations of the species noted. Four active nests were located during 1974 in the following areas: Diamond Valley, Eureka County; Newark Valley, White Pine County; and near Elko, Elko County. In addition, Biale of Eureka has observed four other nests during the past several years in the Eureka area.''

Nappe and Klebenow (1973) reported, ''Possibly more widespread and more common than at present. However, it probably was not an abundant species. Ridgway (1877) reported it as being present near Austin July 2–5, 1868. He (1877) also reported, 'was much less frequently seen than its relative the Common Roughleg. The few observed were sailing majestically overhead.' A generous guess is that now no more than 20 nest in Nevada each year. There are probably only 100 birds in Nevada at any one time.''

Sheldon NWR (1982) reported the Ferruginous Hawk as uncommon spring, summer and fall with occasional records for winter. Some nesting is reported.

One was seen at Badger Camp, northern Washoe County on May 5, 1966 (Alcorn MS).

In Elko County at Wildhorse Reservoir, one was seen August 17, 1971; others were seen near Elko in April, May and June (Alcorn MS).

Ryser (1967) reported several between Lamoille and Elko July 1 to 4, 1967.

LaRochelle (MS, 1967) reported only two sightings of this hawk at Ruby Lake NWR. Both were in February, 1963.

The Ferruginous Hawk is reported by Ruby Lake NWR (1981) as occasional throughout the year.

In the Lovelock area, these birds were recorded in January, March, April, November and December. At nine miles south of Lovelock April 26, 1961, an adult was setting on a nest located 15 feet above ground in a cottonwood tree. The nest was made of heavy sticks lined with salt grass and contained many pieces of dried cow dung. Some of the pieces of dung were about one–fourth of an inch in diameter and some flattened pieces were over two inches across. As we approached the hawk ''flattened out'' low in the nest, then raised its head, flew straight away and was not seen again. The nest contained three eggs (Alcorn MS).

Orr and Moffitt (1971) reported this hawk in the Lake Tahoe region as a rare winter visitant.

Ryser (1969) reported several seen in the Truckee Meadows in December, 1969 and he (1970) for the same area reported one December 6, 1970.

One or two were seen in the Yerington area each year from 1962 to 1984 (Alcorn MS).

Alcorn (1946) reported this hawk in Lahontan Valley (Fallon area) as a winter resident; not abundant. Noted each month from October to March. From 1962 to 1967, inclusive, this bird was seen only on eight occasions in the area. They were observed 3 times in January; twice in March; and 3 times in December. The stomach of one shot six miles northeast of Fallon February 26, 1942 contained the remains of a Townsend Ground Squirrel.

In the area from Fallon to Eureka, one was seen 12 miles northwest of Gabbs on May 29, 1984; one at Eastgate June 2, 1969; one at Brown's Station in Smith Creek Valley November 7, 1962; and two were seen near Austin April 7, 1960 (Alcorn MS).

Biale (MS) reported these birds in the Eureka area 47 times from 1964 to 1967, inclusive. He saw them once in January; 3 times in February; 7 in March; 4 in April; 8 in May; 4 in June; 7 in July; September 30, 1965; October 31, 1964; 5 in November; and 6 times in December. In this same area from 1970 to 1973 inclusive, he reported them 4 times in February; 10 in March; 12 in April; 11 in May; 8 in June; 7 in July; 3 in August; September 26, 1971; October 7, 1973; and twice in November and December.

Biale (MS) reported a nest at least two feet in diameter and two feet deep in a dead juniper tree (6500 feet elev.) on May 4, 1969 in Antelope Valley. The nest was nine feet above ground and made of large dead twigs, some an inch or more in diameter. The nest was lined with large strips of juniper bark and on the top, near the edge of the cup, were a few pieces of cow manure. Four eggs, dirty white with brownish–gray indistinct splotches lay in the nest. A Ferruginous Hawk of the dark phase was incubating the eggs. Biale also reported another Ferruginous Hawk nest in partly dead juniper tree about 11 feet off the ground (6500 feet elev.). This nest was located about a quarter mile from the first nest described. It contained four, dirty white eggs with distinct dark brownish–red spots. A light phase Ferruginous Hawk was incubating the eggs.

Biale (MS) in South Diamond Valley (6100 feet elev.) on May 14, 1969 saw a nest in a juniper tree 10 feet above ground. It was a very large nest made of large dead twigs and lined with juniper bark. Four young approximately one week old were in the nest. The remains of a cottontail rabbit were beneath the nest on the ground.

Biale (PC, 1984) reports in recent years the nesting of this hawk has increased in the Eureka area. In 1983 he saw five nests in the south end of Diamond Valley, all in juniper trees. He reports nests are often built on top of the preceding year's nest and some are five feet deep. He believes in time the nest or nesting activity will kill the tree.

Kingery (1977) reported low numbers of jackrabbits apparently drove Ferruginous Hawks from central Nevada in late August.

In White Pine County, Kingery (1981) reported, "At Ely 27 Ferruginous Hawks produced 69 young."

One was seen near Baker July 5, 1967 and one near Lund April 9, 1962, White Pine County (Alcorn MS).

One was seen near Hiko, Pahranagat Valley November 25, 1961 (Alcorn MS).

Ryser (1967) reported this bird at Overton, Moapa Valley February 15–19, 1967.

Gullion et al., (1959) reported, "Two records: one on October 22, 1949 in Las Vegas Valley between Henderson (1790 feet elevation) and Las Vegas, and a road–killed male recovered from the highway in Pahranagat Valley by Gullion November 16, 1951."

Austin and Bradley (1971) reported for Clark County, "Winter resident in desert scrub and riparian areas. Common from October 22 to April 29."

Lawson (TT, 1976) commented, "The Ferruginous Hawk is a rare bird here in Clark County. An occasional bird will winter in the Moapa Valley also in the Mormon Farm area near Las Vegas. Quite frankly, I think it is becoming more rare all the time."

Rough–legged Hawk   *Buteo lagopus*

Distribution. From Alaska to Newfoundland; south from California and Arizona to Virginia. Also in Eurasia.

Status in Nevada. Winter resident and transient. These hawks are concentrated in areas where populations of meadow mice and other small mammals are available in good supply.

Records of Occurrence. Sheldon NWR (1982) reported the Rough–legged Hawk as uncommon in spring and fall and a common winter resident.

In Elko County these hawks were abundant from October through April (Alcorn MS).

LaRochelle (MS, 1967) reported at Ruby Lake NWR this hawk was, "Very common during winter months, with from five to 19 using the area. The dark form accounts for one–sixth of the birds seen."

In the Lovelock area these birds are common in winter months and have been recorded from October to April 24 (Alcorn MS).

Ryser (1972) reported a few of these hawks were in the Truckee Meadows (Reno area) and the Carson Valley by late October. Ryser (1973) reported there were large numbers of these hawks in the Truckee Meadows in November.

Orr and Moffitt (1971) reported for the Lake Tahoe region this hawk was an occasional visitor, at least in former years.

In the Fallon area from 1961 to 1965, inclusive, these hawks were seen 18 times in January; 11 in February; 8 in March; on April 6, 1961; 8 in November; and 14 times in December (Alcorn MS).

Schwabenland (MS, 1966) reported at Stillwater WMA the Rough–legged Hawk is a common resident during the fall and winter months.

Napier (MS, 1974) reporting for this same area said this hawk was very common from October through March.

From Fallon to Eureka these hawks are often seen at road–killed jackrabbits. One Rough–legged Hawk shot March 1, 1962 along the highway west of Austin was examined and its stomach contained the remains of one Horned Lark and part of a jackrabbit. Both food items were probably road kills (Alcorn MS).

Biale (MS) reported this hawk in the Eureka area from 1964 to 1967, inclusive, a total of 29 times in January; 26 in February; 14 in March; 4 in April; on May 2, 1965; on September 30, 1965; on October 27 and 31, 1966; 14 in November; and 38 times in December. From January 1, 1971 to December 31, 1973 he saw these birds 20 times in January; 13 in February; 10 in March; on April 1, 1971 and 1972; 4 in October (earliest October 21, 1973); 15 in November; and 27 times in December.

In eastern Nevada one was observed at Tippett October 27, and one at Baker October 30, 1967. Others were seen in winter months near Baker (Alcorn MS).

In Pahranagat Valley, lone birds were seen November 25, 1961 and February 28, 1966 (Alcorn MS).

Austin (1968) reported, "I have records for the species from the vicinity of Dike Lake in Pahranagat Valley, Lincoln County as follows: road kills found December 27, 1966 and February 18, 1967 and a sight record for January 22, 1968."

Austin and Bradley (1968) reported, "A bird of light phase observed near Overton on December 22, 1962 appears to be the first record for Clark County."

Lawson (TT, 1976) listed this hawk as a rare to uncommon visitor and winter resident in southern Nevada. "As many as five or six birds have been counted over the Las Vegas Valley. One was seen at Cottonwood Cove, Lake Mohave."

Golden Eagle   *Aquila chrysaetos*

Distribution. From Alaska and Canada; south to central Mexico. Also in Eurasia south to the Himalayas and Japan.

Status in Nevada. Statewide resident. Linsdale (1936) reported these birds occur throughout the state and are especially common about the higher mountain ranges. Present records indicate these birds occur statewide with a majority of the nesting activity taking place in Elko and other nearby northern Nevada counties.

Golden Eagles tend to migrate southward in fall months and are more common in the Fallon, Austin and Ely areas in winter than summer.

Records of Occurrence. No attempt is made to list all of the sight records of these birds. However, records from selected localities are given to show their relative seasonal abundance.

Oakleaf (1974b) stated, "Over 600 sight records of one or more Golden Eagles and approximately 200 nesting sites have been recorded in the state during recent years. In addition to Elko County, nest information has been recorded for 68 other eagle nests. Fifty–five of these were checked with only 31% of the nests being active."

Herron (PC, 1984) estimates 1200 nesting pair of Golden Eagles in the state of Nevada, with 95% of them nesting on cliffs and 5% nesting in trees.

Sheldon NWR (1982) reported the Golden Eagle as a common spring, summer and fall resident; uncommon in winter.

In Elko County from 1959 to 1968, they were observed during visits to this area by the author once in March; 5 times in April; 3 in May; 4 in June; 3 in July and August; 11 in September; 9 in October; and 4 times in November. None were seen in December, January or February, but this was partly due to this author's not being in the area as often, or seeking out the rabbit concentration where the eagles occur (Alcorn MS).

LaRochelle (MS, 1967) reported as many as nine birds have been seen at the Ruby Lake NWR at one time with groups of two to six not uncommonly seen in later winter.

Jerry Page and Donald Siebert (1973) reported, "During 1972, an inventory of these birds in Elko County revealed 88 active Golden Eagle nesting sites. It was found that 93% of the active nests inventoried were on cliffs, 71% were between elevations of 5000–6500 feet and 43% faced east. Locations of 84% of the nests were within two miles of water with desert riparian habitat. The success per nesting attempt was determined from 50 of the 88 active nests with a success ratio of 1.1. young per nest."

Ryser (1974) reported a total of seven Golden Eagles seen in the Carson Valley February 14, 1974.

Alcorn (1946) reported for Lahontan Valley, "Resident; most abundant in winter months. One nest, containing two small young, was seen in a cottonwood tree about two miles north of Stillwater May 13, 1936."

"The following is an account of this nest: 'Today I was walking under some large cottonwood trees when a large Golden Eagle flew from the tree and disappeared to the north. Going 100 yards further, where more large trees were along the ditch, I saw a large nest 4/5 of the way up to the top of the tree. At first glance I thought this nest was a magpie nest, but a closer view and I could see the back of an eagle above the nest. I started up the tree and at about 25 feet, I was startled by a noise and the adult eagle flew off the nest and disappeared to the north, not to be seen again while I was in the vicinity. When the eagle flew I heard a thud on the ground and looking down I saw one downy white eagle that was no bigger than my fist. The fall had killed it, so I looked into the nest and there was another young with much jackrabbit fur and legs in and around the nest. This nest was about 40 feet high.' "

Schwabenland (MS, 1966) reported for Stillwater WMA the Golden Eagle is a year–round resident, with between six and 10 birds seen every winter. No nesting was observed.

Napier (MS, 1974) reported for this same area this eagle is common during winter months with a peak of about five birds. It nests along the Carson River. He saw an active nest in a cottonwood tree in 1972 and 1973. In this area, from 1960 to 1970, inclusive, these birds were seen 7 times in January; 6 in March and April; 4 in May; on August 15, 1969, when Napier saw one carrying a coot in its talons; on September 13, 1967, when Napier reported an immature made a pass at a Great Blue Heron; twice in October; 3 in November; and 5 times in December.

Thompson (PC, 1988) reported three active nests in the Lahontan Valley area in 1987.

Biale (MS) recorded these birds in the Eureka area from January 1, 1965 to December 31, 1967 a total of 21 times in January; 17 in February; 19 in March; 9 in April; 6 in May; 7 in June and July; twice in August; 5 in September; 13 in October and November; and 23 times in December.

In the 3–year period from 1970 to 1972, Biale (MS) reported them 25 times in January; 19 in February; 9 in March; 13 in April; 11 in May; 6 in June; 9 in July; 13 in August and September; 25 in October; 17 in November; and 18 times in December.

Biale (MS) reported that in northeast Antelope Valley (6400 feet elev.) on April 2, 1970 a Golden Eagle nest sat on a rocky outcrop. The nest was easy to walk to and was made of dry sage sticks, juniper branches and greasewood. It was lined with fresh green junipers foliage and juniper bark. Two dirty white eggs, flecked with brown were in the nest.

Biale (MS) also reported a nest found in this same location on March 28, 1971. He stated, "The nest is at best three times as large as it was in 1969 (first year recorded). Lined with fresh juniper twigs, including foliage and freshly pulled juniper bark. Two tan eggs heavily spotted with reddish–brown."

Kingery (1983) reported, "At Eureka, Nevada power poles electrocuted seven immatures that came in with a fall influx of goldens."

In Lincoln County, in the vicinity of Caliente, these birds were seen in April and May. In Pahranagat Valley one was observed November 22, 1962 (Alcorn MS).

Gullion et al., (1959) reported, "These birds occur sparingly over all parts of the desert from as early as August 7 (1954, Mormon Mountains, Lincoln County) to as late as June 27 (1954, same area). Two immature birds on July 9, 1952 in the Gold Butte area constitute the single summer record for the desert, although van Rossem (1936) cites July records for the Charleston Mountains, Clark County."

In Moapa Valley, Clark County, these birds are not commonly seen. From 1960 to 1964, inclusive, they were observed three times in January and February 21, 1964. One was seen at Gold Butte October 30, 1960 and January 26, 1963 (Alcorn MS).

Austin and Bradley (1971) for Clark County reported, "Permanent resident in montane forest areas, winter resident in desert scrub and riparian areas, and transient in woodland areas."

Lawson (1977) says, "In southern Nevada this eagle is an uncommon summer resident and a common winter resident."

Behavior. Along the highway 10 miles southwest of Austin on June 18, 1962, an eagle tried to fly with a road–killed jackrabbit. The eagle flopped along on the ground for about 100 feet, but did not get airborn (Alcorn MS).

James V. Kelleher and William F. O'Malia (1971) reported, "On May 24, 1969, while censusing birds on the Whittell Forest and Wildlife area of the University of Nevada in Little Valley, Washoe County, Nevada, we watched a Golden Eagle capture a male Mallard *(Anas platyrhynchos)* in flight. As we approached a large beaver pond on Franktown Creek, our attention was attracted to a Mallard flying rapidly, within 25 feet of the ground, through a bordering stand of lodgepole pine *(Pinus contorta)* with a Golden Eagle in close pursuit. As the duck maneuvered side to side and up and down, the eagle remained directly behind it. The eagle was able to maneuver through the pines and over the willows *(Salix sp.)* almost as well as the duck. Within seconds the Mallard disappeared downstream in a thick stand of pine, and the eagle rose to about 100 feet in the air, apparently breaking off the chase. Suddenly, the duck reversed direction and reappeared in weaving flight close to the ground. The eagle turned, stooped, snatched the Mallard out of the air, and still holding the duck in its talons, landed within 100 feet of us. As it surveyed the surrounding area, the eagle detected us, released the duck and flew away. After the eagle left we tried to find the duck, but it was gone; all we found were a few bloodstained duck feathers.''

Homer S. Ford and J. R. Alcorn (1946) wrote the following on the observations of Golden Eagle attacks on coyotes: ''The predatory activities of both the Golden Eagle and the coyote *(Canis latrans)* are well known, and it seems to be commonly accepted that each obtains a good share of its food through predation. We know of no written accounts of one preying on the other. However, agents of the Bureau of Sport Fisheries and Wildlife have reported that it is not uncommon for eagles to prey on coyotes in the puppy stage. But observations indicate that at times Golden Eagles will attack mature coyotes. Two instances of this were witnessed by agents in Nevada.''

''On May 23, 1961, while aerial hunting for coyotes on the antelope kidding areas of the Charles Sheldon Antelope Range located in northwestern Nevada, Hayden Purdy and T. C. Barber observed an eagle attacking a coyote. An adult coyote had been spotted standing above a rocky outcrop on a hillside. As the plane approached, the coyote began to move off in a trot. At this point a Golden Eagle flew past the plane in a steep dive and struck the coyote over the hips with both feet and continued on in flight. The coyote was partially knocked to the ground. Recovering, it whirled and jumped, biting in the direction of the eagle which by now was gaining altitude. The men in the plane could see a considerable amount of hair torn from the coyote's back. The plane was then within range and the coyote was dispatched, unfortunately ending the observation. The eagle was not sighted again.''

''On March 5, 1963, another Golden Eagle was sighted making an attack on a coyote in White Pine County, Nevada. James C. Harris and Wendell Ross were engaged in aerial hunting for coyotes on sheep lambing ranges in Long Valley. In sagebrush and sand dune terrain a Golden Eagle was observed making dives on a coyote about 500 yards from the airplane. The plane was turned toward the fight and during the time which elapsed before the arrival at the location, the eagle continued to attack. In the course of two attacks observed at close range the eagle came in contact with the coyote, but did not completely knock it down either time. Blood was evident on the coyote's back and at the time of the second pass the coyote lost a considerable amount of hair. The coyote was traveling at full speed at the time of both attacks and at no time made an effort to fight the bird or take cover. Due to the nearness of the airplane the eagle stopped the attacks and moved on. This may also have interfered with the coyotes willingness to defend itself.''

Ranchers sometimes complain about coyotes attacking their livestock. Government trapper Hayden Purdy (PC, 1966) commented, ''You would be surprised how many antelope kids and domestic sheep and lambs that eagles kill.''

At eight miles northwest of Silver Peak, Esmeralda County on February 12, 1969, the author was standing alongside a vehicle when an eagle was seen about 100 yards above the crest of the hills. It was apparently using the air currents to maintain a relatively stationary position without flapping its wings. I then saw a second Golden Eagle nearby in a similar flight pattern. To my surprise, one eagle suddenly went into a steep dive directly towards me. It was then that I noticed a Short–eared Owl nearby. As it flew from the ground, the eagle was in close pursuit.

When the owl veered sharply to one side, the eagle ascended to prepare for another attack. By now, the second eagle was attacking the owl. The eagles continued taking turns attacking the owl until the owl was captured in flight. The eagle immediately alighted on a large boulder, pulled the owl's head off and partially deplumed it. It then flew away with the body (Alcorn MS).

Scott (1970) reported, "At Ruby Lake NWR on April 9, 1970 two Golden Eagles were locked together and fell to the ground from at least 300 feet in the air. The eagles hit the ground and bounced several feet into the air. One was able to fly away shortly afterward, but the other could not, and subsequently died. Upon examination of this bird (an adult) it was found that the body cavity had been broken open in the fall. It was suspected that the two eagles have become locked together while mating in mid–air."

Family **FALCONIDAE:** Caracaras and Falcons

American Kestrel   *Falco sparverius*

Distribution. Formerly Sparrow Hawk. From Alaska to southern Newfoundland; south through the Americas, including the West Indies to Tierra del Fuego.

Status in Nevada. A statewide summer resident. During winter time, they move out of the colder areas of the state.

Herron (PC, 1984) reports this bird as a common nester statewide.

Records of Occurrence. Sheldon NWR (1982) reported the American Kestrel as common spring, summer and fall; uncommon in winter. This bird is known to nest in the area.

In Elko County these birds were repeatedly seen each month from April 5 through September with a record of one near Tuscarora October 20, 1965 (Alcorn MS).

LaRochelle (MS, 1967) reported these birds were common at Ruby Lake NWR during summer until mid–fall.

In the vicinity of Lovelock from 1960 to 1965, inclusive, these birds were observed 8 times in January; 9 in February; twice in March; 4 in April; twice in May and June; two seen July 28, 1965; August 15, 1965; twice in September; October 6, 1961; 4 in November; and 9 times in December (Alcorn MS).

This bird is frequently seen along the lower Truckee River and Archibald Johnson (1936) reported, "During seven weeks spent last summer (1935) on the Pyramid Lake Indian Reservation, in Nevada, two Paiute families were found keeping Sparrow Hawks as pets. A family near the agency had a young male taken from the nest before it was able to fly. It seemed in the best of condition except for one foot that had been injured accidentally, and was fed on lizards caught by the children of the family."

Thomas D. Burleigh collected nine specimens, now in the USNM, in Reno from 1967 to 1969. They were taken in January, March, April, May, July, October and December.

At Reno in 1942, these birds were seen February 8; 5 times in March; April 2; twice in May; June 23; 3 in July; August 29; and 3 times in September (Alcorn MS).

In the Mason Valley (Yerington area) over 13 years time they were seen 6 times in January; February 5, 1961; 5 in March; 7 in April; 6 in May; 5 in June; 3 in July and August; two seen September 14, 1965; October 11, 1964; 4 times in November; and twice in December (Alcorn MS).

In the Fallon area from January 1, 1960 to December 31, 1967 these birds were seen 102 times in January; 87 in February; 100 in March; 91 in April and May; 74 in June; 78 in July; 126 in August; 114 in September; 102 in October; 104 in November; and 192 times in December (Alcorn MS).

Napier (MS, 1974) reported these birds at Stillwater WMA as, "Present all year. Birds migrate through. A Sparrow Hawk, banded on Stillwater WMA on July 2, 1967 was recovered at San Jacinto, California on November 15, 1969. They nest here, use Wood Duck nest boxes. Feed young many horned lizards as evidenced by remains in nest box."

Biale (MS) recorded these birds in the Eureka area from January 1, 1970 to December 31, 1973 on January 11, 1970; none seen in February; 8 times in March; 47 in April; 70 in May; 32 in June; 45 in July; 35 in August; 42 in September; 15 in October; on November 26, 1972 (one bird); and December 2, 1972. He also said they are seldom seen in winter.

In Pahranagat Valley from 1960 to 1965, inclusive, these hawks were seen twice in January and February; 9 times in March and April; 5 in May; twice in June; 3 in July; 4 in August; twice in September; and 3 times in October, November and December (Alcorn MS).

Gullion et al., (1959) reported, "Desert records primarily in the spring months, from March 22 (1952, Searchlight area) to mid–July. Fall records are most numerous from August 5 (1953 and 1954, Pahrump Valley) to about September 26 (1954, Gold Butte area), with one late fall record, November 28, 1953 from Gold Butte."

J. R. and A. A. Alcorn (MS) in Moapa Valley from 1959 to 1964, inclusive, reported seeing these birds 11 times in January; 4 in February; 5 in March; 8 in April; 3 in May and June; 8 in July; 10 in August; 9 in September; 3 in October; twice in November; and 5 times in December.

Austin and Bradley (1971) reported, "Summer resident in woodland areas. Permanent resident in desert scrub and riparian areas. Transient in montane forests. In woodland and montane forests it is common from April 7 to November 9. In desert scrub and riparian areas it occurs all year."

Lawson (TT, 1976) commented, "This species is simply common the year round in southern Nevada."

Nest. The author found sparrow hawks nesting four miles west of Fallon on May 17, 1942. The nest, in a nest box nailed to a cottonwood tree, contained four eggs. An adult flew from the box when it was touched. On June 16, this same nest still contained four eggs. I wrote, "Judging by the spider webs in the box, the absence of the sparrow hawks and the coolness of the eggs the adult sparrow hawks have deserted their eggs."

At this same location on June 27, 1943, three young, judged to be three–fourths grown were seen perched on a nest box nailed to a cottonwood tree about 25 feet from the ground. This nest box was about 100 yards from the box occupied by sparrow hawks in 1942.

At six miles west–southwest of Fallon on July 14, 1966 three nest boxes were occupied by these hawks. One contained four feathered young and the remains of about eight horned lizards. The second nest contained three young and the remains of horned lizards. The third nest contained an adult kestrel on five eggs. The author wrote, "The adult did not move, so I took a six–inch stick and moved the bird over to one side, then to the opposite side to count the eggs. Then I closed the lid and the hawk still remained on the eggs."

Biale (PC, 1974) reported a nest seen June 4, 1979. It was situated in a wooden box that housed electric switches, on a pole about five feet above ground. No nesting materials, but the box contained four tan, brown–spotted eggs laid in the rubble on the floor of the box (6400 feet elev.). In Antelope Valley on June 22, 1984 in a wooden valve box near a stock watering tank another nest was found. Five nestlings about 10 days old were on the insulation inside the box which was 18 inches above ground (6500 feet elev.).

Merlin   *Falco columbarius*

Distribution. Formerly Pigeon Hawk. From Alaska to Newfoundland; south to South America (Peru).

Status in Nevada. Rare spring and fall migrant. Four races have been reported from Nevada: *F. c. suckleyi, F. c. richardsoni, F. c. columbarius* and *F. c. bendirei*.

Herron (PC, 1984) reports no known nesting occurs in the state.

Nappe and Klebenow (1973) reported, "Was formerly common in Nevada in the winter months. Recorded most frequently in western Nevada, but probably more common statewide than records indicate."

Records of Occurrence. Kenneth W. Voget (MS, 1986) reported the Merlin as a rare transient visitor spring and fall at Sheldon NWR.

In the Santa Rosa Range, Solid Silver Canyon, one was seen October 7, 1961 (Alcorn MS).

In the Reno area in 1942, two were seen February 5, one was seen March 8, and one October 14 (Alcorn MS).

Ryser (1966) reported one was seen in Truckee Meadows that year. He (1969) reported one February 20, 1969. He also reported at least eight different birds were seen in the Truckee Meadows in late February, 1969.

Heilbrun and Rosenburg (1981) reported one on the Truckee Meadows CBC on December 20, 1980.

Alcorn (1946) reported this hawk in the Lahontan Valley, Fallon area to be, "A winter visitant, not common. An example of the race *(bendirei)* was taken November 12, 1931, and another on April 11, 1944. A specimen of *F. c. suckleyi* was taken January 19, 1941 (Alcorn, 1943b). No other records are available."

From 1940 to 1960, although not common, these birds were seen in the Fallon area a few times each winter. Since 1960 they have practically disappeared from this valley. Only two birds have been seen by the author since 1960. One was observed January 9, 1960 and one January 18, 1965 (Alcorn MS).

Napier (MS, 1974) reported in eight years (1967–74) at Stillwater WMA he had only one record of this bird which was at East Lake February 8, 1974.

Mills (PC) reported that at five miles east of Fallon on January 3, 1960 he saw a Pigeon Hawk. "A small Prairie Falcon took after this Pigeon Hawk and they tangled in the air and dove at each other a few times."

Biale (MS) reported seeing these birds in the Eureka area May 2 and December 2, 1965; January 9 and February 13, 1966; February 23, December 21 and 25, 1967. He (PC, 1986) saw one December 22, 1985.

Gullion et al., (1959) reported, "Three records of one bird each: Pahranagat Valley, October, 27, 1951; the Virgin Mesa (1750 feet elevation, on the west slope of the Virgin Mountains, Clark County), April 17, 1953; and seen by Imhof at Frenchman Flat, October 3, 1951."

J. R. and A. A. Alcorn (MS) reported one in the Moapa Valley December 15, 1959 and one February 14, 1963.

Austin and Bradley (1971) reported for Clark County, "Five records; none since 1961."

Lawson (TT, 1976) commented, "I would say it is a rare to uncommon winter resident here in southern Nevada. In the early 1960s we didn't see them at all. We seem to be seeing more of them in recent years."

Peregrine Falcon   *Falco peregrinus*

Distribution. Also referred to as the Duck Hawk. Nearly cosmopolitan, absent from the islands of the eastern Pacific and from New Zealand. From Alaska through the Americas to Tierra del Fuego and the Falkland Islands.

Status in Nevada. Rare. Formerly nested at Pyramid and Walker Lakes.

Herron (PC, 1984) reported no known nesting in Nevada.

Records of Occurrence. Linsdale (1936) reported, "Duck Hawks were seen by Ridgway (1877) at Pyramid Lake on May 23, 1868, and a young male was obtained by him on the Big Bend of the Truckee River on May 23, 1868. A young Duck Hawk, thought to be a female was seen October 17, 1931, flying over the Muddy River near Saint Thomas, Clark County (van Rossem, MS)."

Oakleaf (1974b) stated, "A very limited amount of information is available on the distribution of Peregrine Falcons in Nevada. Only five eyrie sites have been recorded. Three eyries in the western corner of the state (Ridgway 1877; Wolfe, 1937; and Tom Trelease, pers. commun.), and two eyries in Elko County (Porter and White, 1973; and Fred Evenden, pers. commun.). None of the eyries have been recorded as active in the past decade. Only seven sightings were made by Fish and Game personnel during 1973–74 season. Only two of these sightings were made during the breeding season."

Fred G. Evenden, Jr. (1952) reported, "On June 23, 1949, a nest found in a 200 foot cliff along the North Fork of the Humboldt River in Elko County. Both adults remained in the vicinity of the nest."

Sheldon NWR (1982) reported the Peregrine Falcon as a rare summer and fall migrant.

LaRochelle (MS, 1967) reported at Ruby Lake NWR he did not see any of these birds; however, reports of sightings were not uncommon. At this same refuge, the narrative reports for 1971 and 1972 indicated a total of 19 separate observations of these birds.

Heilbrun and Rosenburg (1983) reported one on the CBC at Ruby Lake NWR December 18, 1982.

Herron (PC, 1984) reported three of these birds (two females and one male) were introduced at Ruby Lake NWR and says a minimum of 20 will be introduced in the next four years in an attempt to establish a nesting population there.

Orr and Moffitt (1971) reported a Peregrine Falcon positively identified by Moffitt on the east side of the summit on Mount Rose road July 20, 1926.

Napier (MS, 1974) reported no sighting of this bird at the Stillwater WMA from August, 1966 to August, 1974.

Ryser (1966) reported two reports of this bird in the Stillwater region in December, 1966 and Ryser (1970) reported three in western Nevada in May 29, 1970.

One positive record of this bird in Nevada in recent years is of a male shot by Richard Mills, nine miles north–northeast of Fallon, August 22, 1954. It was examined by the author and prepared as a study skin by A. A. Alcorn.

Mills (PC) reported seeing a Peregrine Falcon at Rattlesnake Reservoir, Churchill County, April 29, 1985.

Biale (MS) reported at Tonkin Ranch, west of Eureka one was seen November 13, 1969. He (PC, 1984) reports this is the only one seen in this area.

Austin and Bradley (1971) reported eight records for Clark County, three since 1954.

Castetter and Hill (1979) and O'Farrell and Emery (1976) reported a few sightings at Yucca Flat.

Lawson (TT, 1976) commented, "This bird is a very rare migrant and occasional winter visitor here in southern Nevada. I have seen this species at Tule Springs, Lake Mead, Moapa and Pahranagat Valleys, but I have not seen one of these hawks in the past two years here in southern Nevada."

Herron (PC, 1988) reported Nevada's first documented nesting Peregrine Falcons at Lake Mead. "They have raised an average of two young to the fledgling stage each year for the past three years."

Comment. Considerable controversy prevails over the reports of Peregrine Falcon sightings in Nevada since 1950. All too often they are confused with the Prairie Falcon. In the past 20 years this author has attempted, without success, to get good field identification of one of these birds in Nevada. I may have seen one in Elko County, March 14, 1969; one in Mason Valley, January 26, 1961; and one near Fallon February 10, 1970. These were not positive identifications. Dr. Ned K. Johnson reported March 11, 1971 he did not believe most of the records in Nevada either.

Prairie Falcon *Falco mexicanus*

Distribution. From southeastern British Columbia to Saskatchewan; south to Baja California, Texas and Aguascalientes, Mexico.

Status in Nevada. Statewide resident. Oakleaf (1974b) stated, "Records of 120 nesting territories and numerous sightings in recent years indicate that breeding populations occur in every county of the state."

Herron (PC, 1984) reported this bird as a common statewide nester with 1200 to 1400 estimated nesting pair. "Nests are found any place with a cliff and a prey base."

Records of Occurrence. Sheldon NWR (1982) reported the Prairie Falcon as common in spring, summer and fall; seen occasionally in the winter. They also nest in the area.

In Elko County this falcon was seen once April 14, 1966; twice in May and June; 3 times in July; August 17, 1971; 4 times in September; twice in October; and November 30, 1965 (Alcorn MS).

LaRochelle (MS, 1967) reported this bird at Ruby Lake NWR to be common in early spring through summer.

Howard (1972) reported on this same refuge one was seen at headquarters May 13, 1972.

The fact that Howard placed it in his narrative report indicates this bird was not commonly seen in 1972.

In the Lovelock area from 1960 to 1965, inclusive, this bird was seen January 11, 1965; 4 times in February; twice in April; May 28, 1962; November 20, 1963; and 3 times in December (Alcorn MS).

Ryser (1967) reported in the Reno area September 10, 1967 a Prairie Falcon was perched on a power pole that appeared to be a favorite perching place, as a Prairie Falcon was there the previous December.

Ryser (1970) reported as many as three at once could be seen along Kleppe and Boynton Lanes in Reno in February, 1970.

Alcorn (1946) reported this bird as resident in Lahontan Valley (Fallon area). Noted on numerous occasions in each month. Since that time there has been a noticeable reduction in the population of these birds in this area. From January, 1960 to December, 1965, inclusive, these falcons were seen 13 times in January; 15 in February; March 22, 1961; April 25, 1963; June 16, 1964; August 24, 1963; 5 in November; and 10 times in December. Near Fallon, the crop of a female was examined March 17, 1945. It contained the remains of a Townsend's Ground Squirrel.

Schwabenland (MS, 1966) reported at Stillwater WMA, "A few birds were seen through the fall and winter months and rarely in the spring and summer. Nesting possible but none seen."

For this same area Napier (MS, 1974) reported, "Occasional birds regularly seen in fall, winter and spring. A rare sighting is made in summer. The first fall I was here (1966) still impresses me with the number of Prairie Falcons I saw. Since that year sightings have been very few."

Biale (MS) for the Eureka area reported from January, 1965 to December, 1967, inclusive, these falcons were seen 3 times in January and February; twice in March; 3 in April; 5 in May; twice in June; 3 in July; twice in August; 4 in November; and 11 times in December. He (MS) knows of two active nests. From 1970 to 1973, inclusive, Biale reported these birds twice in January; 6 times in February; 4 in March; 3 in April; 6 in May; 5 in June and July; 3 in August; 7 in September; 5 in October; 6 in November; and 7 times in December.

On a very high bluff directly west of Baker, White Pine County, an adult was seen perched on a nest May 10, 1972 (Alcorn MS).

At Springdale, north of Beatty on October 7, 1958 a Prairie Falcon dove on seven Green–winged Teal that practically flew into the water as they hastily alighted on a small pond (Alcorn MS).

In Pahranagat Valley from 1961 to 1965, inclusive, these birds were observed January 8, 1962; February 16, 1965; twice in March; April 21, 1961; November 30, 1964; and twice in December (Alcorn MS).

J. R. and A. A. Alcorn (MS) for Moapa Valley, Clark County reported that from December, 1959 to February, 1964, this falcon was recorded twice in February and March; June 7, 1961; twice in August; October 30, 1960; November 23, 1961; and December 15, 1959.

A. A. Alcorn (MS) saw one trying to catch a Redhead duck in flight at Overton WMA June 7, 1961. The falcon flew close to the water and "dove up" in its attack on the duck.

Austin and Bradley (1971) for Clark County reported, "Occurs all year. Permanent resident in riparian area, winter resident in desert scrub and transient in woodlands and montane forests."

Lawson (TT, 1976) commented, "This falcon is an uncommon summer resident and a common winter resident in southern Nevada."

Order GALLIFORMES

Family PHASIAIDAE: Partridges, Grouse, Turkeys and Quail

Gray Partridge   *Perdix perdix*

Distribution. Frequently referred to as the Hungarian Partridge. Introduced from Eurasia into North America. Established from British Columbia to Nova Scotia in Canada and in various parts of the United States (Nevada to Wyoming, South Dakota and Iowa to Vermont).

Status in Nevada. Resident; common locally in northern part of the state. These non–native game birds were introduced into Nevada as early as the early 1930s. However, the 1925 biennial report of the Nevada Fish and Game Commission reported, "25 pair of Hungarian Partridges were purchased and liberated on the recreation grounds and game refuges in Elko and Churchill Counties."

Records of Occurrence. Hunters harvested 9265 of these birds in 1974 in Nevada. In 1977 only 1503 were harvested and in 1981 a total of 8671 were harvested. This illustrates very well the widely fluctuating populations of Gray Partridge in Nevada.

Flocks of eight to 15 birds have repeatedly been seen in the McDermitt, Midas, Tuscarora and Bruneau River areas. In some years, they are very abundant (Alcorn MS).

Marshall and Alcorn (1952) sated, "This species is not listed by Linsdale. It is well established over a wide area in the northern part of the state. Alcorn collected one about 10 miles south–southeast of Denio, Humboldt County January 29, 1942. Two others were shot by Mills April 21, 1942 about 15 miles north of Winnemucca and prepared as specimens by Alcorn. This species has been seen by Alcorn as far south as the southwest side of Smoke Creek Desert, Washoe County and at Kingston Canyon on the east side of the Toiyabe Range, Nye County."

LaRochelle (MS, 1967) reported at Ruby Lake NWR, "Not often seen but are a permanent resident. A group of eight to 15 use the Civilian Conservation Corps camp area in winter and from 25–50 can be found in Narcisse area in summer and fall. Nesting occurs as several broods have been seen."

Biale (PC, 1984) reported at Hay Ranch west of Eureka, he frequently trapped a pair in the 1960s during his bird banding operations. He noted that they come and go in the area. Some years he will see five to 10 birds in a bunch, other years he will not see any. In Diamond Valley he saw nine on November 30, 1980 and three on November 15, 1981.

Gullion et al., (1959) reported, "Records maintained by the Clark County Game Management Board show about 200 birds of this species were released in Clark County (site not specified) during the early 1930s. This plant apparently failed since there have been no subsequent reports of this species in southern Nevada."

Lawson (TT, 1976) commented, "We do not have any of these birds here in southern Nevada."

Himalayan Snowcock   *Tetraogallus himalayensis*

Distribution. Commonly referred to as the Snow Partridge. In the western Himalayas and the high mountain ranges to Kashmir and Ladak.

Status in Nevada. Introduced to Nevada in the 1960s. The Snow Partridge is now established and reproducing in the Ruby Mountains and East Humboldt Range of northeastern Nevada.

Records of Occurrence. These birds were originally imported from Pakistan after work done in 1960 suggested that similarities existed between the Himalayan Mountains and Nevada mountain areas between 7000 feet and 13,000 feet elevation. Glen C. Christensen's 1966 report on this bird explains, " . . . a Reno sportsman, Mr. Hamilton McCaughey, had managed to cut through miles of red tape and secure special permission from the President of Pakistan to enter the remote northern state of Hunza in quest of the famous Marco Polo sheep and he readily volunteered to try and secure some Snow Partridges for the Commission [Nevada Fish and Game Commission]."

In 1961 one bird (a female) of six survived the trip from Pakistan to Nevada.

Christensen continued, "Following Mr. McCaughey's trip to Hunza it was possible to establish a liason with the Mir of Hunza and to arrange for a yearly trapping program and shipment of the captured Snow Partridge to the United States."

In 1962, 27 of 36 birds survived. Of the 27, 19 were released and eight were taken to a game farm. In 1963, only 15 of 60 birds survived the trip and all were sent to a game farm."

"It became apparent that due to the long distance involved in transporting the birds, mechanical problems during the various phases of transport, and disease epidemics, that it would not be feasible to import a sufficient quantity of Snow Partridges on a yearly basis to sustain a wild release program."

"In order to try and alleviate some of these problems and also to obtain additional information about the life history and habitat requirements of the Snow Partridge, a Nevada Fish and Game biologist was sent to Northern Pakistan in the fall of 1964."

In 1964, 46 of 95 birds survived and were taken to a game farm. "Upon completion of the 1964 importation program, it was felt that sufficient Snow Partridges were on hand in Nevada to phase into a full Snow Partridge propagation program and to discontinue the importation projects. A small game farm facility was built at Yerington, Nevada in 1965 for this purpose."

In 1965, 57 eggs were layed by five hens and 16 birds were raised to maturity. Since that time the raising of these partridges has been discontinued at the Mason Valley game farm located at the Mason Valley WMA.

A concentrated effort to establish these birds has been made in the Ruby Mountains and East Humboldt Range, with most success coming from the peaks of the Ruby Mountains. A 1979–80 report on these partridges stated, "A total of 2158 birds have been released into the Ruby Mountains between March, 1963 and September, 1979."

In 1980, the Snow Partridge was felt to be established enough in the Ruby Mountains to have a hunting season. Two birds were harvested in 1980 and three in 1981.

A report from the Nevada Department of Wildlife (1982) stated, "While preferred habitat of this species makes for difficult hunting, reports from hunters confirmed good bird availability. Several hunters commented on the shy nature of the species and said that they had difficulty getting close to the birds."

Stiver (PC, 1986) commented this snowcock is very difficult to hunt. He estimates only 16 birds have been harvested since the 1980 hunting season began. Stiver also noted the name "Snow Partridge" was applied to this bird in the early years because it was felt the name would be more appealing to hunters and game people. The proper name is Himalayan Snowcock.

SeeSee Partridge   *Ammoperdix griseogularis*

Distribution. From southern Russia and Iran to northwestern India.

Status in Nevada. Introduced in 1968. Efforts to establish this bird in Nevada appeared to be partially successful in the Shawave Range of Pershing County and even more successful in the Seaman Range, Lincoln County. However, at last report the birds are not established in Nevada.

Records of Occurrence. In January, 1980 a report by the Nevada Department of Wildlife entitled, "Exotic Game Propagation" stated: "In 1968 the Utah Fish and Game Commission supplied Nevada with 50 SeeSee Partridges to initiate propagation efforts. Also, during 1968, new incubators were obtained and a new brooder house was built in order to accommodate the three species of birds that were being raise at the Nevada State Game Bird Farm (near Yerington)."

"The SeeSee phase of the project proved to be more successful than the tinamou and at least as successful as the Snow Partridge program. The SeeSee propagation effort continued until the Mason Valley propagation facility was shut down in 1979."

"A sufficient number of SeeSee Partridges were reared to enable a total release of 1504 birds during the course of the projects. This large number of birds enabled the Department to make releases in 10 different areas of the state and to adequately test the suitability of various habitat types for this species."

The sites in this report, plus other release sites include the Shawave Range, Lava Beds, and Egbert and Stonehouse Canyons of Pershing County; Buffalo Meadow and Rattlesnake Canyon, Washoe County; Churchill Canyon, Lyon County; Dutch Creek at Pilot Mountain and Cinnabar and Davis Springs, Mineral County; Cave Spring and Silver Peak of Esmeralda County; and White Rock Spring on the Seaman Range, Lincoln County. Releases were made during the spring and fall seasons with best release success in the fall.

The 1979–80 report by the Nevada Department of Wildlife on the SeeSee Partridges seen at the Shawave Range, Pershing County on September 6, 1979 noted these were the only partridges in Pershing County. At the White Rock Spring and Coyote Spring release sites of Seaman Range, Lincoln County, up to 70 birds were seen in a year and it was stated that, "Establishment of this site appears favorable."

Stiver (PC, 1984) reported, "SeeSee Partridge does not appear established in Nevada."

Black Francolin   *Francolinus francolinus*

Distribution. From Cyprus eastward to India. In North America, established in Hawaii, southwestern Louisiana and southern Florida.

Status in Nevada. Attempts to establish this game bird in Nevada in the 1960s appear to have failed. No recent records are available.

Records of Occurrence. At the Overton WMA in Moapa Valley in March and April, 1960, a total of 235 Black Francolin were released. Another 50 were released at this same location in February, 1961. The last ones seen in this area were on July 22, 1965, when two were observed.

Stiver (PC, 1986) commented these birds reproduced very successfully in the wilds of Nevada. However, in their native habitat of India, the climate allowed for two broods to be hatched each year. Stiver believes the inability of the bird to reproduce two broods a year is the reason the bird was unable to establish itself in Nevada.

Gray Francolin   *Francolinus pondicerianus*

Distribution. Native to Pakistan and India, introduced in America. Established on Hawaiian Islands.

Status in Nevada. Attempts to establish this game bird in southern Nevada apparently have failed. No recent records are available.

Records of Occurrence. From 1959 to 1961, inclusive, the Nevada Department of Fish and Game released a total of 1,762 Gray Francolin in southern Nevada. Release sites included Moapa Valley (1,005 released); Virgin Valley (180 released); and Pahranagat Valley (421 released).

On February 24, 1960 in Moapa Valley, 280 of these birds were released and one week later 69 were found dead.

A. A. Alcorn (MS) saw two adults with four young in Moapa Valley June 16, 1962.

Stiver (PC, 1986) commented that similar to the Black Francolin, these birds reproduced successfully in the wilds of Nevada. However, in their native habitat of India, the climate allowed for two broods to be hatched each year. Stiver believes the inability of the bird to reproduce two broods a year in Nevada is the reason the bird was unable to establish itself.

Chukar   *Alectoris chukar*

Distribution. From Eurasia. Successfully introduced into western North America.

Status in Nevada. Comprehensive studies of the Chukar partridge have been undertaken previous to this time. In–depth information on Chukar can be obtained by referring to *The Chukar Partridge in Nevada* by Alcorn and Richardson (**Journal of Wildlife Management**, Vol. 15:3, July 1951) and *The Chukar Partridge in Nevada* by Glen Christensen (Nevada Fish and Game Commission, 1954, 77pgs.).

Common resident in foothills and mountain ranges of north, central and western Nevada.

Records of Occurrence. The Chukar population fluctuates widely, depending upon range conditions. Post–season questionnaires sent out by the Nevada Department of Wildlife indicate from 1974 to 1984 the highest number of Chukar harvested in one year was in 1980 when 218,965 birds were harvested. In contrast, the lowest number harvested was in 1984 when 52,243 were harvested.

These game birds are so widely distributed, no specific records of occurrence are given. Sheldon NWR (1982) reported the Chukar as a common permanent resident.

In the Ruby Valley area these birds are present in limited numbers, but have never become abundant there. A few more occur on the west slope of the Ruby Mountains and huntable numbers were reported in the Green Mountains Creek area in favorable years.

They were frequently seen in the Mason Valley area throughout the year and Colleen Ames and this author saw them often in the Weed Heights area in 1983 and 1984 (Alcorn MS).

Alcorn (1946) reported these birds in the Lahontan Valley as, "Resident; not abundant. Noted frequently in the vicinity of Stillwater and west of Fallon in the cultivated area." Alcorn and Richardson (1951) who reported on the birds west of Fallon discussed their possible weakly established status. Since that time, these birds have disappeared from the area.

Biale (PC, 1984) reports they are abundant in the Eureka area, especially in the Diamond Mountains.

Gullion et al., (1959) reported this bird in southern Nevada was sparingly established in at least four locations: the Virgin Mountains, Clover Mountains, the Flat–Nose Ranch, and near the Comet Mine on the west side of the Highland Range. He reported attempts to establish these birds in various other areas have failed.

In that same year, Gullion et al., (1959) reported, "54 birds of the race *A. g. cypriotes,* the so–called Turkish Chukar, were released in Meadow Valley Wash a few miles south of Caliente March 17, 1955. The success of this release is still not certain."

Lawson (TT, 1976) commented, "We have a few small flocks on the western slopes of Spring Mountain Range. There are also some in the Mount Lincoln area in Lincoln County. I would characterize it as a rare to uncommon bird for Clark County."

Stiver (PC, 1986) reported, "In southern Nevada, common in some years. As a matter of fact, it was more common in Spring Mountains than any other place in 1984."

Barbary Partridge   *Alectoris barbara*

Distribution. Morocco, Algeria and Libya.

Status in Nevada. Attempts to establish this bird in Nevada have failed.

Records of Occurrence. On November 8, 1956, 60 Barbary Partridges were released in Mineral County, by the Nevada Department of Fish and Game. Christensen (1975) reported efforts to establish this bird in Nevada were a failure.

Ring–necked Pheasant   *Phasianus colchicus*

Distribution. Native from southern Russia to Formosa. Introduced birds now established widely in North America.

Status in Nevada. This introduced bird has been released into many parts of the state and has become established in some cultivated valleys. In southern Nevada, the White–winged race was released in the 1960s and is established there.

Records of Occurrence. A cock pheasant was repeatedly heard crowing at Lamance Creek in Santa Rosa Mountains June 10, 1970 (Alcorn MS).

Stiver (PC, 1986) reported some are established in Elko, Eureka, Lander and Humboldt

Counties as far east as Lamoille. They are also along the Humboldt River, in Paradise Valley, Kings River Valley and Quinn River Valley.

In the Lovelock area these birds have been seen during all visits to that valley for the past 25 years. A female with nine small young was seen May 9, 1961 south of Lovelock (Alcorn MS).

In the Fallon area from January, 1960 to December 31, 1965 these birds were recorded 27 times in January; 37 in February; 50 in March; 65 in April; 53 in May; 24 in June; 31 in July; 27 in August; 21 in September; 31 in October; 22 in November; and 31 times in December (Alcorn MS).

W. H. Alcorn saw a dead female pheasant about 15 feet from a pheasant nest containing 15 eggs south of Fallon April 30, 1941. He skinned the pheasant and found a fracture on the side of its head. It was undetermined what had delivered the fatal blow (Alcorn MS).

In the Fallon area two males were seen fighting March 29, 1961. A female flew from a nest containing 10 eggs June 6, 1963 (Alcorn MS). Another female was seen with six half–grown young August 3, 1960. Mills (PC) reported rancher John Gomez said July 16, 1966 the airplane spraying of the alfalfa to control aphids killed most of the pheasants on his ranch southwest of Fallon. He also said this spraying also killed many other species of birds (Alcorn MS).

Biale (MS) reported for the Eureka area 100 birds were released November 3, 1966 in two separate locations in the nearby Diamond Valley.

Biale (PC, 1984) reported these birds have not successfully established themselves in this area.

Between April 10, 1963 and March 13, 1969 a total of 751 of the White–winged race of pheasant were released in Nye, Lincoln and Clark Counties, Nevada.

Stiver (PC, 1986) reported, "In southern Nevada, the White–winged race is established. Ringed–necks released in southern Nevada lose their characteristics to the White–winged race within five years."

In Pahranagat Valley from 1960 to 1965 from one to three pheasants were seen once in January and March; 4 times in April; 3 times in May; and once in June and December (Alcorn MS).

In Moapa and Virgin Valleys from 1960 to 1967, these birds were recorded 5 times in January; twice in February; 4 in March; 5 in April; 6 in May; 4 times in June; twice in July; and once in September and November (Alcorn MS).

Lawson (TT, 1976) reported, "In southern Nevada this bird is a rare species."

Blue Grouse   *Dendragapus obscurus*

Distribution. From Alaska to Mackenzie and south to California. East to South Dakota and New Mexico.

Status in Nevada. Resident in most of the higher mountain ranges, except in southern Nevada. Four races occur in Nevada: *D. o. obscurus, D. o. pallidus, D. o. sierrae* and *D. o. howardi.*

Records of Occurrence. Limited numbers of these birds occur in mostly all of their former ranges. Apparently large numbers occurred in some sections of the state in the early 1800s.

The Nevada Department of Wildlife's post–season questionnaire revealed 3409 Blue Grouse were harvested in 1974. In the 1983 hunting season only 939 were harvested. This decline may appear alarming to some; however the population of these grouse fluctuates widely. In 1985 the 1124 harvested indicate an upward trend.

Henshaw (1877) reported, "The whole pine–timbered region lying along the eastern slope of the Sierras west of Carson Valley, was formerly the home of very great numbers of this fine bird. Some of the stories told by the early settlers of its abundance are almost incredible."

"The sound of the woodman's axe is followed by the almost complete abandonment of a locality and chiefly from this cause and from the persecution they have been subjected to at the hands of settlers and the Indians, the localities are very few where the grouse still exists in abundance."

About three miles south of Jarbidge an adult with four young was seen August 22, 1963. Others were seen in the same area October 12, 1968. Limited numbers, not exceeding 10 per day, were observed at the head of Mary's River in October, 1970 (Alcorn MS).

These birds are not normally seen in the valleys. However, LaRochelle (MS) reported at Ruby Lake NWR seven were flushed from junipers on the west edge of the refuge March 26, 1964. Several males were heard "hooting" at Hunter's Creek, southwest of Reno in April and several were seen in the Mount Rose area August 31, 1959 (Alcorn MS).

Orr and Moffitt (1971) reported, "In the spring and early summer the 'hooting' of the males is a common sound on nearly all timbered slopes surrounding the lake (Tahoe). This phase of the breeding cycle begins when deep snow still covers the ground (earliest date heard, April 20) and is continued until early July. Males have been noted hooting in Jeffrey pine, sugar pine, white fir, and incense cedar. Fairly large trees are generally selected for this purpose and the calling birds characteristically perch on a large limb, not far from the trunk, in the upper third of the tree. When they hoot, their gular sacs are inflated and the head is directed forward and down."

Biale reported seeing these birds in the Toiyabe Range, four at Big Creek and two adults and three immature at Kingston Canyon August 12, 1973. He reported seeing up to 12 birds in the Diamond Mountains, near Eureka in September and October, 1973).

Biale (MS) reported in the Ely area in the Duck Creek Basin of the Schell Creek Range he saw 15 on October 8 and three on October 25, 1972. In the same area he saw 12 September 16, 1973.

Johnson (1973) reported, "On June 20, 1953, LaRivers saw a hen with approximately five chicks in an aspen grove at the base of the cliff near the head of Scofield Canyon and on the same date and in the same area, Carl (referring to Ernest Carl) observed four adults in mountain mahogany near a meadow."

"In 1972 the fresh remains of an adult female which had been killed by a predator were found strewn on a log and about the ground in an aspen grove at 8000 feet in Scofield Canyon."

Lawson (TT, 1976) reported, "We do not have any of these birds in southern Nevada."

Mowbray (MS, 1986) reported the Blue Grouse as uncommon in southern Nevada.

Ruffed Grouse   *Bonasa umbellus*

Distribution. Resident in forested areas from Alaska to Nova Scotia; south to California and North Carolina. Introduced and established in Iowa and Newfoundland.

Status in Nevada. Introduced to Elko County.

Records of Occurrence. Clyde R. Madsen (MS) reported what he thought to be a Ruffed Grouse was seen along the road in the Jarbidge Mountains between Charleston and Jarbidge in 1940. However, Ryser (1985) commented, "There is no hard evidence that it ever occurred naturally within Nevada before its 1963 introduction to northeastern Nevada by the State Fish and Game Commission."

Len Hoskins (PC, 1969) reported in the spring of 1968 he heard three different Ruffed Grouse drumming at "John Day" in Elko County.

Stiver (PC, 1984) reported the population of these birds is now low due to the recent hard winters in northeastern Nevada. He believes their estimated harvest during the hunting season is less than 100 birds a year for the entire state. He (PC, 1986) commented the Ruffed Grouse is well distributed on the west side of the Ruby Mountains. Any aspen forested area seems to be suitable habitat for this grouse.

Sage Grouse   *Centrocercus urophasianus*

Distribution. Resident from British Columbia to southeastern Alberta; south to eastern California and Colorado to northern Mexico.

Status in Nevada. Resident in the northern two–thirds of the state.

Records of Occurrence. These birds have been seen by the author in the mountains and sagebrush–covered valleys in Washoe, Humboldt, Elko, Pershing, Lyon, Churchill, Lander, Eureka, White Pine, and northern Nye Counties. They formerly were more abundant than now, with some reports of the numbers killed in earlier years being almost unbelievable. Their numbers fluctuate considerably from one year to another, but they are still present in considerable numbers over a wide area.

Nevada Department of Wildlife post–season questionnaires sent to hunters show wide variation in the number of birds harvested each year. For example, in the 10–year period of 1974–84, the greatest number in any one year was in 1979—28,228 birds were taken. The lowest number was in 1977—7561 birds taken.

Gullion (1954) stated, "Wildlife technicians are still at a loss to explain the drastic fluctuations in this magnificent grouse. We do know that the numbers of most other species of North American grouse are subject to drastic periodic decreased and subsequent resurgence. These we call grouse cycles. We know that adverse weather conditions and increased losses to disease and parasites often accompany these decreases, but the general concensus of opinion is that these factors are secondary and that the major cause of these declines is still unknown."

Sheldon NWR (1982) reported the Sage Grouse as a common permanent resident that nests in the area.

In Elko County in April 1979, five separate active strutting grounds were observed in an area from Dinner Station to Taylor Canyon (Alcorn MS).

Kingery (1981) reported, "Sage Grouse began drumming at Jarbidge, Nevada by January 19."

LaRochelle (MS, 1967) reported a strutting ground at Ruby Lake NWR. He reported the birds were, "Commonly on the area in spring and fall with broods of young often seen. Up to 75 on the area."

Kingery (1985) reported, "At Ruby Lake, Evans reported . . . Sage Grouse counts, both on the refuge and on the strutting grounds evidenced rapid decline."

Male Sage Grouse strutting. Photo by A. A. Alcorn.

Merrill Van (MS) reported at a dry lake at Bald Mountain, east of Yerington on June 9, 1965 he saw a male Sage Grouse strutting. He was there June 8–10 and saw from eight to 10 strutting birds each day.

In Churchill County, an effort was made to study these birds in the Clan Alpine Mountains and nearby valley to the south and east from 1942 to 1954. In this area it was found these birds spend late winter in the sagebrush–covered flat areas near the base of the mountains. One flock of about 100 birds was consistently found at lower Camp Creek in the Clan Alpine Mountains of Churchill County in December and January. Also a strutting ground was observed every spring in this general area. Their nests were found in this same type of terrain. Nearby at a higher altitude, examination of the meadows at the head of Cherry Creek in early spring revealed no evidence of these birds. However, in June adults and young birds appeared on the meadows and were present at this higher elevation until late fall (Alcorn MS).

One strutting ground, in a flat area covered with a low growth of bud sage, was located five miles north–northwest of Eastgate. This strutting area was used on a yearly basis for many years until a new highway was built through it. A total of 42 males were seen at this place March 28, 1965 and on April 15, 1965 at least 30 were present. On May 5, 1965 Lois Saxton saw 29 males and three females at this location. William Saxton saw six males April 23, 1966. Only three were seen April 3, 1968 (Alcorn MS).

Biale (MS) in the Eureka area recorded these birds many times throughout the years over a wide area in all seasons. He (PC, 1984) noted a decrease in numbers of this bird from 1954 to 1984 in the Eureka area.

Lawson (TT, 1976) reported, ''We don't have this bird in southern Nevada.''

Sharp–tailed Grouse   *Tympanuchus phasianellus*

Distribution. From Alaska to Quebec; south to Oregon, Michigan, the central U.S. and possibly northern Texas.

Status in Nevada. Probably extinct. Has not been seen in Nevada within the past 25 years.

Records of Occurrence. Linsdale (1936) stated, "In early days this bird was found commonly in Elko County, northeastern Nevada. In the northern part of the county, Hoffman, (1881) found it in 1871 in moderate numbers on the Bull Run Mountains. Ridgway (1877) collected a young female from the upper Humboldt Valley, September 16, 1868 and near Trout Creek he found the species abundant in rye–grass meadows at the base of the Clover Mountains."

Linsdale (1951) wrote, "LaRivers (1941) reported he has seen the bird once recently in Nevada, but does not give particulars. In a letter under date of August 23, 1949 he wrote that he saw a flock of six on July 9, 1939. The birds were in a high aspen grove in the Bull Run Mountains about two miles west of the Homer Andre Ranch some nine miles south–southwest of Mountain City."

William Q. Wick (1955) reported, "On September 14, 1952, Grover Freeman and I, observed a flock of 12 Sharp–tailed Grouse. These grouse were found on a spur of the Capitol Range in Humboldt County, Nevada, near the Humboldt–Elko County line, approximately 26 miles northeast of Golconda . . . . By stalking we were able to come within 35 yards of the bird and make a certain identification."

Earl Dudley, Nevada Department of Fish and Game, (PC, 1971) reported, "In Elko County, the Sharp–tailed Grouse were in Ruby Valley in the early 1940s." The last ones Dudley saw in Nevada were four birds in the Lamoille area in 1960. He commented, "These grouse liked to hang around the rosebush thickets."

Ryser (1985) says, "Now it is confined to a few small areas in northeastern Nevada . . . it no longer occurs in northwestern Nevada."

Stiver (PC, 1986) comments that none of these birds have been seen in Nevada in the last 25 years. However, he notes plans are being made to transplant Sharp–tailed Grouse from Idaho. "There is potential for Goose Creek [Elko County] to have a population."

Biale (MS) reported these birds occurred in the Eureka area in the early days and the last ones seen were reported by Floyd Sadler in the 1930s.

Lawson (TT, 1976) reported, "I made a very pointed effort to try to locate some of these birds on my trip to northeastern Nevada this year, with no success at all."

Wild Turkey   *Meleagris gallopavo*

Distribution. From Alberta to Ontario and south across North America to Veracruz.

Status in Nevada. Introduced in western and southern Nevada.

Records of Occurrence. Stiver (PC, 1988) reported on February 25, 1987, 16 Wild Turkeys of the Rio Grande subspecies, imported from Livermoore, California, were released at the Mason Valley WMA. "They did reproduce, but they were naive birds, so a number of them were taken by coyotes."

Stiver (PC, 1988) reported on January 13 and 14, 1988 a total of 114 Wild Turkeys of the Rio Grand subspecies were trapped and transported from Wheeler County, Texas to Nevada. On January 16, 1988 they were released at the following locations: 44 at the Mason Valley WMA, Lyon County; 30 at the John Ghiglia ranch in Weeks, Lyon County; and 40 in the Spring Mountains, Clark County. Stiver expects these turkeys will do better than those released in 1987 because they are less naive and more adept at getting out of reach of predators.

Stiver (PC, 1984) reported a low population is established in the Carson Range and small numbers may be found in the Charleston Mountains of the Spring Range in southern Nevada. He (PC, 1986) comments, "A lack of a consistent crop of food such as pinon nuts is believed to be responsible for low numbers in the Charleston Mountains."

The Nevada Department of Fish and Game released 24 Merriam's Turkeys one mile north of Cold Creek Field Station, Charleston Mountains, February 24, 1960.

A. A. Alcorn (MS) saw two adults with five young near this release site June 24, 1962 and he found turkey feathers at Trough Springs July 19, 1963.

Christensen (1963) reported turkeys were released in two locations: the Bird Spring Range, Clark County in 1960 and 1962 and on the eastern slopes of the Sierra Nevada Mountains, Washoe County, 1963.

These introductions have been moderately successful and limited hunting has been permitted in these areas.

Johnson (1965) reported, "We recorded one turkey in pine–fir three miles north of Charleston Peak, 8900 feet, on June 17 and four in juniper–pine–fir at Macks Canyon, 8100 feet on June 21 in the Spring Range."

Lawson (MS) saw four on March 8, 1976 at Cold Creek Spring, Spring Mountains.

Lawson (TT, 1976) said, "A limited number have been seen here in southern Nevada and I would say it is probably a rare species."

Northern Bobwhite   *Colinus virginianus*

Distribution. From Wyoming to Ontario and Maine; south locally to Guatemala. Introduced in northwestern North America.

Status in Nevada. Introduced in various parts of the state by private citizens.

Records of Occurrence. Mike Wickersham (PC, 1983) reported Bobwhite have not been released by Fish and Game personnel, although individuals have raised and released some birds using hobby permits. One was shot on the south fork of the Owyhee River in 1974.

Stiver (PC, 1984) reports that 1977 was the last year these birds were seen in Ruby Mountains. At this time, six bobwhites, adults and immatures, were shot by Alan Flock and Walt Campbell while hunting for partridge in Huntington Valley. Stiver speculated the birds had been released by a private individual and believes there are no wild birds left in the area.

F. A. Harrigan, a pioneer resident of Fallon, when interviewed January 29, 1941 stated that Tom Dymond introduced Bobwhite Quail into the Fallon area in the early days. Dymond was a gambler from Virginia City who did a great deal of duck hunting in the Fallon area.

The *Fallon Standard* on February 10, 1926 reported the Greenhead Club ordered 25 pair of Bobwhite Quail from Mississippi for release west of Fallon.

In 1984 a few of these birds were frequently seen southeast of Fallon, having been released there in 1982. Frequent introductions of this bird in the Fallon area over a period of many years has not resulted in establishment (Alcorn MS).

Scaled Quail   *Callipepla squamata*

Distribution. From Arizona to eastern Colorado, southwestern Kansas; south in Mexico to the western Tamaulipas.

Status in Nevada. Introduced.

Records of Occurrence. Wickersham (PC, 1983) reported a few were released at Montello (early 1970s) and Buffalo Valley, Lander–Pershing County. He noted none have been seen in that area for a year.

A lone Scaled Quail was observed six miles southeast of Fallon in September, 1933 (Alcorn MS).

Biale (MS) reported Scaled Quail were released at Fish Creek in Eureka County in the early 1960s.

Lawson (TT, 1976) commented, "It is not found here in southern Nevada, but it is apparently common at Sunnyside, White River Valley."

Stiver (PC, 1986) commented before the 1978–79 winter he would observe as many as 200–300 Scaled Quail daily in White River Valley. The harsh winter, however, reduced the population drastically and they have not yet recovered.

Gambel's Quail   *Callipepla gambelii*

Distribution. Resident from east–central California east to western Colorado; south to northwestern Chihuahua and western Texas.

Status in Nevada. Common resident in southern Nevada.

Records of Occurrence. Linsdale (1951) reported these quail as a common resident in southern Nevada, north to Quinn Canyon Mountains.

In Pahranagat Valley, these quail were observed from December, 1959 to June, 1965 a total of 5 times in January; twice in February and March; 5 times in April; 8 in May; once in June; twice in July; once in August; twice September and October; and 3 times in November and December (Alcorn MS).

Near Hiko on April 17, 1960 an adult with 10 young was seen (Alcorn MS).

A. A. Alcorn (MS) reported Gambel's Quail in southern Nevada 194 times between December, 1959 and July, 1965. He reported them in all months and stated, "Downy young were seen at the Gold Butte area April 24, 1961 by park ranger Don Squires."

"Lloyd Marshall at the Overton WMA accidentally stepped on a female quail on her nest May 10, 1961. There were 11 eggs in the nest. This nest was in a clump of barley at the edge of the barley field. The nest was made of short (1/2 inch) pieces of barley leaves and the leaves were about 1 1/2 inches deep in the bottom of the nest."

A. A. Alcorn saw a total of 24 downy young at Gold Butte May 10, 1962. In the Moapa Valley he saw two adults chasing a roadrunner June 6, 1962 and on June 15, 1962 he saw two adults with three young.

A. A. Alcorn (MS) commented, "Installation by Nevada Fish and Game personnel of artificial watering devices (guzzlers) have extended the range of this quail to more arid parts of southern Lincoln County and Clark County. Adult to young ratios are determined at these watering sites for the establishing of hunting seasons. Late spring ratios are usually a good indicator of hunting expectations. Populations very greatly depending upon production of springtime range forage."

"During the early 1960s a 'truck farmer' producing radishes, onions and tomato plants complained that quail had eaten the green plants at the edge of the fields to the extent that he wanted help in reducing his crop losses. Grain was spread along the edge of the fields in an effort to get the quail to eat the grain instead of the greens. It was estimated that the farmer lost 10–15% of his crop."

Lawson (TT, 1976) commented, "Numbers of these quail fluctuate rather widely in any year, but this quail is a very common species in southern Nevada."

Stiver (PC, 1986) commented, "Fluctuating populations are so dramatic the harvest goes from 7000 to 120,000 birds during a population cycle. Winter water supply determines population."

California Quail   *Callipepla californica*

Distribution. Native in the area from southern Oregon, California and western Nevada; south to the Cape region of Baja California. Introduced in other states and British Columbia.

Status in Nevada. Resident locally in the northwestern two–thirds of the state. The first introduction was of 17 birds from Sacramento on May 17, 1862. However, it has been widely introduced into numerous localities. The present populations may represent a mixture of several races.

Starker Leopold (1977) reported, "It's native range includes a small portion of western Nevada . . . ."

Whether native or not, these quail were brought into the area possibly between 1860 and 1870. It was not until these years that mining activity in the western part of the state produced sufficient money and attracted sufficient persons to make it likely that anyone would be interested in transplanting game birds.

A number of quail taken in the Fallon area by this author and Mills were prepared as study skins. They were critically examined in 1940 and compared with other specimens taken in California, Oregon and Nevada. The examination revealed that some individuals showed characteristics of the race *californicus* and others showed characteristics of the race *vallicola*. This would indicate introductions were of both races, or *californicus* was introduced into a native species.

Most of the Nevada "Valley Quail" collected show characteristics placing them in the race *vallicola*. Whether native or introduced, the "Valley Quail" are thriving in numerous locations in Nevada.

Records of Occurrence. The majority of these quail are found in the irrigated valleys of west–central Nevada. However, in good quail years, they are abundant locally at springs and along streams in the foothills and valleys over a wide area, extending from the vicinity of Wellington and Fallon, north to the Oregon line, and from the California line eastward to Owyhee River, Ruby Lake NWR, Battle Mountain and Smoky Valley. Limited numbers occur as far east as Baker (near the Utah line).

Linsdale (1936) reported that E. A. Preble in 1915 found them rather common in the McDermitt area and in the Quinn River Valley. From 1966 to 1968, Lloyd Miller reported these birds extremely abundant in this area.

Sheldon NWR (1982) reported the California Quail as a common permanent resident.

In the foothills of the Santa Rosa Range, flocks of about 30 were seen at Abel Creek October

Male California Quail. Photo by A. A. Alcorn.

5, 1963. Another flock was seen at Stonehouse Creek November 13, 1963. It was also abundant in Paradise Valley (Alcorn MS).

Preble was informed by a man familiar with conditions in Paradise Valley that when the man first arrived in the area in 1895 these birds were abundant, having been brought there from some place in California some years before by William Stocks.

Good–sized flocks have been seen around stock watering ponds and along the streams leading into and out of the Owyhee Desert. In other years, few if any of these birds could be found (Alcorn MS).

LaRochelle (MS, 1967) reported at the Ruby Lake NWR, "48 of these birds were released at Cave Creek by the Nevada Fish and Game on October 20, 1962 where they have survived and reproduced, until now 75–100 are present. The California Quail were reportedly common in the area 10 to 12 years prior to this release."

In the East Range, south of Winnemucca, at Sulphur Canyon and Klondike Pass, these birds were numerous in 1958 (Alcorn MS).

Linsdale (1936) reported in 1902 a number of quail were brought from Paradise Valley to Lovelock to replace stock planted there that had become depleted. In 1975, these quail were abundant in the Lovelock area with sight records available for each trip into the valley.

In western Nevada, these birds are abundant in the irrigated valleys and along the Truckee, Carson and Walker Rivers. In favorable years, good numbers are found around the springs and streams in the lower elevations of the mountains. An example of this was reported by E. Raymond Hall who saw large numbers in Hardscrabble Canyon, west of Pyramid Lake. He estimated several thousand quail there.

In Idlewild Park, Reno in 1942, a resident flock of up to 100 birds were seen on 62 different days in all months (Alcorn MS).

Their abundance in the Fallon area is shown by their being recorded on 797 days while taking a general census of birds in that area from January 1, 1960 to December 31, 1966. They have been established in this area many years (Alcorn MS).

Vernon Bailey collected a male quail, now in the USNM, east of Fallon at Stillwater May 9, 1898.

Long time residents of the Lahontan Valley were interviewed in an effort of determining the status of these birds in earlier years in this area. Mrs. I. H. Kent, a pioneer resident of Fallon, Nevada when interviewed on January 17, 1941 said from 1880–1881, many quail were on their ranch near Stillwater, Churchill County, and they occasionally shot them for "pot pie".

Mr. F. A. Harrigan, another pioneer resident of Fallon said in an interview January 29, 1941 that in 1884, many quail were in the Lahontan Valley, having been introduced in the valley by Dymond, a gambler from Virginia City who did a great deal of duck hunting in the Fallon vicinity.

W. A. Harmon, another pioneer resident of Fallon, said Valley Quail were numerous in Lahontan Valley when he was a boy and in 1878 many quail were in the Stillwater and Sheckler districts near Fallon. Harmon considered the Valley Quail to be native to the area.

Wuzzie George, (PC, 1969) a Paiute Indian woman, reported California Quial were cooked with feathers in ashes or coals, or they were boiled after skinning. The skins, with feathers attached, were used as pillows for baby papooses.

It is now difficult to determine whether Valley Quail already inhabited the area at the time of the first introductions.

On May 21, 1941 at the Bill Farrington Ranch, Smoky Valley, Nye County, Farrington stated California Quail were plentiful in some favorable sections of his ranch.

Biale (PC, 1984) says these quail were released in the Eureka area in the 1950s and 1960s. These quail did well when fed during winter months. None have been seen in the past 10 years.

Lawson (TT, 1976) commented, "We do not have any of these birds in southern Nevada."

Behavior Studies. In the Fallon area, the author began a study of this quail in February, 1938. It was hoped time would permit a complete life history study of this species. However, the study was discontinued in December, 1941. Within this period a series of birds were collected and examined for taxonomic differences, egg development, food habit studies and other information. Some of the results, as well as miscellaneous notes follow:

Food Habits. Dr. A. L. Nelson, in charge of Economic Investigations Laboratory, Patuxent Refuge of the Fish and Wildlife Service agreed to analyze the crop contents of the birds collected. George B. Fell made the analysis with assistance on insects, etc., from others as indicated by Fell's letter which follows in part: "Two crops collected by Vernon Bailey in November, 1920 were included in the series. Two crops collected in 1942 were examined. The first of these had whole grasshoppers in it, whereas only traces of grasshoppers were found in other crops. This could have been because of the year, location or fact that few crops were collected during summer months."

"A few of the crops in February, July, October and December were not tabulated by percent because they were small in volume and would have more effect on the average percent than their real significance."

FOODS OF THE *CALLIPEPLA CALIFORNICA*, CHURCHILL COUNTY, NEVADA
Based on Examination of 193 stomaches

PLANT ITEMS	Percent	Number of Occurrences
Triticum sp.	11.1	55
Salsola kali	9.1	89
Bassia hyssopifolia	8.9	122
Hordeum vulgare	8.9	59
Populus sp.	8.5	20
Leaves of legume (Medicago sativa, etc.)	7.8	95
Melilotus sp.	7.2	102
Medicago sativa	6.7	41
Robinia pseudoacacia	6.4	22
Zea mays	4.9	23
Shepherdia argentea	3.1	28
Setaria viridis	2.8	93
Chenopodium sp.	2.4	72
Leaves of grass	1.4	38
Amaranthus sp.	1.2	79
Sarcobatus vermiculatus	1.0	9
Chrysothamnus sp.	0.9	15
Leaves (unidentified)	0.7	40
Echinochloa crusgalli	0.5	55
Taraxacum sp. (officinale?)	0.5	13
Sporobolus sp.	0.4	11
Capsella bursa-pastoris	0.4	10
Panicum miliaceum	0.4	1
Elymus sp.	0.3	18
Descurainia sp.	0.3	14
Populus sp. flowers	0.3	4
Distichlis sp.	0.2	47
Malva sp.	0.2	25
Artemisia sp.	0.2	22
Setaria lutescens	0.2	22
Polygonum persicaria	0.2	17
Asclepias sp.	0.2	15
Avena sativa	0.2	11
Leaves of Artemisia sp.	0.2	9
Atriplex sp.	0.2	8
Panicum capillare	0.2	7
Rumex sp. (crispus?)	0.1	44
Helianthus sp.	0.1	22
Descurainia sp.? (Capsella?)	0.1	8
Leaves of Atriplex sp.	0.1	4
Physalis sp. (or Solanum)	0.1	2

Amaranthus sp. (blitoides?)	Tr.	24
Polygonum sp.	Tr.	18
Cenchrus sp.	Tr.	13
Scirpus sp. *Scirpus sp. (acutus type)* *Scirpus acutus*	Tr.	12
Panicum sp.	Tr.	10
Oryzopsis hymenoides	Tr.	8
Petalostemon sp. (?)	Tr.	8
Eleocharis sp. (Palustris type)	Tr.	7
Bromus sp.	Tr.	5
Chenopodiaceae seeds and flowers	Tr.	5
Equisetum sp. stem	Tr.	5
Verbena sp.	Tr.	5
Cryptantha sp.	Tr.	4
Grass (unidentified)	Tr.	4
Polygonum hydropiper	Tr.	4
Polygonum sp. (knotweed type)	Tr.	4
Portulaca oleracea	Tr.	4
Ulmus sp.	Tr.	4
Viola sp.	Tr.	4
Carex sp.	Tr.	3
Eragrostis sp.	Tr.	3
Juncus sp.	Tr.	3
Marsilea sp.	Tr.	3
Buds (galls?)	Tr.	2
Compositae (like Solidago)	Tr.	2
Convolvulus arvensis	Tr.	2
Erodium cicutarium	Tr.	2
Flowers of legume	Tr.	2
Galls ? (black)	Tr.	2
Asparagus officinalis	Tr.	1
Bidens frondosa	Tr.	1
Bud of *Salix*	Tr.	1
Bud (unidentified)	Tr.	1
Corns (unidentified)	Tr.	1
Dicot (unidentified)	Tr.	1
Eriogonum sp.	Tr.	1
Franseria sp.	Tr.	1
Hordeum sp.	Tr.	1
Lappula sp.	Tr.	1
Legume (unidentified)	Tr.	1
Lepidium sp.	Tr.	1
Lotus sp.	Tr.	1
Malus sp.	Tr.	1
Malvaceae	Tr.	1

Mentzelia sp.	Tr.	1
Morus sp.	Tr.	1
Poa	Tr.	1
Polygonum lapathifolium	Tr.	1
Polygonum sp. (smartweed type)	Tr.	1
Portulaca sp.	Tr.	1
Rubus sp.	Tr.	1
Secale cereale	Tr.	1
Setaria sp. (immature)	Tr.	1
Sidalcea sp.	Tr.	1
Skin of fruit (similar to apple)	Tr.	1
Trifolium sp.	Tr.	1
Unidentified seed	Tr.	1

ANIMAL ITEMS	Percent	Number of Occurrences
Aphididae (Homoptera)	1.3	18
Lygus sp. (Hemiptera)	0.1	8
Crematogaster sp. (Hymenoptera)	0.1	1
Formica sp. (Hymenoptera) *Formica sp. (fusca?) (Hymenoptera)* *Formica sp. (pallide-fulva?) (Hymenoptera)* *Formica sp. (rufa?) (Hymenoptera)*	Tr.	20
Formicidae (Hymenoptera)	Tr.	17
Hemiptera egg case *(?)*	Tr.	15
Coccinellidae, (Coleoptera)	Tr.	12
Hymenoptera pupa	Tr.	12
Lygaeidae (Hemiptera)	Tr.	10
Coleoptera larva	Tr.	9
Planorbidae	Tr.	9
Lasius sp. (Hymenoptera)	Tr.	7
Lymnaea sp.	Tr.	7
Homoptera	Tr.	6
Insect (unidentified)	Tr.	6
Blapstinus sp. (Coleoptera)	Tr.	5
(?) Cicadula sp. (Homoptera)	Tr.	5
(?) Lygus sp. (Hemiptera)	Tr.	5
Acrididae (Orthoptera)	Tr.	4
Physa sp.	Tr.	4
Carabidae (Coleoptera)	Tr.	3
Elateridae (Coleoptera)	Tr.	3
Harpalinae (Coleoptera)	Tr.	3
Lepidoptera larva	Tr.	3
Prenolepis sp. (Hymenoptera)	Tr.	3
Aphididae (brown) *(Homoptera)*	Tr.	2

(?) Chrysomelidae (Coleoptera)	Tr.	2
Lygaeus sp. (Hemiptera)	Tr.	2
Membracidae (Homoptera)	Tr.	2
Myrmicinae (Hymenoptera)	Tr.	2
Pentatomidae (Hemiptera)	Tr.	2
Acarina	Tr.	1
Acrididae Acridinae (Orthoptera)	Tr.	1
Acrididae Melanoplus sp. (Orthoptera)	Tr.	1
Aeolus dorsalis (Coleoptera)	Tr.	1
Amara sp. (Coleoptera)	Tr.	1
(?) Ameiurus sp. Siluridae	Tr.	1
Aphodius distinctus (Coleoptera)	Tr.	1
Aphodius sp. (Coleoptera)	Tr.	1
Araneida	Tr.	1
(Aufeius sp?) (Hemiptera)	Tr.	1
Camponotus sp. (Hymenoptera)	Tr.	1
Cercopidae (Homoptera)	Tr.	1
Chloropidae (near Meromyza) (Diptera)	Tr.	1
Cicadellidae (Homoptera)	Tr.	1
Coccinella transversoguttata (Coleoptera)	Tr.	1
Collops sp. (Coleoptera)	Tr.	1
Diptera	Tr.	1
Droppings (insect?)	Tr.	1
Eciton sp. (Hymenoptera)	Tr.	1
Edrotes ventricosus (Coleoptera)	Tr.	1
Fulgoridae (Homoptera)	Tr.	1
Gastroidea cyanea (Coleoptera)	Tr.	1
Gastropoda	Tr.	1
Harpaline Platynus (?) sp. (Coleoptera)	Tr.	1
Ichneumonoidea (Hymenoptera)	Tr.	1
Lucilia sp. (Diptera)	Tr.	1
(Lygaeidae or Miridae) (Hemiptera)	Tr.	1
Lygaeus kalmi	Tr.	1
Microtus sp. hair	Tr.	1
Monocrepidius sp. (Coleoptera)	Tr.	1
Muscidae (near Morellia) (Diptera)	Tr.	1
(Myrmicinae?) (Hymenoptera)	Tr.	1
Notoxus sp. (Coleoptera)	Tr.	1
Perimegatoma sp. (Coleoptera)	Tr.	1
Phytonomus sp. (Coleoptera)	Tr.	1
Pteromalidae (Hymenoptera)	Tr.	1
Polistes fuscatus (Hymenoptera)	Tr.	1
(?) Vespoidea (Hymenoptera)	Tr.	1
Vespula sp. (Hymenoptera)	Tr.	1
(Vespula sp.?) (Hymenoptera)	Tr.	1
Xerophloea viridis (Homoptera)	Tr.	1

"Notes on food items: *Scirpus sp.*, *Scirpus sp.* (acutus type), and *Scirpus acutus* can be considered as *Scirpus sp.* acutus type. Leaves of legume (*Medicago sativa,* etc.,) included fragments of unknown leaves which could not be distinguished readily from the legume leaves. This non–legume material probably made up about 10% of the 'leaves of legume' category. Most of the legume leaves were *Medicago.* Some of the leaves (possibly 10–20%) appeared to be *Trigolium sp.* Other legume leaves might have been present but could not be distinguished. It is interesting that most of the leaves were identified as *Medicago sativa* while *Melilotus sp.* seeds were so abundant."

"The snails present in some stomachs were identified by F. M. Uhler as *Lymnea, Physa* and *Planorbidae*, all of which are aquatic species."

"The insects present were mostly fragments, many of which might have been eaten more or less unintentionally. Definite exceptions to this would be the aphids, *Lygus sp.*, and perhaps the ants."

"Most of the species of seeds were identified with the assistance of A. C. Martin, and insect identifications were made by Hewson H. Swift and R. T. Mitchell."

"There were single hairs, probably cattle hairs, in a portion of the crops. Also, feather fragments, presumably from the quail, were sometimes present. Neither of these were recorded. The lead shot present in some was apparently acquired at the time the birds were shot, rather than being picked up as grit. Small bone fragments were sometimes present and were usually listed as grit."

Roosting Habits. In the Fallon area most of the quail roosted in buffalo berry thickets *(Shepherida argentata)* in fall and winter. These thickets often covered from a fraction of an acre to several acres in size and were scattered at locations over the valley. Since this time, cultivation of land acres has resulted in the loss of much of the buffalo berry thicket habitat.

The quail would disperse from the thickets in early morning by traveling on the ground. They ranged out over the feeding areas and did not return to roost areas until after sundown. As evening approached they traveled on the ground toward the roost site. When within 100 yards or more of the site, they would fly as a flock into the thicket. By placing observers on each side of the thicket, one could count the number of quail entering the roost. One thicket of about an acre, was located four miles west of Fallon. It was surveyed each month over a 2–year period (1940–1941). The following information was obtained: in January an average of 189 quail roosted in this thicket, coming in groups of 20–55 birds and occasionally, one or two birds; in February an average of 177 quail roosted, arriving in groups of 20–50, with an occasional single or pair; in March an average of 188 quail roosted, arriving in groups of 20–60 (occasional singles or pair); in April an average of 121 quail roosted, arriving in smaller groups of one to five birds; in May an average of 55 quail roosted, arriving mostly in singles and pairs; in June an average of 33 quail roosted, arriving in singles and pairs (young seen); in July an average of 107 quail roosted, arriving in groups, sometimes with young or as singles and pairs; in August an average of 298 quail roosted, arriving in groups upwards of 25 birds; in September, an average of 380 quail roosted, arriving in groups of 15–20 birds (occasional pair and singles); in October an average of 155 quail roosted in this thicket, arriving in groups of up to 100 (some singles and pairs). (It should be noted here that when hunting season opened at the end of October the numbers roosting in the thicket dropped drastically.) In November, an average of 154 quail roosted, arriving in groups of 20–50 birds and in December an average of 208 roosted in the thicket, arriving in groups of 20–50 birds.

Nesting. During the time quail were collected for food habit studies, the ovum of all the females were examined to determine when egg laying began. The earliest evidence of egg laying was of a quail examined March 26, which layed one egg and other ovum were well developed.

A nest with eggs was seen as early as April 30 and a brood of 14 young were seen May 26, 1963. Re–nesting is evidenced by very small young seen July 31, 1960. Other evidence of re–nesting is of young seen as late as September 16 (1941) (Alcorn MS).

Kingery (1984) reported, "At Carson City, Nevada California Quail thrived–200 young hatched in clutches of 16–20 in one small subdivision."

Mountain Quail   *Oreotyx pictus*

Distribution. From southwestern British Columbia to northern Baja California. (Formerly to New Mexico, where bones have been found in prehistoric cavern deposits in the Guadalupe Mountains.)

Status in Nevada. Sparse resident in western part of the state. Uncommon in northern, central and southwestern part of the state. In good quail years, these birds become abundant in selected areas, often followed by a dramatic drop in the population.

Stiver (PC, 1986) reported, "I suspect they are relatively widely distributed, our most widely distributed native quail." He noted they were locally common, but limited populations exist.

Records of Occurrence. Linsdale (1936) reported the quail as sparse residents in the mountains of the western part of the state as far east as the Toiyabe Mountains. At the present writing these quail occur over their former range in limited numbers.

In the lower canyons of the Santa Rosa Range, west of Paradise Valley, these quail were seen in fall months from 1960–1980. One was taken from a flock of about 30 at Solid Silver Canyon in October, 1963 (Alcorn MS).

Stiver (PC, 1986) reported no sightings in the Santa Rosa Mountains for six to eight years.

About 10 miles northeast of Midas, Elko County, about 10 Mountain Quail were seen October 3, 1964. Several hunters reported these birds were quite common in the area in the 1964–68 period (Alcorn MS).

A flock of over 20 were seen southeast of Mountain City near the junction of Meadow Creek and the Bruneau River September 7, 1961 (Alcorn MS).

Wesley O. Baumann (PC) reported in the fall of 1978 these quail were very abundant in the hills about 20 miles north of Battle Mountain. He reported in the following year they had all disappeared, or at least he did not see any.

Orr and Moffitt (1971) reported these quail in the Lake Tahoe region as a "Common summer visitant, though by no means abundant, a few wintering. Slopes heavily covered with manzanita, ceanothus and similar types of dense brush are especially preferred by members of this species."

This author, in company with G. Hammond Hansen, near the northeast side of Lake Tahoe June 22, 1941 saw an adult female with 12 young. The female kept undercover and issued a "clucking" sound at which time the young were seen to work their way through the brush in the direction of the female. The young were about the size of a Marsh Wren. On this same date another adult was flushed from the general area and it ran downhill with wings hanging and quivering and it appeared to be in great distress. Searching around, this author noted a small nest in the ground at the base of a tree. The nest contained three young quail.

Wendell Wheat (MS) saw eight Mountain Quail in the Stillwater Range in west Lee Canyon in August, 1941. Livestock rancher Hammie Kent, frequently rode this area tending his cattle. He reported these quail were plentiful in the Stillwater Range for 15 years prior to the heavy snow of the winter of 1948–49. He said the Mountain Quail were introduced into the Stillwater Range in the 1930s. He estimated 8000–12,000 in this range in peak years. In 1982, the population of these quail in the Stillwater Range was very low.

Near the head of Cow Canyon in the Reese River Canyon, Toiyabe Range, Nye County, 11 of these quail were seen October 4, 1941. None were seen on recent visits to the area by the author (Alcorn MS).

In the Eureka area Mountain Quail were reportedly very abundant "in the early days." Biale (MS) reported Dale Elliott, Nevada Fish and Game agent, checked one that had been shot by a hunter at Big Pole Creek in the Cortez Range in the fall of 1969.

Johnson (1974) reported, "Occurred locally and in very small numbers in the Grapevine Mountains in 1939–40. I did not search the exact spot where species were found by Miller and his co–workers, for example, the brushy areas near springs; the absence is attributable to insufficient field work in proper places. The probability is high that it was not actually missing during the recent period and that virtually all species resident in 1939–40 were also breeding in 1971 and 1973."

Lawson (TT, 1976) commented, "Three mated pair were found at Kilin's Spring, Mount McGruder, southwest of Goldfield in 1976."

Dennis Rechel (PC, 1988) reported six Mountain Quail six miles southwest of Lida October 3, 1987.

Coturnix Quail   *Coturnix coturnix*

Distribution. Europe, west Asia to Africa and India.

Status in Nevada. Introduced.

Records of Occurrence. The Nevada Department of Fish and game released 64 Coturnix Quail at the Mason Valley WMA April 22, 1957.

None have been seen in recent years and the attempt to establish them in this area apparently was a failure.

Order **GRUIFORMES**

Family **RALLIDAE:** Rails, Gallinules and Coots

Yellow Rail   *Coturnicops noveboracensis*

Distribution. Recorded locally from Mackenzie to New Brunswick; south to California and Florida. South in Mexico to the state of Mexico.

Status in Nevada. Accidental in Nevada.

Records of Occurrence. The only Nevada record is of an individual identified by Linsdale (1936) at a pond five miles southeast of Millett Post Office, Smoky Valley, Nye County the evening of May 19, 1932.

Clapper Rail   *Rallus longirostris*

Distribution. Along the west coast from San Francisco; south to Peru. Along the east coast from Connecticut; south to South America.

Status in Nevada. Hypothetical.

Records of Occurrence. A. A. Alcorn and Al Jonez, both employees of the Nevada Fish and Game Department in 1959, reported (MS) seeing eight Clapper Rails at the Las Vegas Sewage disposal drain ditch, Clark County, September 6, 1959.

A. A. Alcorn (MS) later reported seeing a lone Clapper Rail at this same drain ditch September 19, 1959.

Virginia Rail   *Rallus limicola*

Distribution. From British Columbia to Newfoundland; south to Guatemala. Also found in some parts of South America to the Straits of Magellan.

Status in Nevada. Summer resident in marshes, wet meadows and slow–moving streams.

Records of Occurrence. Sheldon NWR (1982) reported the Virginia Rail as a resident spring, summer and fall. The Virginia Rail is also thought to nest in the area.

LaRochelle (MS, 1967) reported seeing one at Ruby Lake NWR in 1963.

Kingery (1980) reported newly found isolated breeding sites were discovered and recently fledged young were seen. He (1983) reported, "Ruby Lakes produced 600 Virginia Rails."

This dramatic increase in numbers is attributed to the censusing technique of using a tape recording of a rail's voice to elicit responses of these rails. The secretive nature of these birds makes it difficult to census them using observation techniques alone.

One Virginia Rail was seen near Minden November 24, 1941 (Alcorn MS).

Linsdale (1951) reported Marriage found this rail nesting along the Carson River in 1949.

Alcorn (1946) reported these birds in the Fallon area as a resident with records for March, April, July, October and December. Since 1960 these birds have been seen in January, February, May, July and September in this area.

Napier (MS, 1974) reported they were commonly heard at the Stillwater WMA from early spring to early winter, although seldom seen. Adult with young have been observed.

Biale (MS) reported a bird at the Sadler Ranch near Eureka May 9, 1971.

One was heard six miles north–northeast of Beatty June 3, 1957 (Alcorn MS).

In Pahranagat Valley one was seen July 1, 1961 (Alcorn MS).

A. A. Alcorn (MS) saw one at the Overton WMA, Moapa Valley, February 7, 1962.

Austin and Bradley (1968) reported, "A specimen was collected by Bradley at Henderson Slough February 27, 1964. Observations were made at the same locality on March 5, 1964 and

April 7, 1965 and at Tule Springs on March 30, 1964. These are the first records for Clark County.''

Austin and Bradley (1971) for Clark County reported, ''Winter resident in riparian areas, occurring from August 16 to May 8.''

Snider (1971) reported September 7, 1970, ''A tape recorder used near Las Vegas attracted at least 25 Virginia Rails. This is many more than we believed to be in this area.''

Heilbrun and Rosenburg (1981) reported a total of 17 on the Henderson CBC December 20, 1980.

Lawson (TT, 1976) commented, ''This rail is a common resident here in southern Nevada and I have seen some very, very young birds out in the Las Vegas Wash area.''

Sora   *Porzana carolina*

Distribution. From southeastern Alaska, east to Newfoundland; south to Central America. Casual in Bermuda, England and the Outer Hebrides.

Status in Nevada. Resident in limited areas.

Records of Occurrence. Sheldon NWR (1982) reported the Sora as a resident spring, summer and fall.

Taylor (1912) reported he found a nest containing 10 eggs in an open marsh near the Quinn River, Humboldt County May 22, 1909.

LaRochelle (MS, 1967) reported only two sightings at Ruby Lake NWR, both in the fall of 1965.

Kingery (1983) reported, ''Ruby Lake produced 1200 young Soras.''

As with other rails, the secretive nature of these birds makes it difficult to census them using observation techniques alone. The dramatic increase in numbers at Ruby Lakes is attributed to censusing techniques using a tape recording of a rail's voice to elicit responses of these rails.

Burleigh collected an immature female specimen, now in the USNM, in the Reno area September 24, 1967.

Linsdale (1951) reported nesting in the northern and eastern parts of Nevada and reported Marriage found nesting along the Carson River in 1949.

Alcorn (1946) reported this bird as resident in the Fallon area with records for January, July, August, October and November. Mills (PC) saw one near Fallon April 30, 1979.

Napier (MS, 1974) for Stillwater WMA reported, ''These birds are common from early spring to early winter. No production observed, but it is fairly certain they nest here.''

Biale (MS) thinks these birds breed in the Eureka area, but he does not have any breeding records. He saw them August 23, 1964; May 25, 1966; September 2 and 3, 1973; and March 9, 1975.

A. A. Alcorn (MS) reported in Moapa Valley, near Overton, two were seen January 26, 1960; one was found dead on the road September 5, 1963; and one was seen March 10, 1964.

Austin and Bradley (1971) reported in Clark County, ''A winter resident in riparian areas. Usually occurs from September 16 to May 15.''

Snider (1971) reported a tape recorder used near Las Vegas attracted at least 15 Soras September 7, 1970.

Lawson (TT, 1976) reported, ''It is a very common resident here in southern Nevada.''

Heilbrun and Rosenburg (1981) reported five during the Henderson CBC December 20, 1981.

Purple Gallinule   *Porphyrula martinica*

Distribution. From the interior of eastern U.S. to Florida; south to Peru and Argentina. Wanders widely but irregularly to Arizona, Nebraska and Newfoundland.

Status in Nevada. A straggler to Eureka and Clark Counties.

Records of Occurrence. Biale (MS) found one Purple Gallinule dying at Eureka November 25, 1965. It was prepared as a specimen.

Austin (1968) reported for southern Nevada, "Two individuals of this species were observed at Tule Springs Park, 13 miles northwest of downtown Las Vegas in 1966. Single birds were observed there on the 7, 8 and 20 of September and a male was collected September 8, 1966."

Lawson (MS, 1981) reported, "There was an individual at Corn Creek in the spring of 1967." A gallinule in breeding plumage was observed by Lawson March 19, 1978.

Common Moorhen   *Gallinula chloropus*

Distribution. Formerly Common Gallinule. In the Americas, Eurasia and Africa. From California to Nova Scotia; south through Mexico and Central America to Chile.

Status in Nevada. Resident in selected localities. Transient statewide.

Records of Occurrence. LaRochelle (MS, 1967) reported one at Ruby Lake NWR July 14, 1962 and one was seen by Lowell Napier May 13, 1971.

Kingery (1977) reported moorhens at Ruby Lake during May and he (1983) reported, "Nesting Common Gallinules recorded for the first time. Ruby Lakes–a nest found June 8, young hatched June 27. (First latilong nesting record.)"

Kingery (1986) reported, "A Common Moorhen had produced a brood at Ruby Lakes Refuge by April 26."

Ryser (1964) reported a single bird in the vicinity of Kleppe Lane in Reno was seen on several occasions in the spring of 1964.

Alcorn (1946) reported these birds in the Fallon area as transient with records for October and August. Since that time, possibly better censusing techniques or a changed population has caused these birds to be reported in all months with young seen in recent years. From 1961 to 1968, inclusive, these gallinules were seen 3 times in January; once in February; 4 in April; 3 in May; once in June; 5 in July; 8 in August; once in September; twice in October and November; and once in December.

Mills (PC) saw a brood of seven young July 2, 1966; two young were seen July 22, 1968. He (PC) also reported six baby gallinules southwest of Fallon April 15, 1986.

Napier (MS, 1974) reported, "The presence of gallinules is marked by the abundance of water. A small number of gallinules regularly nest in the valley (Fallon area). If one wanted to find one, he could do so in a day."

Lee and Marie White (PC) reported about 10 were seen daily four miles west of Fallon in the winter of 1983–84. The author observed two broods in this same area during the summer of 1984.

Biale (MS, 1988) reported none have been seen in the Eureka area.

Kingery (1986) reported, "A Common Moorhen was at Kirch WMA, July 7."

One was seen in Pahranagat Valley May 29, 1965 (Alcorn MS).

J. R. and A. A. Alcorn (MS) reported from one to six of these birds repeatedly seen in the Moapa Valley in March, May, June, July and August.

Kingery (1977) reported Common Gallinules appeared at Corn Creek March 13.

A. A. Alcorn (MS) observed one in Pahrump Valley June 16, 1961.

Austin and Bradley (1966) reported, "A sight record from the southern tip of the county on January 27, 1934 is the only published record for Clark County (Linsdale, 1936). A specimen was collected at the Las Vegas Sewage Plant by Bradley on May 13, 1962. Sight records throughout the year by Austin from Henderson Slough, Twin Lakes (Las Vegas), and Tule Springs indicate this species is an uncommon but regular resident in the Las Vegas Valley."

In southern Nevada Austin (1968) reported, "Downy young and birds in juvenile plumage were observed at Tule Springs Park and Henderson Slough (north of Henderson) during the summers of 1966 and 1967. This species has not been reported previously as breeding in Nevada."

Heilbrun and Rosenburg (1978) reported 107 seen in the Henderson area and (1981) they reported 102 seen in this same area December 20, 1980 during the CBC.

Lawson (TT, 1976) commented, "This bird is uncommon to common in Clark County."

American Coot   *Fulica americana*

Distribution. Commonly referred to as the mudhen. From east–central Alaska to Nova Scotia; south to South America.

Status in Nevada. Statewide summer resident; winter resident mainly in southern Nevada.

Records of Occurrence. Sheldon NWR (1982) reported the American Coot as common to abundant spring through fall, uncommon in winter. It also nests in the area.

LaRochelle (MS, 1967) reported at Ruby Lake NWR these birds are very common and numerous from ice–out until freeze–up with greatest numbers late summer. From 800–1900 pair raise broods, averaging 2.1 per brood. Five–hundred are banded yearly as a matter of interest and returns from as far as Lake Chipala, Mexico have been made.

Kingery (1985) reported, "At Ruby Lakes, American Coots have steadily increased over the past three years—20,000 adults and young this year."

Orr and Moffitt (1971) reported these birds on Lake Tahoe as, "Common winter visitant from October into April with very few in the area in summer."

At Idlewild and Virginia Parks in Reno, 1942, these birds were seen 9 times in January; 19 in February; 14 in March; 9 in April; 6 in May; 5 in June; 7 in July and August; 8 in September; 9 in October; 3 in November; and 5 times in December. The largest numbers seen were in winter months when over 100 were seen. In summer months usually only one or two were seen (Alcorn MS).

In the Fallon area from 1960 to 1965, inclusive, these birds were seen 15 times in January; 12 in February; 25 in March; 23 in April; 22 in May; 14 in June; 16 in July; 17 in August; 6 in September; 10 in October; 12 in November; and 15 times in December (Alcorn MS).

Schwabenland (MS, 1966) reported at Stillwater WMA these birds, "Are very abundant in the fall and spring months with up to 160,000 birds in fall and 60,000 in spring. Nesting, up to 4800 young raised in 1965. A few birds will stay on the area through the winter."

Napier (MS, 1974) for the same area reported, "Abundant during most of the year. Freeze–up in winter forces all but a few from the valley. Common nester." He reported a nest with six eggs and another with nine eggs pipping June 5, 1970.

Gabrielson (1949) reported, "Common on Walker Lake, Mineral County, October 7 and 8, 1932."

Biale (MS) reported for Eureka from 1970 to 1973, inclusive, these birds were seen January 4, 1970; 3 times in February; 6 in March; 9 in April; 21 in May; 15 in June; 14 in July; 4 in August; 8 in September; 6 in October; 17 in November; and 18 times in December. He (PC, 1984) reports them as common all year in the Eureka area.

A. A. and J. R. Alcorn (MS) for Pahranagat Valley from December, 1959 to March, 1965, inclusive, recorded these birds 6 times in January; 4 in February; 12 in March; 16 in April and May; 6 in June; 4 in July; 3 in August; on September 19, 1960 (2 seen); 5 in October; 6 in November; and 7 times in December. Numerous broods of young were on Upper Pahranagat Lake June 4, 1961; one nest containing five eggs was seen May 23, 1962 and an adult with 10 young was seen near Hiko June 11, 1963.

J. R. and A. A. Alcorn (MS) saw these birds in the Moapa Valley, Clark County from December, 1959 to February, 1966, inclusive, 14 times in January; 12 in February and March; 8 in April; 15 in May; 11 in June; 3 in July; twice in August; 4 in September; and 5 times in October, November and December.

Lawson (TT, 1976) commented, "It is a very common species here in southern Nevada year round. In the winter time there are enormous numbers of them on Lake Mead, on Lake Mohave and below Davis Dam in the Fort Mohave area."

Family GRUIDAE: Cranes

Sandhill Crane   *Grus canadensis*

Distribution. From northeastern Siberia, northern Alaska and Canada to Nova Scotia (casual) and Quintana Roo; south across North America to northeastern Mexico.

Status in Nevada. Summer resident in northeastern Nevada, transient statewide. According to Drewien, Oakleaf and Mullins (MS, 1975), "Two subspecies of Sandhill Cranes occur in Nevada: the lesser subspecies *(G. c. canadensis)* an arctic and subarctic nester, apparently occurs infrequently as a migrant and the greater subspecies *(G. c. tabida)* as a migrant and summer resident. The type specimen of this subspecies was collected on May 19, 1859 in the Valley of the South Fork of the Humboldt River (Peters, 1925) in Elko County."

Records of Occurrence. Since 1970, Nevada Department of Wildlife personnel have noted small groups during spring and fall in northern Washoe and Humboldt Counties.

Sheldon NWR (1982) reported the Sandhill Crane as occasional in spring and summer; uncommon in the fall. They are thought to nest in the area.

Oakleaf (MS, 1975) reported a pair at Duck Flat, Washoe County May 23, 1973.

LaRochelle (MS, 1967) reported at Ruby Lake NWR these birds are common on the area from early spring until early fall. Up to 20 pair nest on the refuge and raise young.

This author's earliest record for this area was March 23, 1966 when seven were observed and the latest record was of three seen September 29, 1965. Late records of these birds in 1965 were recorded by A. A. Alcorn on this refuge. He saw nine on September 27; heard one October 14; saw three on October 28; and saw five on November 18.

Howard's narrative for Ruby Lake NWR stated, "In 1971 approximately 10 pair utilized the refuge and production is estimated at 10 young raised to flight stage. In 1972, peak spring population was recorded on April 11 as 42 birds. An estimated 15 pair nested on the refuge, and the first chicks were observed on May 16. A total of four chicks from three broods were seen, but only one survived to flight stage."

Drewien, Oakleaf and Mullins (MS, 1975) reported, "Cranes are generally present in Elko County from March through October. Extreme dates are February 25 and November 3 (Walkinshaw, 1973). Most cranes arrive during mid–March. An estimated 500 were in Ruby Valley on March 10, 1974. Personnel of Nevada Fish and Game and Ruby Lake NWR report that nesting usually begins in April."

"Principal nesting areas in Elko County include: Ruby Valley, Lamoille–Starr Valleys, Mary's River, Humboldt River and Independence Valley."

"However, most wet–meadow habitat in Elko County can be expected to receive some crane use and observations show that cranes are widely distributed in this habitat type during summer."

"Fall migration occurs in late September through mid–October. Prior to migrating, cranes flock in Ruby and Lamoille Valleys. Presumably, these birds represent summer residents from the surrounding region. On October 11, 1972, 275 cranes were observed in Ruby Valley while up to 75 have been sighted in Lamoille Valley in recent years."

Kingery (1983) reported, "A survey of Sandhills in northeastern Nevada counted 436 and extended the known nesting range southeast from Ruby Valley about 80 miles. The Elko area had 69 Sandhill Crane nests and Ruby Lake had 20 pair with 15 young." He (1984) reported, "Ruby Lake had four nests among 47 birds present; they produced five young."

Kingery (1986) reported, "Sandhill Cranes reached Ruby Lake by February 23 and some had established territories by March 4." He (1986) later reported, "At Ruby Lake 20–30 Sandhill Cranes summered with three on territory August 11, but they produced no young."

Kingery (1987) reported 15 pair of Sandhill Cranes bred at Ruby Lake NWR.

Oglesby (PC) saw three at the Reno airport January 6, 1959. Ryser (1969) reported one repeatedly seen in the Truckee Meadows, east of Sparks from February 16 to February 20, 1969.

Oakleaf (MS, 1975) observed 10 cranes in the Truckee Meadows near Reno October 4, 1974. He believed all were the lesser subspecies.

Alcorn (1946) reported in the Fallon area these birds were frequently seen in March and April, the only months it was recorded. More recent records for this area are of two seen doing a "snake dance" by Mills, four miles west of Fallon on March 14, 1965 and three on October 1, 1968. He also saw four in a field south of Fallon October 2, 1973. Rancher Charles Frey (PC) saw six on his place eight miles south–southeast of Fallon on February 15, 1969.

Thompson (PC, 1988) reported about 10 Sandhill Cranes were seen flying over a ranch near Fallon October 22, 1987.

Biale (MS) in the Eureka area has seen these birds on May 7, 1967, May 9, 1971 and he saw 51 flying over Eureka on November 23, 1983.

Drewien, Mullins and Oakleaf (MS, 1975) reported, "Transient cranes have also recently been noted in Eureka County where Drewien sighted 47 on March 1, 1974."

A pair seen at Comins Lake, south of Ely June 5, 1962 may have been nesting there (Alcorn MS).

Kingery (1986) reported, "Kirch WMA saw 1600 in February in an erratic migration."

Oakleaf (MS, 1975) reported a single crane in the Spring Valley, White Pine County May 23, 1973. He said, "Migrant cranes stop annually each spring and fall in the White River Valley near Lund, White Pine County. They are present from late February through mid–March and again in late September and October. In March 1973, 1003 were counted by Nevada Department of Fish and Game (A. Bassac, PC) and over 600 were observed February 28, 1974 by Drewien and Oakleaf. Both observations and mid–toe track impression measurements obtained in February, 1974 indicate that cranes using the Lund area probably belong to the greater subspecies."

Oakleaf (MS, 1975) reported, "A pair summered in 1973 near Dyke Lake, Pahranagat Valley."

"In addition to the Lund area, the migration route apparently includes the Pahranagat Valley and Hiko area, Lincoln County; although only small numbers usually stop. Refuge manager L. D. Hill, Pahranagat NWR, reports that numerous flights pass over the area each spring and fall but few birds land on the refuge. Hill estimated that over 1000 cranes migrate annually through this area."

"Cottam (1936) also reported migrating cranes in the Pahranagat Valley noting a flock of 15 near Alamo in early November, 1924."

At the Overton WMA in Moapa Valley, three were seen October 4, 1960 and three more March 5, 1961 (Alcorn MS).

Kingery (1977) reported 650 cranes at Lund February 24 and 120 passed through Pahranagat in February. He (1978) reported, "A flock of 80 flew over the Dead Mountains in southern tip of Nevada on January 26; they wintered in nearby Arizona."

Lawson (TT, 1976) commented, "It is an uncommon fall migrant and a common spring migrant. In the last week of February and the first week of March, they sometimes come through in small flocks of 25 to 50 birds, other times they are in flocks of 175 to 300 birds."

Order CHARADRIIFORMES

Family CHARADRIIDAE: Plovers and Lapwings

Black–bellied Plover   *Pluvialis squatarola*

Distribution. Breeds on the arctic tundras from Alaska eastward to Siberia; winters south to South Africa and South America.

Status in Nevada. An uncommon migrant in Nevada.

Records of Occurrence. Linsdale (1951) reported, "Transient. Cogswell (MS) saw one at Little Washoe Lake on July 26, 1950. Alcorn (1941) reported a specimen taken at Soda Lake, Fallon area on September 25, 1940 and others observed two days later. He recorded (1946) another specimen obtained in the same area in October, 1943 and Marshall (1951) recorded flocks there on April 10, April 18 and May 16."

Pete Herlan (PC) collected one Black–bellied Plover at Washoe Lake April 2, 1964.

In the Lahontan Valley (Fallon area) the following records are available: about 20 were seen April 19, 1961; Napier (MS) saw one November 17, 1966; Mills saw five on April 17, 1967; Jerry Alberson (PC) reported about 100 seen April 20, 1971; 204 were in a bare dry alfalfa field south of Fallon April 21, 1971; 394 were in two adjoining alfalfa fields April 24, 1971; and over 400 were scattered out over a field seven miles southeast of Fallon on May 4, 1971 (Alcorn MS).

Biale (MS) reported one of these birds at Bartine's Ranch in the Eureka area April 19, 1970 and May 4, 1975. He (PC, 1986) reported two seen in the same place May 4, 1986.

A. A. Alcorn (MS) saw one in the Moapa Valley April 27, 1962 and five on May 12, 1964.

Grater (1939) reported 10 along the shore of Lake Mead, near the site of Saint Thomas on May 6, 1938.

Snider (1970) reported one at Las Vegas on April 20, 1970 and she (1971) reported one photographed at Overton September 13, 1970.

Lawson (TT, 1976) commented, ''The Black–bellied Plover is a rare to uncommon migrant in spring and fall in southern Nevada. I think there were three individuals at Overton this spring and you would expect to see anywhere from four to seven birds here. Most generally at Las Vegas, Lake Mead or Overton.''

Kingery (1986) reported, ''An oddly–timed one appeared at Las Vegas July 5–6.''

Lesser Golden–Plover   *Pluvialis dominica*

Distribution. Formerly the American Golden–Plover. From the arctic coast of Alaska; south to Argentina. Rarely to the interior of North America. Also, winters from India to New Zealand.

Status in Nevada. Rare migrant.

Records of Occurrence. Kingery (1983) reported, ''An American Golden–Plover at Kirch May 18–21 provided the first eastern Nevada record.''

Ryser (1970) listed this bird, ''Of accidental or irregular occurrence in Nye and Clark Counties.''

Snider (1970) reported one in breeding plumage at Las Vegas, May 16, 1970.

Lawson (1977) reported them at Las Vegas Wash, one male immature collected October 3, 1973 (one of 2 present); one at Mormon Farm on May 16, 1970; one at Overton WMA on April 13, 1975; and one at Pahranagat NWR on October 7, 1976.

Lawson (MS, 1981) reported, ''I would say that this bird is a rare migrant here in southern Nevada, being slightly more common in the fall than in the spring.''

Kingery (1981) reported, ''American Golden–Plover dropped in at Las Vegas on August 10, 1980, the first fall record in seven years.'' He (1982) reported Golden–Plover were at Las Vegas in September, 1981.

Snowy Plover   *Charadrius alexandrinus*

Distribution. From Washington to Kansas; south through Mexico to Venezuela. Also found from England to Siberia and Japan and south to South Africa.

Status in Nevada. Uncommon summer resident and transient.

Records of Occurrence. Evenden (1952) reported the first evidence of this species nesting in Nevada when he reported a downy young seen with a flock of 35–40 adults at the Fernley Sink, Lyon County on August 30, 1949.

Kingery (1987) reported, "Carson Lake near Fallon had 15 on October 10."

In the Fallon–Stillwater area from 1960 to 1963, two of these birds were seen March 27, 1961; from one to eight Snowy Plovers were seen a total of 10 times in April; 5 times in May; twice in June; 6 in July; and four were seen August 23, 1960 (Alcorn MS).

Napier (MS, 1974) reported Snowy Plovers at Stillwater WMA from April into November, with small numbers present during the breeding season.

Marshall and Alcorn (1952) reported a nest containing three eggs June 19 and on June 21, 1951 two additional nests containing three eggs each were seen in the northeast corner of the Stillwater Marsh.

Two adults and a nest with three eggs were seen in this same area July 4, 1971. Two adults with three young about one–third grown were seen on Four–Mile Flat about 20 miles east–southeast of Fallon July 10, 1971. Adults with young have been seen in this same area in each summer since 1971 (Alcorn MS).

Steven J. Herman et al., (1980) reported finding adult birds in the summer for 1980 in the following places: five at Alkali Lake, Washoe County on May 19; 12 at Lower Pitt–Taylor Reservoir, Pershing County on June 20; 112 at Artesia Lake, Lyon County on June 29; and 43 at south end of Walker Lake, Mineral County. He located 671 adults at this location in the last week of June. Nesting was confirmed by eggs and/or pre–flight juveniles for all Nevada sites except at Alkali Lake in Washoe County and at an unnamed pond north of Stillwater, Churchill County.

Biale (MS) reported these birds in the Eureka area on June 28 and July 5, 1964; July 4, 1965; June 19, 1966; July 2, 1967; April 19, 1970; and May 23, 1971. The author's records for the area are of three seen August 28, 1960; four seen July 27, 1961; two seen August 10, 1961; and one on August 15, 1961.

Biale (MS) reported about 40 in Diamond Valley June 22, 1980 and he saw one nest containing two eggs. He (PC, 1986) reported one at Bartine's Ranch May 4, 1986.

Kingery (1983) reported, "Snowy Plovers at Kirch May 1–8 provided the first latilong record."

A. A. Alcorn (MS) saw one at the Overton WMA May 24, 1964.

Grater (1939) reported, "Transient visitant. Observed along the lake near St. Thomas on June 20, 1938; collected from the same area on April 10, 1939."

Austin (1968) reported, "This species is an uncommon transient in southern Nevada. I have records of single birds at Henderson Slough on August 16 and 23, 1966. A female was collected on August 16. These appear to be the first fall records for southern Nevada."

Austin and Bradley (1971) reported for Clark County, "Transient in riparian areas with records on March 22, May 8, June 10 and 20 and August 16 and 23."

Lawson (TT, 1976) reported, "This plover is an uncommon to common migrant in southern Nevada and it is more common in the springtime than in the fall. Around the lakes at Overton on the mud flats it is possible to see from 10 to 15 of these birds in the springtime, in the fall you might see only two or three."

Semipalmated Plover   *Charadrius semipalmatus*

Distribution. From Alaska to Newfoundland; south to Columbia and Chile.

Status in Nevada. An uncommon migrant.

Records of Occurrence. Linsdale (1936) reported the first record of this bird in Nevada was one taken by Streator at Smoke Creek, Washoe County May 11, 1896.

Linsdale (1951) reported, "A recent record is for Truckee Meadows on May 5, 1940, by Christensen and Trelease (1941). On April 27, 1948, Ned Johnson (MS) saw two about a mile east of Reno. In the Lahontan Valley, Alcorn (1946) reported specimens on May 6 and Marshall (1951) has records between August 22 and September 13. At Lake Mead on May 7, 1938, reported by Grater."

Ryser (1968) reported in western Nevada a very few in passage this spring, 1968. Ryser (1970) reported occasional small parties of Semipalmated Plovers during the April migration.

Ryser (1985) reported, "They can be seen quite regularly at the Fernley WMA and Stillwater WMA."

Scott (1968) reported these birds were numerous on the Fernley Marsh and outnumbered the Killdeer on May 1, 1968.

In the Fallon area from 1960 to 1971, inclusive, from one to 18 birds were seen 5 times in April; 11 in May; twice in July; and on September 6, 1961 (2 seen) (Alcorn MS).

Biale (MS) in the Eureka area recorded one bird at Hay Ranch, 11 miles from Eureka May 7, 13 and 14, 1967. He (PC, 1986) reported one at Bartine's Ranch May 4, 1986.

Mowbray (MS, 1986) reported the Semipalmated Plover as uncommon in northeastern Nevada.

Kingery (1986) reported, "Kirch WMA's count of 41 Semipalmated Plovers May 5 topped Regional [Mountain West] totals."

Three were seen near Hiko, Pahranagat Valley on May 4, 1962 (Alcorn MS).

Kingery (1980) reported one at Coal Valley, Lincoln County August 18.

Austin and Bradley (1971) reported for Clark County, "Transient in riparian areas. Usually occur from April 26 to May 15 and from August 16 to September 13."

Lawson (TT, 1976) commented, "This species is an uncommon to common migrant in southern Nevada being found more commonly in the springtime than in the fall."

Killdeer   *Charadrius vociferus*

Distribution. From Alaska to western Newfoundland; south to Peru and Chile.

Status in Nevada. Widespread resident; however, most move out of the northern part of Nevada in mid–winter.

Records of Occurrence. Linsdale (1936) reported these were the most widespread kind of shorebird in the state, occurring in all the valleys and in the meadows in the valleys in the mountains up to 7000 feet altitude.

Sheldon NWR (1982) reported the Killdeer as common spring through fall resident, occasional in the winter.

In Elko County from 1960 to 1968, two of these birds were seen March 23, 1966; others were seen 7 times in April and May; 9 in June; 3 in July; 4 in August; 10 in September; and once October 12, 1965 (Alcorn MS).

LaRochelle (MS, 1967) reported at Ruby Lake NWR these birds were common on the area from spring through fall with a few staying through the winter. "About 25 pair nest on the area."

A nest with four eggs was seen in the Lovelock area April 26, 1961 (Alcorn MS).

Orr and Moffitt (1971) reported for Lake Tahoe area these birds are, "Common in summer and they nest in the area, occasionally noted in the winter."

A specimen in the USNM was taken by Burleigh in Washoe County, north of Lake Tahoe (9000 feet elev.) November 3, 1967.

At Virginia Lake and Idlewild Park of Reno in 1942, these birds were seen February 28; 5 times in March; 6 in April; 8 in May; 7 in June; 10 in July; 7 in August; 9 in September; and 8 times in October (Alcorn MS).

Heilbrun and Rosenburg (1982) reported 37 were seen on the Truckee Meadows CBC December 19, 1981.

Ryser (1965) reported a nest with three eggs close to the ranch building on the Thomas Ranch, Truckee River between Wadsworth and Nixon June 12, 1965.

Their abundance in the Fallon area is indicated by their being recorded from January 1, 1960 to December 31, 1966 a total of 42 times in January; 33 in February; 79 in March; 84 in April; 113 in May; 95 in June; 90 in July; 108 in August; 83 in September; 74 in October; and 39 times in November and December (Alcorn MS).

From three to six miles west of Fallon, nests were seen March 29 (one egg); April 4 (4 eggs); May 5 (4 eggs); May 11 (4 eggs); May 14 (4 eggs); May 24 (4 eggs); May 28 (4 eggs); June 10 (4 eggs); and 4 young May 8, all in 1976 (Alcorn MS).

Mills (PC) reported seeing tracks in the snow in two places near Fallon on December 18, 1972. Killdeer had alighted in the canal, walked a few feet and there they were dead. The snow–covered ground along with below zero Fahrenheit temperatures probably resulted in these birds' inability to find food and they were frozen to death.

Biale (MS) recorded this bird in the Eureka area 176 times between April 27, 1964 and December 24, 1967. Most common in spring, summer and fall. However, it was seen February 12, 17 and 22, 1967.

Kingery (1980) reported one seen near Eureka February 11.

In the Eureka area the Killdeer is not normally seen in November and December. However, from 1964 to 1967 it was seen four times in November and six times in December. From 1970–73 these birds were seen January 3, 1971; 15 times in March; 17 in April; 26 in May; 30 in June; 21 in July; 16 in August; 11 in September; 4 in October; 6 in November; and 10 times in December (Biale MS).

In Pahranagat Valley from 1960 to 1965, these birds were seen twice in January; none in February; 9 in March and April; 13 in May; 5 in June; 3 in July and August; on September 19, 1960 (8 seen); October 4, 1964 (one bird); November 25, 1961 (2 birds); and 3 times in December (Alcorn MS).

J. R. and A. A. Alcorn (MS) recorded this bird in the Overton area of Moapa Valley, Clark County on 51 different days. It was seen in all months. An adult was setting on a nest with four eggs May 24, 1961 and on July 1, 1961 a half–grown young was seen.

In Clark County, Austin and Bradley (1971) reported, "Permanent resident in riparian areas, occurring all year."

Lawson (TT, 1976) commented, "It is a very common species here in Clark County."

Mountain Plover   *Charadrius montanus*

Distribution. From southern Alberta and North Dakota; south into Mexico. Winters in California.

Status in Nevada. An uncommon transient.

Records of Occurrence. Alcorn (1946) reported specimens taken at the Carson Lake area south of Fallon in November, 1940.

Two were seen in the Austin area September 5, 1963 (Alcorn MS).

Biale (MS) reported one seen in the Eureka area December 1, 1967.

Gullion (1952) reported, "On November 1, 1951, a single bird was flushed from the creosote bush—Mohave yucca vegetation at the 3100 foot level on the west slope of the Charleston Mountains, 12 miles southeast of Pahrump, Nye County, Nevada. This seems to be a first record for southern Nevada."

Gullion (1953) reported, "Supplementing an earlier record is an observation of a single bird of this species flushed from the shoulder of U.S. Highway 95 about nine miles southeast of Beatty, Nye County, on October 1, 1952. It was accompanied by a smaller, unidentified shorebird."

Snider (1971) reported one found dead near Las Vegas September 19, 1970.

Lawson (TT, 1976) commented, "Two single individuals were here in Clark County. Both were fall records."

Family **RECURVIROSTRIDAE:** Stilts and Avocets

Black–necked Stilt   *Himantopus mexicanus*

Distribution. From southern British Columbia; south to Peru and Brazil.

Status in Nevada. Summer resident.

Records of Occurrence. Gabrielson (1949) reported, "Six or more birds, obviously nesting, noted on Virgin Creek, Charles Sheldon Refuge, Humboldt County, May 30 and 31, 1948."

Sheldon NWR (1982) reported the Black–necked Stilt as an uncommon migrant in spring; seen occasionally in fall.

LaRochelle (MS, 1967) reported at Ruby Lake NWR only a few sightings have been made of this conspicuous bird. No nesting occurs.

Saake (MS) reported five at Duck Flat, Washoe County on May 28 and 29, 1975.

These birds were observed in limited numbers in the vicinity of Reno, Minden and Yerington each summer from April to August (Alcorn MS).

Alcorn (1946) reported in Lahontan Valley (Fallon area) these birds were frequently seen from the middle of April through August with a sight record for March and one for September. More recently, from 1960 through 1970, the earliest record is of two seen March 24, 1970 and the latest record is of one seen September 3, 1961.

Schwabenland (MS, 1966) reported at Stillwater WMA, "During spring and fall migration

Black-necked Stilts. Photo courtesy of U.S. Fish and Wildlife Service.

population almost reaches 10,000 mark. Never stays through winter. Nesting, up to 2000 young raised in summer.''

At this same refuge, Napier (MS, 1974) stated this bird is a common nester and noted they were present from March to September.

Kingery (1987) reported, ''Black–necked Stilts enjoyed good nesting success at Stillwater, where adults and young swelled the population to 6300 by July 28.''

In the Eureka area this is not a common or abundant bird. Alcorn and Biale (MS) reported over a 9–year period, this bird was seen only twice in April; 6 times in May; 3 in June; once in July; and 3 times in August.

A. A. Alcorn (MS) reported seeing seven and six in the vicinity of Baker, White Pine County May 31, 1967.

Kingery (1986) reported, ''Good numbers (283) of Black–necked Stilts populated Kirch WMA.'' He (1987) reported, ''40% fewer nesting stilts had poor success, a problem related to refuge water levels.''

In Pahranagat Valley one was seen May 11 and two were seen May 17, 1960; two on July 4, 1961; and they were seen June 2, 1962 (Alcorn MS).

In Moapa Valley (Overton area) from 1960 to 1965, inclusive, they were seen 13 times in April; 10 in May; and 5 times in June (Alcorn MS).

Austin and Bradley (1971) reported for Clark County, ''Transient in riparian and aquatic areas. Usual occurrences from April 2 to June 13 and from August 4 to September 30. Unseasonal occurrences in July.''

American Avocet. Photo courtesy of the U.S. Fish and Wildlife Service.

Kingery (1980) reported, "Two young provided the first breeding records for southern Nevada in several years."

Lawson (TT, 1976) commented, "This is a common spring and fall migrant here in southern Nevada. It is also a rare summer resident, with a record of 2–day–old chicks in Las Vegas Wash in 1972."

American Avocet   *Recurvirostra americana*

Distribution. From southeastern British Columbia to Nova Scotia; south to Guatemala.

Status in Nevada. A common summer resident.

Records of Occurrence. Sheldon NWR (1982) reported the American Avocet as an uncommon spring, summer and fall resident. It nests in the area.

Saake (MS) reported 100 at Duck Flat, Washoe County May 28–29, 1970.

LaRochelle (MS, 1967) reported at Ruby Lake NWR this is a common bird arriving in early summer and numbering 200–400 individuals.

Saake (MS) reported about 20 of these birds seen on an aerial survey of the Humboldt Sink February 22, 1971.

These birds are common in summer where marshes and shallow ponds occur in the Reno, Minden and Yerington areas from April through July (Alcorn MS).

Kingery (1978) reported, "The Alves found 8000–10,000 American Avocets April 9 in the Fernley Marshes between Reno and Lovelock. 'The marshes extend for miles and as far as we could see they were full of avocets. We have *never* heard of this many around.' "

Alcorn (1946) reported for the Lahontan Valley (Fallon area), "A summer resident; abundant. Numerous sight records available from the middle of March through October. Also seen frequently in the Carson Sink in November. There is one December record. Five nests containing eggs were seen May 8, 1941. Another nest containing four eggs was seen May 30, 1941."

In the Fallon area from January, 1960 to December, 1966 about 100 were seen February 28, 1963 and 12 were seen February 28, 1964. They were also seen a total of 23 times in March; 22 in April; 29 in May; 20 in June; 31 in July; 37 in August; 21 in September; and 26 times in October (Alcorn MS).

Schwabenland (MS, 1966) reported for Stillwater WMA, "The most abundant shorebird, as 10,000 or more birds have been on the area in late summer. Has been known to winter, otherwise arrives in early spring and leaves in late fall. Nesting raises up to 4000 young in summer."

Napier (MS, 1974) reported for this same refuge, "Occurs each month of the year, fairly hardy bird forced out by freeze–up. Generally arrives about February and last depart late in November, early December. Abundant nester. In past eight years (1967–74) population fluctuated, but in 1974 production dropped significantly from loss of habitat."

Kingery (1986) reported early arrivals at Stillwater WMA February 26, 1986. He (1987) reported 51,000 American Avocets at Stillwater WMA July 28, "Even though they had poor nesting success."

The combined records of Biale and Alcorn (MS) for the Eureka area reveal over a 12–year period, these birds were seen twice in March; 14 times in April; 45 in May; 8 in June; 12 in July; 10 in August; and 4 times in September. At one time there were over 30 nests at Bartines Ranch west of Eureka.

Saake (MS) reported three at Railroad Valley WMA May 28 and 29, 1975.

Kingery (1986) reported, "Kirch had 33 nesting pair on June 23, its peak—a drop from last year, but they had better success this year."

In Pahranagat Valley 28 were seen May 11, 1960; one March 8, 1961; one May 5, 1962; 10 on May 25, 1962; about 10 on June 2 and 11, 1963; and in 1965 from one to 30 were seen once in March and 3 times in April and May (Alcorn MS).

In Moapa Valley (Overton area) from 1960 to 1965, inclusive, these birds wee seen 9 times in March; 8 in April; 7 in May; 4 in June; twice in July; and 3 times in August (Alcorn MS).

Austin and Bradley (1971) reported for Clark County, "Transient in riparian and aquatic areas. Usual occurrences from March 10 to June 9 and from July 21 to October 4. Unseasonal occurrences on July 9 and December 1."

Kingery (1980) reported, "Southern Nevada had its first nesting record of American Avocets, two pair with young near Las Vegas."

He (1984) reported, "The American Avocet January 11 gave southern Nevada its first winter record."

Lawson (TT, 1976) reported, "The avocet is a common spring and fall migrant. The fall migration from mid–July through October." He also reported having seen 300–400 in a flock at Lake Mead and large flocks in Las Vegas Wash and Moapa Valley.

Family **SCOLOPACIDAE:** Sandpipers, Phalaropes and Allies

Greater Yellowlegs   *Tringa melanoleuca*

Distribution. From Alaska to Newfoundland; south to southern South America.

Status in Nevada. Transient and winter visitant.

Records of Occurrence. Linsdale (1951) reported, "Transient; winter visitant to southern end of the state."

Sheldon NWR (1982) reported the Greater Yellowlegs as an occasional spring and fall migrant.

This author has seen Greater Yellowlegs in many areas of Nevada. One was seen near Elko April 12 and 13, 1966; one was seen near Lovelock December 12, 1960; and one was at Reno October 16, 1962. It was recorded in the Minden area in February, March, September, October and November; and in the Yerington area in March and April (Alcorn MS).

Alcorn (1946) reported for Lahontan Valley (Fallon area), these birds were seen in all months, except February and May. From January of 1960 to September of 1966, inclusive, these birds were recorded 7 times in January; on February 28, 1963; on March 30, 1960; 6 times in April; 3 in May and July; on August 1, 1966; 4 times in September; 6 in October; and 4 times in November and December.

Napier (MS, 1974) reported at Stillwater WMA these birds were recorded during most months, usually not present in June and July. Sightings were generally of solitary birds or small groups of up to five individuals.

At Bartine's Ranch west of Eureka, two were seen August 28, 1960 and three on August 29, 1961 (Alcorn MS). Biale (MS) reported this bird at this same location April 19, 1970.

Kingery (1986) reported, "A Greater Yellowlegs stopped at Eureka, Nevada, fugitive from a late snowstorm."

Miller and Baepler collected a Greater Yellowlegs at Peavine Reservoir, Nye County April 21, 1973.

Kingery (1986) reported, "Greater Yellowlegs set their usual mystifying dates: one to two at Kirch June 23–July 7."

In Pahranagat Valley three were seen April 27, 1961 and in Moapa Valley from one to three were seen on three different days in March, 1962 (Alcorn MS).

Grater (1939) reported, "Transient visitant. Recorded at St. Thomas in May, September and November, 1938.

Austin and Bradley (1971) reported for Clark County, "Winter resident in riparian areas. Usually occurs from August 4 to April 28. Unseasonal occurrences in May and on June 30."

Lawson (TT, 1976) reported, "The Greater Yellowlegs is a common migrant in southern Nevada with an occasional bird wintering over in the Las Vegas Wash."

Lesser Yellowlegs   *Tringa flavipes*

Distribution. From Alaska to Quebec; south to Argentina.

Status in Nevada. An uncommon transient.

Records of Occurrence. Linsdale (1936) reported this bird as, "Transient; recorded on the basis of two females reported by Ellis (1935) as in his collection from west side of Ruby Lake, three miles north of the Elko County line, taken August 2 and 8, 1927."

Sheldon NWR (1982) reported the Lesser Yellowlegs as an occasional spring and fall migrant.

Alcorn (1946) reported for Lahontan Valley, "Transient. Obtained on April 23, 1941 and also taken by Mills on April 24, 1941. Probably more numerous than records indicate."

In the Fallon area these birds were seen over a 10–year period (1960–1970) twice in March; April 21, 1970 (2 birds); 5 times in September; 3 times in October and November; and twice in December (Alcorn MS).

Marshall (1951) reported, "I have but one additional record of this uncommon species, a male taken on July 29, 1949 in the Indian Lakes [Churchill County]."

Thompson (MS, 1988) reported two Lesser Yellowlegs seen in the Carson Lake area January 4, 1987.

Biale (MS) reported a chick seen on the Tonkin Ranch, Eureka County August 7, 1969.

Kingery (1986) reported, "A Lesser Yellowlegs at Kirch WMA February 23."

Gullion et al., (1959) reported for southern Nevada, "One record: one bird on a lawn being watered with a sprinkler at Desert Game Range Headquarters, Corn Creek (2900 feet elevation, Clark County) on July 30, 1951."

One was observed at the Overton WMA in Moapa Valley March 10, 1964 (Alcorn MS).

Austin and Bradley (1971) reported for Clark County, "Transient in riparian areas. Occurred on April 21 and 26, July 30 to September 13. Unusual occurrence on November 13."

Lawson (TT, 1976) commented, "I would say it is an uncommon migrant here in Clark County. Usually see two or three at a place."

Spotted Redshank   *Tringa erythropus*

Distribution. Breeds from Scandinavia to northern Siberia; south from central Russia to Kamchatka. Winters from the Mediterranean region to equatorial Africa.

Status in Nevada. Hypothetical.

Records of Occurrence. Lawson (MS, 1975) reported Nevada's only record of the Spotted Redshank in Vegas Wash August 16–20, 1975. The record was published in *American Birds* (Kingery, 1976).

Solitary Sandpiper   *Tringa solitaria*

Distribution. From Alaska to Labrador; south to Argentina.

Status in Nevada. An uncommon transient of irregular occurrence.

Records of Occurrence. Linsdale (1936) reported these birds were recorded during seasons of migration and irregularly in summer.

Sheldon NWR (1982) reported the Solitary Sandpiper as an occasional spring and fall migrant.

LaRochelle (MS, 1967) reported for Ruby Lake NWR up to 45 use the area in early summer.

Alcorn (1946) reported, "Transient in the Lahontan Valley with a specimen taken by Mills in September, 1940. One was found dead at Soda Lake on July 27, 1941 and others were seen April 25 and June 25, 1941."

In this Fallon area from 1960 to 1965, inclusive, single birds were seen July 23 and September 4, 1960; April 29, 1961; May 3, 1962; and from one to three were seen on five different days in October, 1965 (Alcorn MS).

Biale (MS) reported one at Hay Ranch, west of Eureka April 25, 1971.

Gullion et al., (1959) reported, "Two records for southern Nevada. Four birds in Pahrump Valley on August 5, 1953; and several birds at Pahranagat Lake, August 18, 1954."

Austin and Bradley (1971) reported for Clark County, "Transient in riparian areas. Usually occurs from April 29 to May 2 and from August 29 to September 13."

Lawson (TT, 1976) reported, "This bird is a rare to uncommon transient in southern Nevada. It is possible sometimes to see six or seven in a single day during migration."

Willet   *Catoptrophorus semipalmatus*

Distribution. From Oregon to Nova Scotia; south to northern South America.

Status in Nevada. Summer resident and transient.

Records of Occurrence. Sheldon NWR (1982) reported the Willet as uncommon in spring and fall, occasional in summer. It also nests in the area.

In Elko County these birds were seen in limited numbers from April through September 26 (Alcorn MS).

LaRochelle (MS, 1967) reported at Ruby Lake NWR about 45 arrive in mid–spring and quickly begin nesting. They leave in early fall.

Johnson (1956) reported, "On May 14, 1954, at the north end of Washoe Lake, 5000 feet, Washoe County, Ryser, John S. Spencer and the author saw an adult Willet fly from a nest containing two eggs which was built on a grassy knoll near the lake. To my knowledge this is the first report of a nest of this shorebird."

Willets were repeatedly seen in the Minden area from June, 1941 to April, 1963. Three downy young were seen with an adult on June 28, 1943 (Alcorn MS).

Alcorn (1946) reported the Willet in Lahontan Valley (Fallon area) as, "Summer resident but not common. It is recorded from April to September, inclusive." Records between 1959 and 1965 for this area indicate this bird is present from April through July with an unseasonal record of one seen October 17, 1965.

Thompson (PC, 1988) reported in the Lahontan Valley, "The numbers are not very high. They probably should nest, but we haven't found any nests."

Mills (PC) reported a family group of Willets seen flying by him at Rattlesnake Reservoir on July 7, 1975.

The combined reports of J. R. Alcorn and Biale (MS) of the Eureka area for 10 years recorded this bird 3 times in April; 15 in May; 6 in June and July; and twice in August.

Kingery (1983) reported, "Two Willets stopped at Kirch WMA May 17."

In Pahranagat Valley, near Hiko, three birds were seen May 7, 1965 (Alcorn MS).

Snider (1970) reported four Willets near Las Vegas April 19, 1970.

Austin and Bradley (1971) report for Clark County, "Transient in riparian areas. Usually

Spotted Sandpiper. Photo by A. A. Alcorn.

occurs from April 24 to May 11 and from July 17 to August 25. Unseasonal occurrences on June 5 and July 1.''

Lawson (TT, 1976) commented, ''The Willet is a common spring and fall transient–migrant here in southern Nevada.''

Spotted Sandpiper   *Actitis macularia*

Distribution. From Alaska to Newfoundland; south to South America.

Status in Nevada. Summer resident and transient.

Records of Occurrence. Sheldon NWR (1982) reported the Spotted Sandpiper as an uncommon spring, summer and fall resident, nesting in the area.

In Elko County single birds were frequently seen along the south fork of the Owyhee River as Robert Quiroz and J. R. Alcorn floated down this river July 11 and 12, 1967.

LaRochelle (MS, 1967) reported at Ruby Lake NWR up to 45 use the area, arriving the second or third week of March. He found a nest with five eggs in the spring of 1965 and subsequently saw a female with young.

In the Reno area in 1942, these birds were seen April 29 and 30; 3 times in May; 5 in June; 8 in July; twice in August; on September 10 (2 birds); and lone birds were seen October 1 and 30 (Alcorn MS).

Kingery (1977) reported one at Reno January 26.

Alcorn (1946) reported for the Lahontan Valley, "Summer resident, frequently seen in all months from April to September, inclusive." Recent records are similar to those reported above, except an early record of one seen by Napier (MS) on February 3, 1968; and others seen October 11 and 18, 1964 and one October 4, 1965.

A nest three miles northeast of Fallon was seen July 14, 1959. It was lined with sticks and was situated in the sand at high water level of the reservoir. An adult left the nest at it was approached. The nest was partially under an overhanging bank and contained four spotted eggs (Alcorn MS).

The combined records of J. R. Alcorn and Biale (MS) for 10 years revealed this bird was recorded in the Eureka area 35 times in May; 8 in June and July; 6 in August; and an unseasonal record of one on October 2.

In eastern Nevada, one was seen at Comins Lake June 5, 1962 and A. A. Alcorn (MS) reported one from near Baker May 11, 1967.

J. R. and A. A. Alcorn (MS) reported for Pahranagat Valley from 1961 to 1965 one was seen April 23, 1964; from one to two birds were seen 8 times in May; and lone birds were observed August 9, 1961 and August 8, 1963.

In Moapa Valley from 1962 to 1965, inclusive, these birds were seen 8 times in May and twice in August (Alcorn MS).

In Virgin Valley one was seen August 29, 1963 and in the Pahrump Valley two were seen August 22, 1961 (Alcorn MS).

Cruickshank, for the Henderson CBC, reported one December 21, 1969 and one December 18, 1971.

Austin and Bradley (1971) reported in Clark County, "Summer resident in riparian areas, usually occurs from April 20 to October 20. Unseasonal occurrences on February 8 and December 22."

Lawson (TT, 1976) reported, "I have recorded this species in every month of the year in Clark County and suspect that it may breed somewhere here. It is a common species."

Upland Sandpiper *Bartramia longicauda*

Distribution. Breeds locally from Alaska to southern New Brunswick; south to South America.

Status in Nevada. Hypothetical.

Records of Occurrence. Linsdale (1936) listed this bird on his hypothetical list based on a set of eggs in the British Museum taken at Soda Lake.

Lawson (TT, 1977) commented on Linsdale's 1936 report, "I wrote the British Museum and had them examine that particular set of eggs and they informed me that they are not Upland Sandpiper eggs, but that they are avocet eggs."

Lawson (1977) reported one seen in Las Vegas Wash, near Las Vegas April 19, 1970.

Kingery (1985) reported, "Las Vegas had an Upland Sandpiper May 2."

Whimbrel *Numenius phaeopus*

Distribution. From northern Alaska to Siberia; south to Chile and Brazil. Also, from Mediterranean to China; south to southern Africa and New Zealand.

Status in Nevada. Rare transient.

Records of Occurrence. Ryser (1970) reported this bird of accidental or irregular occurrence in Nevada.

In the Carson Lake Pasture area, southeast of Fallon, two were seen on May 1, 1971 in the company of a small group of Long–billed Curlews (Alcorn MS).

Kingery (1977) reported a single Whimbrel occurred at Fallon May 12.

Thompson (PC, 1986) reported one seven miles west–southwest of Fallon May 20, 1986.

Biale (MS) reported one of these birds in Antelope Valley about 24 miles southwest of Eureka April 6, 1975.

Gullion (1952) reported, "On July 31, 1951, while driving across the Joshua tree–creosote bush desert of northern Clark County, Nevada, I flushed a crippled Hudsonian Curlew [Whimbrel] from the shoulder of the highway about 1/2 mile east of Cactus Springs. The bird, dangling one leg as it flew, moved about 20 yards off the road and settled among the cactus where it remained nearly motionless for several minutes while I carefully examined it with binoculars. The head striping and short bill were so prominent that there can be no doubt about the species. The bird was standing still among the cactus when I left. This is apparently the first record for this species in the state of Nevada."

Snider (1970) reported three near Las Vegas April 19, 1970 and a single with four Long–billed Curlews at Tule Springs Park, Las Vegas April 16, 1972.

Lawson (TT, 1976) reported, "The Whimbrel is a rare migrant, spring and fall here in Clark County. Scovill has photographed this species at Boulder Beach, Lake Mead in late August. Others have been seen at Moapa Valley, Lake Mead, and Mormon Farm area in May, 1975."

Kingery has consistently reported from one to three Whimbrels in the Las Vegas area since 1980.

Long–billed Curlew   *Numenius americanus*

Distribution. From south–central British Columbia to Manitoba; south in western United States to Guatemala.

Status in Nevada. Summer resident in meadows and pastures. Occasional winter bird.

Records of Occurrence. Sheldon NWR (1982) reported the Long–billed Curlew as an uncommon spring, summer and fall resident; nesting in the area.

LaRochelle (MS, 1967) reported at Ruby Lake NWR about 75 pair use the area with the first arriving about the first week of April. No nests have been found but pairs with flightless young have been seen.

Ryser (1973) reported a flock of about 50 were seen south of Reno April 29, 1973.

Alcorn (1946) reported this curlew in Lahontan Valley as, "Present in summer in limited numbers. This bird reportedly nests in the vicinity of Carson Lake. Recorded in April, May, June and July. Also there is one sight record for December 10, 1941."

Kingery (1987) reported a Long–billed Curlew with a bad leg at Carson Lake until November 26.

Mills (PC) reported two young curlews at Sheckler Reservoir June 5, 1975.

In the Fallon area from 1961 to 1966, inclusive, five Long–billed Curlews were seen by Mills February 7, 1966; two were seen March 22, 1966; they were observed 11 times in April;

4 in May; 6 in June; 3 in July; twice in October; 3 in November; and on December 12, 1965 (5 seen) (Alcorn MS).

Mills and Schwabenland (MS, 1966) reported the most seen in one day at the Carson Lake Pasture was on July 12, 1965 when over 100 were seen.

A lack of records for August and September is believed to be due to this author's failure to take a census in the habitat where they occur and not from the absence of these birds during these months.

Mills (PC) saw a downy baby curlew at Willow Lake, Stillwater Marsh June 10, 1973; also he saw 26 adults out in a bare alfalfa field five miles west of Fallon on March 28, 1974; and he reported one group of 45 and another of 12 individuals in a field west of Stillwater April 11, 1975.

Napier (MS, 1974) reported for Stillwater WMA, "Common during the summer. Nesting population probably does not exceed 25 pair. Also present in small flocks during spring and fall."

In the Eureka area, J. R. Alcorn and Biale (MS) recorded these birds on May 16, 1960; May 13 and 31, 1964; May 11, 1965; June 19, 1966; July 4, 1970; and May 6, 1973 when Biale saw 36 at the Hay Ranch. Biale (PC, 1984) reported seeing three adults and a young about a week old at Fish Creek Ranch June 8, 1980.

A. A. Alcorn (MS) repeatedly saw from one to two of these curlews in the Baker area, White Pine County in May and June, 1967 and he thought they were nesting in that area. He also saw them in the Spring Valley east of Ely and thought they also nested in that area.

A. A. Alcorn (MS) in Pahranagat Valley, two miles south of Hiko, saw 10 on April 10, 1965. He also reported one in the Virgin Valley April 19, 1964. In Moapa Valley four were seen April 4 and one was observed August 7, 1960.

Austin and Bradley (1971) reported for Clark County, "Transient in riparian areas. Usually occurs from March 10 to May 20 and August 16. Unseasonal occurrence on January 15 (early) and on June 12 (late)."

Lawson (TT, 1976) reported, "It is a rare to uncommon spring and fall migrant."

Hudsonian Godwit   *Limosa haemastica*

Distribution. From Alaska to Hudson Bay; south to South America and the Falkland Islands.

Status in Nevada. Rare visitor to southern Nevada.

Records of Occurrence. Lawson (1977) reported, "There are no previous sightings of this species in Nevada. I collected two of five birds present May 15, 1976, one male and one female . . . . The presence of these birds had been discovered earlier in the day by Mowbray."

Marbled Godwit   *Limosa fedoa*

Distribution. From Alberta to northern Ontario and along coasts to Peru and Chile.

Status in Nevada. Transient.

Records of Occurrence. Kingery (1983) reported, "Marys River north of Deeth, Nevada had 30 on April 26."

In the Yerington area two were seen March 27, 1962 and one October 14, 1943 (Alcorn MS).

Alcorn (1946) reported for Lahontan Valley (Fallon area), specimens taken or birds seen in April, July, August, September, November and December.

In the Fallon area from 1960 to 1969, inclusive, these birds were recorded 5 times in April; twice in May and June; 3 in July; 4 in August; 3 in September, October and November; and twice in December (Alcorn MS).

Biale (MS) saw one in Diamond Valley May 15, 1983.

Gullion et al., (1959) reported, "Two records: Pahranagat Lake, August 18, 1954, seven birds on a mud flat; and Boulder Dry Lake (with water in it), Eldorado Valley, one bird September 11, 1957."

Snider (1971) reported 20 photographed at Overton September 13, 1970.

Austin and Bradley (1971) reported for Clark County, "Transient in riparian areas. Usual occurrences from August 16 to October 4 and unusual occurrence in June, July 9 and November 11."

Lawson (TT, 1976) commented, "It is an uncommon to common migrant and occasional winter resident in southern Nevada."

Kingery (1983) reported one in Las Vegas July 5. He (1987) reported 75 at Overton April 26.

Ruddy Turnstone   *Arenaria interpres*

Distribution. From northern Alaska eastward to the Bering Sea; winters from central California to South America. Also, the British Isles, the Mediterranean and south to southern Africa and New Zealand.

Status in Nevada. A straggler to Nevada.

Records of Occurrence. Kingery (1979) reported two visited Pyramid Lake August 29 and he (1980) reported two Ruddy Turnstones at Pyramid Lake August 20.

Lawson (MS, 1981) reported two seen at Washoe Lake August 29, 1970 by G. Bolander and Dr. B. Parmeter.

Alcorn (1942) reported one found dead at Soda Lake July 26, 1941. It apparently died, with other birds, of botulism.

Another Ruddy Turnstone was obtained by Dave Marshall, 17 miles northeast of Fallon August 27, 1952. It was prepared as a study skin by A. A. Alcorn (Alcorn MS).

One was seen by Napier and Paullin at Swan Lake in the Stillwater Marsh July 31, 1970 (Alcorn MS).

Thompson (MS, 1988) reported one male Ruddy Turnstone in breeding plumage at Carson Lake May 7, 1986.

Kingery (1987) reported, "22 Ruddy Turnstones at Stillwater July 30."

Austin (1968) reported a specimen collected near Logandale September 13, 1966.

Lawson (MS, 1981) reported seeing one at Bowman Reservoir, Moapa Valley on November 7, 1973.

Kingery (1982) reported two immature Ruddy Turnstones visited Las Vegas September 4, 1981 and he (1983) reported, "Nevada's first Ruddy Turnstone in two years, in alternate plumage, stayed at Las Vegas July 30–August 6."

Red Knot   *Calidris canutus*

Distribution. From Alaska to Victoria Island and Newfoundland; south to South America. Also occurs in the Old World to the arctic mainland.

Status in Nevada. Uncommon irregular transient.

Records of Occurrence. Marshall (1951) collected one of two birds seen at the Carson Sink of Churchill County May 16, 1950.

Richard Mewaldt (MS, 1979) reported at Lake Tahoe, a Red Knot was seen September 23, 1977 and another April 23, 1978. Two were seen at Topaz Lake August 12, 1976.

Osugi saw four Red Knots six miles north of Fallon on May 14, 1973. She promptly reported the sighting and with Napier and the author returned to the small pond and observed these four birds.

Kingery (1975) reported, "Red Knots visited Overton April 17."

Lawson (TT, 1976) reported, "The Red Knot is a rare migrant in Clark County and several records from Lincoln County."

Kingery (1982) reported two Red Knots were seen at Las Vegas April 12. He (1987) reported two others at Las Vegas September 26.

Sanderling   *Calidris alba*

Distribution. Over parts of North and South America and in the Old World from the Mediterranean to China, South Africa and Australia.

Status in Nevada. An uncommon transient in western and southern Nevada.

Records of Occurrence. One was seen in the Yerington area August 16, 1942 (Alcorn MS).

Alcorn (1946) reported for Lahontan Valley (Fallon area) specimens were taken May 10, July 26, September 22, 1941 and May 18, 1942. Others were seen September 22 and October 2, 1941.

Stillwater WMA personnel reported the following records: 15 of these birds October 18, 1950; two on April 28, 1953; and six on August 10, 1954.

Paul Springer (PC) reported two Sanderlings in winter plumage with other shorebirds in a pond west of Hazen September 29, 1978.

Kingery (1979) reported two at Stillwater September 20.

Austin (1968) reported he saw a single bird at Tule Springs Park, Clark County May 1–3, 1966.

Lawson (TT, 1976) commented, "In southern Nevada the Sanderling is a rare to uncommon transient visitor, primarily in the fall. We have had as many as six individuals at Lake Mead although more commonly we see lone individuals at Pahranagat, Moapa, Las Vegas Wash or at Lake Mead."

Kingery (1979) reported one Sanderling at Lake Mead April 4.

Semipalmated Sandpiper   *Calidris pusilla*

Distribution. From northwestern Alaska to Quebec; south to Chile.

Status in Nevada. Hypothetical.

Records of Occurrence. Mowbray (MS, 1986) reported the Semipalmated Sandpiper as rare in southern Nevada. He (PC, 1988) reported it was first seen in southern Nevada August 30, 1970 by Lawson and himself. He commented, "They've been seen almost every year since then, mostly in the fall."

Kingery (1987) reported one Semipalmated Sandpiper in Las Vegas May 8.

Western Sandpiper   *Calidris mauri*

Distribution. On the coasts of Alaska and along Pacific coast; south to Ecuador and Peru.

Status in Nevada. A common statewide spring and fall transient.

These sandpipers are more abundant than records indicate. Due to the difficulty of separating and identifying the various kinds in the field, most wildlife biologist census reports do not list these birds separately.

Records of Occurrence. One was seen at Quinn River near Orovada April 22, 1942 (Alcorn MS).

Ryser (1966) reported "small parties" in the Fernley Marsh in late April and early May, 1966 and he (1968) reported many were in passage in western Nevada during spring, 1968. He also (1970) reported the April migrations were noticeable through western Nevada.

Alcorn (1946) reported numerous records were available in the Lahontan Valley (Fallon area) for May, July and August. A flock of over 20 were seen May 1, 1961.

Schwabenland (MS, 1967) reported at Stillwater WMA these birds are very abundant in spring and fall migration.

Alcorn (MS) in the Eureka area observed these birds at Bartines Pond April 24, 1960.

Biale (PC, 1986) reported these sandpipers are more common in the Eureka area during spring migration than records indicate.

A. A. Alcorn (MS) saw a flock of 15 at Pahranagat Valley April 22, 1981.

Gullion et al., (1959) reported for southern Nevada, "One spring record: one bird, Las Vegas Valley on April 30, 1951. Four fall records, from August 8 (1953, Meadow Valley) to August 18 (1954, Pahranagat Valley)."

Austin and Bradley (1971) reported for Clark County, "Transient in riparian areas. Usually occurs from March 26 to May 12 and from August 4 to October 4. Unseasonal occurrences on June 1 and 30, and on July 9 and 11, in November, and on December 21 and 30."

Kingery (1982) reported 25 at Las Vegas November 11, 1981.

Lawson (TT, 1976) reported, "I have seen enormous flocks of them at Overton, Las Vegas Wash and Fort Mohave, so it is a common to very abundant species in spring and fall in southern Nevada."

Least Sandpiper   *Calidris minutilla*

Distribution. From Alaska to Newfoundland; south along coasts to Peru and Brazil.

Status in Nevada. A common visitor to western Nevada and a winter resident in the southern part. A statewide transient.

Records of Occurrence. Linsdale's (1951) total report is given as follows: "Transient; collected in May, July and September. Marshall (1951) on July 13, 1949, collected a female in the Stillwater marshes. Reported by Grater (1939) in Clark County in April, in May and August to December."

This bird is more common than records indicate. Many of these sandpipers are not easily identified in the field and are not listed as two separate species on census records.

Cruickshank (1970) reported one Least Sandpiper was reported during the Truckee Meadows CBC December 21, 1969.

Ryser (1968) reported in western Nevada many were in passage in spring, 1968.

A small flock of about 10 was seen near Minden April 26, 1944 (Alcorn MS).

Alcorn (1946) reported in Lahontan Valley (Fallon area), specimens were taken or birds seen in April, May, August, September, October, November and December. About 30 were seen three miles east–northeast of Fallon July 23, 1960.

Schwabenland (MS, 1966) at Stillwater WMA reported, "This bird is bunched with the Western Sandpiper when censused, as identification is hard in the field. Between the two species several thousand migrate through this area each fall and spring. Nesting—none observed. A few birds stay through summer."

For the same area, Napier (MS, 1974) reported, "One of 'peep' sandpipers usually grouped along with westerns. Present in large numbers during spring and fall migrations. A few seem to stay during summer months. No known nesting. In winter, a few stay along drains and open ditches."

In the Eureka area, these birds were seen April 24, 1960 (Alcorn MS). Biale (MS) reported them May 9, 1965. He (PC, 1986) reported these sandpipers are more common during spring than his records indicate.

Cruickshank (1970) reported during the Henderson CBC from one to 26 were consistently seen in late December and January.

Austin and Bradley (1971) reported, "Winter resident in riparian areas. Usually occurs from August 4 to May 9."

Kingery (1977) reported 35 to 79 wintering at Davis Dam, and he (1978) reported several dozen wintered at Davis Dam and one to two wintered at Las Vegas.

Lawson (TT, 1976) reported for southern Nevada, "This sandpiper is a common to abundant spring and fall migrant. We have large numbers of these birds just about anywhere there is water—they are found during the spring and fall."

Baird's Sandpiper   *Calidris bairdii*

Distribution. From Alaska to Baffin Island and into northeastern Siberia. Winters locally from El Salvador to Argentina.

Status in Nevada. An uncommon transient.

Records of Occurrence. So few records are available of this bird in Nevada that all of Linsdale's (1936) account is given. "Transient. During the spring migration this sandpiper was found about Pyramid Lake in May, 1868 (Ridgway, 1877). Between August 9 and 14, 1872, Nelson (1875) saw several flocks on the small sandbars along the Humboldt River near Elko and he obtained one specimen. Ridgway (1877) on August 26, 1867, obtained a female on the Humboldt Marshes, Pershing County. A male was shot on August 20, 1932 at the margin of an alkali lake three miles southeast of Rogers in Smoky Valley, Nye County. One other individual was observed there (Linsdale, MS)."

Johnson and Richardson (1952) reported, "On September 14, 1950, female was taken from a group of three at the north shore of Washoe Lake, Washoe County."

Ryser (1973) reported "small parties" were seen migrating through western Nevada in the spring of 1973, "Down compared to the large flights of 1972."

The author (MS) observed Baird's Sandpipers at Four Mile Flat, about 20 miles southeast of Fallon on May 2, 1971 and June 1, 1973.

Monson (1972) reported three at Las Vegas April 28 and two at Tule Springs Park near Las Vegas April 29, 1972.

Cruickshank (1972) reported four on the Henderson CBC December 18, 1971.

Austin (1968) reported, "Hayward et al., gives the only previous report of this species for southern Nevada. There is an unpublished record for Corn Creek on April 22, 1946. I have additional records for these birds 1.5 miles west of Logandale on September 13, 1966 (female) and for one at Tule Springs Park on May 2, 1967."

Austin and Bradley (1971) reported for Clark County, "Transient in riparian areas, recorded on April 27, May 23 and September 13."

Lawson (TT, 1976) reported, "This bird is a rare to uncommon migrant spring and fall."

Pectoral Sandpiper   *Calidris melanotos*

Distribution. From Alaska to Hudson Bay and the arctic coast of Siberia. South to South America, Australia and New Zealand.

Status in Nevada. An uncommon transient.

Records of Occurrence. Burleigh collected a female specimen, now in the USNM, in the Reno area September 30, 1967.

James L. Hainline (1974) reported three were feeding with a small group of Western Sandpipers at Stillwater WMA September 11, 1972.

Austin and Bradley (1968) reported, "There is one record for the Nevada Test Site, Nye County, Nevada (Hayward, 1963). Two were observed at Overton Beach on October 4, 1964, constituting the first record for Clark County and the second record for Nevada."

Snider (1969) reported four Pectoral Sandpipers near Las Vegas on September 22, 1968; Snider (1970) reported three at Las Vegas April 24, 1970; one at Las Vegas May 2, 1970; three near Las Vegas July 26, 1970; six near Las Vegas August 30, 1970; six near Boulder City August 30, 1970; and one to three were near Las Vegas August 13–15, 1971.

A Pectoral Sandpiper specimen, now in the UNLV collection, was taken at Henderson Slough October 11, 1973.

Lawson (TT, 1976) reported, "This sandpiper is an uncommon migrant in southern Nevada, being more common in fall than in spring. You might see as many as eight or 10 birds in Las Vegas Wash and the same number at Moapa.''

Kingery (1977) reported one at Davis Dam December 14.

Sharp–tailed Sandpiper   *Calidris acuminata*

Distribution. Found in Asia; casually to North America.

Status in Nevada. Hypothetical.

Records of Occurrence. Mowbray (PC, 1988) reported the only record of the Sharp–tailed Sandpiper in Nevada is of one seen by him at the sewage ponds near Lake Mead Marina September 18, 1975.

Dunlin   *Calidris alpina*

Distribution. From Alaska to the arctic coast of Siberia; south to Baja California and Florida. Also from the British Isles to China and Japan south to northern Africa.

Status in Nevada. Transient visitor. Most abundant in spring migrations.

Records of Occurrence. Linsdale's (1951) entire report states, ''Transient: four records, three of them in May. A specimen from Smoke Creek, Washoe County and sight record at Pyramid Lake. R. G. Miller shot one in 1947 on Washoe Lake (Richardson, MS). Alcorn (1946) took a specimen on May 2, 1941 at Soda Lake.''

Ryser (1966) reported several ''small parties'' were present at the Fernley Marshes in late April and early May, 1966; he (1968) reported in western Nevada a number were in passage this spring (1968); and he (1970) reported the April 1970 migration was noticeable.

Leroy W. Giles collected a specimen, now in the USNM, at Stillwater WMA on January 14, 1955.

In the Fallon area from 1961 to 1975, these birds were consistently seen in April and May. Most sightings were of three to 12 birds, but 100 were seen at the Carson Lake Pasture May 11, 1971 (Alcorn MS).

Napier (MS, 1974) reported the Stillwater WMA occasionally has flocks of 25 to 50 birds in spring migrations. ''They are present for only several days.''

Snider (1970) reported at Stillwater WMA 200 were seen May 1, 1970.

Kingery (1972) reported a flock of 800 stayed at Stillwater, the largest flock ever seen there, and perhaps in Nevada, April 16–22.

Biale (MS) saw these birds at the Bartine Ranch, west of Eureka, April 25 and May 23, 1971. He (MS) also observed two on an airstrip runway in Eureka May 4, 1975. He (PC, 1986) reported 10 at Bartine's Ranch May 4, 1986.

Snider (1966) reported eight at Overton January 29, 1966.

Ryser (1972) reported four at Las Vegas July 26, 1972 and he (1977) reported one Dunlin at Las Vegas December 4, 1976.

Kingery (1982) reported Dunlins at Las Vegas in the fall of 1981. He (1984) reported 18 at Las Vegas on April 18 and seven or eight at Kirch WMA April 23–30.

Lawson (TT, 1976) reported, ''The Dunlin is an uncommon to common migrant in southern

Nevada. Normally at each of the various water sites where we go to observe shorebirds we will find one or two individuals during the spring or fall.''

Stilt Sandpiper   *Calidris himantopus*

Distribution. From Alaska to Nova Scotia; south to Argentina.

Status in Nevada. An occasional transient.

Records of Occurrence. Kingery (1979) reported one Stilt Sandpiper at Ruby Lakes August 7.

J. W. Slipp (1942) reported one between Hazen and Mahala, Churchill County, June 30, 1939.

One was reported by Lehman Caves National Monument personnel, White Pine County, August 26, 1957 (Alcorn MS).

Hayward et al., (1963) observed seven at the Nevada Test Site, Nye County October 8, 1961.

Monson (1972) reported one at Las Vegas Wash April 28, 1972.

Kingery (1979) reported one to four at Las Vegas September 10 to October 3.

Lawson (1977) reported, ''I collected one bird 17 September 1976, an immature male . . . . Vince Mowbray observed two birds on two separate occasions, 16 September 1973 and 7 September 1976.''

Lawson (MS, 1981) photographed one August 31, 1977 at Las Vegas Wash, in the tailing ponds.

Buff–breasted Sandpiper   *Tryngites subruficollis*

Distribution. From northern Alaska across Canada to Nova Scotia; south to northern Argentina.

Status in Nevada. Hypothetical.

Records of Occurrence. Kingery (1986) reported, ''The Buff–breasted Sandpiper at Las Vegas August 26 probably will not make the state list, as a single observer report, despite a good description.''

Short–billed Dowitcher   *Limnodromus griseus*

Distribution. From Alaska to Manitoba; south to Peru and Brazil.

Status in Nevada. Rare.

Records of Occurrence. Ryser (1985) says, ''Specimens have been collected in the western part [Great Basin] at the Fernley WMA, Washoe Lake, Lahontan Valley and Goldfield.''

Lawson (TT, 1976) reported, ''This is a rare migrant here in southern Nevada. We do have May records and fall records from late July on.''

Kingery (1987) reported singles at Las Vegas September 26 and 27.

Long-billed Dowitchers. Photo by A. A. Alcorn.

Long–billed Dowitcher   *Limnodromus scolopaceus*

Distribution. From Alaska to Mackenzie into Siberia; south to California along the Gulf coast and south through Mexico and Guatemala.

Status in Nevada. Transient.

Records of Occurrence. Sheldon NWR (1982) reported the Long–billed Dowitcher as a migrant of occasional occurrence spring, summer and fall.

LaRochelle (MS, 1967) reported at Ruby Lake NWR this bird is a rarely seen transient and summer visitor. Kingery (1981) reported, "Ruby Lake had 300–500 in late October, 1981."

Mills (PC) reported three large flocks totaling thousands of birds flying over his place four miles west of Fallon the evening of May 8, 1960. On this same date at six miles west of Fallon, a flock of about 700 birds was seen flying northward in the evening.

In the Fallon area from 1960 to 1963, inclusive, these birds were seen twice in March; 10 times in April; 8 in May; 4 in June; on July 16, 1963 (1000 birds at Carson Lake Pasture); twice in August; on September 10, 1960 (12 birds); 5 times in October; and on November 21, 1960 (10 birds) (Alcorn MS).

At the Carson Lake Pasture, over 1000 were seen July 12, 1965 and Mills (PC) reported seeing about 14,000 at this place August 15, 1966 (Alcorn MS).

Napier (MS, 1974) reported at Stillwater WMA this dowitcher is abundant in large flocks from March to May and August to November. "A few seen through the summer."

Kingery (1987) reported 50,000 Long–billed Dowitchers in Lahontan Valley in September.

Biale and Alcorn (MS) recorded these birds in the Eureka area over an 11–year period 4 times in March; 9 in April; 16 in May; on July 27, 1961 (10 birds); on September 22, 1960 (one bird); and on October 2, 1960 (one bird).

Kingery (1984) reported, "Kirch WMA reported the top count of Long–billed Dowitchers was 600 on May 7."

J. R. and A. A. Alcorn (MS) reported for Pahranagat Valley 10 were seen May 4, 1962; two on March 16; 25 on March 19; and two on May 7, 1965.

J. R. and A. A. Alcorn (MS) reported for Moapa Valley these birds were seen on April 27, 1962; May 4 and 5, 1962 (10 birds); on March 10, 1964 (4 birds); and April 22, 1964 (6 birds).

Gullion et al., (1959) reported for southern Nevada, "Three records: nine birds, Las Vegas Valley on March 26, 1951; many birds, Pahranagat Valley, October 26, 1951; and one bird on the shoreline of Lake Mead, near Overton, August 12, 1953."

Austin and Bradley (1971) reported for Clark County, "Transient in riparian areas. Usual occurrences from March 2 to May 8 and from August 4 to August 30. Unseasonal occurrences on January 6, June 1 and October 4 and 26."

Kingery (1977) reported them from Davis Dam on January 5, 1977.

Lawson (TT, 1976) commented, "It is a common spring and fall migrant and a rare winter resident. Sometimes in mild winters we will have a few."

Common Snipe   *Gallinago gallinago*

Distribution. From Alaska to Labrador; south to Venezuela and Brazil. Also found from the British Isles to Siberia; south to the Philippines and Africa.

Status in Nevada. Resident, common in summer in the north; winter resident in south.

Records of Occurrence. Linsdale (1951) reported this snipe as, "Present at all seasons: most numerous in fall and least numerous in winter. Present in summer, south at least to 39 degrees [latitude] in Smoky Valley."

Sheldon NWR (1982) reported the Common Snipe as an uncommon permanent resident spring, summer and fall with occasional occurrences in winter. Nesting is reported in the area.

In spring and early summer at all wet pastures and meadows from Fallon northwards, these birds were heard making their courtship flights. Farther south, in the Fallon, Austin and Smoky Valley areas, they were heard only at major pastures and meadows (Alcorn MS).

LaRochelle (MS, 1967) reported up to 100 at Ruby Lake NWR at times during the summer. "A few stayed over the winter of 1963–64."

Alcorn (1946) reported in the Lahontan Valley (Fallon area) these birds were, "Frequently recorded each month from October through April of each year. Also seen less often in May, August and September. Not seen in June or July."

In the Fallon area from January, 1960 to December, 1964, these birds were recorded 10 times in January; twice in February; 7 in March; 3 in April; twice in May; on August 31, 1960; on September 9, 1964; 3 times in October; 5 in November; and 4 times in December (Alcorn MS).

In this area they are most abundant in spring and fall migrations. Mills (PC) saw a group of about 30 in a small area five miles west of Fallon March 26, 1974. He reported he has seen young birds in the Harmon Pasture east of Fallon and also at the Carson Lake Pasture south–southeast of Fallon. Mills also reported a nest containing three newly hatched young and one egg on May 31, 1977 at Harmon Pasture. He saw two young at Sheckler Reservoir July 5, 1975.

Napier (MS, 1974) reported at Stillwater WMA they probably nest at the Stillwater Marsh, but none were observed.

Biale (MS) saw these birds in the Eureka area from November, 1964 to 1973, inclusive, twice in January; 6 times in March; on April 25, 1971; twice in May; 3 times in September; twice in October; 10 times in November; and 15 times in December. In 1984 he reported that he suspected nesting in the area, but had not found a nest.

In Pahranagat Valley one bird was seen October 31, 1960; two on March 11, 1962; two on November 30, 1964; one December 18, 1964; one February 16, one April 14, and four May 6, all in 1965 (Alcorn MS).

J. R. and A. A. Alcorn (MS) reported in the Moapa Valley, Clark County two were seen October 7, 1961; two on March 14, one October 1, and one November 6, all 1962; four on January 23, two on January 28 and one February 19, all 1963.

One was seen in the Virgin Valley near Mesquite April 22, 1963 (Alcorn MS).

At Potosi Spring about 30 miles southwest of Las Vegas, one was noted October 3, 1959 (Alcorn MS).

Austin and Bradley (1971) reported for Clark County, "Winter residents in riparian areas. Usually occurs from August 9 to May 15."

Lawson (TT, 1976) reported, "This species is a common migrant, spring and fall, and an uncommon winter resident here in southern Nevada."

American Woodcock   *Scolopax minor*

Distribution. From southern Manitoba to Newfoundland; south to Louisiana and Florida

Status in Nevada. Accidental.

Records of Occurrence. A banded Woodcock, recovered by Frank Dodge along the shoreline of Comins Lake, White Pine County on March 13, 1972 was one of several released by the California Department of Fish and Game just south of Sacramento on January 14, 1972. When found, the bird was presumed to have been dead for at least two days. It was a female with slightly enlarged ovaries in extremely emaciated condition, weighing only 100 grams. The original stock for the California introduction came from Louisiana.

Wilson's Phalarope   *Phalaropus tricolor*

Distribution. From British Columbia to southwestern Quebec; south to Argentina.

Status in Nevada. Summer resident and transient.

Records of Occurrence. Sheldon NWR (1982) reported the Wilson's Phalarope as a common resident spring, summer and fall, with nesting in the area.

Wilson's Phalarope. Photo by Norm Saake.

This is not an abundant or widespread species in Elko County. One was seen May 27 and 28, and on June 16, 1965 in the vicinity of Halleck (Alcorn MS).

LaRochelle (MS, 1967) reported at Ruby Lake NWR small groups of 15 to 45 birds have been seen on three occasions in mid–summer.

Schwabenland (MS, 1966) reported at Stillwater WMA in Lahontan Valley, "For census purposes, Wilson's Phalaropes are bunched with Northern Phalaropes. It is the more abundant of the two birds. During spring and fall migration between 10,000 and 12,000 of the two species have been on the area. Nesting; some must occur, although none has been observed."

Napier (MS, 1974) reported for the same area, "Common from April to October. Abundant during May into June and July. They nest here as I have found nests and eggs."

Alcorn (1946) reported for Lahontan Valley (Fallon area), "Summer resident. Seen frequently from April 15 through August." In the Fallon area from 1960 to 1965, inclusive, these birds were seen 5 times in April; 10 in May; twice in June; 6 in July; on August 23, 1960 (3 birds); and on October 11, 1961 (10 birds). Once in May and twice in July, groups of over 100 birds were observed; otherwise, from three to 10 were seen.

Kingery (1987) reported, "Wilson's Phalaropes at Stillwater reached a peak of 67,000 July 2, an indication of successful nesting."

Biale and Alcorn (MS) over a 12–year period recorded these birds in the Eureka area 6 times in April; 66 in May; 23 in June; 18 in July; 9 in August; and 3 times in September. Biale is sure they nest in the area, but none have been found.

A. A. Alcorn (MS) saw about 30 on a small pond near Baker, White Pine County, on May

4 and eight were seen May 25, 1962; about 10 were seen July 11, 1963; and a lone bird was observed May 31, 1965.

Kingery (1986) reported Wilson's Phalaropes May 5 (1322 birds present) at Kirch, plus another 1000 to 1500 in Railroad Valley.

In Moapa Valley (Overton area) from 1961 to 1964, inclusive, seven were seen August 10, 1961; two on April 27, 1962; five on April 20, five on May 21 and six on May 30, 1963; and 10 or more seen April 22 and 23, 1964. In Pahrump Valley one was seen June 16, 1961 (Alcorn MS).

Austin and Bradley (1971) reported for Clark County, "Transient in riparian and aquatic areas. Usual occurrences from April 23 to June 1 and from August 4 to October 4."

Lawson (TT, 1976) reported, "It is a common migrant spring and fall."

Kingery (1985) reported, "Las Vegas had an unusually high count of 100+ Wilson's Phalaropes all summer." He (1987) reported, "Wilson's Phalaropes declined at Las Vegas and Kirch (a drop off 90–95%), with a peak of 22 August 25."

Red–necked Phalarope   *Phalaropus lobatus*

Distribution. Formerly Northern Phalarope. From Alaska to Bering Sea. Winters at sea in the Western and Eastern Hemisphere.

Status in Nevada. Spring, summer and fall visitant.

Records of Occurrence. Sheldon NWR (1982) reported the Red–necked Phalarope as a rare migrant in spring and fall.

LaRochelle (MS, 1967) reported at Ruby Lake NWR a group of 23 were seen August 29, 1964, the only record for the area.

Ryser (1964) reported, "The overcast night sky of May 13, 1964, rained Red–necked Phalaropes on Reno and its environs. Untold thousands of migrating Red–necked Phalaropes trying to set down, were attracted to ground lights. Flock after flock was attracted to the neon glare of downtown Reno. Phalaropes were wildly flying between buildings and lights and colliding with obstacles by the scores. All the dead phalaropes were seen as far afield as Mount Rose, a peak in the Carson Range to the southwest of Reno."

Alcorn (1946) reported, "Transient in the Lahontan Valley, frequently seen in July, August, September and October. Obtained or seen on three occasions by Mills in May, 1942."

In the Fallon area from 1960 to 1963, inclusive, they were seen on April 29, 1963; 5 times in May; 4 in July; 3 times in August; and 30 or more were seen September 10, 1960. The latest fall date is October 3, 1967 when over 10 were seen.

Kingery (1986) reported, "The sizable count of Red–necked Phalaropes came from Stillwater September 12, with 16,200."

Biale (MS) saw these birds in the Eureka area May 13 and 24, 1964; May 23 and 24, 1965; May 8, 1966; May 4, 1969; May 23, 1971 and May 4, 1986.

At Overton WMA in Moapa Valley, three were seen May 30 and June 1, 1963 (Alcorn MS).

Austin and Bradley (1971) reported for Clark County, "Transient in riparian and aquatic areas. Usual occurrences on May 6 and from August 16 to September 15. Unseasonal occurrence on October 20."

Kingery (1984) reported, "Las Vegas had a late Red–necked Phalarope December 9."

Lawson (TT, 1976) stated, "This is an uncommon to common migrant in spring and fall in southern Nevada."

Red Phalarope   *Phalaropus fulicarius*

Distribution. Circumpolarly along the northern coast and offshore islands of North America, Europe and Asia, mainly above the Arctic Circle. Winters at sea chiefly in the Southern Hemisphere.

Status in Nevada. Accidental and irregular occurrences in southern Nevada.

Records of Occurrence. Snider (1971) reported four near Las Vegas June 5, 1971 and the first specimen record for Nevada was taken at the time.

Ryser (1972) reported on November 1, 1972 at Las Vegas Wash, five or six were seen, including a male still in breeding plumage.

Lawson (1973) reported, "On July 26, 1972 a male Red Phalarope was collected from a flock of about 250 phalaropes at Las Vegas Wash, 10 miles southeast of Las Vegas. The flock consisted of four or five Red Phalaropes in winter plumage, 225 Wilson's Phalaropes and 25 Northern Phalaropes."

Kingery (1976) reported one Red Phalarope at Las Vegas September 18. He (1977) reported, "The fall brought five Red Phalaropes to Lake Mead September 19 and a single to Las Vegas August 7.

Lawson (TT, 1976) commented, "It is a rare migrant spring and fall here in Clark County."

Family **LARIDAE:** Skuas, Gulls, Terns and Skimmers

Parasitic Jaeger   *Stercorarius parasiticus*

Distribution. From northern North America and in the Old World and northern Russia; south offshore in the Pacific and Atlantic Oceans to the coast of Africa.

Status in Nevada. Rare.

Records of Occurrence. Kingery (1978) reported two immature birds October 27 and one November 14 and 26 at Pyramid Lake.

Fred Wright (PC, 1985) reported seeing a Parasitic Jaeger in "intermediate phase" in the Humboldt Sink on August 4, 1953.

Kingery (1986) reported one in Nevada flying along the highway near Carson City September 1."

Kingery (1987) reported, "One well–observed Parasitic Jaeger with long tail streamers at Carson City September 14."

Kingery (1981) reported, "A juvenile Parasitic Jaeger picked up and released in Spring Valley, White Pine County, Nevada in September, 1980 received a definite identification this spring from photographs."

Lawson (TT, 1976) reported one at Pyramid Lake on September 6, 1976. He photographed one at Las Vegas Bay, Lake Mead September 13, 1975.

Lawson (1977) reported in 1976 he saw one at Davis Dam September 17 and 19, and one at Lake Mead November 14.

Long–tailed Jaeger   *Stercorarius longicaudus*

Distribution. Breeds in North America from western Alaska to northern Quebec; in the Old World from Ireland to the Sea of Okhotsk. Migrates through central and eastern United States to South America.

Status in Nevada. Hypothetical.

Records of Occurrence. Kingery (1980) reported one seen at Pyramid Lake October 4, 1979.

South Polar Skua   *Catharacta maccormicki*

Distribution. Breeds in the South Shetland Islands and along the Antarctic Coast. Ranges from the Gulf of Alaska south to California, Hawaii and Japan; in the North Atlantic along the east coast from Massachusetts to North Carolina.

Status in Nevada. Accidental.

Records of Occurrence. Norm Saake of the Nevada Department of Wildlife reported (PC, 1988) a leg and leg band of a South Polar Skua that had been dead for some time were found by Alex Williams at Soda Lake January 1, 1988. The bird had been banded 17 years before in Antarctica.

Saake commented, "Soda Lake gets some oddball records. The bird probably was blown off course in a storm and died there."

Laughing Gull   *Larus atricilla*

Distribution. From Nova Scotia; south to South America. Occasionally to California in winter.

Status in Nevada. Accidental.

Records of Occurrence. Lawson (MS, 1981) reported, "A Laughing Gull was reported and photographed at Davis Dam March 19–20, 1979 with three adult Mew Gulls. This was a first winter bird."

Franklin's Gull   *Larus pipixcan*

Distribution. From east–central British Columbia to Manitoba. South to Chile.

Status in Nevada. Summer resident in western Nevada; transient.

Records of Occurrence. Linsdale (1951) had no record of this bird in Nevada except for a bone reported by Howard (1939) that was found at prehistoric Lovelock Cave, West Humboldt Mountains.

Alberson (PC) reported one seen in a field four miles south–southeast of Fallon on May 8, 1977.

Kingery (1981) reported, "Franklin's Gulls reportedly breed now at Fallon, Nevada, a westward extension of the range of that expanding species."

In the Lahontan Valley at Carson Lake, these birds were repeatedly seen in spring and summer months from 1970 to 1975. The earliest spring record was of four seen April 24, 1971 and the latest record was of 20 seen September 9, 1971 (Alcorn MS).

At the Carson Lake June 16, 1971 about 100 adults and 50 nests were seen in a marsh area where the water was from one to three feet deep. Bullrush and other vegetation was utilized as a platform for these nests. Some of the nests contained eggs, other contained young and three adults were seen carrying nest material (Alcorn MS).

At this location, Osugi (PC) reported seeing two Franklin's Gull nests June 30, 1975. One contained five eggs and one was empty. They saw four Franklin's Gulls during this visit.

Biale (MS) reported seeing these birds in the Eureka area on May 15, 1960 and April 19 and May 30, 1970.

Kingery (1987) reported two Franklin's Gulls at Dyer May 23.

Lawson (1977) reported this bird was photographed near Las Vegas July 13, 1970.

Monson (1972) reported them at Lake Mead from October 8 through November 17, 1971. He (1972) reported 13 were at Las Vegas Bay April 8, 1972.

Kingery (1979) reported one at Davis Dam March 22 and four at Las Vegas May 3.

Lawson (TT, 1976) reported, "The Franklin's Gull is an uncommon to common migrant here in southern Nevada, being more common in spring than in fall."

Bonaparte's Gull   *Larus philadelphia*

Distribution. From Alaska to Ontario; south in Mexico to Yucatan.

Status in Nevada. Transient.

Records of Occurrence. Kingery (1983) reported, "Summer Bonaparte's Gulls were reported from Elko June 13 and Las Vegas July 15."

Kingery (1978) reported 15 at Pyramid Lake in November and in the same year he reported one at Reno June 3.

Lawson (1977) reported two Bonaparte's Gulls at Washoe Valley in October, 1975.

Alcorn (1946) reported in Lahontan Valley (Fallon area), "Transient with specimens taken in May, June, August and November. Others were seen in May, June, July, September, October, November and December."

In the Fallon area from 1960 to 1971, lone individuals were recorded in April, May, June and November (Alcorn MS).

Kingery (1981) reported, "Bonaparte's Gulls dotted the Region [Mountain West] with three at Fallon on June 19 and one at Hawthorne July 21."

In the Eureka area three were seen May 16, 1960 (Alcorn MS) and Biale (MS) reported one in the Eureka area on May 26, 1968.

Kingery (1983) reported, "Far from water a Bonaparte's Gull winged north along a highway near Baker, Nevada, May 11.'' He (1984) reported, ''Observations of Bonaparte's Gulls were topped by 10 at Kirch WMA May 14.''

Kingery (1986) reported, ''Three immature Bonaparte's Gulls summered at Kirch, spending much of their time on mudflats.''

At Overton WMA, Moapa Valley, about 10 were seen April 22, 1964 (Alcorn MS).

Grater (1939) reported Bonaparte's Gulls as common along the lake shore near St. Thomas May 7, 1938.

Austin and Bradley (1971) reported for Clark County, ''Transient in riparian and aquatic areas. Usually occurs from April 12 to May 15. Unusual occurrences on October 26 and December 30.''

Lawson (1977) has the following records: Bonaparte's Gull, 15 fall records of 52 birds; all in October and November. Also five winter records, four for December and one for January; one at Yucca Lake, Nevada Test Site, October 26, 1951; and one at Pahranagat on May 8, 1964.

Lawson (TT, 1976) commented, ''Uncommon migrant, spring and fall, and a rare winter resident. An occasional individual will over–winter in Las Vegas Wash or at Davis Dam.''

Kingery (1982) reported 30 Bonaparte's Gulls at Overton, Nevada April 24.

Heermann's Gull   *Larus heermanni*

Distribution. From southern British Columbia along the islands of the Gulf of California; locally along the coast from Oregon to Guatemala.

Status in Nevada. Accidental.

Records of Occurrence. Michael Wooten and David B. Marshall (1965) reported, ''On June 19, 1961, an adult Heermann Gull was observed on Anaho Island NWR, Pyramid Lake, Washoe County, Nevada. The bird was watched closely and photographed over a period of several hours . . . . It was found in the midst of a large colony of California Gulls and returned to the same spot each time it was disturbed. Approaching California Gulls were driven off in determined fashion. There was no evidence of a second bird, but it seems highly probable that nesting would have occurred had a suitable mate been available.''

Lawson (TT, 1976) reported, ''Skip Prange observed an immature individual on Lake Mohave last fall.'' He (1977) reported a Heermann's Gull seen at Davis Dam September 17, 1976.

Mew Gull   *Larus canus*

Distribution. From Alaska to Saskatchewan; from the British Isles to Siberia south to the Mediterranean Sea; and in eastern Asia to Formosa.

Status in Nevada. Transient.

Records of Occurrence. Kingery (1981) reported, ''Mew Gulls seen at Pyramid Lake on March 26 were the first record for northern Nevada.''

Kingery (1986) reported, ''A Mew Gull joined Ring–billed Gulls in a Reno park February 9.''

Monson (1972) reported, ''January 9 to February 27, 1972 a Mew Gull was present at Las

Vegas Bay and the city dump. It was photographed. San Diego Museum of Natural History checked the photos and vouch for the authenticity of the record. It is a new species for Nevada."

Lawson (1973) reported an immature Mew Gull resting on the beach at Las Vegas Wash Marina, Lake Mead January 9, 1972. The bird was last seen February 27, 1972.

In November, 1972, another juvenile of this species was sighted at Boulder Beach, Lake Mead by Mowbray.

Kingery (1978) reported, "An immature visited Lake Mead January 5 to February 20, a different one appeared February 26, and one was at Lake Mohave January 25."

Kingery (1979) reported one Mew Gull September 22 and November 1 through 28 and noted that this matched patterns of previous years for southern Nevada.

Lawson (TT, 1976) reported, "For the Mew Gull we have a few fall and winter records."

Kingery (1982) reported, "Las Vegas had a Mew Gull January 30 through February 7." He (1983) reported, "Unusual inland in the spring, four adult Mew Gulls stopped at Las Vegas April 3."

Ring–billed Gull   *Larus delawarensis*

Distribution. From southern British Columbia to Newfoundland; south to El Salvador.

Status in Nevada. Resident in western Nevada, transient statewide. A few winter in the southern part of the state.

Records of Occurrence. A few of these birds have been recorded from all sections of the state. Many wildlife biologists have difficulty in distinguishing the Ring–billed Gull from California Gull in the field. On many census sheets they are lumped together as "gulls."

Sheldon NWR (1982) reported the Ring–billed Gull as an uncommon migrant, spring and fall; occasional in the summer.

LaRochelle (MS, 1967) reported at Ruby Lake NWR from one to three of these gulls arrive two to three weeks later and leave earlier than the California Gulls.

In the Truckee Meadows area these gulls are seen in all seasons (Alcorn MS).

Alcorn (1946) reported for Lahontan Valley (Fallon area), "Recorded in every month except June. Most abundant from October through December. No known nesting colony is situated in this area."

Thirteen nests were seen located on an island in the eastern part of Lahontan Reservoir on May 16, 1959 (Alcorn MS).

Thompson (PC, 1986) reported 150 Ring–billed Gull nests on this small island in Lahontan Reservoir June 18, 1986.

The author, in company with A. A. and M. E. Alcorn visited this same area at Lahontan Reservoir on May 17, 1987. The west island had 50 Ring–billed Gull nests and the east island had about 250 Ring–billed Gull nests (Alcorn MS).

Schwabenland (MS, 1966) reported for Stillwater WMA, "Permanent resident. Common in summer and winter. Has nested in the past, but none observed in recent years."

Napier (MS, 1974) for the same area reported these gulls are present in small numbers during most of the year. "No nesting at Stillwater from 1967 to 1974."

Biale (MS) reported one bird found dead at Bartine Ranch April 24, 1960.

Gullion et al., (1959) reported, "Two records away from Lakes Mead and Mohave (where

Ring-billed Gulls feeding behind a tractor in Lahontan Valley. Photo by A. A. Alcorn.

they are of regular occurrence): about 40 birds on March 31 and three birds on May 23, 1951 in the Pahranagat Valley.''

Austin and Bradley (1971) reported for Clark County, ''Winter resident in riparian and aquatic areas. Usually occurs from August 9 to May 26.''

Lawson (TT, 1976) reported, ''This gull is an abundant winter resident here in southern Nevada. Sometimes as many as 6000 or 7000 individuals, mostly at Lake Mead, Lake Mohave and Davis Dam.''

Food. In winter months, these gulls have been seen feeding at several dump sites in the Fallon area and are at these dumps on a daily basis.

About 500 were present daily at the Nevada Cattle Feeding Company feedlot southeast of Fallon in December, 1974. Others have been seen feeding on dead fish in the canals and limited nuumbers patrol the school yards and parks of Fallon in search of discarded food.

On numerous occasions in this area Ring–billed Gulls have been seen feeding in flight. This behavior, similar on all occasions, is described by the author (Alcorn 1943) as follows: ''Flight was slow and irregular and often birds within a flock fly in opposite directions. Repeatedly it was noted that an individual would cease flapping its wings and sail upward. Then it appeared to hesitate momentarily, as it probably captured an insect and then dropped downward for a short distance as normal flight was resumed. A flock of about 50 gulls were seen feeding in this manner two miles east of Lahontan Dam September 12, 1942. They were flying and feeding from 100 to 500 feet from the surface of the ground. The distance between the two outside individuals of the flock did not exceed 600 feet.''

"Another flock of about 30 gulls was seen feeding in flight 3 1/2 miles west–southwest of Fallon October 5, 1942. Mills obtained two of the gulls from this flock. Examination of their stomachs, gullets, and mouths revealed many insects. These insects were saved and later identified by Herbert T. Dalmat of Cornell University, as ants of the species *Lasius (Acanthomyops) murphyi, Lasius (A.) latipes* and the host of these two *Lasius (Lasius) niger americana*; there also were two stink bugs *(Pentotomidae).*''

"Another flock, consisting of about 100 gulls, was seen four miles west of Fallon October 17, 1942. These were feeding about 60 to 300 feet above the ground. Examination of one gull that was shot revealed ants (queens of *Lasius sp.*) in its mouth.''

California Gull   *Larus californicus*

Distribution. From southeastern Alaska to Manitoba; south across much of the United States to Mexico.

Status in Nevada. Common summer resident in western Nevada, transient statewide. A few winter in Clark County and in recent years, Ryser (1985) reported them overwintering in the Reno area. In the northern part of the state a few individuals have been seen at major bodies of water in summer months.

Records of Occurrence. Sheldon NWR (1982) reported the California Gull as an uncommon spring migrant with occasional occurrences in summer and fall.

LaRochelle (MS, 1967) reported at Ruby Lake NWR from two to seven of these gulls arrived the third or fourth week in March and stay until early fall.

California Gulls have been seen in the Lovelock area eating insects and worms in cultivated fields. Occasionally, they will eat small mammals. One such occurrence was on November 1, 1967 at the Brinkerhoff Range near Lovelock, when about 100 gulls were seen capturing live Meadow Mice flushed from their burrows because of flood irrigating. A gull would pick up a live mouse, take flight and swallow the mouse while in flight (Alcorn MS).

Kingery (1980) reported 167 California Gulls at South Lake Tahoe December 16, 1979.

Heilbrun and Kaufman (1977) reported 307 California Gulls on the Truckee Meadows CBC.

These gulls nest at Anaho Island, Pyramid Lake and on islands in eastern part of Lahontan Reservoir.

Kingery (1977) reported, "California Gulls had 3000 nests at Anaho Island NWR and 1700 nests at Stillwater.''

The abundance of these gulls in the Fallon area is indicated by their being seen 1263 times from 1960 to 1967, inclusive. They forage for food in the many cultivated fields in the area where they eat many insects and worms (Alcorn MS).

The majority of these gulls arrive in late March or early April and are common and abundant until the first part of August. They frequently are not separated by wildlife census takers from the smaller Ring–billed Gulls which are not as numerous in the Fallon area (Alcorn MS).

Marshall (1951) reported 13 California Gull nests on a small island at Stillwater Point Reservoir on the Stillwater WMA June 2, 1950.

Schwabenland (MS, 1966) reported, "Several hundred of these have been seen with

Ring–billed Gulls during the summer. Some birds come to feed from Anaho where they nest. Have nested at Stillwater in the past although not at present.''

Napier (MS, 1974) reported, ''Common in small numbers. No nesting at Stillwater from 1967 to 1974. Abundant nester at Anaho Island with no significant change in population in recent years.''

Thompson (MS, 1988) reported California Gulls nested at Pelican Island, Stillwater WMA, in 1986 and 1987.

Biale (PC, 1984) reported, ''Seven birds were observed at Bartine Ranch, Antelope Valley May 11, 1980.''

In Pahranagat and Moapa Valleys of southeastern Nevada, gulls judged to be mostly California Gulls with an unknown number of Ring–billed Gulls were seen 8 times in January; 7 in February; 23 in March; 6 in April; 11 in May; twice in July, August and October; once in November; and 4 times in December (Alcorn MS).

Austin and Bradley (1971) reported for Clark County, ''Winter resident in riparian and aquatic areas. Usual occurrences on December 21 and May 15. Unseasonal occurrence on July 24.''

Heilbrun and Stotz (1980) reported 15 California Gulls on the Henderson CBC December 15.

Lawson (TT, 1976) commented, ''This gull is a common winter resident in southern Nevada. You might see as many as 50 individuals at Lake Mead, but you could see as many as 200 at Davis Dam. They winter primarily from early November through February with the spring migrations in early March.''

Kingery (1983) reported, ''Las Vegas again this year had very low numbers—half of the next lowest year and a very few first-year birds.''

Nesting. These gulls have nested for many years on one or both of two islands in the eastern part of Lahontan Reservoir, Churchill County. A total of 65 partly grown young were banded at this location June 20, 1939. Sixty–one young were banded on June 22, 1940. Since that time, observations on their nesting activity have been undertaken intermittently up to 1987. On May 12, 1941 there were 172 nests with 388 eggs. On May 26, 1942 there were 380 nests of which 301 contained eggs and 29 contained both eggs and at least one newly hatched young. On July 5, 1942 all young appeared over half–grown. On May 22, 1943 there were 246 nests, all containing from one to three eggs, except one that had two young and a pipped egg. On May 29, 1944 there were 364 nests of which seven contained young or pipped eggs. On May 10, 1946 there were 342 nests that contained 833 eggs; also three newly constructed nests did not yet have eggs.

Thompson (PC, 1988) reported 500 nests at Lahontan Reservoir June 18, 1986. The author (MS) reported over 400 nests in this same area May 17, 1987.

Herring Gull   *Larus argentatus*

Distribution. From Alaska, Canada, Europe and Asia, south to Panama, central Africa and the Philippines.

Status in Nevada. Uncommon occurrence in Nevada.

The author conducting a census of nesting California and Ring-billed Gulls on an island in Lahontan Reservoir, Churchill County. Photo by A. A. Alcorn.

Records of Occurrence. Kingery (1980) reported three Herring Gulls at South Lake Tahoe December 16, 1979.

Heilbrun and Kaufman (1979) reported six Herring Gulls at Truckee Meadows in December, 1978.

Leukering and Rosenburg (1987) reported six Herring Gulls at Carson City December 27, 1986 during the CBC.

Alcorn (1940) reported a specimen collected by Earl J. Alcorn at the north end of Walker Lake, Mineral County November 13, 1939.

Alcorn (1946) reported one taken in the Lahontan Valley (Fallon area) January 13, 1942.

Biale (MS) reported one immature seen at the Eureka dump on August 31, 1982.

Austin and Bradley (1968) reported individuals at Overton Beach December 22, 1962 and February 1, 1964.

Cruickshank (1972) reported 10 Herring Gulls on the Henderson CBC December 18, 1971.

John P. Hubbard (1973) reported 18 Herring Gulls on the Henderson CBC December 16, 1972.

Austin and Bradley (1971) reported for Clark County, "Winter resident in riparian and aquatic areas. Occurred on February 1 and 18 and December 22 and 30."

Heilbrun and Kaufman (1977–78) reported Herring Gulls in Henderson during the CBC December 15, 1979 (3 gulls) and December 20, 1980 (2 gulls).

Lawson (TT, 1976) reported, "This gull is a rare to uncommon winter resident here in southern Nevada."

Thayer's Gull   *Larus thayeri*

Distribution. From Alaska to Greenland; south to California and Florida.

Status in Nevada. Uncommon.

Records of Occurrence. Ryser (1985) reported, "There have been a few winter sight records for Thayer's Gulls from Reno."

Kingery (1986) reported, "First–winter Thayer's Gulls at Reno, Nevada November 23, 1985."

Heilbrun and Kaufman (1978) reported four Thayer's Gulls on the Henderson CBC. They (1980) reported three during the Henderson CBC December 15, 1979 and they (1981) reported one on this same route December 20, 1980.

Lawson (1977) reported the following records: one at Lake Mohave December 4, 1975; two at Lake Mead in November, 1974 and February, 1975; two at Davis Dam (also wintered); one at Mormon Farm April 20, 1975; one at Lake Mead March 10, 1975; two at Overton WMA April 26, 1976; and fall of 1976 records; all reports from Lake Mead are immatures.

Lawson (TT, 1976) commented, "Thayer's Gull is a rare to uncommon winter resident here in Clark County. We will have from one to four individuals winter in the vicinity of Las Vegas Wash at Lake Mead, another couple of individuals on Lake Mohave and maybe one or two at the Davis Dam area."

Western Gull   *Larus occidentalis*

Distribution. From southwestern British Columbia; south along the Pacific coast to Sinaloa and Nayarit in Mexico.

Status in Nevada. Prehistoric record with three recent records.

Records of Occurrence. Howard (1939) reported a bone the size of this species from Lovelock Cave, Humboldt Mountains. (This cave is situated in the West Humboldt Mountains on present day maps.)

Ryser (1985) reported, "There is a sight record of the Western Gull from Virginia Lake in Reno."

Kingery (1981) reported, "A Western Gull stayed briefly at Lake Mead October 31, 1980." On this same bird, Mowbray commented, "In view of the presence of the yellow–legged race on the lower Colorado River and Salton Sea, I would have anticipated that race to show up here, but this bird was very clearly a pink–legged bird with no hint of yellowish or orange."

Lawson (MS, 1981) reported one at Boulder Beach, Lake Mead from December 5, 1980 to January 28, 1981. It was a third year bird and a photograph was taken December 15, 1980.

Glaucous–winged Gull   *Larus glaucescens*

Distribution. From the Bering Sea to southeastern Alaska; south to California and Mexico.

Status in Nevada. Accidental in western and southern Nevada.

Records of Occurrence. Kingery (1979) reported, "A Glaucous–winged Gull banded July 11, 1972 at a location eight miles northwest of Tofine on the west coast of Vancouver Island, British Columbia was found dead one mile from the east shore of Lake Tahoe near Glenbrook, Douglas County, Nevada in late July, 1972. The band was recovered but the skeletal remains were left at the discovery site."

Wesley O. Baumann (PC) reported seeing seven Glaucous–winged Gulls at Lahontan Reservoir, Churchill County in the summer of 1979.

Monson (1972) reported from October 30 through November 29, 1971, one to three were identified at Las Vegas Bay, Lake Mead for a new Nevada bird. Subsequently, from one to three individuals were observed until February 27, 1972. During November and December, 1972 another juvenile appeared at the same site.

Mowbray (1973) reported in the Las Vegas area a Glaucous–winged Gull was seen December 16–28.

Kingery (1978) reported one to two appeared January 17 to 22, and he (1982) reported that Las Vegas had a Glaucous–winged Gull arrive November 19, 1981.

Heilbrun and Rosenburg (1981) reported one on the Henderson CBC December 20, 1980.

Lawson (1977) reported these records for the Glaucous–winged Gull: one at Davis Dam January 9, 1976; two at Lake Mead November 30, 1974; one at Davis Dam February 18, 1975; one at Lake Mojave December 4, 1975; and one at Lake Mead November 14, 1976. He reported, "It occurs in small numbers that is one or two individuals almost every winter—either at Las Vegas Wash or at Fort Mohave in the Davis Dam area."

Glaucous Gull   *Larus hyperboreus*

Distribution. Arctic coast and islands from Alaska to northeast Asia. In America casually to Baja California and Georgia.

Status in Nevada. Hypothetical.

Records of Occurrence. Lawson (1973) reported a juvenile resting ashore at Boulder Beach, Lake Mead November 30, 1972. "The bird was last seen December 16, 1972."

Lawson (TT, 1976) commented, "I don't know of any other record of this gull in Nevada."

Black–legged Kittiwake   *Rissa tridactyla*

Distribution. From Alaska to Newfoundland; south to Baja California. Also in northern Europe, Siberia, Japan and the Mediterranean Sea to northwest Africa.

Status in Nevada. Accidental in western and southern Nevada.

Records of Occurrence. Ryser (1972) reported several at Virginia Lake, Reno in 1972.

Lawson (1977) reported one at Davis Dam on February 17, 1975; one at Lake Mohave December 4, 1975; two at Davis Dam December 10 and February 25, 1976; and one on Lake Mead November 10, 1976.

Heilbrun and Rosenburg (1981) reported seven during the Henderson CBC December 20, 1981.

Kingery (1981) reported, "Several Black–legged Kittiwakes wintered at Lake Mead, peaking at 11 immatures December 9."

Red–legged Kittiwake *Rissa brevirostris*

Distribution. From northern North Pacific Ocean to the Gulf of Alaska; accidental in Yukon, Oregon and Nevada.

Status in Nevada. Accidental.

Records of Occurrence. Lawson (MS, 1981) reported a little girl picked one live bird up at Tule Springs July 3, 1977. The bird died the same day.

Lawson (MS, 1981) also reported a specimen of an immature bird from Las Vegas Wash, Lake Mead, was taken December 8, 1980.

Sabine's Gull *Xema sabini*

Distribution. From Alaska to Greenland and the New Siberian Islands; south to Peru. Also in the Atlantic.

Status in Nevada. Transient in western and southern Nevada.

Records of Occurrence. Ryser (1972) reported several were at Virginia Lake in Reno. Clark et al., (1976) reported one in June, 1974 at this same lake.

Kingery (1978) reported one at Topaz Lake October 6, one October 15, and four on November 4.

Kingery (1982) reported one at Lake Lahontan September 28, 1981.

Lawson (1977) reported a specimen in the University of Nevada Museum of Biology taken at Soda Lake September 26, 1940.

Fred Wright (PC, 1984) reported sighting a Sabine's Gull in Pintail Bay, Stillwater WMA October 7, 1954.

Lawson (TT, 1976) reported one September 13, 1975 at Lake Mead and he (1977) reported the following records: Cottonwood Basin, Lake Mohave, September 15, 1976; Davis Dam, September 17; Logandale, October 6.

Kingery (1982) reported a Sabine's Gull at Las Vegas September 21, 1981.

Lawson (MS, 1981) reported, "One December 9, 1980 at Las Vegas Wash Arm on Lake Mead. Very warm, very late, usually pass through mid–September through mid–October."

Kingery (1986) reported one Sabine's Gull at Las Vegas September 28.

Caspian Tern *Sterna caspia*

Distribution. Found locally in the interior from central Mackenzie to Newfoundland; south to Baja California and Louisiana. Also found in parts of Europe, Asia, Africa, Australia and New Zealand.

Status in Nevada. Summer resident in western Nevada and transient.

Nesting Caspian Terns. Photo courtesy of the U.S. Fish and Wildlife Service.

Records of Occurrence. Information on this bird in Nevada is limited, therefore, all of Linsdale's (1936) account is given as follows: "Ridgway (1877) reported this tern as common in May at Washoe Lake and near Pyramid Lake and in September at Humboldt Marshes. Henshaw (1877) listed it as present in small numbers late in the fall on Washoe Lake. One to three individuals were noted by Hall (MS) at Pyramid Lake on July 6, 7 and 9, 1926."

Sheldon NWR (1982) reported the Caspian Tern as an occasional migrant in spring and fall.

Ruby Lake NWR reported (1981) the Caspian Tern is occasional spring, summer and fall.

Napier (MS, 1974) reported at Anaho Island, Pyramid Lake, there was no known nesting of these terns from 1967 through 1973. However, on May 21, 1974 he saw about 70 of these terns in a gull colony on Anaho Island. "They appeared to be nesting, some adults were sitting on scooped out depressions. Adults were bringing fish to other birds at the site. Upon investigation no eggs were found. No further trips were made to this island, but I feel reasonably sure they did nest."

Kingery (1984) reported, "A Caspian Tern stopped at Carson City, September 9."

Ryser (1970) reported a solitary Caspian Tern hunted on Virginia Lake, Reno and had been present for the last few days of April, 1970.

Alcorn (1946) reported in the Lahontan Valley these birds were frequently seen from April 15 through September. "A few nest each year on a small island in the Lahontan Reservoir." A visit to this island colony of nesting terns on July 12, 1939 revealed about half of the eggs had hatched. At this same island 25 young, swimming in the water, were banded and about 25 nests containing eggs or downy young were observed July 5, 1942.

Marshall (1951) reported a colony of 11 nests at Stillwater WMA June 15, 1949. On June 12, 1950 on an island in the Stillwater Point Reservoir 107 nests were found.

Schwabenland (MS, 1966) for this same are reported, "A few birds have been seen in the spring and summer months. No nesting occurs presently at Stillwater."

Napier (MS, 1974) reported, "About 25 birds arrive at Stillwater in April and are commonly seen in April and May. During the rest of the summer through August a few are occasionally seen."

Kingery (1987) reported 475 Caspian Tern nests at Stillwater. Thompson (PC, 1988) reported 110 nests at Stillwater in 1987.

In the Eureka area one was seen April 7, 1960 (Alcorn MS).

Kingery (1986) reported a Caspian Tern, "Appeared at Kirch April 21, two remained to the end of the season."

In Pahranagat Valley one was seen near Hiko May 4, 1962 (Alcorn MS).

In Moapa Valley (Overton area) one was seen May 4, 1962 and two were observed May 30, 1963 (Alcorn MS).

Kingery (1975) reported, "Six Caspian Terns visited Overton April 17 with four remaining April 24." He (1983) also reported, "13 Caspian Terns appeared at Mesquite June 15 and two at Davis Dam June 23."

Lawson (TT, 1976) reported, "The Caspian Tern is a rare to uncommon spring and fall migrant in southern Nevada. The two places we find them most are at Lake Mead and at Overton—sometimes six to 12 birds in a flock."

Food. The stomach of a male taken near Fallon April 25, 1945 contained one black bullhead *(Ameriurus melas)* that was 7 1/4 inches in length.

Common Tern   *Sterna hirundo*

Distribution. From Alberta to Mackenzie to Newfoundland, Europe and Siberia; south to South America, Africa and New Guinea.

Status in Nevada. An irregular migrant. Linsdale (1951) had no record of this bird in Nevada. Since 1964, however, these birds have been reported from Pyramid Lake, Washoe Lake, Lake Tahoe, Lahontan Reservoir, Stillwater WMA, Walker Lake and various places in southern Nevada.

Records of Occurrence. Lawson (1977) reported, "On 15 September, David Winkler found one at Washoe Lake, Washoe County and one at Lahontan Lake, Churchill County."

Lawson (1973) reported one Common Tern seen at Stillwater WMA May 18, 1971.

Snider (1971) reported, "The fall of 1971 was a good fall for this tern in southern Nevada. At Las Vegas Bay, 25 in all were observed between September 13 and October 16, with a maximum of 14 on September 18. On April 8, 1972 two were at Las Vegas Bay and on April 23, 1972 four were at Las Vegas Wash."

Lawson (1973) reported, "On August 15, 1972, an adult was collected at the tailing ponds in Las Vegas Wash about 10 miles southeast of Las Vegas. First Nevada specimen. Scovill (1968–69) photographed this species in Iceberg Canyon of Lake Mead (1968) and Lake Mohave (1969)."

Kingery (1977) reported 43 at three locations in southern Nevada August 7 to October 10, including 14 on September 26.

Kingery (1978) reported Common Terns at Las Vegas April 29 and again (1981) he reported, "Las Vegas noted 58 Common Terns waiting out a heavy rainstorm September 9, 1980."

Kingery (1986) reported five Common Terns in Las Vegas June 27.

Forster's Tern   *Sterna forsteri*

Distribution. From southeastern British Columbia to Manitoba; south to Florida and Guatemala.

Status in Nevada. Summer resident in selected localities; transient statewide.

Records of Occurrence. Sheldon NWR (1982) reported the Forster's Tern as an occasional migrant in spring and fall.

LaRochelle (MS, 1967) reported at Ruby Lake NWR about 75 to 120 use the area from late spring through early fall. "Nesting seems likely, but none observed."

Kingery (1983) reported, "Forster's Terns had successful nesting at Ruby Lakes (250 nests)."

Alcorn (1946) reported for Lahontan Valley these terns were summer residents frequently seen in May, June and July. Other records are for April and August.

In the Lahontan Valley from 1960 to 1971, inclusive, these birds were observed from April to July. The earliest record was April 10, 1968 and the latest was July 16. One June 16, 1971 at Carson Lake, numerous nests were seen associated in the Franklin Gulls' nesting area.

Schwabenland (MS, 1966) reported for Stillwater WMA the Forster's Tern, "Is a common summer resident. Up to 1000 birds use the area. Comes in late spring and leaves in early fall. Nesting, 400 young were raised in 1965 in four colonies. Nesting in greater numbers in the 1950s.

Napier (MS, 1974) reported for this same area, "Arrive in April with Caspian Terns. Stay until late September. Breeding population has decreased in recent years, particularly in 1974, due to loss of habitat."

In the Eureka area two were seen May 16, 1960 (Alcorn MS) and Biale (MS) saw these birds June 5, 1966, June 1, 1975 and May 4, 1986.

Kingery (1987) reported Forster's Terns nested at Kirch.

In the Overton area of Moapa Valley four were seen May 5, 1962 (Alcorn MS).

Austin and Bradley (1971) for Clark County reported, "Transient in riparian and aquatic areas. Usually occurs from April 13 to August 4. Unseasonal occurrence on October 4 and 8."

Lawson (TT, 1976) commented, "Forster's Tern is a common migrant, spring and fall, in southern Nevada. We see large numbers of them at Moapa and Lake Mead in the spring. They seem to move through in flocks in the springtime but straggle through in the fall."

Food. In the Fallon area, 15 of these terns were examined to determine their food habits. Five examined July 19, 1942 revealed that each bird contained from one to three largemouth black bass that were from three to 3 1/2 inches in length. Four examined July 3, 1943 revealed that each contained from one to three largemouth black bass that were from 1 3/4 to two inches in length. One examined July 16, 1943 contained five largemouth bass that were from 2 1/4 to three inches long. Five examined June 27, 1946 all contained Sacramento perch from two to three inches in length.

Least Tern   *Sterna antillarum*

Distribution. Locally, from central California to Nova Scotia and south to Brazil; also from the British Isles, to India, China, Japan, West Africa and Australia.

Status in Nevada. Accidental in southern Nevada.

Records of Occurrence. Grater (1939) reported this bird near the site of Saint Thomas May 7, 1938 and at Hemenway Wash June 6, 1938. Both sightings were on Lake Mead, Clark County.

Lawson (MS, 1981) reported one seen August 5, 1966 by Nora Poyser at Las Vegas Wash Arm, Lake Mead.

Kingery (1986) reported, "A Least Tern spent several hours over Las Vegas Sewage ponds and mudflats on July 6; harassed by Cliff and Northern Rough–winged Swallows, it dove into the water several times and caught a fish once."

Black Tern   *Chlidonias niger*

Distribution. From British Columbia to Nova Scotia; south to South America and Africa. Also found in Eurasia.

Status in Nevada. Summer resident in northern and western Nevada, transient statewide.

Records of Occurrence. Sheldon NWR (1982) reported the Black Tern as an uncommon resident in spring and fall, occasional in summer. They are thought to nest in the area.

Along the Humboldt River in the vicinity of Halleck, Elko County, these birds were repeatedly seen in May and June. In the Ruby Marsh they were conspicuous in July (Alcorn MS).

LaRochelle (MS, 1967) reported at Ruby Lake NWR about 125 of these birds use this area at times beginning in late spring and extending through early fall with no known nesting.

Kingery (1983) reported, "Black Terns produced 450 young at Ruby Lake."

Alcorn (1946) reported these birds were frequently seen in the Lahontan Valley from April 30 through August, with nesting in the Carson Lake area each year. A visit to this area June 16, 1971 revealed three to five nests (total of 11 nests) in three colonies of Eared Grebes. No extensive search was made for other nests and the total number nesting in the area is unknown.

Thompson (PC, 1986) reported 20 nests in this same Carson Lake area June 2, 1986. Two of the nests contained one and two eggs.

At the Reese River, west of Austin, four were seen May 16, 1960 and one was observed August 28, 1960 (Alcorn MS).

Biale and Alcorn (MS) recorded this bird in the Eureka area from 1960 to 1973 11 times in May; 6 in June; twice in July; and on August 29, 1961. Biale (PC, 1984) reports these birds are common each summer at reservoirs in this area.

A. A. Alcorn (MS) saw three near Baker May 12, 1967.

In Pahranagat Valley from 1962 to 1965, two were seen May 25, 1962; four were seen August 8 and 9, 1963; two on May 7,1965; and five on May 14, 1965 (Alcorn MS).

Gullion et al., (1959) reported, "Two records: eight to 10 birds in migration over Stonewall

Flat (4500 feet elevation, Esmeralda County), August 4, 1951; one bird Pahranagat Valley, August 18, 1954.''

A. A. Alcorn (MS) reported one in Moapa Valley June 16, 1960 and five on May 10, 1965.

Austin and Bradley (1971) reported for Clark County this bird is, "Transient in riparian and aquatic areas. Usually occurs from May 3 to August 20.''

Lawson (TT, 1976) commented, "The Black Tern is a rare to uncommon migrant, being more common in the springtime than in the fall. They straggle through in the fall, one or two at a time. They come through in small flocks in spring.''

Family ALCIDAE: Auks, Murres and Puffins

Ancient Murrelet   *Synthliboramphus antiquus*

Distribution. From southern Alaska; south in west and central North America to Baja California and Formosa.

Status in Nevada. Accidental in Nevada.

Records of Occurrence. Gullion (1956) reported on November 14, 1955, Frank Lespade found an Ancient Murrelet in the yard of a local lumber dealer in Elko. "Later on the same day it was learned that Mike Coboz, an associate of Lespade's saw another bird of the species on the Humboldt River in Elko.''

V. K. Johnson obtained one specimen from Pyramid Lake March 4, 1961.

Bill Smith (1966) reported, "On 27 November 1965, Albert Taylor captured an Ancient Murrelet, which he had observed floundering about on the main street in Carson City, Ormsby County, Nevada. Attempts to get the bird airborne proved futile . . . . The murrelet was set free overnight, but the following morning Taylor found it dead and frozen in a vacant lot near his home. He promptly presented the specimen to the Nevada State Museum where it was prepared as a study skin.''

One female was found by C. J. Chamberlain in a dry field four miles west of Fallon April 5, 1970. It was alive, but unable to fly when found. This author prepared it as a study skin (Alcorn MS).

Lawson (1977) reported one at Lake Mead March 24, 1964 present until March 26.

Order COLUMBIFORMES

Family PTEROCLIDIDAE: Sandgrouse

Chestnut–bellied Sandgrouse   *Pterocles exustus*

Distribution. Formerly Common Sand Grouse. Throughout northern Africa, Arabia and Syria east to India.

Status in Nevada. Efforts to establish this bird in southern Nevada have failed.

Records of Occurrence. A. A. Alcorn (MS) reported these birds were introduced into the Pahranagat and Moapa Valleys. Near Hiko, Pahranagat Valley on July 26, 1960, 200 of these birds were released. One of these birds was recovered in Moapa Valley in October, 1960. In April and May, 1961 in the same area, 391 were released. In Moapa Valley June 14, 1960, 400 were released on the Overton WMA. In April and May, 1961 a total of 639 Chestnut–bellied Sandgrouse were released in this same area.

A. A. Alcorn (MS) reported, "On May 13, 1961 a flying sandgrouse was observed being pursued by a raven. The raven unsuccessfully tried to down the sandgrouse twice and on a third attempt the grouse fell into a ditch. The raven alighted, carried the sandgrouse to a ditch bank, and began eating it. The raven was frightened away and examination revealed the raven had eaten the crop and neck region of the sandgrouse. I feel that if the sandgrouse had been in good flying condition it would have been able to out–fly the raven. The sandgrouse had been released the previous day, as indicated by band number."

A. A. Alcorn (MS) reported, "During 1960 and 1961, 2030 Common Indian Sandgrouse were released in southern Nevada. These game birds were introduced because of their potential of becoming adapted to the game–void areas of southern Nevada. They were introduced under a federal aid project and the Nevada Fish and Game Commission."

Imperial Sandgrouse   *Pterocles orientalis*

Distribution. From northern Africa, the Middle East and northern India.

Status in Nevada. Introduction of this bird during the early 1960s in an attempt to establish it was unsuccessful.

Records of Occurrence. According to a Nevada Fish and Game report, in 1959–60, four birds were trapped in India; two survived the trip to Nevada. In 1960–61, 37 were captured, 32 of which were shipped from Jodhpur, India. Twenty–three survived the trip to Nevada and were sent to the Nevada Game Farm near Yerington.

A 1962 report by the Fish and Game covering the period July 1, 1961 to June 30, 1962 stated, "The sandgrouse releases in Nevada have all disappeared. Following a recovery of two banded sandgrouse in Sonora, Mexico, it appears the birds released in Nevada have moved substantial distances south."

Family **COLUMBIDAE:** Pigeons and Doves

Rock Dove   *Columba livia*

Distribution. Resident in parts of Europe, Asia and Africa. Now established widely in the Americas and other parts of the world from feral domestic stock.

Status in Nevada. Found statewide, but most common in and near larger towns and cities.

These birds have been observed in all parts of Nevada. They are most frequently seen in areas where cattle feedlots and other sources of food are available, near nesting sites such as cliffs or large buildings.

Records of Occurrence. D. E. Lewis collected a female specimen, now in the USNM, at Cave Creek, Ruby Lake NWR June 5, 1968.

These birds are well established in the city of Reno where they nest on the ledges of some of the larger buildings. Along the Truckee River from Reno to Wadsworth these birds nest under the railroad bridges and colonies have prevailed there for many years (Alcorn MS).

In the Fallon area they are common in the vicinity of several cattle feedlots and nesting occurs under bridges and at rocky outcroppings in the nearby foothills (Alcorn MS).

Biale (PC, 1984) reported these birds are common but not abundant in the Eureka area.

Lawson (TT, 1976) reported, "They are a very common bird around town (Las Vegas)." He has seen Rock Doves at cliffs in the areas of Lake Mohave, along the Colorado River and Red Rock Canyon.

Band–tailed Pigeon   *Columba fasciata*

Distribution. From British Columbia to Colorado; south in the mountains to Nicaragua.

Status in Nevada. Transient statewide. Uncommon resident in limited areas of southern and western Nevada.

Records of Occurrence. Bill Pogue (PC) reported seeing 15 at Alder Creek, Humboldt County October 6, 1962.

Lewis collected an adult male at the Ruby Lake NWR October 25, 1966 and Howard observed one at this same refuge July 14, 1971.

Alan Foster (PC) saw a flock of about 24 Band–tailed Pigeons near Hunter Lake, Carson Range September 12, 1964.

Others were reported by Ryser (1972) during July and August, as singles, pairs and flocks of 6 to 11. He (1972) reported about 100 birds in back of Bowers Mansion, west side of Washoe Valley.

Ryser (1985) commented, "The largest recorded winter invasion occurred in early 1972; flocks totaling well over a 1000 birds roosted at several sites in the Carson Range, including forested areas to the west of Washoe Valley and along the southeast shore of Lake Tahoe. At their roosts in conifer forests, the ground was littered with feathers and droppings. During the day flocks could be seen flying to the east and southeast to forage in the pinon–juniper woodlands in the Virginia and Pine Nut ranges—the pigeons which the late Pete Herlan and I collected as museum specimens had their crops loaded with pinon pine seeds. Another invasion occurred in the Carson Range in early 1978."

One was seen near Yerington September 3, 1982 at the McGowan residence (Alcorn MS).

Alcorn (1941) reported a lone individual taken four miles west of Fallon October 17, 1940. At the Wolf Ranch, seven miles north–northeast of Fallon, one of these birds was observed eating the fruit of a mulberry tree July 12, 1971. The most recent record by the author is of three seen six miles west–southwest of Fallon May 22, 1983 (Alcorn MS).

Ned Kendrick phoned this author to report a Band–tailed Pigeon eating berries from his honeysuckle bush at his residence in Fallon on July 5, 1972. Thirty minutes later we checked the area and found a lone individual perched on a limb about 30 feet from us (Alcorn MS).

Near Ellsworth, east side of Paradise Range, one was seen September 28, 1967 (Alcorn MS).

Biale (MS) reported one in the Eureka area October 6, 1969, three on June 17, 1973 and one on July 10, 1984.

Curran (PC) reported seeing about 60 of these birds at Cucamunga Springs, McGruder Mountains in the fall of 1969. These birds were hanging around in an area where a good crop of pinon pine nuts was available.

Johnson (1965) reported, "One was noted flying high over Hidden Forest Canyon, 7500 feet on June 5 and juvenile female was taken at Hidden Forest Cabin, 7900 feet on June 11, in the Sheep Range. In the Spring Range, the following records were obtained: an adult was seen in circling flight over a canyon and later heard calling in pine–fir timber three miles north of Charleston Peak, 9500 feet on June 17; an individual was heard calling in scattered ponderosa pine, white fir and juniper in Macks Canyon, 7800 feet on June 21; and a female accompanied by a male was taken from the top of a pine snag in virgin pine–fir in Macks Canyon, 8500 feet on June 24. Although fat, this female was ready to lay. In the Clover Mountains, Lincoln County, two were seen in ponderosa pine and Gambel Oak at 7200 feet on the east side of Ella Mountain on June 27."

Johnson (1970) reported, "A single adult was flushed from the top of a fir at 9000 feet on the north face of West Virgin Peak, Virgin Mountains on May 21, 1970."

Austin and Bradley (1971) reported for Clark County, "Summer resident in woodland and montane forest areas and transient in riparian areas. Usually occurs from May 1 to August 12. Unseasonal occurrence on October 8 and 30, November 1 and December 3."

Kingery (1979) reported two at Las Vegas August 31 and September 5, and one at Lida Junction October 28. He (1979) also reported one at Las Vegas June 23.

Lawson (TT, 1976) reported, "This species is a rare to uncommon resident here in southern Nevada." He (MS, 1981) reported 50–100 in winter in the Spring Range and commented they were probably resident.

Ringed Turtle–Dove   *Streptopelia risoria*

Distribution. Native country unknown; long domesticated throughout the world. Naturalized in a wild state locally in California, Texas and Florida.

Status in Nevada. Uncommon.

Records of Occurrence. Cruickshank (1968) reported one seen during the Truckee Meadows CBC December 23, 1967.

Each summer from 1974 to 1984, these birds have been heard and seen by this author's sister, Velma Watkins, at her residence four miles west of Fallon (Alcorn MS).

Don Johnson (PC) reported seeing two Ringed Turtle–Doves four miles west–southwest of Fallon April 23, 1984 and this author saw one in the city of Fallon June 5, 1984.

Lawson (TT, 1976) commented, "We have seen some banded birds (once captive) that were released in southern Nevada. However, we have also seen one or two which were not banded, which would indicate there is some reproduction here in Las Vegas Valley."

Spotted Dove   *Streptopelia chinensis*

Distribution. From southern China, East Indies and the Philippines. Introduced and established in Los Angeles and other areas of California and Hawaii.

Status in Nevada. Birds are occasionally released or escape in southern Nevada.

Records of Occurrence. Lawson (PC, 1984) reported these birds are brought to southern Nevada as pets by people and they are released or escape. No establishment has occurred.

White–winged Dove   *Zenaida asiatica*

Distribution. From southeastern California, southern Nevada and New Mexico; south to the Rio Grande Valley in Texas. South through Mexico and Central America to northern Chile.

Status in Nevada. Summer resident in the southern part of the state. Stragglers have been reported as far north as Ruby Lake and Fallon.

Records of Occurrence. These birds are normally restricted to the hot bottomlands of the southern part of the state. However, a lone bird was observed by A. A. Alcorn (MS) at Ruby Lake NWR May 22, 1966.

Howard Young shot one of these birds in the Fallon area while hunting Mourning Doves in September, 1946. The wing was saved and later identified by the author.

A. A. Alcorn (MS) saw one near Baker June 4, 1967.

Gullion et al., (1959) reported, "This dove is a common summer resident in the hot, humid bottomlands of the Mohave Valley."

J. R. and A. A. Alcorn (MS) reported seeing these birds in Moapa Valley from 1961 to 1964 a total of 6 times in May; 9 in June; 6 in July; 4 in August; and on September 6 and 21, 1963.

Austin and Bradley (1971) reported for Clark County, "Usually occur from March 12 to August 31."

George T. Austin and Amadeo M. Rea (1976) reported, "Young fledged from several nests in Paradise Valley (Las Vegas) during 1969 and 1970."

Lawson (TT, 1976) reported, "In the extreme southern tip of Nevada it is a very common bird. In the Las Vegas area it is a rare to uncommon bird."

Mourning Dove   *Zenaida macroura*

Distribution. From Alaska to Nova Scotia; south through Mexico and Central America to western Panama.

Status in Nevada. Common statewide, summer resident; uncommon winter resident. The majority migrate into the area in late March to early April and leave in late August or early September. A few winter as far north as Elko County. In most years, hunters harvest over 100,000 Mourning Doves in Nevada.

Records of Occurrence. Sheldon NWR (1982) reported the Mourning Dove as a common resident from spring to fall. This dove also nests in the area.

In Elko County one of these birds was observed March 23, 1966;; others were seen 3 times in April; 11 in May; 10 in June; 5 in July; 11 in August; 18 in September; on October 15, 1965 (3 seen); on November 28, 1965 (4 seen); and one December 12, 1965 (Alcorn MS).

LaRochelle (MS, 1967) reported at Ruby Lake NWR these doves usually arrive in the first half of April and leave in mid–September. In this area, "Young are raised mainly in the juniper and pinon belt along the west edge of the refuge. Several stayed through winter of 1962–63 and 1965–66. Up to 900 were banded each year and most of these were immature birds."

Although a common and widespread summer resident, limited numbers winter in western Nevada. On January 9, 1942 one was seen in Reno; January 10, 1965, 15 were seen in the Lovelock area; and in the Fallon area they have been reported in all winter months of each year (Alcorn MS).

Their abundance in the Fallon area is indicated by their being recorded from 1960 to 1966, inclusive, a total of 22 times in January; 29 in February; 62 in March; 78 in April; 131 in May; 108 in June; 88 in July; 103 in August; 91 in September; 24 in October; and 20 times in November and December. They are usually heard cooing beginning the last week of February. One was heard cooing in Fallon as early as February 18 (1972) (Alcorn MS).

Biale (MS) recorded these birds from 1971 to 1973, inclusive, 14 times in April; 49 in May; 50 in June; 36 in July; 61 in August; 40 in September; and 21 times in October.

In Pahranagat Valley from 1960 to 1965, inclusive, these doves were observed twice in January; once in February; twice in March; 10 times in April; 13 in May; 4 in June, July and August; twice in September and October; and once in December (Alcorn MS).

In Moapa Valley (Overton area) from 1960 to 1965, these birds were observed on February 1, 1963; 8 times in March; 18 in April; 27 in May; 26 in June; 23 in July; 20 in August; 18 in September; 4 in October; and on November 4, 1961 (Alcorn MS).

Austin and Bradley (1971) reported for Clark County, "Occurs from March 3 to November 24. Unseasonal records on February 6 and mid–December."

Lawson (TT, 1976) reported, "The Mourning Dove is an extremely common to abundant bird here in Clark County   spring, summer and fall."

Nest. In the Fallon area, many Mourning Dove nesting records are available. Nests have been observed in trees, sarcobatus bushes, on the ground and on rocky outcroppings; however, the majority of nests have been seen in willow or cottonwood trees (Alcorn MS).

The earliest nesting record for the Fallon area is of two young doves out of the nest reported by Mills (PC) April 1, 1986. Another early record is a nest seen containing two eggs April 4, 1964. A late nesting record for the Fallon area is of a nest containing two eggs September 6, 1966. Re–nesting is evidenced by a nest observed throughout the summer of 1966 that produced three broods. It is assumed the same pair produced each brood.

The author has never observed more than two eggs in any Mourning Dove nest (Alcorn MS), but Mills (PC) reported a late dove nest containing four eggs two miles north of Fallon on August 23, 1972.

On April 18, 1988 a nest was seen at Donald and Norma Cooper's residence in Fallon. Located in an evergreen tree, the dove had placed sticks on an old robin's nest and was setting on two eggs.

Biale saw a nest containing two eggs on a horizontal mahogany tree limb June 18, 1967 (7200 feet elev.). The nest was nine feet above ground and was constructed of small twigs and grass. On June 6, 1972 he reported another nest eight feet above ground in a boxelder tree. The nest was made of weeds and coarse grass and contained two white eggs (6250 feet elev.). Biale (PC, 1984) reports many other nests on the ground, including one in Antelope Valley containing two young ready to leave the nest June 22, 1984.

In southern Nevada near Overton, four nests were seen by A. A. Alcorn to contain one or two eggs each on May 23, 1960. They nest commonly in this area (Alcorn MS).

A Mourning Dove with two young in nest built on top of a robin's nest. Photo by A. A. Alcorn.

Passenger Pigeon   *Ectopistes migratorius*

Distribution. It is now extinct. Formerly bred from central Montana to Nova Scotia; south to Kansas and Georgia. Wintered from Arkansas to Florida. Was casual or accidental from British Columbia to Labrador and south to Veracruz, Valley of Mexico, Cuba and Bermuda. Also accidental in Scotland, Eire and France. Last specimen obtained in the wild was taken at Sargento, Pike County, Ohio, March 24, 1900; last living individual known died in captivity in the Cincinnati Zoological Gardens, Cincinnati, Ohio, September 1, 1941.

Status in Nevada. Extinct.

Records of Occurrence. Ridgway (1877) reported, "On September 10, 1867 I collected a young female Passenger Pigeon in the West Humboldt Mountains. This bird, the only one seen, flew rapidly past me and alighted upon a stick at the edge of a stream. Its stomach contained dogwood berries."

Inca Dove   *Columbina inca*

Distribution. From extreme southeastern California to central Texas; south over to lowlands of Mexico, Costa Rica and Nicaragua.

Status in Nevada. Rare in southern Nevada.

Records of Occurrence. Gullion (1952) reported, "On October 17, 1952, I saw a dove of this species in the center of Logandale, Clark County. It was not possible to attempt to collect this bird, but prolonged observation at very close range left no doubt as to the species identification. This dove has not been reported from Nevada previously."

Monson (1972) reported one at Boulder City, October 24–31, 1971.

Lawson (1977) reported two or three at Boulder City December 17 to March 31 (1972–73), and singles at Tule Springs Park February 18, 1973, March 5, 1973, and May 15, 1975. He also reported one at Cottonwood Cove, Lake Mohave, October 10–17, 1976.

Lawson (TT, 1976) commented, "It is irregularly rare and more commonly seen in fall and winter than any other season here in southern Nevada."

Kingery (1982) reported, "Las Vegas birdwatchers turned up a flock of 15–20 wintering Inca Doves which have reportedly inhabited a trailer court for years, and probably nested. The species has been reported from Nevada only sporadically." He (1982) reported, "Las Vegas recorded the first Nevada Inca Dove nest; found March 9, it blew down after a severe windstorm; the doves disappeared soon after April 11."

Kingery (1983) reported, "Las Vegas' nesting flock of Inca Doves left: the people who fed it moved."

Kingery (1987) reported Inca Doves in backyards in Boulder City, Nevada.

Common Ground–Dove *Columbina passerina*

Distribution. From southern California to South Carolina to South America.

Status in Nevada. Irregular occurrence in southern Nevada.

Records of Occurrence. William H. Behle (1976) reported, "Wauer (1969) observed one at Overton, Nevada on 19 October 1965 and took an immature male there on 20 July 1966. Prior records for southern Nevada were provided by Hardy (1949) who observed one at Corn Creek Ranch, Desert Game Range, Clark County, Nevada on 27 and 29 June 1945, and by Gullion (1953) who obtained a specimen four miles east of Las Vegas, Nevada . . . . Gullion (1953) gave a sight record of this species on 17 October 1952 at Logandale, Clark County, Nevada."

A. A. Alcorn (MS) in the Overton area, Moapa Valley reported two of these birds seen January 18, 1961; one June 17, 1961; one July 24, 1961; and two January 4, 1964.

Snider (1966) reported one collected at the Overton WMA July 20, 1966.

Austin (1968) reported a female found dead at Corn Creek, Desert NWR June 9, 1967.

Monson (1972) reported three were at Mormon Farm, near Las Vegas from December 18 to 25, 1971 and one was collected December 25, 1971.

A specimen, now in the UNR Museum, was taken from the Mormon Farm near Las Vegas February 25, 1972.

Lawson (TT, 1976) who recorded this bird from Mormon Farm, Tule Springs, Boulder City, and along the Colorado River down to Fort Mohave reported, "I would say it is irregularly rare here in southern Nevada."

Lawson (1977) reported the Ground Dove records for 1976 were: two at Pahranagat NWR August 4; one September 26, two on October 3 and one October 10, all at Lake Mohave.

Kingery (1982) reported the first Ground Dove since 1976 at Las Vegas May 29.

Order **PSITTACIFORMES**

Family **PSITTACIDAE:** Lories, Parakeets, Macaws and Parrots

Black–hooded Parakeet   *Nandayus nenday*

Distribution. From southern South America. Widely reported in the United States and Puerto Rico as escapees. Breeding reports in southern California and small population at Coney Island, Brooklyn, New York.

Status in Nevada. Rare. Only one record for Nevada, probably an escaped captive bird.

Records of Occurrence. One female was taken six miles west–southwest of Fallon July 27, 1975. It had been seen flying with other wild birds for several days before it was collected. It is now in the Museum of Vertebrate Zoology at Berkeley, California.

   NOTE. Parakeets of various kinds are popular as pets and sometimes escape into the wild. One of these escapees was a small bird of unknown species, seen rapidly flying southward about 10 miles north of Austin. It was flying about 20 feet above the pinon and juniper trees, and was very vociferous.

Order **CUCULIFORMES**

Family **CUCULIDAE:** Cuckoos, Roadrunners and Anis

Yellow–billed Cuckoo   *Coccyzus americanus*

Distribution. From southern British Columbia to Newfoundland; southward through the United States, Mexico and Central America to South America.

Status in Nevada. Rare summer resident and transient. Found mostly in western and southern Nevada. Breeding populations in Nevada are possibly extinct or in any event, threatened with extinction.

Records of Occurrence. Voget (MS, 1986) reported at Sheldon NWR, "Single bird observed two consecutive mornings (23 and 24 June, 1986) at Thousand Creek."

   Refuge personnel found a dead Yellow–billed Cuckoo on the headquarters lawn of Ruby Lake NWR June 16, 1972. It was sent to the USNM for identification, and was a new record for northeastern Nevada.

   Linsdale (1936) reported Ridgway (1877) observed cuckoos on the lower Truckee River in July, 1867 and Hall (MS) saw a single bird two miles west of Sutcliff, Washoe County, June 12, 1924.

   Alcorn (1946) saw these birds each year in May, June, July and August in Lahontan Valley. He observed individuals on nine separate occasions in July, 1941 near Fallon. Practically all of those seen were in large cottonwood trees. Since 1946, these cuckoos have become rare in the Lahontan Valley with the last ones seen by this author June 30, 1964 and August 20, 1977.

However, they were heard calling west of Fallon in Lahontan Valley on June 17–22, 1978 and June 24 and 25, 1979 (Alcorn MS).

Mills (PC) heard a cuckoo calling July 19, 1969 in the Fallon area.

Karen Platou (PC, 1986) reported a Yellow–billed Cuckoo at Carson Lake June 24, 1986. Photographs provided positive identification.

Oakleaf (1974a) reported, "Surveys to document the present distribution of cuckoos in Nevada were conducted by first locating area containing preferred habitat of thick riparian growth 300 meters in length and 100 meters wide as described by Gaines (1974). Only six isolated areas containing the required habitat size were located on the Truckee, Carson and Walker Rivers. Two of the areas and adjacent marginal habitat were intensely surveyed using taped calls, a technique described by Gaines (1974). No cuckoos were heard or seen. In addition, Don Klebenow has established a breeding bird transect (one mile in length) on the Lower Truckee River. This transect has been worked 12 times in the last three years and cuckoos have yet to be recorded."

Biale (MS) for the Eureka area reported a cuckoo August 7, 1976 and another was found dead June 22, 1976. Since that time he has not seen or heard of any of these birds in the area (1984).

Cottam (1941) reported, "Because of the relatively few records of this bird in this state, it seems appropriate to record that the writer observed a cuckoo in the town of Alamo, Clark [sic] County, Nevada on August 28, 1940. The bird was observed in a cottonwood tree about 30 feet from the ground."

This author saw one in a large cottonwood tree near Overton July 11, 1963 (Alcorn MS).

Austin (1971) reported these cuckoos on, "May 23, June 27, July 3 to 24, and August 14; they are transient in riparian areas."

Kingery (1979) reported six to seven pair probably breeding at Beaver Dam Wash, and he (1980) reported one at Las Vegas October 22 and 27, and he (1984) reported one July 14 near Las Vegas singing in atypical habitat.

Kingery (1978) also reported three Yellow–billed Cuckoos at Corn Creek August 3.

Lawson (TT, 1976) reported, "The Yellow–billed Cuckoo is a rare riparian area bird here in southern Nevada."

Greater Roadrunner *Geococcyx californianus*

Distribution. Resident from California to Louisiana; south to Veracruz, Mexico.

Status in Nevada. Resident in southern Nevada.

Records of Occurrence. The northernmost records of these birds in Nevada are of one seen between Walker Lake and Schurz, Mineral County by Robert Cress May 3, 1970; one reported near Baker, White Pine County by A. A. Alcorn in the summer of 1966; and one reported at Sunnyside, northeastern Nye County, September 15, 1968 by Dean Doell.

Kingery (1984) reported a Greater Roadrunner seen at Kirch WMA July 16.

In Pahranagat Valley, limited numbers occur and these birds were recorded in all months (Alcorn MS).

Gullion et al., (1959) reported, "Although apparently a permanent resident of all southern Nevada desert areas, this bird is most abundant in the valley areas. The northernmost record in Nevada is of a bird seen near Panaca on August 9, 1951. The curious courting antics of this bird are under way by mid–March and a nest with newly hatched chicks and some eggs was found in cholla *(Opuntia bigelovii)* at about 1700 feet elevation on the southeast slope of the Dead Mountains on June 7, 1952.''

J. R. and A. A. Alcorn (MS) recorded these birds in southern Nevada from 1959 to 1965, inclusive, a total of 112 times; seen in all months.

Austin and Bradley (1971) reported for Clark County, "Occurs all year and is a permanent resident in riparian areas, winter resident in desert scrub and visitant in woodland areas."

Lawson (TT, 1976) commented, "They are almost abundant in the Las Vegas Wash in the vicinity of sewage plants and Mormon Farm. It is possible to see them at Tule Springs, Corn Creek, Cold Creek, Red Rock Canyon   just about everywhere."

Groove–billed Ani   *Crotophaga sulcirostris*

Distribution. From southern Baja California and southern Sonora east to southern Louisiana; south through Mexico and Central America to Peru and British Guiana.

Status in Nevada. Accidental in Clark County.

Records of Occurrence. Pauline Long and Florence E. Poyser (1965) reported, "A Groove–billed Ani was seen by us at Boulder City, Clark County, Nevada, on December 7, 1964, and was kept under observation for about half an hour. The bird seemed unafraid and we were able to approach to within a few feet of it and note the grooves in the heavy beak and the zygodactylous feet. When discovered it was perched in a trellis in a garden eating morning glory seeds; later it foraged to arbor vitae and fed on fruits of the myrtus."

Lawson (TT, 1976) reported Mowbray saw one at Tule Springs in the fall of 1973 and another in the Moapa Valley in the fall of 1974.

Order **STRIGIFORMES**

Family **TYTONIDAE:** Barn–Owls

Common Barn–Owl   *Tyto alba*

Distribution. Nearly cosmopolitan, from British Columbia to Newfoundland; south to South America. Also in Europe, Asia, Africa and Australia.

Status in Nevada. Resident. Herron et al., (1985) reported, "Barn–Owls have statewide distribution, with highest densities found in agricultural and riparian areas."

Records of Occurrence. Sheldon NWR (1982) reported the barn owl as a fall transient.

One was found dead along the highway at Dunphy, northern Eureka County March 15, 1969; another was found dead along the highway to Ryepatch Reservoir, Pershing County (Alcorn MS).

Mills (PC) reported two Barn–Owls five miles south of Lovelock June 2, 1974. He (PC) reported two young Barn–Owls in a hole in a ditch bank June 15, 1975.

Burleigh collected a male specimen, now in the USNM, in the Reno area in February of 1969.

Kingery (1984) reported six at Reno during the 1983–84 winter.

In the Fallon area these owls are present in limited numbers throughout the year with nesting reported each year. Napier (MS, 1974) reported a nest with an adult setting on four eggs May 8, 1968.

Thompson (PC, 1988) noted he has seen "quite a few" Common Barn–Owls in the cattails of Dry Lake and Foxtail Lake. "They appear to be year–round, but no nests have been found."

Barn–owls are unreported by Biale in the Eureka area.

Gullion et al., (1959) reported, "Three records: one bird, Searchlight area, August 17, 1951; one bird, Dead Mountains (3200 feet elevation, at the head of Empire Wash), July 11, 1952; and one bird in the Gold Butte area, September 20, 1954."

Kingery (1986) reported, "One or two were seen at Kirch June 9 and July 21."

One was found dead at Overton WMA, Moapa Valley, Clark County September 22, 1961 (Alcorn MS).

Austin and Bradley (1971) for Clark County reported, "Transient in desert scrub and riparian areas. Usually occurs from July 11 to October 19. Unseasonal occurrence in March."

Prange reported a good population at Lake Mohave existed; almost every abandoned mine shaft has a nesting pair and he has found up to 10 road kills a year.

Heilbrun and Rosenburg (1981) reported one during the Henderson CBC December 20, 1980.

Lawson (TT, 1976) commented, "Barn–Owls are rare to uncommon here in southern Nevada except during the winter time when they are abundant in the Fort Mohave and Lake Mohave areas."

Kingery (1986) reported, "Three Common Barn–Owls summered at Las Vegas; two adults and one immature; they might have nested."

Food. Alcorn (1941) reported, "On February 22, 1941 this writer and Miss Laura Mills visited the north end of Soda Lake, Churchill County, Nevada. Here, in a south–facing vertical bank, Miss Mills had seen Barn–Owls on several previous occasions. In the bank, less than 30 feet from the water's edge we found an opening about 15 inches in diameter. Seven feet from the base, the hole went straight into the bank for about two feet, then made a 90 degree turn to the left and extended for about three feet farther. With the aid of a mirror placed at the turn, so as to reflect light into the hole, we could seen an adult Barn–Owl standing near the end of the burrow. Miss Mills had earlier in the day seen two Barn–Owls fly from this location."

"Examination of the ground at the base of the bank below the opening revealed numerous pellets judged by us to have been ejected by the Barn–Owls."

"A total of 89 that were reasonably fresh were picked up for study. Each of the food items listed in the table were represented by the major part of the cranium."

FOOD OF THE COMMON BARN–OWL, FALLON, NEVADA

MAMMALS		Number of items	Percent of items
Ground Squirrel	*Citellus townsendi*	2	3
Pocket Gopher	*Thomomys bottae*	20	34

Kangaroo Rats	*Dipodomys sp.*	24	41
	Dipodomys ordii	8	14
	Dipodomys merriami	2	3
Meadow Mouse	*Microtus montanus*	3	5
TOTAL		59	100

BIRDS		Number of items	Percent of items
California Quail	*Lophortyx californica*	2	14.3
Virginia Rail	*Rallus limicola*	1	7.1
Brewer Blackbird	*Euphagus cyancephalus*	11	78.6
TOTAL		14	100.0

"I am grateful to Dr. Alden H. Miller and Harvey I. Fisher, of the Museum of Vertebrate Zoology for identifying the remains of birds."

Family **STRIGIDAE:** Typical Owls

Flammulated Owl   *Otus flammeolus*

Distribution. From southern British Columbia south through the mountains west of the Great Plains to the Highlands of Mexico and Guatemala; also along Gulf coast to Florida.

Status in Nevada. Uncommon summer resident in mountain forests and transient in valleys.

Herron et al., (1985) commented, "Once considered a rare species, studies of its life history have shown that it is a retiring species that is difficult to locate since it remains so still while perched. Many biologists believe that this bird is much more common than was first believed."

Records of Occurrence. Linsdale (1951) reported one record of these owls in Nevada: a female collected May 8, 1930 at the east base of the Toiyabe Mountains northern Nye County.

Kingery (1979) reported two visited Jarbidge September 18 and he (1983) reported three at Elko.

Lewis collected a male at Cave Creek, Ruby Lake NWR May 16, 1968.

Ned K. Johnson and Ward C. Russell (1962) obtained the first record of this owl in western Nevada 1 1/2 miles north of Crystal Bay, Lake Tahoe, Washoe County on June 23, 1961. Eight individuals were called in using tape recorded calls.

A female was found dead six miles west–southwest of Fallon May 9, 1963 (Alcorn MS).

Biale (MS) photographed a Flammulated Owl at Eureka October 16, 1981.

Johnson (1973) reported, "In the Quinn Canyon Mountains, the Flammulated Owl was numerous in rather open–mixed growth of white firs, aspen, mahogany, and scattered ponderosa pines in Scofield Canyon from 7800–8000 feet elevation. Although no specimens were taken, from one to three individuals were called in on the evenings of June 18 to 20, 1972."

"In the Highland Range, a male was called in on the evening of June 22, in Water Canyon, 7500 feet, in mixed white fir and mahogany and on the following evening another male responded to imitated calls in 40 to 60 foot white firs in Anderson Canyon, 8200 feet. I have reported previously the discovery of this species in the Clover Mountains in 1963 (Johnson

1965). In 1972 in the ponderosa pine zone on the west slope of Sawmill Peak, Clover Mountains, several were heard and tape recordings of vocalizations were obtained.''

Johnson (1965) reported, "This owl was common in the Sheep Range in ponderosa pine and white fir in Hidden Forest Canyon from 7200–8200 feet, above which no nighttime hunting was attempted. Hansen found a dead male floating in Wiregrass Spring when we arrived at Hidden Forest Cabin on June 3; it was prepared as a skeleton. As many as five males were called in by the author during a 2–hour evening hunt. In the Spring Range the Flammulated Owl was common in a canyon three miles north of Charleston Peak, between 8700 feet and 8900 feet, from June 16 to 19, where it occurred chiefly in ponderosa pines, bristlecone pines and white firs. As many as four males were heard in one evening. In Macks Canyon we obtained only one certain record, that of a very responsive male calling in heavy white fir and ponderosa pine at 8100 feet on June 25 . . . . In the Clover Mountains, Lincoln County, a male was taken in ponderosa pine and Gambel oak at 7200 feet on the east side of Ella Mountain on June 26 . . . . Three other Flammulated Owls were found that evening, two of which repeatedly returned to a probable nest cavity in the dead top of a pine at the edge of a manzanita field.''

Snider (1962) reported May 20, 1962, Nevada's second Flammulated Owl was collected at the Corn Creek Field Station.

Austin and Bradley (1971) reported for Clark County, "Summer resident in montane forest areas, transient in riparian areas. Usually occurs from May 20 to June 25.''

Lawson (TT, 1976) commented, "It is a rare to uncommon bird here in southern Nevada in the very high mountain forests.''

Western Screech–Owl   *Otus kennicotti*

Distribution. The scientific name of the Western Screech–Owl was formerly *Otus asio*. Resident, locally from southeastern Alaska to northern Idaho and western Montana; south to the state of Jalisco in Mexico.

Status in Nevada. Resident in small numbers throughout the state. Linsdale (1951) reported, "An adult of *O. a. macfarlanei* was obtained on June 10, 1936 by D. C. Smiley on Smoke Creek (see Linsdale, 1938). This is near the southern limit of the range of the bird. The race *O. a. cineraceus* is resident in small numbers in the southeastern part of the state; specimen from Ash Springs, Pahranagat Valley and skins from the Grapevine Mountains, Nye County, and Potosi Mountain, Clark County, are in the Museum of Vertebrate Zoology.''

"A Screech Owl from Fallon was identified by Hall (1938) as belonging to the race *O. a. inyoensis*. Alcorn (1946) reported additional specimens. Richardson (MS) found it at Reno.''

"The race *O. a. yumanensis* is resident in the Colorado River Valley to at least the extreme southern tip of the state; three specimens in winter (see Miller and Miller, 1951).''

Herron et al., (1985) commented, "Although it ranges throughout the state, the owl is considered to be uncommon in Nevada. The species is generally non–migratory.''

Records of Occurrence. Sheldon NWR (1982) reported the Western Screech–Owl as an occasional spring, fall and winter resident; uncommon in summer. Nesting is also reported.

Heilbrun (1982) reported one seen at Ruby Lake NWR December 19.

Kingery (1987) reported Western Screech–Owls roosted in an open garage in Ruby Valley "on a January night.''

Western Screech-Owl. Photo by A. A. Alcorn.

One was seen near Wellington, Lyon County January 11, 1961 (Alcorn MS).

Alcorn (1946) reported for the Lahontan Valley (Fallon area), "Common but not abundant. Specimens were taken in August, September, October and December."

At present (1988) these owls are seen in limited numbers in all seasons in the Lahontan Valley.

At five miles southwest of Fallon in mid–day on May 19, 1937, a young Screech–Owl was seen sitting on a limb in a buffalo berry thicket. The owl was captured and placed on this author's finger where it sat for about 15 minutes as I walked through the thicket in search of a nest. I then walked out into the sagebrush toward my car and the young owl flew from my finger and went back into the thicket (Alcorn MS).

One was seen near Ellsworth, Paradise Range, Nye County December 17, 1965 (Alcorn MS).

Biale (MS) found young large enough to leave the nest in a hole in an aspen tree on the southeast slope of the Diamond Mountains, near Eureka June 27, 1965. He also saw two in Diamond Valley February 13, 1974.

One was heard near Hiko, Pahranagat Valley, August 20, 1964 and another was seen at the same location March 19, 1965 (Alcorn MS).

Austin and Bradley (1971) for Clark County reported, "Occurs all year; summer resident in woodlands and montane forest areas; and transient in desert scrub and riparian areas."

Lawson (TT, 1976) commented, "The Screech–Owl is an uncommon resident here in southern Nevada."

Food. Mills (PC) reported, "I am fairly well convinced that these owls would rather eat frogs than almost anything else they can get."

Two owl stomachs were examined and one contained the remains of one leopard frog and some unidentified insects. The second contained mainly beetles and grasshoppers with a few other unidentified insects (Alcorn MS).

Although the leopard frog was a major food item of various birds, including the Screech–Owl, in the Lahontan Valley at one time, it has virtually disappeared from the area and is an endangered species (Alcorn MS).

Great Horned Owl   *Bubo virginianus*

Distribution. From Alaska to Newfoundland; south through the Americas (except West Indies and other islands) to Tierra del Fuego.

Status in Nevada. A common statewide resident. Four races have been reported from Nevada. Linsdale (1951) reported, "The race *B. v. occidentalis* is resident over most of the state, except extreme southern and west–central portions. Gabrielson (1949) reported that a specimen *B. v. lagophonus* obtained by him on October 9, 1932 on the Pine Creek Ranch near Potts Post Office, Nye County, is identical with birds from northeastern Oregon and southeastern Washington. The race *B. v. pacificus* is resident along the central part of the western border of the state in the vicinity of the Sierra Nevadas. In southern Nevada *B. v. pallescens* is resident in the Charleston Mountains and along the Colorado River."

Records of Occurrence. Sheldon NWR (1982) reported the Great Horned Owl as an occasional permanent resident, with nesting reported in the area.

LaRochelle (MS, 1967) reported at Ruby Lake NWR about 100 use this refuge. Their hooting is common year round in evenings, also in daytime during mating season. "They take a considerable number of ducks, but we feel they are beneficial because of the pressure they put on the crow and magpie populations."

Kingery (1984) reported, "Undeterred by the winter, Great Horned Owls had hatched young at Reno February 26."

In the Fallon area these owls are present in all months of the year (Alcorn MS).

Biale (MS) recorded these owls in the Eureka area in all months of the year. He reported a nest at the Hay Ranch April 28, 1969 with three young nearly ready to leave the nest. Near Eureka, a nest in an old hawk nest on a head frame above a mine shaft contained two young on April 23, 1979. There were five freshly killed cottontail rabbits lying on the edge of the nest.

A. A. Alcorn (MS) recorded these owls in the Baker area, White Pine County in the winter and summer of 1966 and 1967.

Austin and Bradley (1971) reported for Clark County, "A permanent resident in riparian and woodland areas, winter resident in desert scrub areas and summer resident in montane forests."

Lawson (TT, 1976) reported, "This species is common here in southern Nevada. We do have some influx during the wintertime, probably from the mountains—I don't know, but from two to four appear in late October in Boulder City and roost in the same trees all winter long, then disappear again about mid–July.

Nesting and Food Information. Alcorn (1942) reported, "On February 26, 1942 an adult Horned Owl was seen on a nest near the south bank of the Carson River, four miles west of Fallon, Churchill County, Nevada. This nest had been used the previous year by a pair of Swainson Hawks. The nest, situated near the top of a cottonwood tree *(Populus fremonti)* was estimated to be about 30 feet from the ground. I was unable to examine the contents of the nest for the limbs supporting it was judged to have insufficient strength to support a person's weight. Frequent visits were made to the area in the nesting period at which time an adult owl was usually seen on the nest and another was often seen perched in a nearby tree.''

"The principle vegetation in the area is sagebrush *(Artemisia tridentata)* and many scattered cottonwood trees. California Quail *(Lophortyx californica)* were frequently seen near the nest site.''

"On April 3, 1942 an adult owl was flushed from the nest and one small downy young was seen protruding above the edge of the nest.''

"On other occasions when the adult flew from the nest, due to my presence, it usually returned in about 12 minutes, from a different direction than taken when it flew from the area. On April 19, the young was not in the nest, but was found on the next visit to the area, April 27, at which time it was seen perched in a small tree about 10 feet high. This small tree was about 200 feet east of the nest. Two adult owls and the young that was yet unable to fly were seen repeatedly in the area until May 5 at which time the young was judged to be fully grown. On May 20, the date of this author's last visit to the nest area, the young was not located, but one adult was seen.''

"Repeated search for pellets in the area resulted in the finding of 55 under trees near the nest. None were under the nest. In consideration of the length of time that the area was under observation, this number of pellets is low compared with the findings of Bond (1940) in the Pahranagat Valley, Nevada. Conceivably, this is due to the number of trees in the area that afford numerous places for the owls to perch. This may scatter the pellets over a larger area where they would be more difficult to find.''

"Only the cranium and lower jaws, or their parts were saved from the pellets for study. Each part found was recorded as one individual prey item. For example, one cranium, and a lower jaw were recorded as two prey items. Results of the pellet examination were as follows:

FOOD OF THE GREAT HORNED OWL, FALLON, NEVADA

ANIMALS		Number of items	Percent of items
Ground Squirrel	*Citellus townsendi*	10	31.3
Pocket Gopher	*Thomomys bottae*	5	15.6
Meadow Mouse	*Microtus montanus*	3	9.3
Cottontail Rabbit	*Sylvilagus nuttallii*	12	37.5
California Quail	*Lophortyx californica*	2	6.3
TOTAL		34	100.0

"On April 3, a dead California Quail was seen lying on the edge of the nest. On three occasions, California Quail feathers were found under trees where pellets were picked up. Feathers and wings of a Long–eared Owl *(Asio wilsonianus)* also were picked up from under a tree near the Horned Owl nest. It is not known whether the quail or the Long–eared Owl were captured or eaten by the Horned Owls. In addition to the mammal and bird remains found in the pellets, the mandible of one Jerusalem Cricket *(Stemopelmatus)* was found."

Snowy Owl   *Nyctea scandiaca*

Distribution. Breeds on the tundras from northern Alaska to Greenland; irregular winter migrant southward. Also found in northern Eurasia.

Status in Nevada. A rare migrant to Nevada.

Records of Occurrence. Scott (1961) reported, "A Snowy Owl was reported staying at Ruby Lakes this fall by Clair Aldous. This is a very rare bird in Nevada."

Lawson (TT, 1976) reported one south of Reno in January, 1973. He (1977) reported one 16 kilometers south of Reno January 27, 1972. Guy McCaskie found feathers of a Snowy Owl on a fence post in the same area two weeks later.

Linsdale (1951) reported one Nevada record. A specimen, obtained December 1, 1929 at Indian Springs, Clark County.

Monson (1960) reported a Snowy Owl at Boulder City, November 5, 1959.

Northern Pygmy–Owl   *Glaucidium gnoma*

Distribution. From southeastern Alaska through the Rocky Mountains and California to Guatemala.

Status in Nevada. Uncommon resident in the Carson Range of the Sierra Nevada Mountains and in mountains in the southern part of the state.

Records of Occurrence. Johnson (1956) reported finding this species from July 24 to September 19, 1953–54 at Zephyr Cove, Glenbrook, and Sand Harbor on Lake Tahoe, Daggett's Pass, Douglas County, and Marlette Creek, Washoe County.

Burleigh collected a female specimen, now in the USNM, at Galena Creek, 5400 feet, Carson Range on January 16, 1969.

Ryser (1967) reported one at Holcomb Lane, Reno, during February, 1967; Scott (1967) reported one near Reno February 23, 1967; and Ankers (1971) reported one in the Truckee Meadows (Reno area) December 26, 1970 during the CBC.

Kingery (1980) reported one seen at South Lake Tahoe December 16, 1980.

Johnson (1965) reported, "A male was taken by Miller in Hidden Forest Canyon, 7500 feet, on June 11, in mixed ponderosa pine and pinon. This is the only verifiable record of this species in Nevada away from the Carson Range in the Lake Tahoe area and it represents a locality well isolated from breeding populations in adjacent states."

Austin and Bradley (1971) reported for Clark County, "Summer resident in montane forest areas, occurs in June."

Lawson (TT, 1976) commented, "Here in southern Nevada it is a very rare species. I have personally seen it only once and that was in upper Kyle Canyon, Spring Mountains in the fall of 1975."

Elf Owl   *Micrathene whitneyi*

Distribution. Desert areas and oak zone from southeastern California to southern Texas and central Mexico.

Status in Nevada. Uncommon in southern Nevada.

Records of Occurrence. Kingery (1978) reported at Fort Mohave a Screech Owl took over the cavity in which Elf Owls nested during the pervious year; and an Elf Owl was there April 7 but not April 20.

Kingery (1980) reported the stand of cottonwoods which hosted nesting Elf Owls at Fort Mohave in 1975 to 1978 is dying; no Elf Owls have been there since 1978.

Herron (PC, 1984) reported the population decline of Elf Owls in southern Nevada is due to the burning of cottonwoods and the removal of cactus.

Lawson (1977) reported three Elf Owls at Fort Mohave June 5, 1975. He returned June 12 and collected one of five or six birds present. Also at the same site a pair was seen by van Remsen April 21, 1976.

Lawson (TT, 1976) reported, "In suitable habitat here in southern Nevada the Elf Owl is actually reasonably common, but the habitat is so restricted and so limited that probably it should be listed as rare."

Burrowing Owl   *Athene cunicularia*

Distribution. Plains and unforested areas from southern British Columbia to Manitoba; south to Central and South America. Casually from Quebec to New Brunswick (sighting) along the east and Gulf coast.

Status in Nevada. Summer resident over most of the state and winter resident in the southern part. Kingery (1984) commented that these owls suffer as more land goes into cultivation.

Herron et al., (1985) reported, "Burrowing Owls are a breeding species throughout the state. The majority of the nesting population migrates from northern Nevada during the winter months, but observations of this owl have been recorded throughout Nevada during all months of the year."

Records of Occurrence. Sheldon NWR (1982) reported the Burrowing Owl as an uncommon spring and fall resident, occasional in summer, with nesting reported in the area.

LaRochelle (MS, 1967) reported at Ruby Lake NWR, "An uncommon permanent resident. Nesting is evidenced by downy young."

Gabrielson (1949) reported one near Minden, Douglas County, August 12, 1933.

Alcorn (1946) reported for Lahontan Valley the Burrowing Owl is a common summer resident, recorded each month from March to October. Since 1946, these birds have been recorded in all months in this area; however, the total number of birds has diminished, possibly by as much as 50% in recent years. Thompson (PC, 1988) noted, "Burrowing Owls are seen fall, winter and spring, but "they are rare."

Napier (MS, 1974) reported these owls are present during most of the year at Stillwater WMA, but many move out during the cold winter months. A slight downward trend in

Burrowing Owl. Photo courtesy of the U.S. Fish and Wildlife Service.

population was noted by him during his stay in the area (1967 to 1974). He reported pellet examination revealed that scorpions are heavily used for food during summer months.

Mills (PC) reported four miles west of Fallon an adult was seen flying with a leopard frog in its talons. He also reported, "The Burrowing Owls in early years caught leopard frogs out of the ditches by the dozen. Of course, these frogs are easy to take out in the fields after the hay has been mowed."

In recent years, the population of this frog has diminished and one is seldom seen. At this time (1988) it is not an important source of food for Burrowing Owls (Alcorn MS).

Nine miles south of Fallon, a Burrowing Owl nest contained seven white eggs on May 17, 1939. Laura Mills and Lois Saxton (PC) reported seeing seven of these owls, possibly five young with two adults, 14 miles south of Fallon June 26, 1963. The young were about the same size as the adults. When they stopped their auto to look at these birds, some of the owls ran into the burrow.

Biale (MS) reported these birds in the Eureka area in May, June, July, August and October. He reported that adults flew out of a burrow opening in Antelope Valley, Eureka County on June 1, 1984. He (PC, 1986) reported 10 young out of the burrow in Newark Valley June 13, 1985 and in south Diamond Valley on June 6, 1986, Biale saw another female with 10 young.

In eastern Nevada one was shot at the Lehman Caves National Monument August 25, 1965 (Alcorn MS).

One was seen in Delamar Flat, about 20 miles east of Hiko May 4, 1962 (Alcorn MS).

Austin and Bradley (1971) reported Burrowing Owls are seen all year and are a permanent resident in desert scrub areas, transient in riparian areas in Clark County.

Kingery (1979) reported Burrowing Owls nested at gravel pits in southern Nevada.

Lawson (TT, 1976) commented, "The Burrowing Owl is rare to uncommon here in southern Nevada. We see it most generally as a migrant; however, there are one or two pair in the Mormon Farm area near Las Vegas and we have evidence of young being produced there on at least one occasion."

Food. Bond (1942) picked up a dozen pellets for food habit examination three miles northeast of Yerington on June 8, 1939. Food items consisted of four species of small mammals, one snake, one frog, one scorpion, and 12 various species of insects.

Spotted Owl   *Strix occidentalis*

Distribution. Locally from southwestern British Columbia to California; along the southern Rocky Mountains to Mexico (Michoacan).

Status in Nevada. Rare.

Records of Occurrence. Heilbrun and Rosenburg (1981) reported one Spotted Owl in the Truckee Meadows (Reno area) December 20, 1980 during the CBC.

Herron (PC, 1984) reported one trapped and release by him 15 miles north of Reno on December 12, 1983.

Great Gray Owl   *Strix nebulosa*

Distribution. In the New World from tree limit in Alaska to Ontario; south in the mountains to central Sierra Nevada and California. Also occurs in Russia and Siberia.

Status in Nevada. Hypothetical.

Records of Occurrence. Lawson (MS, 1981) reported one seen October 14, 1980 in Hidden Valley, Washoe County.

Ryser (1985) reported, "There is a summer record for Little Valley in the Carson Range on the western rim of the Basin."

Long–eared Owl   *Asio otus*

Distribution. From Alaska to Nova Scotia; south to Central Mexico. Also in Europe, Siberia, India and southern China to Japan.

Status in Nevada. Statewide summer resident, winter resident in western and southern parts of the state. Nesting occurs mostly in the northern part of the state.

Herron et al., (1985) commented, "In Nevada this owl normally inhabits juniper woodlands, riparian areas and coniferous forests at the higher elevations. The majority of Nevada's Long–eared Owls are considered to be non–migratory. However, individuals of the northern continental populations will winter in the state."

Records of Occurrence. Linsdale (1936) reported, "Probably the most numerous kind of owl in this state."

Sheldon NWR (1982) reported the Long–eared Owl as an occasional permanent resident year round.

In the King River Valley, about three miles north–northwest of the mouth of Thacker Pass Canyon, Humboldt County April 22, 1942, these owls were found nesting in a buffalo berry *(Sheperdia argentea)* thicket. One nest contained four young about one–third grown; another nest contained six young about one–fourth grown; also one two–thirds grown young was seen perched on a limb in this thicket. Food items on the edge of or in the nests consisted of one Meadow Mouse *(Microtus)*, a Kangaroo Rat *(Dipodomys)*, two Pocket Mice *(Perognathus)*, and a Sagebrush Vole *(Lagurus)*. All the nests were situated on old abandoned magpie nests.

A female was collected from a nest containing one egg, five miles north of Red House, Humboldt County April 25, 1942. This bird contained one soft–shelled egg (Alcorn MS).

LaRochelle (MS, 1967) reported at Ruby Lake NWR this owl was a permanent resident that nests on the area, but is not often seen.

Orr and Moffitt (1971) listed this owl in the Lake Tahoe area as a probable resident, but they had no recent records for this area.

Six miles west–southwest of Fallon on December 10, 1972, two of these owls were seen about two feet apart perched in a Russian olive tree. They appeared puffed out and roundish, but as this author approached within 20 feet, they elongated their bodies, pulled their feathers close to their bodies and looked like tree limbs.

On March 31, 1936 at five miles southwest of Fallon an owl nest was made in an old abandoned magpie nest containing six owl eggs. On April 6, this same nest contained three newly hatched young and two eggs. On April 13, examination of this nest revealed a young jackrabbit and a Kangaroo Rat that had been brought to the nest as food for the young owls. The young owls were all dead in the nest and their heads had been pulled off (Alcorn MS).

On April 27, 1942, at four miles southwest of Fallon, an adult flew from a nest containing six eggs. This nest was situated on an abandoned magpie nest. On May 9, 1961 an adult and one nearly full–grown young were seen 14 miles west of Fallon. In the same area, Mills reported an adult on a nest May 4, 1962.

Biale (MS) recorded these birds in the Eureka area May 23, 1965 and May 19 and 24, 1966. An adult owl was seen on a nest May 19, 1966. In Antelope Valley on March 20, 1984, Biale saw two eggs in a nest located in a juniper tree.

Kingery (1979) reported one juvenile at Lida May 30 and 31—the only spring record in seven years.

Johnson (1973) reported, "Although this owl is widespread in northern and central Nevada, the breeding distribution in the southern part of the state remains poorly known. On the evening of June 23, 1972, I attracted a Long–eared Owl to imitated calls in a mixed white fir and mountain mahogany in Anderson Canyon, 7900 feet, Highland Range."

Austin and Bradley (1971) reported, "Permanent resident in Clark County in riparian areas, transient in woodlands and montane forest areas. Occurs in February and from June to December at the Desert Game Range."

Johnson (1965) reported, "Calls attributable to this owl were heard in Hidden Forest Canyon, 7000 feet, Sheep Range, on June 6 and at three miles north of Charleston Peak, 8700 feet, Spring Range, on June 18. Although this species breeds in Las Vegas Valley, there is still no conclusive evidence of nesting in either the Spring or Sheep Ranges."

Gullion et al., (1959) reported this owl as a permanent resident in the Las Vegas Valley with fledgling noted there May 23, 1952.

Lawson (TT, 1976) commented, "This Long–eared Owl is most commonly seen as a migrant in the fall here in southern Nevada."

Food. Johnson (1954) reported finding a pair of adult Long–eared Owls roosting in a single needle pinon tree at 4800 feet, Virginia Range, 11 miles southeast of Reno, Washoe County on March 5. He visited the area again on March 26, April 30 and May 21. One young was present at the latter date. Pellets (131) gathered and analyzed were found to contain 18 pocket mice *(Perognathus parvus)*, 18 kangaroo rats *(Dipodomys panamintinus)*, two pocket gophers *(Thomomys talpoides)*, three unidentified gophers *(Thomomys sp.)*, 15 harvest mice *(Reithrodontomys megalotis)*, 21 deer mice *(Peromyscus (maniculatus?))*, 34 meadow mice *(Microtus montanus)*, two jackrabbits *(Lepus californicus)* and a Western Meadowlark *(Sturnella neglecta)*.

Owl pellets gathered from four miles southwest of Fallon on February 28, 1941 revealed the owls had eaten 10 pocket gophers *(Thomomys bottae)*, five meadow mice *(Microtus montanus)*, 27 kangaroo rats *(Dipodomys sp.)*, three House Finches *(Carpodacus mexicanus)* and one Northern Oriole *(Icterus galbula)* (Alcorn MS).

Comment. In past years, many owl nests, eggs or young were destroyed by magpie bounty hunters. Magpies are now protected by federal law and all magpie bounties have been discontinued. At present, the greatest threat to this owl is the destruction of its nesting and daytime habitat. These owls frequently spend the daytime in buffalo berry thickets and come out to hunt for food at night. In many northwestern valleys of the state within the past 20 years, these thickets have been destroyed to make way for cultivated fields.

Short–eared Owl   *Asio flammeus*

Distribution. From northern Alaska to northern Siberia; south in winter to Mexico. Also found in parts of Europe, Asia and Central and South America.

Status in Nevada. Resident in northern half of the state and transient elsewhere. These birds apparently move into the areas where mice populations are high.

Herron et al., (1985) commented, "The species exists in suitable habitat throughout the state."

Records of Occurrence. Sheldon NWR (1982) reported the Short–eared Owl as an uncommon permanent resident except in winter when it is occasionally seen.

Gabrielson (1949) reported, "A single individual noted in Ruby Valley, Elko County, August 17, 1933; one at the head of Owyhee River, Elko County, August 19, 1933."

Johnson and Richardson (1952) reported, "Recorded as resident in Nevada apparently on the basis of a specimen from Ruby Lake taken on June 27, 1928. A nest with week–old young found in Washoe Valley on May 5, 1950, affords proof that this species is resident."

In northern Humboldt County, large numbers of these owls were seen in September, 1969. From 50 to 60 were counted at nighttime along a 30 mile stretch of road near the north fork of the Little Humboldt River (Alcorn MS).

Ranger Steve Scott (PC) reported at the Mountain City Ranger Station on July 29, 1965,

"The previous year the Meadow Mice came in great quantities and the owls came with them. This year (1965) there aren't so many mice and I haven't seen the owls either."

Page (PC) reported he saw three nests in a crested wheatgrass field eight miles southeast of Halleck, Elko County in June, 1965. The nests were "dished out" places on the ground and one contained four eggs; one contained one egg and four young; and one contained three young.

LaRochelle (MS, 1967) reported at Ruby Lake NWR, although reportedly a nester and permanent resident on this refuge, he had seen the bird only twice.

Schwabenland (MS, 1966) reported at Stillwater WMA these owls were commonly seen during the winter.

Napier (MS, 1974) reporting on the same area, says they are present all year with lowest numbers in summer. Highest population from November to February. He reported five young in a nest on salt grass, covered over with grass and a side entrance on June 23, 1970.

Biale (MS) reported these birds in the Eureka area in all seasons of the year. On May 30, 1965 he reported two nests on the ground in a stubble field. No nest material was used, just a shallow depression in the ground. One nest contained five young of different sizes and judged to be different ages. The second nest contained two young and three eggs.

On May 11, 1969, Biale reported a nest on the ground in an unused field. Again, no nesting material, just a shallow cup in the dirt. The nest contained four young of different ages and five white eggs. One of the remaining eggs was rotten and the other seemed ready to hatch.

Austin and Bradley (1968) reported an individual sighted at Henderson Slough March 5, 1964 as the first record for Clark County.

Lawson (TT, 1976) commented, "The Short–eared Owl is an uncommon winter resident. We do get them out in the Las Vegas Wash area and sometimes we see as many as four in the wintertime."

Northern Saw–whet Owl   *Aegolius acadicus*

Distribution. From Alaska to Nova Scotia; south to Mexico.

Status in Nevada. Uncommon summer resident in some mountain ranges and transient. Probably more common than records indicate.

Herron et al., (1985) reported, "In Nevada, distribution is considered to be statewide, but the majority of observations have been in coniferous forests and riparian vegetation . . . . Currently there are only random sightings and limited nesting observations within the state, indicating that the species exists and is present during the breeding season."

Records of Occurrence. Ridgway (1877) reported one adult female of this species found in Thousand Springs Valley, Elko County on September 24, 1868. He commented, "This one was captured alive by Mr. O. L. Palmer, who found it asleep and placed his hat over it. It was perched on the edge of an old Robin's nest, in a dense willow thicket near the camp."

Linsdale (1951) reported, specimens from Grapevine Mountains, Nye County, and Thousand Springs Valley and Kyle Canyon, Charleston Mountains."

Sheldon NWR (1982) reported the Northern Saw–whet Owl as an uncommon permanent resident thought to nest in the area, although no nests have been found.

Lewis collected a male specimen for the USNM at Ruby Lake NWR October 28, 1966.

Orr and Moffitt (1971) reported in the Lake Tahoe region this owl is known to breed in the area and possibly is resident the year round.

Burleigh collected a female specimen, also for the USNM, near Reno on March 12, 1969.

Ryser (1965) reported one in the Red Rock region north of Reno April 1, 1965.

Herron (PC, 1984) reports one seen in the northwest part of the Pine Nut Range.

Alcorn (1946) reported observations and specimens from September to January in Lahontan Valley. He (1940) discovered another in December, 1938 near Eastgate.

A specimen was found dead on the highway one mile north of Fallon on January 8, 1960. Another was seen by Mills January 15, 1966 in the Fallon area (Alcorn MS).

Biale (MS) recorded this owl in the Eureka area March 20, 1966, when one was caught in a trap two miles south of Eureka.

Herron (PC, 1984) reported one in the Ely area and also from Mt. Charleston.

Miller and Russell (1956) reported, "Juveniles were taken on Indian Creek at 7400 feet, Esmeralda County, on June 1 and 2."

Johnson (1974) reported, "Occurred locally and in very small numbers in the Grapevine Mountains in 1939–40. I did not search the exact spots where these species were found by Miller and his co–workers. The absence is attributable to insufficient field work in the proper places."

"The probability is high that it was not actually missing during the recent period and that virtually all species resident in 1939–40 were also breeding in 1971 and 1973."

Johnson (1965) reported, "van Rossem reported June records from Kyle Canyon in the Spring Range. We did not record this species, although in both the Sheep and Spring Ranges we observed that small birds were generally more responsive to imitated calls of the Saw–whet Owl than to imitated Pygmy Owl notes. This may indicate the greater abundance or more widespread occurrence of the former species, although the population of both owls are undoubtedly small."

Austin and Bradley (1971) for Clark County reported, "Summer resident in montane forests, transient in riparian areas, occurs in June and from October 10 to November 11."

Lawson (TT, 1976) commented, "In southern Nevada, we see very few in the springtime. We do see a few in the summer in the mountains and then we get a fair number of them passing through here in the fall migrations. People are calling me quite frequently about small owls that are bothering their parakeets that are out on patios and so forth, and it is invariably the Saw–whets."

Order CAPRIMULGIFORMES

Family CAPRIMULGIDAE: Goatsuckers

Lesser Nighthawk   *Chordeilus acutipennis*

Distribution. From the interior of central California to southern Texas; south through Mexico and Central America to southern Brazil.

Status in Nevada. Summer resident in the southern part of the state.

Records of Occurrence. Ryser (1985) reported the only sight record for Elko County. He also (1985) reported, "Jack and Ella Knoll saw six Lessers skimming over the vegetation hawking insects in Fernley Marshes, on a cloudy morning in early July, 1982."

A. A. Alcorn (MS) saw nighthawks in the Hiko area of the Pahranagat Valley August 20, 1964; May 5, 29 and 30, 1965; and June 14 and 27, 1965.

Kingery (1980) reported at Dry Lake and Coal Valley north of Pahranagat NWR, these nighthawks were 30 miles north of their cited northern limit August 9.

Gullion et al., (1959) reported, "Common summer resident in all desert valley areas. Earliest spring arrivals were recorded on April 10 for three consecutive years, at Las Vegas in 1951, Boulder City in 1952 and Overton in 1953. Most of the birds have departed by mid–August, but one late record, on October 31, 1953, was obtained in Las Vegas Valley."

"During the hotter summer days, when the air temperature exceeds 108–110 degrees Fahrenheit in the Las Vegas Valley, these birds leave their daytime roosts and course lazily back and forth, low over the mesquite bosques. These are apparently not foraging flights, but simply cooling efforts. As heat diminishes later in the day, the nighthawks disappear into dense thickets, where they roost until their feeding forays commence at dusk."

J. R. and A. A. Alcorn (MS) for the Moapa Valley from 1960 to 1964 reported they were seen 3 times in April; 5 in May; 11 in June; 15 in July; 7 in August; and twice in September.

Austin and Bradley (1971) reported in Clark County these birds are, "Summer resident in woodland areas and transient in montane forests. Usually occurs from June to September 13."

Lawson (TT, 1976) reported, "The Lesser Nighthawk is a common summer resident here in southern Nevada."

Common Nighthawk   *Chordeiles minor*

Distribution. Summer resident from Alaska to Greenland; south in Mexico to Chiapas. It winters in South America.

Status in Nevada. Summer resident in the northern part of the state and transient.

Linsdale (1951) reported these birds were most numerous in the high valleys and flats in the northern part of the state from May to September.

Records of Occurrence. Sheldon NWR (1982) reported the Common Nighthawk as uncommon spring and summer with probable nesting in the area.

In Elko County, this bird was recorded from June to September of most years with the earliest record of several seen June 7 (Alcorn MS).

LaRochelle (MS, 1967) reported the Common Nighthawk at Ruby Lake NWR from late spring until early fall.

This author saw 18 Common Nighthawks flying in a group near Yerington July 26, 1982.

This author noted a Common Nighthawk six miles west–southwest of Fallon May 7, 1977 for an early record.

Alcorn (1946) reported in Lahontan Valley (Fallon area) a summer resident from the latter part of May to the middle of September.

In this area from 1960 to 1966, inclusive, these birds were seen 3 times in May; 25 in June; 37 in July; 78 in August; and 9 times in September. In this period the earliest record was May 28, 1960 and the latest record was September 7, 1966. An adult flew from a nest on the ground containing two eggs seven miles west–southwest of Fallon on July 3, 1962.

Over 200 of these birds were scattered over an area of several miles west of Fallon July 25, 1984 as the author traveled along the highway at dusk (Alcorn MS).

Biale (MS) in the Eureka area from 1970 to 1972, inclusive, saw these birds 3 times in May; 32 in June; 59 in July; 62 in August; and 10 times in September. The earliest record was May 28 and the latest was September 16, 1972. Biale reported a nest on bare ground containing two eggs on June 22, 1965. An unusually large number of about 40 were seen daytime feeding June 26, 1962. He (PC, 1986) reported a nest on the ground in the Ruby Hill Mine area July 18, 1985.

Kingery (1977) reported a Common Nighthawk at Las Vegas October 5.

Lawson (TT, 1976) reported, "This is an uncommon summer resident in the mountains and a common migrant, spring and fall, in southern Nevada."

Common Poorwill   *Phalaenoptilus nuttallii*

Distribution. From British Columbia to Nebraska; south to central Mexico.

Status in Nevada. Summer resident and transient.

Records of Occurrence. Sheldon NWR (1982) reported the Common Poorwill as occasional spring and summer with nesting thought to occur in the area.

In northern Washoe County at Badger Camp, these birds were repeatedly heard and seen May 3, 4 and 5, 1966 (Alcorn MS).

In Elko County these birds were heard in many localities in September and one was found dead on the road near Wildhorse Reservoir September 3, 1960 (Alcorn MS).

Kingery (1980) reported one at Ruby Lake NWR December 19, 1979. It was flushed from a road after sunset.

LaRochelle (MS, 1967) reported at Ruby Lake NWR, "Very few sightings of this bird and no known nesting."

Raymond Evans (1967) reported, "On August 2, 1965, a nest of a Poorwill was found in Little Valley, Nevada at an altitude of 7300 feet. Little Valley is 25 miles south of Reno in the Carson Range. The nest, which was in a slight depression in pine needles and which contained two eggs, was on an east facing slope. The dominant tree of the area is Jeffrey pine *(Pinus jeffreyi)* and the most common shrub of the immediate area is manzanita *(Arctostaphylos patula)*. In the course of taking daily weights of the Poorwill, I found that the nest site was frequently shifted. On August 7, the parent bird flushed exposing the young, 14 feet west of the original site. On August 8, the young were found 20 feet north of site number two. The nest site was in the same place on August 9, but on August 10, the nestlings were found 35 feet west of site number three. On August 11, they were found 17 feet south of site number four. Because of inclement weather, the nest area was not checked on August 12 or 13, but on August 14, the nest was found seven feet west of site number five. The bad weather persisted through April 15, 16 and 17, and on August 18, the young birds could not be found."

Alcorn (1946) reported these birds in Lahontan Valley (Fallon area) as not common with records in May, July, August, September and October.

More recent records for this area are of one seen dead on a road on July 10, 1966 and one seen on a street at 4:30 a.m. on May 7, 1970 in Fallon. It was approached on foot to within eight feet before flying. A Poorwill was also seen six miles west–southwest of Fallon April 17, 1987 and another was seen in Fallon April 29, 1987 (Alcorn MS).

The evening calling of these birds was repeatedly heard in the Toiyabe Range south of Austin in June, July and August. On June 20, 1962 over 10 were heard in Kingston Canyon of this range (Alcorn MS).

Kingery (1978) reported 52 along a 30 mile stretch of road near Ely.

Biale (MS) recorded these birds in the Eureka area on May 17, 1970, May 17, 1972 and May 9, 1973. They were repeatedly seen through June, July, August, September and on October 1, 1966, October 7, 1971 and October 13, 1972.

Biale (PC, 1986) reported a nest on the ground and partially under a black sage (elev. 6200 feet) in Newark Valley, contained two eggs on June 14, 1985.

A. A. Alcorn (MS) reported finding one dead on the road in Las Vegas Valley, Clark County April 17, 1962. Another bird was seen at Cold Creek, Charleston Mountains July 24, 1962.

Austin and Bradley (1971) reported for Clark County, "A summer resident in desert scrub, woodlands and montane forest areas, transient in riparian areas. Occurs from April 5 to November 23, with unseasonal occurrences on January 24 to February 25."

Kingery (1977) reported Poorwills at Lake Mohave November 15 through 22. He (1984) reported a Common Poorwill found dead at Las Vegas February 27.

Lawson (TT, 1976) commented, "The Poorwill is a common migrant spring and fall and an uncommon summer resident in southern Nevada."

Chuck–will's–Widow   *Caprimulgus carolinensis*

Distribution. From extreme southern Ontario; south to Middle America to the east coast.

Status in Nevada. Accidental.

Records of Occurrence. Kingery (1984) reported, "Mowbray commented that people would have viewed a Chuck–will's–Widow in Nevada with skepticism, were the record not a specimen. He picked up Nevada's first record at 6:00 a.m. June 12 under a telephone line crossing the Desert Wildlife NWR."

Whip–poor–will   *Caprimulgus vociferus*

Distribution. Intermittently from Saskatchewan to Nova Scotia; south through Mexico to Central America. Winters to Costa Rica.

Status in Nevada. Uncommon summer resident in southern Nevada; transient.

Records of Occurrence. Kingery (1975) reported, "Reno had its first Whip–poor–will in years May 28."

Johnson (1965) reported one heard in the early morning on June 5, 1963 near Wiregrass Spring the Sheep Range, and noted, "Christman fired at a calling Whip–poor–will perched in a pine snag" that same evening. The following day Johnson collected a male with dried blood on its feathers, believed to be the one shot the night before. On June 9, 1963, another bird was heard calling. Johnson wrote, "Therefore, at least two individuals were present in Hidden Forest Canyon during our visit, indicating the presence there of a small breeding population. This surprising occurrence extends the known breeding range of this species approximately 130 miles north–northwest from the nearest previously reported population in the Hualapai Mountains, Mohave County, Arizona (Phillips et al., 1964).''

Austin and Bradley (1971) reported, ''Summer resident in montane forest regions of Clark County. Occurred in June and September 3.''

Kingery (1975) reported, ''Three Whip–poor–wills called and flushed on Mount Charleston near Las Vegas May 29.''

Lawson (TT, 1976) reported, ''I would say there are three or four pair in Kyle Canyon in the summertime.''

Order APODIFORMES

Family APODIDAE: Swifts

Black Swift   *Cypseloides niger*

Distribution. From southeastern Alaska to Montana; south to Costa Rica. Winters in tropic America.

Status in Nevada. An uncommon transient in western and southern Nevada.

Records of Occurrence. Ridgway (1877) reported finding the remains of a Black Swift on May 31, 1868. ''They had been left by a hawk or owl on a log in the woods at Truckee Reservation near Pyramid Lake.'' He continues, ''On June 23, 1868 this species was observed abundantly in the Carson River Valley, seven miles above Fort Churchill, Lyon County.''

Linsdale (1951) reported the Black Swift from the Grapevine Mountains.

Lawson (MS, 1981) reported one at Sutcliff, Washoe County May 25, 1978.

Biale (MS) saw one flying at Eureka May 15, 1981.

Lawson (MS, 1981) reported six at Mount Charleston September 10, 1978.

Mowbray (MS, 1976) reported the Black Swift as rare in southern Nevada.

Vaux's Swift   *Chaetura vauxi*

Distribution. Distribution discontinuous; from Alaska and northern British Columbia to Central America and northern Venezuela.

Status in Nevada. An uncommon transient.

Records of Occurrence. Linsdale (1951) reported, ''Recorded several times in May and June. Localities in Washoe, Eureka, Nye and Clark Counties. Richardson (MS) reported that in the vicinity of Reno, Ned K. and V. K. Johnson found this bird in 1948 in April at Mayberry Ranch on May 1 and on Hunter Creek September 25 (about 75 birds). Grater (MS) reported two on October 8 and 17 in Hemenway Wash.''

One was seen along the Truckee River near Wadsworth on May 8, 1961 (Alcorn MS).

Mills (PC) saw five in the Fallon area May 20, 1959; two on May 4, 1960; and two on May 8, 1961. He also saw one about 30 miles north of Fallon on May 14, 1970 as it flew northward about 20 feet above the ground, and he saw another four miles north of Fallon May 14, 1970 as it flew northward about 25 feet above ground level.

One was seen September 7, 1985 at six miles west–southwest of Fallon (Alcorn MS).

Biale (MS) for the Eureka area saw one of these birds May 26, 1965 and another August 30, 1970. No others have been seen by him.

Austin (1968) reported, "Formerly considered a rare transient in Clark County, this species has been found to be fairly common in recent years."

Austin and Bradley (1971) for Clark County reported, "Transient in desert scrub, riparian and montane forest areas. Usually occurs from April 29 to June 10 and from August 12 to October 17."

Kingery (1978) reported a surprising combination of 200 Vaux and 1200 White–throated Swifts migrating at Davis Dam February 4.

Lawson (TT, 1976) commented, "This bird is an uncommon migrant here in southern Nevada. Most normally we see them in the first 10 days in May. We find them in the vicinity of Mormon Farm, in Las Vegas Wash, Tule Springs and Corn Creek. Usually we find them on the valley floor."

White–throated Swift   *Aeronautes saxatalis*

Distribution. From southern British Columbia to South Dakota; south through Pacific and northwestern states to Guatemala and El Salvador.

Status in Nevada. A common summer resident and transient. Found around major projections of cliffs in canyons and mountain areas.

Records of Occurrence. Gabrielson (1949) reported, "About a dozen coursing over the meadows at head of Humboldt River, Elko County, August 19, 1933, and approximately half that many about the Duferine Ranch House, and Thousand Springs Creek Canyon on Charles Sheldon Refuge, Humboldt County, May 30 and 31, 1948."

Sheldon NWR (1982) reported the White–throated Swift as occasional spring and fall and uncommon in summer with nesting in the area.

About 30 were flying about a cliff in Rattlesnake Creek Canyon, west side Ruby Mountains May 31, 1968. Others have been recorded in Elko County from April 20, 1966 through September (Alcorn MS).

Lewis collected a male, now in the USNM, at the Ruby Lake NWR October 8, 1967.

Cruickshank (1967, 1968) reported about 25 in the Carson City area December 28, 1966 and December 28, 1967 during the CBC.

Biale (MS) reported these swifts in the Eureka area in May, June, July and September 1, 1973. In earlier years from 1964 to 1967, he reported them from May into August.

In eastern Nevada, near Baker, A. A. Alcorn saw two on June 10, 1967.

In Pahranagat Valley from one to 10 were repeatedly seen in April, May and June (Alcorn MS).

Gullion et al., (1959) reported, "A common summer resident in the rougher desert mountain ranges, tending to wander widely over the desert (as singles and flocks) after the nesting season. Our earliest spring arrival date is February 27 (1954, Overton area), although most of the birds do not arrive until April. Our latest fall record is September 28 (1954, Gold Butte area)."

J. R. and A. A. Alcorn (MS) recorded these birds 25 times in southern Nevada from 1960 to 1965, inclusive. The earliest date was January 10, 1964, when 12 were seen at Overton WMA. One was seen February 28, 1966 and a lone bird was seen at the same place March 18, 1964. On March 13, 1962 a lone bird was seen in Moapa Valley. Others were seen in April, May, June, July, on October 11, 1962 (30 birds), on November 24, 1961 (6 birds), and on December 17, 1964 (12, 3 and 10 birds).

A. A. Alcorn (MS) saw about 20 birds at or near the top of Box Canyon, Sheep Springs, Virgin Mountains, Clark County June 13, 1961.

Austin and Bradley (1971) reported, "Permanent resident in riparian areas, summer resident in desert scrub, woodlands and montane forest areas in Clark County. They are found all year in desert scrub and riparian areas; in woodlands and montane forest they usually occur from May 12 to September 28."

Kingery (1978) reported, "A surprising combination of 200 Vaux and 1200 White–throated Swifts were seen February 4 at Davis Dam."

Lawson (TT, 1976) reported, "It is a common summer resident and uncommon winter resident. I have recorded this species here in Las Vegas Valley every month of the year in large numbers."

Family TROCHILIDAE: Hummingbirds

Broad–billed Hummingbird   *Cynanthus latirostris*

Distribution. From southwestern United States through Mexico to northern Veracruz.

Status in Nevada. Hypothetical.

Records of Occurrence. Kingery (1986) reported, "Nevada recorded its first Broad–billed Hummingbird at a feeder at Alamo September 22–23, documented with a good description of a closely–observed bird."

Magnificent Hummingbird   *Eugenes fulgens*

Distribution. Formerly Rivoli's Hummingbird. From western Colorado to Nicaragua.

Status in Nevada. Accidental.

Records of Occurrence. Kingery (1987) reported, "Extremely rare in Nevada, a Magnificent Hummingbird showed up at a feeder in Jarbidge, on the Idaho line. It stayed for only three days, June 12–14, but long enough for good photographs to document the occurrence."

Lawson (TT, 1976) reported, "My son Karl and I observed one male of this species on the 28th of August, 1975 in Macks Canyon. Essentially this appeared to be an adult male, in that it was just a large, all black, hummingbird. The bird did do one thing that I have observed only in the Rivoli's Hummingbird, and that is it sailed on set wings for a distance of about 50–75 feet as it flew down the canyon."

Black–chinned Hummingbird   *Archilochus alexandri*

Distribution. From British Columbia to northwestern Montana; south to Veracruz, Mexico. Casually to southern Louisiana and Florida.

Status in Nevada. Summer resident.

Records of Occurrence. The Black–chinned Hummingbird is reported by Ruby Lake NWR (1981) as uncommon spring, summer and fall with nesting in the area.

Ridgway (1877) reported collecting a nest containing two eggs on the Truckee Reservation near Pyramid Lake June 1, 1868. A male was taken by Streator at Pyramid Lake May 2, 1896.

Kingery (1981) reported, "A very late Black–chinned Hummingbird stopped at Reno November 12, 1980."

Wilson and Norr (1949) reported young raised in a nest in Minden in 1949.

One was seen at a feeder near Yerington May 12, 1983 (Alcorn MS).

Saxton (MS) repeatedly saw one at her feed station in Fallon the last week of May and the first week of June, 1970.

Biale (PC, 1984) reported, "One bird was seen in Eureka August 3, 1976, feeding on hollyhock flowers."

Two dried bodies of Black–chinned Hummingbirds were picked up by Hall on the Chiatovich Ranch, Fish Lake Valley, in Esmeralda County June 1, 1928.

A. A. Alcorn (MS) saw three of these birds at Rosebud Springs, Charleston Mountains July 16, 1963.

Austin (1971) listed them as summer residents in southern Nevada in riparian areas. Kingery (1979) reported, "A pair summered at Las Vegas raising two young."

Kingery (1986) reported a Black–chinned Hummingbird February 10 at Boulder City.

Lawson (TT, 1976) reported, "It is a very common migrant and common summer resident here in southern Nevada."

Anna's Hummingbird   *Calypte anna*

Distribution. From south–coastal Alaska to central British Columbia; south in western North America to Mexico.

Status in Nevada. Rare, of accidental or irregular occurrence.

Records of Occurrence. Mewaldt (MS) reported one seen at Lake Tahoe August 13, 1962.

Ryser (1970) listed this hummingbird as accidental or irregular in occurrence in the Carson River Basin in western Nevada. Ryser (1985) reported, "During August, 1983 at least three different Anna's Hummingbirds began visiting feeders that Jack and Ella Knoll maintain in northwest Reno. One individual believed to be a young male, was still present as of December 4."

Kingery (1986) reported, "An Anna's Hummingbird appeared in Carson City, Nevada March 23."

Austin (1971) reported, "Sight records by Hansen (personal communication) at Corn Creek on 19 April, 1964 and Jaeger (1927) require substantiation."

Lawson (TT, 1976) reported, "It is rare, but regular. We most normally see them in the fall in southern Nevada."

Kingery (1978) reported one near Lake Mead January 18–22. He (1981) also reported, "One used a Las Vegas feeder from mid– to late November, 1980."

Kingery (1985) reported, "Las Vegas reported two to three Anna's Hummingbirds March 3 to April 16." He (1986) reported, "An Anna's Hummingbird February 9 to a warm canyon near Las Vegas where manzanita had already started to bloom."

Kingery (1987) reported, "Pine Creek Canyon near Las Vegas produced an impressive total of eight Anna's Hummingbirds May 7; having been seen for some years at city feeders, they may breed in the foothills."

Costa's Hummingbird   *Calypte costae*

Distribution. From central California, southern Nevada and southwestern Utah; south into Mexico. Casually north to southwestern British Columbia, Washington and Oregon, central Nevada and northern Utah.

Status in Nevada. Uncommon summer resident in southern half of the state.

Records of Occurrence. Linsdale (1936) reported, "The northernmost record of this bird in Nevada was of a male seen June 19, 1930 in Kingston Canyon, 7100 feet, Lander County by A. H. Miller, (MS)."

"Two specimens were taken on June 18 and 19, 1928 at Cave Spring, Esmeralda County. Specimens in the USNM are from Charleston Mountains, Clark County, Panaca, Lincoln County and Ash Meadows, Nye County."

A nest seen May 3 in Vegas Wash with two full–fledged young was reported by Linsdale (1936). He also reported a nest with two fresh eggs seen on May 3 at the bend of the Colorado River and another nest with two fresh eggs seen May 4, at Bitter Springs, Muddy Mountains.

Austin (1971) reported, "A summer resident in desert scrub, riparian, woodland and montane forest areas. Sight records are from February to September."

Kingery (1977) reported, "Costa's arrived at Lake Mohave on February 14 and stayed feeding on creosole brush and ocotillo nectar."

Lawson (TT, 1976) commented, "Costa's Hummingbird is an uncommon to common summer resident here in southern Nevada."

Calliope Hummingbird   *Stellula calliope*

Distribution. From central British Columbia east to Alberta to western Colorado and southwestern Texas to Mexico.

Status in Nevada. Common summer resident in mountains, mainly in the northern half of the state. Limited nesting in some areas and transient.

Records of Occurrence. Sheldon NWR (1982) reported the Calliope Hummingbird as an uncommon spring and summer resident, thought to nest in the area.

One was seen at Lye Creek, Humboldt County, July 4, 1968 (Alcorn MS).

Kingery (1979) reported the Calliope arrived in Jarbidge May 4.

Ryser (1967) reported this hummingbird observed July 1 to 4, 1967 at Lamoille Canyon and Harrison Pass in the Ruby Mountains.

LaRochelle (MS, 1967) reported them at Ruby Lake NWR in the summers of 1962–66. They were regularly seen a short time each summer and were banded in 1963.

Ryser (1973) saw two males in the Carson Range, Washoe County May 7 at Davis Creek Campgrounds and one male at Genoa in June of the same year.

Colleen Ames (MS) saw them daily at a feeder near Yerington during the summer of 1981.

Saxton (MS) reported one or two at her feed station in Fallon all summer in 1970.

Watkins (PC) reported one nest four miles west of Fallon each summer in 1968 and 1969. At this same location on June 20, 1977, an adult was seen feeding two young on the ground. On July 23, there was no sign of the young and the adult was found dead in a tree.

Two young in a nest were seen at Big Creek Campgrounds south of Austin July 4 (Alcorn MS).

Linsdale (1936) reported, "Two males were obtained June 21 and 24, 1930 at Birch Creek, 7000 feet, Toiyabe Mountains, Lander County."

Biale (MS) reported seeing them in the Eureka area September 1 and 7, 1972.

Johnson and Richardson (1952) reported, "A specimen, identified by Dr. A. H. Miller, was taken by us in extreme southern Nye County at Ash Meadows on June 24, 1956."

Austin (1968) for southern Nevada reported two sight records, one at Corn Creek April 19, 1964 and one in Lee Canyon, 8400 feet, Spring Mountains, June 3, 1966. Also, one specimen, a male was collected in Lee Canyon, 8900 feet, May 23, 1966.

Austin (1970) reported, "A rare transient in the montane forests in southern Nevada."

Kingery (1983) reported, "On April 11 the first Calliope huddled in a snowstorm on Mt. Charleston near Las Vegas."

Lawson (TT, 1976) reported, "I have seen it a few times here in southern Nevada."

Broad–tailed Hummingbird   *Selasphorus platycercus*

Distribution. From north–central Idaho, northern Nevada, to eastern Colorado; south to Texas and Guatemala.

Status in Nevada. Common summer resident. Widespread occurrence of this bird in the state is indicated by the numerous sightings and specimens collected. Ryser (1970) reported this hummingbird as a common summer resident in mountains along streams and rivers in east, central and southern Nevada.

Records of Occurrence. Sheldon NWR (1982) reported the Broad–tailed Hummingbird as an uncommon spring and summer resident, thought to nest in the area.

Kingery (1981) reported these birds stayed at feeders at Jarbidge through September 21–22, 1981.

Observations were made by Ryser (1967) who observed a female on a nest containing one egg at Lamoille Canyon, Elko County May 28 to 29, 1967.

LaRochelle (MS, 1967) reported this hummingbird observed each summer and banded in 1963 at Ruby Lake NWR, Elko County.

Burleigh collected one female north of Lake Tahoe, 9500 feet, August 8, 1967.

In central Nevada, Biale (MS) saw one at the north fork of Allison Creek, Monitor Range, June 18, 1972. He (PC, 1984) reports seeing them repeatedly in the Eureka area in the summer.

The Broad–tailed Hummingbird was also recorded at Lehman Caves National Monument May 28, 1958 and July 18, 1961 (Alcorn MS). Dean E. Medin (1987) reported seeing these hummingbirds every day on Wheeler Peak from July 11–27, 1985.

Gabrielson (1949) reported, "One collected at Wheeler Peak, White Pine County, August 16, 1933."

A. A. Alcorn (MS) saw one at Trough Springs, Charleston Mountains July 19, 1963.

Austin (1971) reported for southern Nevada, "Transient in desert scrub and riparian areas; summer resident in woodland and montane forest areas."

Lawson (TT, 1976) reported, "Common migrant both spring and fall and actually abundant in the fall and it is a common breeding species in all the mountains in southern Nevada."

Rufous Hummingbird   *Selasphorus rufus*

Distribution. From southern Alaska through Great Lakes area to Nova Scotia; south in Mexico to Guerrero and Veracruz.

Status in Nevada. Common transient over most of the state in spring and fall; summer resident.

Records of Occurrence. Linsdale (1936) listed specimens collected in the East Humboldt Mountains, Elko County; Soldiers Meadows and Santa Rosa Mountains, Humboldt County; Snake Mountains, White Pine County; at Lake Tahoe; the lower Truckee River; and at Pine Grove, Mineral County. All specimens were either taken in summer or fall.

Sheldon NWR (1982) reported the Rufous Hummingbird as an uncommon spring and summer resident, thought to nest in the area.

Kingery (1979) reported, "First Rufous arrived at Jarbidge, June 27."

LaRochelle (MS, 1967) reported this bird is less often seen at Ruby Lake Refuge, Elko County than either the Broad–tailed or Calliope.

Burleigh collected an immature female at Reno September 6, 1967 and a female at north Lake Tahoe June 28, 1968.

Kingery (1977) reported a Rufous at Reno on October 30.

Ryser (1970) lists this hummingbird as a common summer resident and transient in mountains (meadows, chaparral, rivers and streams) statewide.

Alcorn (1946) obtained two of these birds near Fallon on August 7, 1939.

Biale (MS) recorded this bird near Eureka August 13, 1960; August 4 and September 12, 1966; a lone individual on five different days in late August, 1972; and one July 28, 1974. Another, a male, was seen in Faulkner Canyon, Monitor Range on July 4, 1981.

Gabrielson (1949) reported, "One collected on Wheeler Peak, White Pine County, August 16, 1933."

Austin (1971) reported for southern Nevada, "Transient occurring in desert scrub, riparian, woodlands and montane forest areas. In woodlands and montane forests they usually occur from June 21 to September 19 and in desert scrub and riparian they occurred April 20 and 21, May 25 and from August 18 to September 19.''

Lawson (TT, 1976) commented, ''Generally, I would say anywhere from the last half of July on through the first week of September it is abundant.''

Allen's Hummingbird   *Selasphorus sasin*

Distribution. From southern Oregon, California, Arizona, Baja California and northeastern Mexico. Accidental in Washington.

Status in Nevada. Rare.

Records of Occurrence. Lawson (TT, 1976) commented, ''We do have perhaps 10 good solid records of this species here, so I would say it is a very rare species in southern Nevada as a spring visitor.''

Order CORACIIFORMES

Family ALCEDINIDAE: Kingfishers

Belted Kingfisher   *Ceryle alcyon*

Distribution. Over most of the North American continent except extreme northern parts; south to northern South America.

Status in Nevada. A common but not numerous resident over the entire state. Usually seen near streams, rivers and lakes.

Records of Occurrence. Linsdale (1936) reported these birds at many places and considered them, ''Resident, found near the larger streams and as a straggler, at other places in the state.''

This writer has recorded them in all seasons and all areas of the state from Jarbidge in Elko County south to Moapa Valley in Clark County and from Reno east to Baker near the Utah state line (Alcorn MS).

Sheldon NWR (1982) reported the Belted Kingfisher as a rare migrant spring, summer and fall.

LaRochelle (MS, 1967) at Ruby Lake NWR reported the Belted Kingfisher as a generally permanent resident, but no known nesting.

A. A. Alcorn (MS) saw lone individuals once or twice each month in September, October, November and December of 1965 at this same refuge.

In 1942 in the city of Reno, this kingfisher was recorded in all seasons along the Truckee River; normally only one bird was seen each day (Alcorn MS).

In the Fallon area in 1966 when bird observations were repeatedly made, this bird was seen in every month. Most often seen in the winter (Alcorn MS).

Biale (MS) in the Eureka area from April of 1965 to April of 1967, saw individual birds on 11 different days, usually in the spring.

Austin (1971) reported, "Transient in southern Nevada in riparian areas and visitant in montane forest areas. Usually occur from March 3 to May 8 and from August 7 to October 21, with unseasonal occurrences on June 21 and December 30."

J. R. and A. A. Alcorn (MS) observed Belted Kingfishers in Pahranagat Valley May 11, 1960; April 22, 1961; and May 5, 1962. In Moapa Valley, Belted Kingfishers were observed on September 24, 1960 (2 birds); and singles were seen October 31, 1960; January 10, 1961; October 7, 1961; January 26, 1962; August 16, 1963; September 14, 1963; and April 13 and 14, 1965.

Lawson (TT, 1976) reported, "It is a common migrant, spring and fall, and an uncommon winter resident and a rare summer resident in southern Nevada."

Family PICIDAE: Woodpeckers and Allies

Lewis' Woodpecker   *Melanerpes lewis*

Distribution. From southern British Columbia to southern Manitoba; south in the west to Sonora, Mexico.

Status in Nevada. Summer resident, mainly in the northern half of the state; not regular in occurrence. Common summer resident at the Sheldon Refuge and in the Jarbidge area. Most abundant in the lower canyons of the Ruby Mountains and the East Humboldt Range of Elko County; widespread in the Carson Range of western Nevada.

Records of Occurrence. Sheldon NWR (1982) reported the Lewis' Woodpecker as an occasional spring migrant.

Ridgway (1877) reported collecting a male from upper Humboldt Valley, Elko County, September 12, 1868 and he noted a few individuals in the lower canyons of the East Humboldt Mountains.

Ryser (1967) reported Lewis' Woodpeckers as very abundant July 1–4, 1967 in the Harrison Pass area of the Ruby Mountains.

LaRochelle (MS, 1967) reported they are often seen in late summer and fall at Ruby Lake NWR.

Kingery (1979) reported one at Ruby Lake in December, 1978. They were frequently seen from May to September in the canyons on the west side of the Ruby Mountains, and one was reported flight feeding at Rattlesnake Creek on May 31, 1968 (Alcorn MS).

One was seen near Lovelock May 9, 1961 (Alcorn MS).

At Mason Valley, a lone individual was seen June 11, 1941 and another near Wilson Canyon July 12, 1981. Also, two were seen at Wilson Canyon May 26, 1984 (Alcorn MS).

Alcorn (1946) reported transient individuals seen May, August, September and October in the Lahontan Valley near Fallon. Solitary birds were seen in September and October of most years from 1960–70.

At the Potts Ranch, Monitor Valley, a lone individual was seen on a fence post in September, 1972 (Alcorn MS).

Biale (MS) reported seeing this woodpecker in the Eureka area on October 3, 1960; May 8, 1965; April 17, 1966; September 20, 1966; and September 11, 1973.

Kingery (1987) reported, "Lewis' Woodpeckers may have begun a slow recovery from diminished numbers. At a Eureka, Nevada ranch, six nesting pair increased from only two last year."

A. A. Alcorn (MS) reported one seen by park personnel at Lehman Caves National Monument, Snake Range May 6, 1958.

Alcorn (MS) saw one at Overton WMA October 31, 1960.

Austin (1971) reported, "Transient in riparian and montane forest areas of southern Nevada, with sightings from March to May and from September to October."

Lawson (TT, 1976) commented for southern Nevada, "This is one of those species that seems to erupt occasionally and goes along and you only see two or three in a year in the spring, another five or six in the fall and then one year will come along and you will see 40 or 50 in a single hour, flying southward or flying northward. For what reason I do not know."

Acorn Woodpecker   *Melanerpes formicivorus*

Distribution. From south–central Washington west of the Sierra Nevadas and west–central Texas; south to western Panama.

Status in Nevada. Rare occurrences in southern Nevada.

Records of Occurrence. Linsdale (1936) reported, "A single specimen taken September 18, 1930 in the Hidden Forest, Sheep Mountains, Clark County, has been identified by van Rossem (MS) as of the race *bairdi*."

Lawson (TT, 1976) reported an individual collected in 1972 in Las Vegas Valley.

Lawson (1977) reported at Boulder City one was seen October 31, 1971. One collected at Sunrise Mountain, just east of Las Vegas October 27, 1972 was identified at the Museum of Vertebrate Zoology.

Kingery (1983) reported, "An Acorn Woodpecker visited Las Vegas May 6."

Gila Woodpecker   *Melanerpes uropygialis*

Distribution. Resident from southeastern California to southwestern New Mexico; south in Mexico to Jalisco.

Status in Nevada. Permanent resident in riparian areas of the southern part of the state.

Records of Occurrence. Linsdale (1936) reported, "This woodpecker was noted as common in the timbered bottomlands of the Colorado River in the extreme southern tip of the state by Hollister (1908). On December 29, 1932 a male was taken in Clark County, opposite Fort

Mohave by Miller. Ten specimens were collected in January and February, 1934, in the same neighborhood."

Linsdale (1951) reported, "In the description of this race [*M. u. oropygialis*] van Rossem (1942) ascribed it to extreme southern Nevada in the riparian cottonwood, willow and mesquite."

Austin (1971) reported for southern Nevada, "Gila Woodpeckers occur all year and are a permanent resident in riparian areas."

Lawson (TT, 1976) commented, "As far as I know it occurs only in the extreme southern tip of southern Nevada in the Fort Mohave area. In its very limited habitat it is a common resident and you can find it any time of the year."

Yellow–bellied Sapsucker   *Sphyrapicus varius*

Distribution. From extreme eastern Alaska to Newfoundland; south to western Panama.

Status in Nevada. Resident statewide in most mountain ranges. According to Linsdale (1951) two races occur in Nevada: *S. v. nuchalis* and *S. v. daggetti*.

Records of Occurrence. Linsdale (1936) reported these birds from most areas of the state with specimens and sight records.

Sheldon NWR (1982) reported the Yellow–bellied Sapsucker as an occasional spring migrant.

Two female specimens of Yellow–bellied Sapsuckers, now in the USNM, were taken from Ruby Lake NWR. One was taken September 22, 1966 and the other November 6, 1967.

The following accounts, unless otherwise noted, pertain to the Yellow–bellied Sapsucker variety.

LaRochelle (MS, 1967) reported at Ruby Lake NWR he had only two records of this bird, one of which was banded in 1963.

Ryser (1967) reported a Red–breasted variety July 1 through 4, 1967 at Lamoille Canyon, Elko County.

A female was collected by Burleigh at Reno May 1, 1967 and another female was taken by Burleigh on Mount Rose, 8000 feet, on June 9, 1968.

Susan Roney Drennan and Gary Rosenburg (1985) reported one Yellow–bellied Sapsucker at Carson City December 30, 1984 during the CBC.

Alcorn (1946) reported the only record for the Fallon area was of a bird taken April 23, 1941. Another was observed six miles west–southwest of Fallon April 27, 1976.

Alcorn (MS) reported seeing two adult Yellow–bellied Sapsuckers carrying food to young at Wilson Creek, Toquima Range, July 4, 1969. The nest was about six feet from the ground in a hole in an aspen tree.

Biale (MS) recorded this bird in the Eureka area only 15 times in the 9–year period 1965–73. He had no winter records. He also reported young in a nest in an aspen tree about 10 feet above the ground on June 27, 1965. He reported two nests in Faulkner Canyon July 4, 1980. The nests were 12 feet and five feet above the ground and contained young (elev. 7800 feet). Another nest was found on the north fork of Allison Creek, Monitor Range, July 18, 1982. The nest was 30 feet above ground and contained young (elev. 7800 feet). He reported two nests, containing young, in Faulkner Canyon, Monitor Range on July 11, 1984.

A. A. Alcorn (MS) reported Yellow–bellied Sapsuckers seen by Lehman Caves National Park employees in November and December, 1958, September 10, 1960 and July 9, 1961.

Cruickshank (1970) reported one at the Desert Game Range, Clark County, December 22, 1969 during the CBC.

Johnson (1973) reported, "Breeding specimens of this woodpecker were collected in mixed white fir between 8200 feet and 8700 feet elevation, Quinn Canyon Mountains, on June 20, 1953, and on June 18 and 19, 1972. In the Highland Range, a pair was taken at a nest hole in a white fir in Anderson Canyon, 8300 feet, on June 23, 1972. This species has been known previously in eastern Nevada from the Snake Range and from the Spring and Sheep Ranges in southern Nevada, and the new records provide intermediate localities of breeding occurrence."

Johnson (1965) found breeding hybrids in southern Nevada which were close to the race *daggetti* in June, 1963.

Austin and Bradley (1965) collected a male at 8500 feet in Lee Canyon, Charleston Mountains on September 16, 1964 which, although nearest *daggetti*, had traces of *nuchalis* characteristics. They considered it to be a resident of the area.

In Moapa Valley one was seen November 24, 1961 and another April 15, 1964 (Alcorn MS).

Austin (1971) reported, "Summer residents in montane forests, transient in woodlands and winter resident in riparian areas in southern Nevada."

A. A. Alcorn (MS) found one of the Red–breasted variety dead at Trough Spring, Charleston Mountains July 16, 1963.

Williamson's Sapsucker   *Sphyrapicus thyroideus*

Distribution. From southern British Columbia eastward to Montana and southward in mountains to Jalisco, Mexico. Casual or accidental east to southern Alberta, Saskatchewan, Oklahoma and west–central Texas.

Status in Nevada. Resident in some mountain ranges. Linsdale (1951) reported two races in Nevada: *S. t. thyroideus* and *S. t. nataliae*.

Records of Occurrence. Evenden (1952) reported, "A pair with a nest hole at a height of about 28 feet on the north side of a tall alder was found near Twin Bridges on the South Fork of the Humboldt River, Elko County, on June 21, 1949."

Gullion (1957) reported, "On July 28, 1955, a small group was encountered in a stand of limber pines and subalpine firs at 9000 feet elevation on the southeast slope of Divide Peak, in the Jarbidge Mountains of northern Elko County."

Ridgway (1877) collected a male and female November 27, 1867 and a female March 10, 1868 from the pines of the Sierra Nevada Mountains near and west of Carson City.

Cruickshank reported the sighting of two birds at Reno December 29, 1958 and lone birds at Truckee Meadows (Reno area) December 21, 1967; December 20, 1970; and December 26, 1971 during the CBC.

Biale (PC, 1984) reported a female Williamson's Sapsucker 15 miles south of Eureka April 13, 1975.

Gabrielson (1949) reported, "A single bird at Potts Post Office, Nye County October 10, 1932."

In the southern part of the state, Linsdale (1936) reported specimens from Irish Mountain, Silver Canyon Mountains, also from the Sheep and Spring Mountains.

A. A. Alcorn (MS) saw one at Trough Springs, Spring Mountains August 7, 1963.

Johnson (1973) reported, "In 1972 specimens of this sapsucker were taken at nests with young at 9000 feet in bristlecone pines, Scofield Canyon, Quinn Canyon Mountains and at 8600 feet, one–half mile northeast of Mount Wilson. No new records resulted from the recent fieldwork on Mount Irish, and Orr's specimen represents the only evidence for summer residency there."

Austin (1971) for southern Nevada reported, "Usually occurs from April 2 to November 3 and is a summer resident in montane forests, transient in woodlands and visitant in riparian areas."

Lawson (TT, 1976) commented, "In the 9000 foot area of Mount Charleston it is reasonably common. The latest I have seen them here is in October, apparently they do move out in the wintertime."

Ladder–backed Woodpecker   *Picoides scalaris*

Distribution. From southeastern California to southwestern Kansas; south through Mexico to British Honduras.

Status in Nevada. Resident in southern Nevada and transient. Usually found in mesquite thickets, Joshua tree forests and desert scrub.

Records of Occurrence. Kingery (1980) reported, "One September 17, 200 miles north of its normal range in Spruce Mountains in a stand of white fir and limber pine."

One was seen at Hiko March 6, 1961 and one at Gold Butte, June 22, 1962 (Alcorn MS).

Others have been seen in Moapa Valley in March, May and June, and an adult was seen feeding a feathered young June 14, 1962 (Alcorn MS).

Gullion et al., (1959) reported for southern Nevada, "Records from the head of Pennsylvania Wash (6500 feet elevation) in the Clover Mountains, eight miles northeast of Elgin, on August 31, 1951, and from the Garden Spring area (4500 feet elevation) on the north edge of the Tule Desert, 15 miles southeast of Elgin, on June 26, 1954, as well as the recent record for the Hiko area (4100 feet elevation, Lincoln County) by Johnson (1956), indicated that these birds range to the northern limits of the Mohave Desert in Nevada."

Austin and Bradley (1971) reported for Clark County that this bird is a permanent resident.

Lawson (TT, 1976) commented, "It is an uncommon to common species in southern Nevada. There is some movement of this species away from nesting grounds, sort of down mountains as it were, into the lower portions of the county. We seldom see them in the Fort Mohave area except in the winter time."

Downy Woodpecker   *Picoides pubescens*

Distribution. From southeastern Alaska to Newfoundland; south to southern California and Florida.

Status in Nevada. Limited numbers resident in the northern, western and eastern part of the state. Transient elsewhere. Linsdale (1951) reported two races occur in Nevada: *P. p. leucurus* and *P. p. turati*.

Records of Occurrence. Linsdale (1936) listed specimens that various collectors have taken of this bird from the upper Humboldt Valley, Elko County; Pine Forest and Santa Rosa Mountains, Humboldt County; Snake Mountains, White Pine County; from Wadsworth, Washoe County; and west Walker River, Lyon County. He also reported their presence in small numbers in the trees along the Truckee, Carson and Walker Rivers.

Sheldon NWR (1982) reported the Downy Woodpecker as uncommon spring and fall, occasionally a summer resident.

Alcorn (MS) saw several in the Jarbidge area of Elko County in September, 1959.

LaRochelle (MS, 1967) reported at Ruby Lake NWR, "Occasional birds were seen nearly any month. No known nesting."

In Idlewild Park, Reno, a total of nine solitary birds were seen in January, February, May, June, August, September and November, 1942 (Alcorn MS).

Burleigh collected four Downy Woodpeckers from Reno and one from nearby Galena Creek from 1967–69.

The author has seen solitary birds among the cottonwood trees along the lower Carson River over a period of 30 years. This bird has never been abundant, but its numbers have been considerably reduced in recent years (Alcorn MS).

Alcorn (1946) collected a transient *P. p. leucurus* December 25, 1941 and a resident *P. p. turati* December 12, 1940. Both birds were collected in the Fallon area. In this same area he reported seeing an adult carrying food to a nest containing young in June, 1944.

Biale (MS) reported seeing this bird in the Eureka area November 22, 1964, September 23, 1973, December 28 and 29, 1973 and March 11, 1974. From 1974 to 1984 he reports they were seen several times each year in the Eureka area.

At Lehman Caves National Monument, White Pine County they were seen July 11 to 14, 1961 (Alcorn MS).

Gullion et al., (1959) saw a Downy Woodpecker in the mesquite bosque (1800 feet elevation) in Las Vegas Valley, four miles east of Las Vegas March 17, 1951. Two other sightings of this bird in southern Nevada were reported by Snider (1969).

Lawson (1969) saw one at Lee Canyon, Charleston Mountains December 23, 1968. He (TT, 1976) commented, "The Downy Woodpecker in southern Nevada is a very rare bird."

Hairy Woodpecker   *Picoides villosus*

Distribution. From Alaska, eastward to Newfoundland throughout most of North America; south to Panama.

Status in Nevada. Resident in mountain ranges statewide, but not abundant. Mostly confined to mountain areas in summer and more widespread into valleys in winter. According to Linsdale (1951) three races occur in Nevada: *P. v. orius*, *P. v. monticola* and *P. v. leucothorectis*.

Records of Occurrence. Linsdale (1936) recorded this bird from all sections of the state.

Sheldon NWR (1982) reported the Hairy Woodpecker as an occasional summer resident, uncommon spring and fall.

LaRochelle (MS, 1967) reported seeing this bird on five different occasions at the Ruby Lake NWR, Elko County in 1966.

Ryser (1967) reported seeing them in Lamoille Canyon, Ruby Mountains July 1–4, 1967.

Burleigh collected single female birds in western Nevada near Reno, on September 13, 1968 and March 13, 1969. Four others were taken at nearby Galena Creek in 1968. All specimens are in the USNM.

Alcorn (MS) saw lone individuals on nine different days at Idlewild Park in Reno in 1942. They were seen in January, March, June and November.

Gabrielson (1949) reported, "A male collected near Eastgate, Lander County, October 14, 1933."

Biale (MS) reports lone individuals in the Eureka area in all seasons.

Austin (1971) reported, "Permanent resident in montane forests and winter resident from October to April in riparian and woodland areas in southern Nevada."

Lawson (TT, 1976) reported, "It is a common resident in the Spring Mountains and Sheep Mountains, here in southern Nevada. By and large, they are found only in the mountains, but occasionally we do get an individual down in the valley."

White–headed Woodpecker   *Picoides albolarvatus*

Distribution. From south–central British Columbia, north–central Washington, and northern Idaho; south to southern California and western Nevada.

Status in Nevada. Resident in small numbers in the Carson Range along the western edge of the state.

Records of Occurrence. Linsdale (1936) reported, "Ridgway (1877) found the White–headed Woodpecker to be common throughout the winter of 1867–68 in the mountains west of Carson City, Ormsby County. He took specimens on March 10 and April 25, 1868. The birds kept entirely among the pines, though they sometimes came down to the lower edge of the woods. Henshaw collected a male on September 16, 1876 at Lake Tahoe, Nevada. Another was taken on May 28, 1889 by Keeler at Glenbrook, Douglas County."

Burleigh collected a male for the UNSM at Galena Creek (5400 feet elevation) January 24, 1969.

Orr and Moffitt (1971) reported the White–headed Woodpecker as a common resident of the Lake Tahoe region.

Kingery (1982) reported, "A White–headed Woodpecker was reported at Carson City January 21."

Three–toed Woodpecker   *Picoides tridactylus*

Distribution. Formerly Northern Three–toed Woodpecker. From the limit of trees in Sweden, Lapland, northern Russia, northern Siberia, Alaska to Quebec; south to Arizona and New England. Also in western China.

Status in Nevada. Resident in the coniferous forests of the Snake Range in the eastern part of the state.

Records of Occurrence. Linsdale (1936) reported, "On May 17, 1934 a male was shot on Lehman Creek, 9000 feet, in the Snake Range. A male and female were taken September 18, 1934 in the same vicinity."

A. A. Alcorn (MS) reported an adult female was in the Baker area January 17, 1959.

Biale (MS) reported one at Lehman Creek, Snake Range, July 23, 1972.

Black–backed Woodpecker   *Picoides arcticus*

Distribution. Formerly Black–backed Three–toed Woodpecker. From central Alaska to Newfoundland; south across most of the northern United States.

Status in Nevada. Rare, reported only from limited areas in extreme western parts of the state.

Records of Occurrence. Linsdale (1936) reported, "Ridgway (1877) obtained an adult female, the only one of the species seen by him, in the pines of the Sierra Nevadas, near Carson City February 19, 1868. The bird was hammering on the trunk of a dead pine tree near the foot of the mountains. Two males from Lake Tahoe were taken August 27 and September 18, 1876."

Orr and Moffitt (1971) listed this bird as resident in small numbers at the high elevations around Lake Tahoe. However, they give no records for the Nevada side of the lake.

Kingery (1980) reported two at South Lake Tahoe December 16, 1979.

Northern Flicker   *Colaptes auratus*

Distribution. Formerly the Red–shafted Flicker. From Alaska to Labrador and Newfoundland; south in Mexico to the Isthmus of Tehuantepec.

Status in Nevada. Three varieties of this flicker occur in Nevada: the Yellow–shafted Flicker, an uncommon transient in Nevada; the Red–shafted Flicker, a common statewide resident; and the Gilded Flicker, reported rarely only from southern Nevada.

Records of Occurrence. Sheldon NWR (1982) reported the Northern Flicker as a permanent breeding resident, common in spring, summer and fall; uncommon in winter.

The Yellow–shafted variety was reported by Ryser (1966) who saw one at the Ruby Lake NWR October 16, 1966. He (1973) also reported a female seen along the Truckee River.

In the Fallon area, a yellow–colored flicker was seen on the average of once each winter. Of two birds collected, both were "hybrids" (Alcorn MS).

Biale (MS) reported seeing one of the Yellow–shafted variety October 3, 1971 in the Schell Creek Range, White Pine County.

A. A. Alcorn (MS) reported one of the Yellow–shafted variety October 19, 1964 and another December 15, 1964 near Hiko, Lincoln County.

Ryser (1967) reported one Yellow–shafted variety at Paradise Valley near Las Vegas, Clark County February 15–19, 1967.

Northern Flicker. Photo by Roy Nojima.

Lawson (MS) reported Yellow–shafted Flickers in southern Nevada are very rare and one is seen every three or four years.

The Red–shafted variety has been seen in all localities of the state. So many sight records are available, that no attempt is made to list all of them.

LaRochelle (MS, 1967) reported at Ruby Lake NWR, Elko County from 1962–1967, these flickers were seen each month of the year, and nested each summer.

Alcorn (MS) recorded these birds in Reno each month between November of 1941 and December of 1942.

Alcorn (MS) recorded 1899 sightings (a total of 4349 birds) in the Fallon area from January, 1960 to December, 1971.

Biale (MS) recorded this flicker 424 times in the Eureka area between June, 1964 and December, 1973. He reported frequent nesting in the area.

A. A. Alcorn (MS) recorded the Red–shafted Flicker variety 60 times in Moapa Valley, Clark County between December 13, 1959 and April 14, 1965. He saw them in all months except May, June and August.

Austin (1971) reported for southern Nevada, "Winter resident in desert scrub and riparian areas and permanent resident in woodlands and montane forest areas."

Snider (1971) reported one Gilded Flicker at the Desert Game Range October 4, 1970.

Lawson (TT, 1976) reported, "We have all three subspecies of this bird in southern Nevada. The Red–shafted Flicker is common resident in mountains in summer, moving to valleys in winter. The Gilded Flicker is here primarily in the fall with four individuals recorded here in the last three years."

Food. Alcorn (MS) reported in the Fallon area on November 9, 1941 a flicker was seen to pick from a tree and eat seven Russian olive berries. On December 15, 1941, seven flickers perched in one Russian olive tree and all were eating the berries from the tree. One was seen to pick and swallow 10 berries. On December 18, 1941, one flicker picked from a tree and swallowed whole, 14 Russian olive berries, then flew 20 feet to alight on the side of a power pole. After about two minutes on the pole, it flew back to the tree and ate five more berries, then flew north out of sight.

On January 14, 1971, a female was seen from a distance of about 15 feet as it ate about 12 kernels of wheat, only after breaking each kernel into small pieces. Each kernel was struck with its beak to break the kernel, then the broken pieces were eaten (Alcorn MS).

Pileated Woodpecker   *Dryocopus pileatus*

Distribution. From British Columbia to Nova Scotia; south to California and along the Gulf Coast to Florida.

Status in Nevada. Nests near the Nevada/California border at Lake Tahoe. However, the author knows of no records for Nevada.

Records of Occurrence. Lawson (PC, 1983) reported he saw one nesting in California within one quarter mile of the Nevada border in the mid– or late 1960s. He also reported that in 1972 a friend hunting near Incline Village saw a pair.

Lawson (PC, 1984) commented, "We still have no Nevada records for this woodpecker, but I think it is just a matter of searching for them in the Lake Tahoe area."

Order **PASSERIFORMES**

Family **TYRANNIDAE:** Tyrant Flycatchers

Olive–sided Flycatcher   *Contopus borealis*

Distribution. From northern Alaska to central Newfoundland; south to South America (Brazil).

Status in Nevada. Uncommon summer resident in some of the higher mountain ranges. Transient statewide in migrations.

Records of Occurrence. Sheldon NWR (1982) reported the Olive–sided Flycatcher as an occasional spring and summer resident.

Specimens in the USNM include a female collected from Ruby Lake NWR June 14, 1967; a female collected from Reno June 5, 1967; a male from Reno May 21 and a female from Reno May 22, 1968; and a male from Carson City June 12, 1967.

Alcorn (1946) reported a specimen taken in the Fallon area May 30, 1942. Since that time, lone individuals were seen April 30, 1960; May 22 and 26, 1960; and June 12, 1960. Mills reported one May 30, 1961. Between March 25, 1961 and May 24, 1965 they were recorded eight times in this area.

Biale (MS) reported one in the Eureka area August 15, 1965; one three miles south of Eureka September 7, 1975; one in the Ruby Hill Mine area September 19, 1976; and one five miles south of Eureka August 25, 1979.

In southern Nevada, Gullion et al., (1959) reported four May records.

Johnson (1973) reported, "Breeding localities for this species are rare over most of Nevada; therefore, the following records are of interest: June 19, 1972, Quinn Canyon Mountains, one sang suggesting probable establishment for nesting; May 18, 1972, Sawmill Peak, Clover Mountains, 6900 feet, one sang repeatedly from the top of a snag in residual ponderosa pine forest."

Austin (1973) reported for southern Nevada, "In woodlands and montane forest, it usually occurs from May 2 to August 23. In riparian areas it usually occurs from April 29 to May 20, and from September 7 to 18, with an unseasonal occurrence on June 9. Transient in riparian and woodland areas. Summer resident in montane forests."

Western Wood–Pewee *Contopus sordidulus*

Distribution. From east–central Alaska, east and southward to central Manitoba; south in the mountains to South America (Peru).

Status in Nevada. Summer resident in wooded areas along streams and rivers.

Records of Occurrence. Linsdale (1936) reported specimens taken in Washoe, Douglas, Humboldt, Elko, Lander, White Pine, Esmeralda, Nye, Lincoln and Clark Counties.

Sheldon NWR (1982) reported the Western Wood–Pewee as an occasional resident spring, summer and fall.

Fifty–three specimens were collected for the USNM from Ruby Lake NWR, Lake Tahoe, Mount Rose, Galena Creek, Carson City and Reno area between August of 1960 and August of 1968.

They were observed in the Jarbidge area of Elko County September 3, 1960 and September 7–8, 1961, and at Harrison Pass August 22, 1965 (Alcorn MS).

They were frequently seen in May and June in the Fallon area, but no nesting has been reported. On May 12, 1963 traveling by auto from Fallon to six miles westward, over 20 migrating flycatchers, mostly Western Wood–Pewee, were seen (Alcorn MS).

Biale (MS) for the Eureka area reported seeing this bird 28 times from May 3, 1964 to September 6, 1967. They were seen 9 times in May; 15 in June; once in August; and 3 times in September. He (MS, 1986) reported a nest in an aspen tree with an unknown number of small young at Hunter Ranch, three miles east of Eureka on July 7, 1968.

Kingery (1975) reported, "A stray Western Wood–Pewee reached Diamond Valley May 19, perished May 23, perhaps from effects of storms on its food supply."

Kingery (1987) reported 100 Western Wood–Pewee at Dyer May 22–23 during a cold spell.

J. R. and A. A. Alcorn (MS) recorded this bird in Pahranagat Valley May 31, 1965 and June 2, 1963. In Moapa Valley they were repeatedly seen in May and June.

Austin (1971) reported, "In woodlands and montane forests it usually occurs from April 29 to September 16. In riparian areas, it usually occurs from April 19 to June 11 and from August

30 to September 21. Unseasonal occurrence in October. Transient in riparian and woodland areas. Summer resident in montane forests.''

Willow Flycatcher   *Empidonax traillii*

Distribution. Formerly Traill's Flycatcher. From central British Columbia to Nova Scotia; south to Mexico and Argentina.

Status in Nevada. Linsdale (1951) reported three races in Nevada: *E. t. brewsteri*, *E. t. adustus* and *E. t. extimus*.

The Willow Flycatcher (song "fitz–bew") and the Alder Flycatcher *(Empidonax alnorum)* (song "fee–bee–O") are virtually indistinguishable in the field. As such, the two constitute a superspecies as listed by the *American Ornithologist's Union Checklist of North American Birds* (6th ed., 1983).

Records of Occurrence. Sheldon NWR (1982) reported the Willow Flycatcher as an occasional migrant spring and fall.

Specimens in the USNM were collected from Ruby Lake NWR on June 12 and 15, 1967 (male and female) and May 23 and 30, 1968 (2 males).

Burleigh collected three specimens, now in the USNM, from the Reno area. A male was taken August 2, 1968 and a male and female were collected September 3, 1968.

Stanley Jewett reported two specimens from Mason Valley taken June 9, 1945.

Alcorn (1946) reported one taken August 26, 1940 near Fallon.

Austin (1971) reported for southern Nevada, "Usually occurs from April 25 to September 22. Summer resident in riparian areas. Transient in the woodland and montane forest areas.''

Lawson (TT, 1976) commented, "The Willow Flycatcher is a very common migrant here in southern Nevada.''

Philip Unitt (1987) reported, "*E. t. extimus* has been recorded at only three Nevada localities: Indian Springs (three nesting pair in 1932, (two specimens collected on 8 and 11 July, Linsdale 1936), Corn Creek (one specimen on 16 May 1962), and Colorado River at the southern tip of the state (one specimen on 9 May 1953, A. R. Phillips, pers. comm.).''

Least Flycatcher   *Empidonax minimus*

Distribution. From southwestern Yukon to Nova Scotia; south through Mexico and Central America to Panama.

Status in Nevada. Accidental in Nevada.

Records of Occurrence. Allan Phillips et al., (1964) reported one specimen taken at Boulder City, Nevada September 6, 1950.

Hammond's Flycatcher   *Empidonax hammondii*

Distribution. From Alaska to Alberta; south through the western United States to northern Nicaragua.

Status in Nevada. Summer resident in mountain forests.

Records of Occurrence. Linsdale (1936) reported specimens from Humboldt, Elko, Lander, White Pine, Churchill, Esmeralda and Clark Counties.

Sheldon NWR (1982) reported Hammond's Flycatcher as an occasional migrant spring and fall.

Specimens from Nevada, now in the USNM, include a female taken June 1, 1967 at Ruby Lake NWR; a male, taken by Burleigh from north of Lake Tahoe August 16, 1967; and a female collected by him near Reno October 6, 1967.

Ryser (1985) reports, "It is also an uncommon breeding species in the mountainous eastern and western rims of the Basin and in some of the mountain ranges within the Basin. Breeding specimens have been collected from the Carson Range in the west and the Snake Range . . . in the east."

Austin (1971) reported for southern Nevada, "Usually occurs from April 20 to May 17 and from August 23 to 27, and on September 19. Unseasonal occurrence July 13. Transient in riparian and montane forest."

Dusky Flycatcher   *Empidonax oberholseri*

Distribution. From southwestern Yukon east to Saskatchewan; south to southern California to Oaxaca in Mexico.

Status in Nevada. Common summer resident in most mountain ranges.

Records of Occurrence. Linsdale (1936) reported specimens collected from Washoe, Humboldt, Lander, White Pine, Mineral, Esmeralda and Nye Counties.

Sheldon NWR (1982) reported the Dusky Flycatcher as an occasional spring, summer and fall resident, nesting occasionally in the area.

Fifteen specimens in the USNM were taken from Ruby Lake NWR, North Lake Tahoe and Reno in May, June and August of 1967 and 1968.

Alcorn (1946) reported specimens collected in the Lahontan Valley (Fallon area) August 11, 1941 and May 30, 1942.

Johnson (1974) stated, "In 1974 *E. oberholseri* was definitely breeding in the Grapevine Mountains."

Johnson (1974) also reported that at Potosi Mountain in 1971 he found at least five singing males in pine fir between 6800 feet and 7500 feet elevation.

Austin (1971) writes, "The Dusky Flycatcher in southern Nevada is found in woodlands and montane forests usually occurs from May 1 to September 19; riparian areas usually occurs from April 5 to May 12 and from September 5 to 13. Transient in riparian areas. Summer resident in woodlands and montane forests."

Gray Flycatcher   *Empidonax wrightii*

Distribution. From south–central Washington east to Colorado; south to central Mexico.

Status in Nevada. Common summer resident in various localities.

Records of Occurrence. Linsdale (1936) reported specimens from Washoe, Humboldt, Lander, Mineral, Esmeralda, Nye and Clark Counties.

Sheldon NWR (1982) reported the Gray Flycatcher as an uncommon summer resident with breeding in the area.

Specimens in the USNM (all males) were taken at Ruby Lake NWR May 30, 1968; and the Reno area May 9, 1967 and April 26, 1969.

Alcorn (1946) reported a specimen taken May 2, 1942 in the Fallon area.

Biale (MS) reported them in the Eureka area August 8, 1965; June 18, 1967; and July 16, 1967. He (PC, 1986) reported a nest with two young partially feathered out in the Silverado Mountains 10 miles east of Eureka on June 19, 1985.

Kingery (1987) reported 13 Gray Flycatchers at Dyer May 22–23 during a cold spell.

Austin (1971) reported for southern Nevada, "Found in riparian areas from April 14 to May 15, with unseasonal occurrence on April 5. Transient in riparian and montane forest areas. Summer resident in woodlands."

Western Flycatcher   *Empidonax difficilis*

Distribution. From southeastern Alaska south to western South Dakota; south in the mountains to Central America.

Status in Nevada. Uncommon summer resident at middle altitudes on the mountain ranges; transient elsewhere.

Records of Occurrence. Linsdale (1936) reported specimens from the mountains in Humboldt, Elko, Lander, Nye, Esmeralda and White Pine Counties.

Sheldon NWR (1982) reported the Western Flycatcher as an occasional spring and fall migrant.

A female specimen was collected by Lewis at Ruby Lake NWR June 1, 1967.

Burleigh collected a female specimen near Reno on June 1, 1967.

Biale (MS) observed single birds in the Eureka area May 24, 1965, June 1, 1967 and September 9, 1973.

Johnson (1973) in his report of avifaunas in southeastern Nevada reported breeding season records from Scofield Canyon, 8000 feet, Quinn Canyon Range; in Water Canyon, Highland Range; on Mount Wilson; and in the Virgin Mountains.

Austin (1971) reported, "In woodland and montane forest it usually occurs from May 23 to September 16. In riparian areas it usually occurs from April 20 to May 26, and from September 5 to 13, with an unseasonal occurrence on April 5. Transient in woodland and riparian areas. Summer resident in montane forests."

Black Phoebe   *Sayornis nigricans*

Distribution. From California to Texas; south to South America (Argentina). Casually to British Columbia, Washington and Oregon.

Status in Nevada. Uncommon resident in the southern part of the state. Visitant at least as far north as Reno, Fallon and Eureka.

Records of Occurrence. Kingery (1978) reported one at Reno May 24.

One was seen at the upper end of Lahontan Reservoir in Lyon county December 21, 1959 (Alcorn MS).

Alcorn (1946) collected one in the Lahontan Valley on February 11, 1945.

Mills (PC) reported for Lahontan Valley (Fallon area) birds seen December 30, 1958; May 24 and 30, 1961; January 1, 1969; and November 30, 1970. Alcorn (MS) reported additional single birds were seen August 15, 1971 (one) and August 6, 1976 (one) in this same valley (Alcorn MS).

Biale (MS) saw one in the Eureka area August 14, 1960.

Kingery (1983) reported, "A Black Phoebe stayed at Baker, Nevada April 28–30."

A. A. Alcorn (MS) reported Black Phoebes from Pahranagat Valley, Lincoln County, on March 11 and May 4, 1962 and April 13, 1965.

J. R. and A. A. Alcorn (MS) observed Black Phoebes in the Moapa Valley December 15, 1959; April 4, 1960; September 29, 1961; March 14 (3 birds), March 15 and 21, 1962; and June 12 and 15, 1962.

Johnson (1956) reported, "Found breeding at Meadow Valley Wash, 4300 feet elevation, four miles south of Caliente, Lincoln County, on April 11, 1954. Both members of the pair and their nearly–completed nest were collected. The female was adding mud to the sides of the nest as I approached. I am unaware of other breeding records in Nevada."

Gullion et al., (1959) reported for southern Nevada, "Probably an uncommon permanent resident, with records available for all seasons."

Austin (1971) reported for southern Nevada, "Occurring all year. Winter resident in desert scrub and riparian areas. Transient in montane forest areas."

Lawson (TT, 1976) commented, "It is an uncommon to common summer resident and winter resident in southern Nevada."

Eastern Phoebe   *Sayornis phoebe*

Distribution. From British Columbia to Nova Scotia; south into Mexico to Veracruz.

Status in Nevada. Accidental in the southern part of Nevada.

Records of Occurrence. Lawson (MS) reported an immature female taken at Hansen Springs, just behind Corn Creek Field Station, December 30, 1973. The bird was present for about a month and was first seen in late November. It is now a specimen at the University of Nevada, Reno.

Kingery (1978) reported another seen at Corn Creek October 10, 1977 and he (1980) reported one seen by Mowbray at Las Vegas October 27, 1979.

Say's Phoebe   *Sayornis saya*

Distribution. From northwestern Alaska to southwestern Manitoba; south into Mexico (Puebla and central Veracruz).

Status in Nevada. Summer resident, statewide. In winter it occurs mainly in the southern half of the state. Linsdale (1951) reported two races in Nevada: *S. s. saya* and *S. s. quiescens*.

Records of Occurrence. This bird has been reported from all parts of the state. The author has seen these birds at Vya and Badger Cabin in northern Washoe County, at Orovada and Winnemucca in Humboldt County; and at Owyhee, Midas, Tuscarora, Elko and Ruby Valley in Elko County. Farther south, they were seen numerous times in the Lovelock, Reno, Yerington and Fallon areas. Eastward, they were seen in the Austin, Eureka and Ely–Baker areas. In southern Nevada, they were seen in the Hiko–Alamo and Overton areas.

Sheldon NWR (1982) reported the Say's Phoebe as uncommon to common spring, summer and fall, with nesting in the area.

LaRochelle (MS, 1967) reported for Ruby Lake NWR this phoebe is a common summer resident, but no nesting has been observed.

A fledgling was seen at the residence of the late Ed and Helen Spence, five miles north of Yerington on July 18, 1983.

In the Fallon area from March, 1960 to March, 1968, this bird was observed on 40 different days. It has been recorded in all months in this area (Alcorn MS).

Biale (MS) reported this bird 69 times in the Eureka area from April, 1964 to September, 1967. His records were from March through September, not seen from October through February. Biale (PC, 1984) reported finding a nest in Antelope Valley on June 4, 1984. It was located in the mouth of a mine tunnel and contained four white eggs (6800 feet elev.).

J. R. and A. A. Alcorn (MS) reported this bird 20 times in the Hiko–Alamo area of Pahranagat Valley from October, 1960 to June 14, 1965. In this area it was seen January 8, 1962; 5 times in March; 3 in April and May; twice in June and July; once in August and October; and twice in December.

J. R. and A. A. Alcorn (MS) reported this bird most common in the Overton area from February through June. From September 20, 1959 to June 14, 1965, they were recorded all months except November.

Austin (1971) reported for southern Nevada, "Occurs all year and is a permanent resident in desert scrubland and riparian areas and transient in woodlands."

Lawson (TT, 1976) commented, "This bird in southern Nevada is uncommon in the summertime and we find more of them in the spring, fall and winter."

Nesting. Biale (MS) reported nesting activity in the Eureka area. On July 22, 1964, two eggs were found on a ledge in a cave. On July 26, 1964, a nest in an old Barn Swallow nest of an old building contained three young. On June 11, 1967, Biale saw a nest in an old building that contained three eggs. The nest was about nine feet above floor level and nest material consisted of string, wool and spider webs.

On July 9, 1967 another nest, containing four young was in an old building on the window trim. Nest material consisted of grass, sage bloom (dry) and wood. Another nest was seen May 24, 1970, five miles west of Hay Ranch. The nest was on a timber under a bridge six feet above ground and was made of webs, wool and small pieces of vegetation. The nest contained five white eggs (6050 feet elev.).

Biale found another nest July 3, 1970 at Clear Creek Ranch (Little) Fish Lake Valley. It was located on the ledge of the porch of a deserted building, seven feet above ground. It was made of wool, webs and dry vegetation and contained four young ready to leave the nest (7000 feet elev.).

Another nest containing four white eggs was examined May 6, 1973 at the Hay Ranch (elev. 6000 feet). The nest was in a pump shed and was made of fine vegetation and lined with hair.

At Lehman Caves National Monument west of Baker, Park Service personnel reported nesting phoebes under the eve of the superintendents' house in May and June, 1959. A pair also nested in 1961 on the porch of the visitor center. On May 13 and June, 1961 a pair nested under the eve of the superintendent's house.

A. A. Alcorn (MS) at the Overton WMA in Moapa Valley on April 19, 1961 found a Say's Phoebe nest. The nest contained four young birds with unopened eyes and downy feathers on their bodies and heads. The nest was located in the rafters of the shop building on a board nailed to the underside of the rafters. The nest was made of mud, sticks and feathers and was quite large for the size of the phoebe. On May 9, 1961, A. A. Alcorn examined the nest and found it empty.

Vermilion Flycatcher   *Pyrocephalus rubinus*

Distribution. From the southwestern United States south to Guatemala, Honduras and Panama. In South America from Galapagos Islands to southern Argentina. Casually in California, Colorado and central and southern United States.

Status in Nevada. Resident in the southern part of the state; accidental as far north as Fallon.

Records of Occurrence. One of two Vermilion Flycatchers seen six miles southwest of Fallon was taken in October, 1948 (Alcorn MS).

Ryser (1981) reported, "In mid–May of 1981, a male and female Vermilion appeared on Valley Road in Reno and were seen there for several weeks."

Scott (1964) reported, "There was a Vermilion Flycatcher at Gabbs, Nevada April 2 and for several days following. This is farther north in Nevada than this bird usually is reported."

One was seen nine miles north–northeast of Beatty April 2, 1960 (Alcorn MS).

These birds were seen in the vicinity of Hiko, Pahranagat Valley February 17, 1960 (4 seen); April 21, 1960 (one seen); April 22, 1960 (2 pair seen); and May 4, 1962 (one seen) (Alcorn MS).

J. R. and A. A. Alcorn (MS) reported seeing this bird in Moapa Valley 37 times from October 10, 1959 to April 20, 1965. They were recorded once in January and February; 3 times in March; nine in April; 14 in May; 5 in June; and once in September, October and December. One was seen building a nest April 15, 1964.

Austin (1971) reported for southern Nevada, "Occurs all year. Winter resident in desert scrub, permanent resident in riparian areas."

Kingery (1985) reported, "A female Vermilion Flycatcher, formerly regular at several sites in southern Nevada, wintered at Las Vegas for the first occurrence their in several years."

Dusky–capped Flycatcher   *Myiarchus tuberculifer*

Distribution. Formerly Olivaceous Flycatcher. From southern California and Nevada to southwestern New Mexico; south to South America, Argentina and Brazil.

Status in Nevada. Rare in southern Nevada.

Records of Occurrence. Snider (1971) reported on November 15, 1970, "A calling Olivaceous Flycatcher was reported from the Desert Game Range; perhaps the first Nevada record, although one was collected to the west in Death Valley, California."

Kingery (1983) reported one at Overton August 1.

Lawson (TT, 1976) reported, "The first mention we had of this flycatcher was in May, 1970 when John Watson reported an individual up in the oaks in the Mount Charleston area. We discounted it and didn't think anything more about it for a couple of years, then Mowbray reported one in the fall at Corn Creek and then another and another."

Ash–throated Flycatcher   *Myiarchus cinerascens*

Distribution. From northern Oregon, eastward to western Colorado and northern Texas; south to El Salvador. Casually to southern British Columbia and Montana; east to northeastern Atlantic region and Gulf coast.

Status in Nevada. Statewide summer resident.

Records of Occurrence. Linsdale (1936) reported, "This bird is common in southern parts of the state, extending as far north as Pyramid Lake and the Ruby Mountains, but less frequent in the higher northern areas. Also present in winter along the Colorado River."

"On June 16, 1898, Obherholser and Bailey (MS) found a nest near Hastings Pass, Ruby Mountains, on a slope which descends to Ruby Valley. It was in a natural cavity in the trunk of a living juniper and contained three eggs."

Sheldon NWR (1982) reported the Ash–throated Flycatcher as an occasional spring and summer migrant.

Ryser (1965) reported nesting along the Truckee River between Wadsworth and Nixon.

Alcorn (1946) in the Lahontan Valley reported, "Frequently seen in May, June, July and August."

The Ash–throated Flycatcher has declined in numbers in some parts of Nevada over the past 30 years. Recorded only five times in the past 15 years in the Lahontan Valley area, it is seen only on rare occasions (Alcorn MS).

Biale (MS) has seen six or eight in the Eureka area, including one August 28, 1974 and one July 6, 1975.

Biale (PC, 1984) reported one seen in Sand Springs Valley, Nye County on June 12, 1984. He (PC, 1986) reported, "They are uncommon in the Eureka area, seen sparingly in late spring and summer. I am almost sure they nest here."

Austin (1971) reported for southern Nevada, "Usually occurs from March 28 to September 5. Unseasonal occurrences in January, February and on December 31. Summer resident in desert scrub, riparian and woodland areas. Transient in montane forest."

Kingery (1981) reported, "A burned mesquite grove at Ft. Mohave, Nevada, had only one bird December 13—an unusual Ash–throated Flycatcher."

Lawson (TT, 1976) commented, "This bird is a common summer resident in suitable habitat and a common migrant in spring and fall in southern Nevada."

Brown-crested Flycatcher   *Myiarchus tyrannulus*

Distribution. Formerly Wied's Crested Flycatcher. From southeastern California to New Mexico and southern Texas; south to South America (Brazil).

Status in Nevada. Uncommon summer resident in southern part of the state.

Records of Occurrence. Gullion et al., (1959) reported, "A single bird seen in a mesquite bosque in Ash Meadows August 25, 1951. This apparently constitutes the second record for this species in Nevada and the first outside of Colorado River drainage system."

Warren M. Pulich (1952) reported, "On July 19, 1951, while in the southeastern part of Clark County, Nevada, at the tip of the state bordering on California and opposite the site of old Fort Mohave, Arizona, a pair of crested flycatchers were heard. These birds were gradually working overhead, some 30–40 feet high, through the desert foliage of a cottonwood–willow association. When a glimpse of them was obtained they appeared larger than the Ash–throated Flycatcher, a common summer resident of the area. One of the two birds was collected and proved to be an Arizona Crested Flycatcher. It was an adult male . . . . Another Crested Flycatcher was obtained on August 18, 1951, in an area at least a mile and a half from that in which the birds were first found on July 19 . . . . It is believed that at least two pair of crested flycatchers were breeding in the area. It is remarkable that this species has been reported so seldom in the lower Colorado River Valley. Its occurrence has been reported only from south of the Fort Mohave area and to the east of it, in the big Sandy Valley of Mohave County, Arizona."

Lawson (TT, 1976) commented, "It is a rare to uncommon summer resident in southern Nevada."

Kingery (1987) reported Las Vegas' first Brown–crested Flycatcher in two years appeared August 8.

Sulphur–bellied Flycatcher   *Myiodynastes luteiventris*

Distribution. From southeastern Arizona; south to northern South America. Casual to southern California.

Status in Nevada. Hypothetical.

Records of Occurrence. Kingery (1987) reported, "An inconspicuous Sulphur–bellied Flycatcher allowed about an hour of observation near Las Vegas June 13 for Nevada's first record."

Tropical Kingbird   *Tyrannus melancholicus*

Distribution. From southeastern Arizona south to Bolivia and Argentina. Casually along the Pacific coast from southern British Columbia to southern California.

Status in Nevada. Hypothetical.

Records of Occurrence. Kingery (1980) reported, "A Tropical Kingbird fed near Davis Dam in the sagebrush, greasewood and mesquite habitat."

Kingery (1983) reported, "Detailed notes of plumage, viewed through a scope, accompany the report of a Tropical Kingbird at Las Vegas November 7."

Cassin's Kingbird   *Tyrannus vociferans*

Distribution. From California to Oklahoma; south to Guatemala. Casual to Oregon, Ontario and Massachusetts.

Status in Nevada. Uncommon summer resident in the southern part of the state and transient.

Records of Occurrence. Ryser (1985) reported, "There is a September record for Minden, Nevada; one kingbird was found dead along a highway, with a live one nearby."

Grater (1939) reported two individuals at Saint Thomas, June 20, 1938, for the first southern Nevada record.

Lawson (TT, 1976) reported, "This kingbird is an uncommon summer resident in southern Nevada. It also migrates through in spring and fall. I believe it is found as far north as Eureka in suitable habitat. I have also seen this species in the pinon–juniper around 6000 feet in the past few years."

Kingery (1981) reported, "A Cassin's Kingbird had reached Overton, Nevada by February 7."

Western Kingbird   *Tyrannus verticalis*

Distribution. From British Columbia to Newfoundland; south to Costa Rica.

Status in Nevada. Common summer resident over entire state. Occurs mainly in the valleys, but also recorded up to 7500 feet elevation.

Records of Occurrence. Due to the number of records of this bird in all areas of the state, no attempt will be made to list all of them.

Following are observations by the author from selected localities: Jarbidge area September 3, 1960 (one); and the Midas–Tuscarora area August 22, 1965 and July 11, 1966; Elko area 22 times between August 22, 1963 and May 31, 1968; 6 times in the Ruby Lake NWR area between May 27, 1965 and May 19, 1966; 14 times in the Lovelock area between April 26, 1961 and July 13, 1966; 5 times in the Reno area from May 21, 1962 to July 18, 1964; 52 times in the Mason Valley area from July 11, 1941 to May 4, 1947 and 9 times in the same area from June 16, 1961 to July 23, 1965. They usually arrive in April and depart by September. The first one seen in 1984 was on April 23.

Sheldon NWR (1982) reported the Western Kingbird as an occasional spring and summer resident with nesting reported in the area.

They were recorded 803 times in the Fallon area between April 22, 1960 and April 21, 1969. The earliest sighting of this kingbird in the Fallon area from 1960 to 1970 was April 1, 1964 by Mills (Alcorn MS).

They were recorded three times in the Austin area between May 16, 1961 and June 10, 1965 (Alcorn MS).

At six miles west–southwest of Fallon, a newly constructed nest contained no eggs June 20, 1964. The same nest contained three eggs June 23 (Alcorn MS).

Western Kingbirds are normally seen on power lines flying about and catching insects to eat. However, Mills (PC) reported three Western Kingbirds eating chokecherries at his residence in

Fallon on June 29, 1987. "The cherries were pretty ripe and the robins were eating them. I guess the kingbirds decided they wanted them too."

Biale (MS) saw these birds in the Eureka area from May 1, 1964 to September 11, 1967 a total of 291 times. Spring arrivals were in May, except one early record of April 30, 1966. It normally left the area by the last of August, but it was recorded as late as September 11. Biale saw one building a nest June 22, 1965 and a nest in an elm tree contained young July 26, 1965. He (PC, 1986) reported a nest in a juniper tree with five eggs on June 7, 1985.

Kingery (1982) reported that Western Kingbirds appeared at Eureka April 7—an early record.

In the Hiko area they were recorded by the author 33 times between April 17, 1960 and June 28, 1965.

J. R. and A. A. Alcorn (MS) recorded these kingbirds 115 times in southern Nevada from 1960 to 1965. It arrived in April (early April to the middle of the month) and normally left the area by the last of August or the first week in September. Late records include: one bird at Overton WMA December 11, 1960 and another at Virgin Valley November 4, 1961.

A. A. Alcorn (MS) reported at Pahrump Ranch, Nye County on May 16, 1961, "I observed a kingbird flying toward a building carrying a three–foot length of string. This string was trailing behind as the bird flew."

Lawson (TT, 1976) reported for southern Nevada, "This is a species that arrives in late April, very quickly establishes nest sites, nests, then departs. They are not at all common after the first of July, they have already migrated southward."

Eastern Kingbird   *Tyrannus tyrannus*

Distribution. From southeastern Alaska to Nova Scotia; south through Nevada to southern Florida. In South America from Peru to Bolivia.

Status in Nevada. An infrequent summer visitant in the northern part of the state; transient elsewhere.

Records of Occurrence. Sheldon NWR (1982) reported the Eastern Kingbird as an occasional summer and fall transient.

Linsdale (1936) reported two females collected at Big Creek Ranch in the Pine Forest Mountains, Humboldt County on July 8 and 9, 1909 by Taylor. He also reported Hanna (1904) observed a few individuals in Humboldt County in June, 1903.

Ridgway (1877) reported two or more pair lived in large cottonwood trees in the Truckee Valley when he made observations there in 1867–68.

More recent records are of one seen perched in an aspen tree at the north end of Wildhorse Reservoir April 8, 1972 (Alcorn MS).

Howard (MS, 1972) observed one perched on the Ruby Lake NWR boundary fence May 17, 1972.

One was observed by Ryser (1966) in the Lovelock area, Pershing County.

In the Fallon area, one was seen June 30, 1968 by the author.

Napier (MS, 1974) reported seeing one near Cattail Lake, Stillwater Marsh June 4, 1970.

Mills (PC) reported one seen one mile southeast of Soda Lake July 29, 1977 and this author saw one six miles west–southwest of Fallon June 19, 1979.

Biale (MS) recorded one in the Eureka area August 31, 1967; one at Hunters Ranch, five miles east of Eureka June 25, 1969; one June 7 and August 30, 1970; one September 8, 1974; and one in Eureka August 25, 1979.

Kingery (1987) reported an Eastern Kingbird at Dyer May 22–31.

A. A. Alcorn (MS) saw one at Hiko, Pahranagat Valley June 2, 1965 and at Crystal Springs in the same valley he saw two Western Kingbirds fighting an Eastern Kingbird.

Grater (1939) reported one at Saint Thomas April 13, 1938 and Austin (1969) reported one observed at Corn Creek September 3, 1966.

Kingery (1984) reported Eastern Kingbirds in Las Vegas in May and from June 16–23. He (1985) reported, "A late Eastern Kingbird was at Las Vegas November 15."

Lawson (TT, 1976) commented, "It is a rare species in southern Nevada, with only three or four sightings from Moapa and Pahranagat Valleys."

Scissor–tailed Flycatcher   *Tyrannus forficatus*

Distribution. From eastern New Mexico to the Atlantic coast; casually throughout most of North America.

Status in Nevada. A rare transient.

Records of Occurrence. Biale (MS) picked up a nearly dead male at the south end of Diamond Valley near Eureka May 21, 1965. This author prepared it as a study skin.

D. A. Zimmerman (1962) reported, "On May 1, 1962 one was seen at Corn Creek Field Station and another was collected there May 20. These are apparently the first published records of the species for the state, although a bird was seen at this same place two years ago."

Austin (1971) reported them on the hypothetical occurrence list as follows: "Four records: Spring, 1960; 1961; 20 May, 1962, Corn Creek (Banks and Hansen, 1970; Zimmerman, 1962); and 23 June, 1966, Desert NWR (Snider, 1966)."

Lawson (TT, 1976) reported, "There aren't too many records of this species, perhaps eight or 12. I am not certain because I don't have Hansen's records from Corn Creek and he did elude to the fact that the species did occur there almost every summer, although only an individual would appear. It's been recorded at Mormon Farm in Moapa Valley. It would have to be considered a rare species here in southern Nevada."

Kingery (1984) reported, "All birdwatchers in southern Nevada saw one at Las Vegas October 3–21, the first local Scissor–tailed Flycatcher since 1971." He (1985) reported, "Wandering Scissor–tailed Flycatchers visited Las Vegas September 7."

Family **ALAUDIDAE:** Larks

Horned Lark   *Eremophila alpestris*

Distribution. Northern parts of both hemispheres; south to Italy, northern China and Japan. In the Americas south to South America (Andes).

Status in Nevada. Common resident statewide. Gabrielson (1949) commented, "Horned Larks are the most common passerine bird in much of Nevada."

Linsdale (1951) reported, "*E. a. lamprochroma* is summer resident in the western half of the state, south to Fish Lake Valley, Esmeralda County. In other seasons, present farther east and south. This and other races of Horned Lark in Nevada are treated fully by Behle (1942)."

"The race *E. a. utahensis* is resident in eastern Nevada from Elko and White Pine Counties south to Smoky Valley in Nye County. Present in other seasons farther south and west."

"In the southwestern part of the state, *E. a. ammophila* is resident; recorded from 19 1/2 miles southeast of Goldfield and Smith Ranch, Charleston Mountains."

"In extreme southwestern Nevada, *E. a. leucansiptila* is resident; a specimen from Arden reported by Behle (1942)."

Records of Occurrence. These birds have been seen statewide so many times and at so many places that no attempt is made to list them all.

In northern Washoe County it was seen during each visit to Vya (Alcorn MS).

Sheldon NWR (1982) reported the Horned Lark as abundant, common year round with breeding in the area.

In Humboldt County, the Horned Lark was recorded frequently in Paradise Valley and Winnemucca.

In the Elko area it was recorded 22 times at Tuscarora from 1964 to 1967 and 19 times in Elko from 1964 to 1968 (Alcorn MS).

In Pershing County from December, 1960 to December, 1967, the Horned Lark was seen 33 times in the Lovelock area. Robert Ferraro phoned to report Horned Larks doing damage as they were eating pollinating leaf cutter bees in the alfalfa fields in Dixie Valley on July 8, 1966 (Alcorn MS).

In Lyon County, in the Yerington area, they were recorded frequently. At Silver Springs on December 18, 1961 this author noted a 40–acre rye field had been heavily damaged by over 2000 Horned Larks (Alcorn MS).

In Churchill County from January 4, 1960 to December, 1967 they were recorded 319 times. Flocks averaging 33 birds were seen flying over alfalfa stubble only during winter months (October to February) (Alcorn MS).

Napier (MS, 1974) reported Horned Larks common in all months in the Stillwater WMA, "Probably more abundant in winter than summer."

In Lander County from April 7, 1960 to October 4, 1966 the Horned Lark was seen 63 times in the Austin area and 39 times in the Upper Reese River area near the Lander/Nye County boundary (Alcorn MS).

Biale (MS) reported Horned Larks in the Eureka area a total of 312 times from May 1, 1964 through December 29, 1967. They were seen in all months. He saw a nest two miles north of Eureka on the ground under a small sage July 4, 1984. It contained three young ready to leave the nest. He (PC, 1986) reported a nest on the ground near Eureka with three eggs June 3, 1984 and a nest with three eggs at south Diamond Valley April 19, 1985.

In White Pine County from 1960 to 1967 they were seen 20 times (Alcorn MS).

In Lincoln County they were seen 15 times in the Hiko area from 1960–64 (Alcorn MS).

A. A. Alcorn (MS) reported them in Moapa Valley, Clark County in all months except February and September from 1960 to 1964.

Gullion et al., (1959) reported, "Common winter visitor and less common, but widespread summer resident. Pulich found young just out of the nest near Boulder City on July 7, 1952."

Austin and Bradley (1971) reported Horned Larks, "Occurring all year in southern Nevada. Permanent resident in desert scrub and riparian areas. Summer resident in woodlands."

Family **HIRUNDINIDAE:** Swallows

Purple Martin *Progne subis*

Distribution. From British Columbia to Nova Scotia; south in the western and eastern United States to South America.

Status in Nevada. Rare transient.

Records of Occurrence. Ridgway (1877) reported the Purple Martin as not common at Carson City; a single bird at Virginia City June 18, 1868.

Gullion et al., (1959) reported, "Four records: one bird in Mohave Valley May 14, 1951; 14 birds in Muddy Valley (1700 feet elevation, Clark County) on August 28, 1951; a 'flock' of birds near Carp (2600 feet elevation, Meadow Valley Wash, Lincoln County) on August 29, 1951; and three birds in Pahrump Valley on August 5, 1954."

A. A. Alcorn (MS) reported about 10 flying in Moapa Valley May 6, 1964.

Austin (1969) reported for southern Nevada, "Additional records for April 19, 1967, Davis Dam area; September 2, 1966, Henderson Slough; and five birds on September 13, 1966 at Logandale. Also, the remains of a male were found at Corn Creek in May, 1964 by Hansen."

Austin (1971) reported for southern Nevada, "Transient in riparian areas."

Monson (1972) reported two observed April 22, 1972 at Corn Creek and six found at Las Vegas Wash April 28, 1972.

Kingery (1980) reported single birds at two different places in Las Vegas May 27 and 29. He (1984) reported, "The Purple Martin at Las Vegas August 11 (1983) provided the first fall record since 1971."

Tree Swallow *Tachycineta bicolor*

Distribution. From Alaska to Newfoundland; south to Central America.

Status in Nevada. Summer resident and transient. Most of these swallows are seen as they migrate through the state in early spring and in late summer. However, nesting does occur in parts of the northern half of the state along streams, rivers and lakes where timber is available. They usually nest in abandoned woodpecker holes.

Records of Occurrence. Sheldon NWR (1982) reported the Tree Swallow as an uncommon spring and occasional nesting summer resident.

LaRochelle (MS, 1967) reported Tree Swallows arrive at Ruby Lake NWR in early April and nest in tree cavities. They usually leave the area in August, but a late record for near Elko is of one seen September 26, 1962.

Orr and Moffitt (1971) reported them nesting at Lake Tahoe.

Ridgway (1877) reported these swallows nest at Carson City and Truckee Bottoms (Reno).

At Idlewild Park in Reno, the earliest arrival was of four birds seen March 5, 1942. They were repeatedly seen in April, May, June and July of the same year and were nesting there. None were seen after August 15 (Alcorn MS).

Kingery (1977) reported one at Reno February 4.

A pair, seen June 14, 1962 along the Truckee River near Wadsworth, were thought to be nesting (Alcorn MS).

Ryser (1971) reported Tree Swallows several miles below the Big Bend of the Truckee River February 17.

Ryser (1973) reported a migrating flock of about 200 seen March 26, 1973 in the Truckee Meadows.

In the cottonwood trees along the Carson River, five miles east of Fort Churchill, numerous Tree Swallows were seen June 12, 1960. One pair was seen at a dead tree with a nest hole revealing a feather–lined nest with five swallow eggs. The feathers lining the nest were mostly wild duck feathers (Alcorn MS).

In the Fallon area, approximately 600 migrating swallows were seen in two flocks March 13, 1960. One flock contained about 100 Tree Swallows and the other about 500 (Alcorn MS).

The majority of migrating birds arrive in March after winter weather has moderated. However, at six miles southwest of Fallon the author saw two Tree Swallows on February 27, 1962. "They were perched on a wire, side–by–side (facing opposite directions). The temperature reached to minus one degree Fahrenheit the previous night. The two Tree Swallows were cold and allowed me to approach within 10 feet. They did not fly, but one took its head out of its feathers when I made a noise."

Other early arrivals were of two seen February 22, 1962 in the Austin area (Alcorn MS).

Biale (MS) for the Eureka area reported them February 26, 1967 and he recorded them numerous times in this same area in March, April and May, from 1965 to 1975.

Kingery (1980) reported 10,000 swallows, primarily Tree Swallows, at Comins Lake near Ely.

A. A. Alcorn (MS) saw two in the Baker area May 22 and July 4, 1967. In the Hiko area he saw them in March, April, May and August from 1963 to 1965.

In Moapa Valley these swallows from 1957 to 1965 were seen 4 times in March; 3 in April; once in May; and twice in August (Alcorn MS).

Cruickshank (1965) reported three of these swallows at Overton, Moapa Valley, December 29, 1964 during the CBC.

Austin (1971) reported for southern Nevada, "Usually occur from February 10 to May 15 and from July 10 to September 3–6. Transient in desert scrub and riparian areas."

Kingery (1977) reported Mowbray saw 2000–3000 Tree Swallows winter at Davis Dam and migration brought the total up to 50,000 by February 27. Kingery (1980) also reported, "Davis Dam had normal numbers wintering with about 1200 counted on each of three trips."

Violet–green Swallow   *Tachycineta thalassina*

Distribution. Western North America from Yukon River Valley south and east to South Dakota; south to Central America.

Status in Nevada. Summer resident mostly in mountains and transient elsewhere.

Records of Occurrence. Sheldon NWR (1982) reported the Violet–green Swallow as uncommon spring and fall; common breeding summer resident.

LaRochelle (MS, 1967) reported for Ruby Lake NWR, "One of the early arriving swallows. Banded several in 1963. Nests around Civilian Conservation Corps Camp."

The author frequently saw Violet–green Swallows at the Ruby Lake NWR in July and August, 1965.

Ryser (1967) reported them in Lamoille Canyon June 1 to 4, 1967.

Ridgway (1877) encountered this swallow nesting at Pyramid Lake.

They were frequently seen in the Reno area in late August and early September (Alcorn MS).

In the Fallon area the Violet–green Swallow is often seen in March, April and May of each year. It is also occasionally seen in late August. No known nesting in the immediate area (Alcorn MS).

Biale (MS) recorded this swallow in the Eureka area 112 times between May 2, 1964 and September 4, 1967. He also reported definite nesting in the area near rocky cliffs where they utilized the holes and crevices in the rocks. Spring arrivals in this area showed the first week of May (except one on April 17, 1966) of each year. Recorded many times in August, but most leave the area by the last of August. Only two September dates: September 2, 1966 and September 4, 1967.

A. A. Alcorn (MS) saw them in the Baker area in May, June, July and early September of 1967. Also, farther south at Hiko in 1965, he saw them twice in March; 5 times in May; on June 27; and on July 1.

Kingery (1975) reported, "Ryser reports a group of Violet–green Swallows tried to enter burrows of Bank Swallows during a heavy snow and hail storm on May 19 at Pahranagat Refuge."

J. R. and A. A. Alcorn (MS) recorded this swallow from Moapa Valley 24 times between April 25, 1957 and June 27, 1965. It was seen from March to July of each year (Alcorn MS).

Gullion et al., (1959) reported for southern Nevada, "Spring and fall migrants on the desert."

Austin (1971) reported them in southern Nevada, "Present in woodlands and montane forests, it usually occurs from March 22 to September 14. In desert scrub and riparian areas it usually occurs from February 23 to May 25 and from July 18 to October 18. An unseasonal occurrence on June 21, July 6, November 26 and December 30. Transient in desert scrub, riparian and woodland areas. Summer resident in montane forests."

Kingery (1977) reported Vince Mowbray saw six at Davis Dam February 17 and he (1978) reported 50 at Davis Dam February 18.

Northern Rough–winged Swallow   *Stelgidopteryx serripennis*

Distribution. From southeastern Alaska to New Brunswick; south to Panama.

Status in Nevada. Common summer resident and transient statewide.

Records of Occurrence. Sheldon NWR (1982) reported the Northern Rough–winged Swallow as an uncommon spring, summer and fall resident with nesting in the area.

Quiroz and this author saw these birds along the south fork of the Owyhee River on July 11 and 12, 1969 while traveling by boat down the river. All were lone birds thought to be nesting in the area.

At Ruby Lake NWR about 10 were seen March 23, 1966. The latest record is of one seen August 22 (Alcorn MS).

At this same refuge, LaRochelle (MS, 1967) reported, "This bird arrives later than Violet–green and Tree Swallows. Nests in gravel pit banks."

Burleigh collected three specimens in the Reno area: a female on June 14, 1967; and a male and female on June 4, 1968.

At Idlewild Park in Reno these birds were seen April 18 and 30, on May 28, and July 1 and 14–17, 1942 (Alcorn MS).

In Lahontan Valley (Fallon area) these swallows regularly nest in the banks of ditches (Alcorn MS).

Biale (MS) saw these birds from June of 1964 to August 25, 1967, 55 times in the Eureka area. He reported them common in May, June and July. His 10–year record from this area reveals the earliest arrival was March 29, 1966 and the latest fall record was September 16, 1964. He saw a nest in Diamond Valley July 3, 1983. It was located in a hole in a dirt bank and contained an unknown number of young. He (PC, 1986) reported a nest in a dirt bank in south Diamond Valley June 20, 1985.

A. A. Alcorn (MS) saw these birds at Baker April 20 and May 13, 1967 and he saw them near Hiko in the Pahranagat Valley May 5, 1962 and April 23, 1964.

J. R. and A. A. Alcorn (MS) reported early arrivals in the Moapa Valley on February 21, 1963 and March 9, 12 and 13, 1962. In this area they were seen infrequently in April and May with one record of July 9, 1963.

Gullion et al., (1959) reported for southern Nevada, "One nesting record: Meadow Valley, May 26, 1951 and another, possibly, in Pahranagat Valley May 24, 1951."

Austin (1971) reported for southern Nevada, "Summer resident in riparian areas and usually transient in desert scrub. From March 10 to October 13, with unseasonal occurrences on November 5 and December 30."

Kingery (1978) reported, "More Rough–winged Swallows than ever wintered with 35–50 at Davis Dam and one to four at Lake Mead."

Kingery (1987) reported two Northern Rough–winged Swallows at Las Vegas February 15.

Bank Swallow   *Riparia riparia*

Distribution. From Alaska to northern Siberia; south to South America, Africa, India and Japan.

Status in Nevada. Summer resident; transient statewide.

Records of Occurrence. LaRochelle (MS, 1967) reported for the Ruby Lake NWR, "A mid–spring arrival nests in gravel pits on the area."

In the Reno area it was observed regularly from April 30 to September 4, 1942 (Alcorn MS).

At Lahontan Reservoir, Lyon County in June, 1959 about 30 Bank Swallows were seen flying in and out of some small openings in a sandy bank. Nests lined with feathers were at the end of the horizontal burrows that were 18 to 24 inches deep. A visit to the area on July 11, 1959 revealed not a single live swallow although one dead was found in a nest. One or more badgers had honeycombed the area apparently to feed on the eggs or young birds and had destroyed most of the nests (Alcorn MS).

Alcorn (1946) reported these swallows as regularly nesting at the north end of Soda Lake in the Fallon area. Bank Swallows are regularly seen in June, July and August in the Fallon area.

Biale (MS) noted this bird is not common in the Eureka area. He observed them April 11, May 26 and July 4, 1965.

At lower Pahranagat Lake, two were seen May 25, 1962 (Alcorn MS).

A. A. Alcorn (MS) saw one in the Virgin Valley April 22, 1963 and J. R. and A. A. Alcorn (MS) saw about 10 birds in the Moapa Valley April 22, 1964.

For southern Nevada, Austin (1971) reported, ''These birds usually occur from April 3 to May 15 and from August 6 to September 29. Unseasonal occurrence on March 10. Transient in desert scrub and riparian areas.''

Cliff Swallow   *Hirundo pyrrhonota*

Distribution. From central Alaska to Nova Scotia and southern Greenland; south to South America (Brazil).

Status in Nevada. Summer resident and transient. Alcorn (1946) reported two races in Nevada: *H. p. hypopolia* and *H. p. albifrons*.

Ryser (1985) commented, ''Not only is the Cliff the most abundant and widespread swallow in the Great Basin today, but it was so even in Ridgway's day.''

Records of Occurrence. These swallows have been seen in all areas of the state. Linsdale (1936) reported them from 12 of 17 Nevada counties. This author's records represent all 17 counties in Nevada with nesting colonies in many areas.

Sheldon NWR (1982) reported the Cliff Swallow as uncommon spring and fall and a common breeding summer resident.

Nesting sites follow: on the south fork of the Owyhee River, Elko County, J. R. Alcorn and Quiroz reported seeing numerous colonies of Cliff Swallow nests on the cliffs above the river. Each group of nests seen July 11, 1967 contained from 40–100 nests. On July 12, 1967 they saw four more colonies with about 40, 80, 50 and 30 nests (Alcorn MS).

About 100 nests were seen under the highway bridge at the south end of Wildhorse Reservoir, Elko County on May 30, 1971 (Alcorn MS).

Over 100 birds were observed nesting under a bridge near Jiggs, Elko County June 27, 1962 (Alcorn MS).

At the Quilici Brothers Ranch, six miles west of Fort Churchill, Lyon County on June 14, 1941, these swallows were building nests under the water tower tank. Of the 21 nests seen only two appeared completed. The other 19 were under construction. A revisit to the site on July 2, 1941 revealed all nests apparently completed and as this author walked under them, Cliff Swallows were looking out of every nest (Alcorn MS).

These swallows have also been repeatedly seen nesting in Wilson Canyon, Lyon County. Some nests are situated on the rocky cliffs, others are under the highway bridge (Alcorn MS).

These swallows regularly nest on the control tower and under the arches of the road at Lahontan Dam, Churchill County. Examination of this nesting site on March 23, 1960 revealed no swallows, but on April 8, 1960 about 50 Cliff Swallows were flying about and perching at this site with no nesting activity. On May 14, 1960 about 90 nests were being constructed at the small north tower and about 40 nests were partially constructed at the north arch. The earliest spring arrival for Lahontan Valley in 1962 was April 12. The latest fall record was August 26, 1962 (Alcorn MS).

Biale (MS) reported these birds as a summer resident and he saw nests in Newark Valley east of Eureka on the porch of a house. He (PC, 1986) reported 50 active Cliff Swallow nests in a concrete culvert under the highway in north Antelope Valley June 6, 1986.

Kingery (1984) reported, ''Cliff Swallows face an uncertain future when they nest around homes and condominiums: owners dislike their droppings and resort to various means, from brooms to garden hoses, to dislodge them. At Eureka, Nevada where Eyre doesn't mind them, the tale of seven nesting pair seems to follow the script of the 'perils of Pauline': They failed to finish one, House Sparrows took three, starlings destroyed one, and two survived.''

Gullion et al., (1959) reported Cliff Swallows as an uncommon migrant in southern Nevada.

J. R. and A. A. Alcorn (MS) recorded 28 nests in Moapa Valley, southern Nevada on May 31, 1963. These birds were seen yearly from 1961 through 1965. The earliest spring record was March 3, 1965; though they were normally not seen until April. The latest fall record was July 4, 1961.

Austin (1971) reported for southern Nevada, ''Summer resident in riparian areas and transient in desert scrub, usually occurring from March 18 to October 18.''

Kingery (1981) reported, ''Cliff Swallows were building nests at Davis Dam on February 28, extremely early.'' He (1984) reported ''very early'' Cliff Swallows at Davis Dam February 7.

Barn Swallow   *Hirundo rustica*

Distribution. In both Hemispheres. From Newfoundland and Greenland; south to southern South America. In the Eastern Hemisphere from northern Scandinavia to Siberia; south to Africa, the Philippines and Micronesia.

Status in Nevada. A common summer resident in the valleys of the northern half of the state; transient statewide.

Records of Occurrence. These birds have been seen in all sections of the state. Their widespread abundance represents many localities.

Sheldon NWR (1982) reported the Barn Swallow as an abundant summer resident, uncommon spring and common in fall.

LaRochelle (MS, 1967) reported, ''A common mid–spring arrival that surely nests, though not as yet observed.''

Barn Swallows were frequently observed in the Reno area from March 29 through the summer months to as late as September 17 (Alcorn MS).

Their abundance in the Fallon area is evident from this author recording them 984 times from March, 1960 to March 25, 1969. The earliest spring arrival and the latest fall record for eight years in the Fallon area are of two seen March 7 and October 18 when one was seen (Alcorn MS).

Many of their nests are located under the bridges which span the canals in this area. These nests are located a few feet above the canal water and seldom can one find more than three or four at one bridge (Alcorn MS).

Biale (MS) reported these birds in the Eureka area 204 times between May, 1964 and September, 1967. Their greatest abundance during this period was June, July and August. He reported only four records for April and the earliest was April 3, 1966. His latest record was September 30, 1965 in this period. Later, he reported seeing one as late as October 17, 1969.

Biale (MS) reported nests in the Eureka area: July 10, 1964, nest in a pump house contained three eggs; and July 26, 1964, six nests in an old building, four nests contained four eggs each, one contained five eggs, and one contained five young. He also reported two nests July 24, 1967 in a highway underpass. They were made of mud and lined with grass, hair and feathers.

One nest contained four young and the other three young. The nest was about seven feet above ground.

J. R. and A. A. Alcorn (MS) in Moapa Valley from 1960 through 1965, recorded them on occasion in April and May, with one observed November 24, 1961.

Austin (1971) reported for southern Nevada, "Transient in desert scrub, riparian and woodland areas, usually occurring from April 5 to June 4 and from September 1 to October 18."

Gray Jay   *Perisoreus canadensis*

Distribution. From Alaska to Nova Scotia; south in the United States to Arizona in the west and New England in the east.

Status in Nevada. Hypothetical.

Records of Occurrence. Linsdale (1936) reported, "This jay has been reported by Para-menter (1924) as observed on August 17, 1923 near Glenbrook, Douglas County. Mr. Parameter wrote as follows. 'We were on an elevated point some 300 feet above the lake (Tahoe) or at about 6500 feet above sea–level. The jay was perched on top of a large pine tree and we observed him at rest for more than 10 minutes.'"

Mowbray (PC, 1988) reported the Gray Jay as suspect on Mt. Charleston.

Steller's Jay   *Cyanocitta stelleri*

Distribution. From Alaska southeast to Alberta and South Dakota. Through southern California and Mexico to Nicaragua. Accidental in Quebec.

Status in Nevada. Resident in some mountain ranges. Linsdale (1951) reported three races in Nevada: *C. s. frontalis, C. s. macrolopha* and *C. s. percontatrix*.

Records of Occurrence. George K. Tsukamoto (1966) reported, "Hoskins records two separate sightings of this species in his field notes. A flock of eight birds was observed on McDermitt Creek in Cherry Creek Mountains on 18 February 1959. On 30 October 1960 a single bird was seen above the W. Payne Ranch, east of Pequop Summit, Pequop Mountains. These are the first records of this species in Elko County and the northernmost occurrence for eastern Nevada."

These birds have been seen repeatedly in the Carson Range near Lake Tahoe, Hunters Lake, Mount Rose, Reno, Carson City and southeast to Pine Grove (Alcorn MS).

Burleigh collected eight specimens for the USNM from the Mount Rose and Galena Creek areas. They were taken in January, April, July, September and November, from 1967 to 1969.

Evenden (1952) reported, "Two were seen on the east slope of the Desatoya Mountains, Lander County, on June 24, 1949."

Over the past 40 years, the author has spent many days in the mountains Evenden discussed and has never seen this jay in this range. It is thought the birds Evenden saw were stragglers.

Biale (MS) reported one seen in the Duck Creek Basin near Ely October 25, 1972 and one at Mt. Hamilton, White Pine County October 14, 1972.

Steller's Jay. Photo by Roy Nojima.

At Lehman Caves National Monument they were recorded October 25, 1960; August 26, 1964; September 1, 1964 (2 seen); and June 23, 1965 (Alcorn MS).

Johnson (1972) reported, "In 1972, the species was seen regularly in ponderosa pines on the west slope of Sawmill Peak, Clover Mountains, and a population of moderate size probably breeds there. A significant extension of known range for the species was recorded May 26, 1972, Sawmill Canyon, Quinn Canyon Mountains (7900 feet elevation) when an apparently mated male and female in post–breeding condition were taken in an open stand of mature ponderosa pine, bristlecone pine and white fir."

J. R. and A. A. Alcorn (MS) saw four of these jays at Gold Butte, Clark County October 30, 1960.

Austin (1971) reported for southern Nevada, "These jays occur all year as a permanent resident in montane forests and as a visitant in riparian and woodland areas."

Blue Jay   *Cyanocitta cristata*

Distribution. From extreme east–central British Columbia to Newfoundland; south to central and southeastern Texas, the Gulf coast and southern Florida. Casually west to southwestern British Columbia, western Washington and the western United States.

Status in Nevada. Rare straggler.

Records of Occurrence. Cruickshank (1958) listed this bird at Ruby Lake NWR December 28, 1958 during the CBC.

Kingery (1978) reported two Blue Jays stayed at Incline Village, Lake Tahoe March 24, 1978.

V. K. Johnson (PC, 1988) reported a Blue Jay stayed in his yard in Fernley through the 1987–88 winter.

Herron (PC, 1988) reported a Blue Jay was seen in the Reno area in March, 1988. He speculated that it was the same Blue Jay Johnson had observed in his yard in Fernley over the winter.

A. A. Alcorn (MS) saw two Blue Jays four miles west of Fallon December 14, 1976. One was obtained as a specimen.

Biale (MS, 1987) reported one Blue Jay at his feeder in Eureka from November 28 through December 4.

Scrub Jay   *Aphelocoma coerulescens*

Distribution. Formerly the Woodhouse Jay. From southwestern Washington to Colorado; south in Mexico to Oaxaca and Veracruz. Also in central Florida. Casual in British Columbia and southern midwest.

Status in Nevada. Resident in all sections of the state. Most commonly found in lower portions of mountain ranges, but moves into the valleys in winter. Linsdale (1951) reported three races in Nevada: *A. c. superciliosa, A. c. nevadae* and *A. c. woodhouseii.*

Records of Occurrence. Combining this author's records with published material, these birds have been recorded from all 17 counties in the state. However, these birds are not numerous and often only one is seen at a place.

Sheldon NWR (1982) reported the Scrub Jay as a common permanent resident spring and fall; uncommon in winter.

Specimens in the USNM were taken from Ruby Lake NWR, Reno, Galena Creek near Reno, and Carson City.

Gabrielson (1949) reported records from, "Pine Nut Mountain, Douglas County, August 22, 1933; near the top of Silver Peak, Esmeralda County, where a male and female were taken on November 22, 1934; several, Potts Post Office, Nye County, October 11, 1932; several on the lower slopes of Wheeler Peak, White Pine County, August 16, 1933; two on the eastern slopes of Ruby Mountains, Elko County, August 18, 1933."

In the Fallon area they are not a consistent winter visitor. One Scrub Jay was observed six miles west–southwest of Fallon September 27, 1972. None were recorded in summer months. Two Scrub Jays were observed at the same place November 19, 1972. One was feeding on Russian olive berries, but this author did not determine whether the bird was eating whole berries or just the pulp surrounding the hard seed. Later both birds were seen eating kernels of corn from the cob. This was sweet corn just left on the ground for bird feed and the kernels were not yet completely dry and hard. Scrub Jays repeatedly fed on a hunk of beef fat tied to a platform in our front yard in the winter of 1973 and 1974. One was seen April 2, 1980 (Alcorn MS).

Scrub Jay. Photo by A. A. Alcorn.

Linsdale (1936) reported, "Woodhouse Jays nest early but not as early as most other species in this family. In the Toiyabe Mountains, near the center of the state, nests were found on June 3 and 4, 1932 (Orr, MS). One was in a juniper and had just been vacated by the young; the other was in a pinon and held two nearly grown young and an unhatched egg."

Biale (MS) recorded this jay in the Eureka area a total of 304 times from June 29, 1964 to December 29, 1967. One nest containing three eggs was seen seven feet above ground on a horizontal branch of a pinon tree on May 13, 1966. One other nest containing three eggs was seen April 28, 1982.

A. A. Alcorn (MS) reported this bird at the Spring Creek Fish Rearing Station in the Baker area in the months of April, May, June, July, October, November and December.

Two of these jays were seen in the Pahranagat Valley near Hiko on March 1, 1962 (Alcorn MS).

Combined records of J. R. and A. A. Alcorn (MS) reveal this jay observed in the Moapa Valley area in all months from 1959 to 1965.

Johnson (1965) reported up to 10 individuals, including fully grown juveniles, in the Sheep and Spring Ranges in June.

Austin (1971) says, "Occurs all year in desert scrub and riparian areas. In woodlands and montane forests, it usually occurs from August 11 to April 25 with an unseasonal occurrence on June 13. Winter resident in desert scrub and riparian areas. Permanent resident in woodland. Transient in montane forests."

Pinyon Jay   *Gymnorhinus cyanocephalus*

Distribution. From central Oregon to South Dakota; to northern Chihuahua, Mexico.

Status in Nevada. Common resident of the pinon–pine, tree–covered part of mountain ranges. Transient elsewhere over the entire state.

Records of Occurrence. LaRochelle (MS, 1967) reported for Ruby Lake NWR these jays are found in raucous flocks of 2550 in fall and winter. He observed no nesting.

About 35 were seen by this author in this area April 15, 1966. On October 24, 1971 a flock of about 100 were seen west of the Fort Ruby Ranch at the south end of Ruby Valley (Alcorn MS).

In western Nevada, they were observed frequently in the Virginia City area and in the pinon pine forests south and east to Montgomery Pass (Alcorn MS).

Specimens in the USNM were taken by Burleigh from the Reno vicinity January 24, April 30 and July 19, of 1967; December 13, 1968; and December 25, 1967.

Ridgway (1877) reported he found near Carson City 100 or more nests in one grove April 21, 1868. All nests were vacated except one nest with four young ready to leave.

Transient birds have been seen at irregular intervals in the Fallon area in spring and fall months. In the mountain ranges from Eastgate, Churchill County, eastward to the Snake Range in easternmost Nevada, these birds were frequently seen in tight flocks numbering 30–200 individuals (Alcorn MS).

On December 13, 1970 in Buffalo Canyon, Desatoya Mountains, Churchill County, three vociferous flocks numbering approximately 200, 80 and 40 individuals were seen. On January 23, 1971 in the same canyon, three flocks were heard, then seen (Alcorn MS).

On April 16, 1967 Biale spent about four hours searching an area of about 19 acres for nests of these jays. He found 21 nests in a very, very loose colony. All nests were found built in pinon trees six to 12 feet above the ground. Most nests were built close to the main trunk of the tree. However, some were built on horizontal branches a foot or more away from the trunk. All nests found were made nearly entirely of juniper twigs and were lined with cheat grass and juniper bark. A few nests also had feathers and some horse hair mixed into the lining.

Of the 21 nests found and examined by Biale, 11 were old nests from the previous year or years; six were new nests without eggs; one nest contained a broken egg; one nest contained four eggs, two of which were pipped; one nest contained three eggs; one nest contained four young, one of which was dead.

In 1984, the earliest nest found by Biale was April 5, which contained three young and one egg and the latest nest found was June 1. It contained six young. In 1985, the earliest nest seen was March 30 and it contained four eggs.

Gabrielson (1949) reported, ''Seen near Potts Post Office, Nye County October 11, 1933; numerous on Wheeler Peak, White Pine County, August 16, 1933; very common, Pine Nut Mountains, Douglas County, August 22, 1933. On November 22, 1934, on Silver Peak, Esmeralda County, I saw the largest flock of Pinyon Jays in my experience. Several thousand birds were spread over a large area in a loose aggregation of flocks that filled the pinon pines for some distance. Two males and a female were taken from this group.

Bond (1940) reported, ''No occupied nest was found, but young were still being fed by

Pinyon Jay. Photo by Art Biale.

parents on May 20. Several times small flocks were seen in the Delmar [Delamar] Valley in the Joshua trees.''

A. A. Alcorn (MS) in the area west of Baker reported seeing these birds almost daily from March through September, 1964.

A. A. Alcorn (MS) reported in the Hiko area of Pahranagat Valley a flock of about 50 was seen flying December 27, 1964.

Johnson (1965) reported, "On June 14, a group of six was found in scattered pinon woodland at 6200 feet in Hidden Forest Canyon, Sheep Range. This species was recorded by van Rossem only as a common visitant during the fall and winter . . . . There is a female from the north side of Potosi Mountains, 7000 feet, taken on June 14, 1940. This specimen suggests that this species is a summer resident in the Spring Range, although it could be a wanderer inasmuch as the species breeds early.''

Gullion et al., (1959) reported them in southern Nevada as widely scattered spring, summer and fall residents, but of erratic occurrence.

Austin (1971) reported, "A permanent resident in woodlands, transient in riparian and montane forests, winter resident in desert scrub in southern Nevada. Usually occurs from February 3 to December 30.''

Food. Ronald M. Lanner (1981) noted, "The Pinyon Jay eats many kids of insects, berries and seeds, including pinon nuts. But the way it uses the latter is extraordinary . . . . Cones are pecked loose from their branches and carried to secure perches where the birds noisily hammer them open and pick them apart scale by scale. The seeds are removed intact, up to 20 being stored temporarily in the jay's elastic esophagus. These seeds are then carried off to the flock's traditional nesting area, as much as six miles away, where they are placed in the ground. Typically, the jay thrusts several seeds into the litter of dead needles and twigs that makes up the woodland floor. Some are deposited on the south sides of pinon trees, where deep snow will accumulate and where snowmelt will come early in spring. The caching of nuts continues through the fall, as the cones dry and open, the jays are able to continue the harvest by picking seeds from intact cones still attached to the tree.''

Clark's Nutcracker   *Nucifraga columbiana*

Distribution. From central Alaska to Manitoba; south to Texas.

Status in Nevada. Resident on mountain ranges; usually moves to lower elevations after nesting in early spring; transient in valleys.

Records of Occurrence. In Elko County, these birds were repeatedly seen in the Jarbidge and Ruby Mountains from 1964 to 1974 (Alcorn MS).

LaRochelle (MS, 1967) reported at Ruby Lake NWR they were regularly seen in fall and winter along the western edge of the refuge.

Kingery (1978) reported Clark's Nutcrackers common all spring at Ruby Lake.

Clark's Nutcrackers were very abundant in Little Valley of the Carson Range. Three specimens collected by Burleigh in April, August and November of 1968 from this range are in the USNM.

This author observed one in Idlewild Park, Reno January 23, 1942.

The Audubon CBC from 1966 to 1972 consistently revealed one to six in the Reno area (Truckee Meadows).

In the Fallon area one was observed by Don Johnson in the summer of 1970. Another lone bird was observed eight miles south–southwest of Fallon March 16, 1973 by Napier and Aldous.

This author, in summer and autumn, repeatedly saw these birds in all the major mountain ranges in the area from Eastgate, Churchill County to the Snake Range of White Pine County.

Biale (MS) reported them in the Eureka area 37 times from August 30, 1964 to August 27, 1967. He saw them twice in January; once in April and June; 3 times in August; 8 in September; 12 in October; and 10 times in November.

Johnson (1974) reported, ''In 1971 I saw or heard nutcrackers from May 18 to 20. On the latter date, a group of 10 perched together in pinons and the cool windswept ridge just below Wahguyhe Peak. This species nests early and thus it is unlikely that breeding residency could be proved by May or June records unless juveniles were recorded. Their presence at this date, however, strongly suggests an earlier nesting in the Grapevine Mountains and the flock observed is assumed to have been composed of local postbreeding individuals. The species was not seen in 1973, when my visit was later than in 1971.''

A. A. Alcorn (MS) reported these birds in southern Nevada from the Charleston Mountains on February 26, 1963 at Deer Creek (two seen); on July 16, 1963 at Rosebud Springs (2 seen); on July 18, 1963 at Lee Canyon (two seen); and October 31, 1961 in Pahrump Valley (one seen).

Gullion et al., (1959) reported, ''This species seldom occurs on the deserts. However, in

1951 a small group of nutcrackers spent most of July in Boulder City and were noted there as late as August 23. During their stay in this area, maximum daily temperatures reached 109 degrees Fahrenheit and average about 89 degrees Fahrenheit.''

Austin (1971) reported, ''A permanent resident in montane forests in southern Nevada and transient in riparian and woodland areas.''

Black–billed Magpie   *Pica pica*

Distribution. From Alaska to Manitoba; south to California and Kansas. Also, in Europe, Russia, China and Africa.

Status in Nevada. A common summer resident over the entire state except southern Nevada.

Records of Occurrence. Magpies, because they are a conspicuous bird, appear more often on census reports than some of the less conspicuous, smaller species. This writer has recorded them in 14 Nevada counties. There is no point in listing all of their occurrences. Their abundance is represented by the fact that they were seen 2211 times statewide from 1960 to 1971. Following are some observations from areas throughout the state.

Sheldon NWR (1982) reported the Black–billed Magpie as a permanent common resident spring, summer and fall; uncommon in winter.

LaRochelle (MS, 1967) reported for Ruby Lake NWR, ''Common throughout the year and nests in willow and aspen thickets in great numbers. Exerts appreciable depredation on nesting waterfowl and is controlled in late winter when it is vulnerable to several controls.''

Alcorn (1946) reported for Lahontan Valley (Fallon area), ''Common and widespread over the whole area. Nine nests containing two to eight eggs or young birds were seen May 12, 1936.''

From 1960 to 1968, these birds were observed 1271 times in the Fallon area (Alcorn MS).

Schwabenland (MS, 1966) reporting on these birds at Stillwater WMA stated, ''Abundant on the area at all times of the year, although more during the summer than winter. Presents a possible depredation problem to waterfowl nests. Nesting commonly observed.''

Napier (MS, 1974) reporting for Stillwater says, ''Common along river and brushlands, seldom seen in the marsh. Doubtful that 'control' of winter flocks has much influence on breeding population in local areas. If bird depredation control is needed, should control magpies during breeding season.''

Apparently little movement of magpie populations exists, as indicated by one young banded eight miles northwest of Fallon May 18, 1939. It was later shot in the Fallon area by J. C. Savage on December 2, 1939. Another young banded eight miles southwest of Fallon on June 1, 1939 was shot by Walter Davis in the same area January 21, 1940 (Alcorn MS).

Biale (MS) reported magpies many times in the Eureka area in all months from 1965 to 1975. Their abundance is indicated by his reporting them 337 times from 1965 to 1967, inclusive. Several nests were reported by him. On May 20, one mile north of Eureka, a nest was found in pinon pine about six feet above ground. The nest was made of coarse twigs roofed over and semi–lined with mud and entirely lined with roots. Six young, one to three days old, were seen (6400 feet elev.). On May 14, 1972 at Roberts Creek Ranch area, a nest eight feet up in a juniper tree and made of dry twigs, mud and grass (roofed over) was found. One newly hatched young and five rotten eggs were in the nest (6400 feet elev.). Biale (PC, 1984)

Immature Black-billed Magpie. Photo by Roy Nojima.

reported a considerable increase in the number of magpies since these birds were put on the protected list.

Magpies were not seen by this author in Pahranagat, Virgin or Moapa Valleys, even though they were looked for during each trip to these areas.

A. A. Alcorn (MS), stationed at Key Pittman WMA near Hiko and the Overton WMA from July, 1959 to July, 1965 did not see any magpies in these areas.

Kingery (1980) reported eight Las Vegas observations.

Food Habits. Magpies have been seen eating, or their stomachs have been examined and found to contain the following: Russian olive berries and seeds (up to 60% of stomach contents), barley seed (as many as 10 seeds), wheat (52 rain–softened kernels were seen eaten by one magpie), asparagus seeds (up to 22 seeds), grape seed, dock seed, grain husk (comprising up to 95% of stomach contents), sunflower seeds (six seeds cracked and only the kernels eaten), honeysuckle berries, eggs (shells and egg parts; also one magpie seen flying with a quail–sized egg), mammal hair, meat (up to 80% of stomach contents), fresh water shrimp (up to 50% of stomach contents), fish (carp, up to 55% of stomach contents), snails (up to 80% of stomach contents), and a ground squirrel that was barely alive when examined that was dropped by a magpie when the bird was frightened by a shot. Magpies are also frequently seen on the highway at the road kills of rabbits or any domestic or wild animal that dies.

At the McGowan residence near Yerington, magpies consistently visited a bird feed station there and selected the sunflower seeds from a mixture of five kinds of seeds. They then held the sunflower seeds on a hard surface with their foot and pecked the kernel out of its shell, then it was swallowed whole.

Pinon pine nuts were also a favorite food of these birds. Often, the magpies filled their gullets with nuts and flew away to a nearby place to stash the nuts under some leaves or other litter. In this manner the magpies may help reforest the area with new trees.

Population Management. Sentiment against these birds has been very strong, especially many years ago. An article in the **Yerington Times,** dated February 19, 1910 announced, "From now until the 15th of September, Fred Burner offers a shotgun to the one who brings him the greatest number of magpies. The gun is a Baker make, 16 gauge hammerless and is a fine fowling piece. This is a good opportunity for our hunters to get in and display their marksmanship, for the lucky one who will not only win a good gun, but rid the country of a pest that is rapidly exterminating our game and their eggs or killing the defenseless young."

Up until the 1960's many sporting clubs or counties offered bounties for magpies. In the Fallon area, bounty hunters were offered 10 to 25 cents per head.

Ryser (1985) commented, "Mammalogists have been troubled by magpies following along their small–mammal traplines and flying off with their snap traps which have caught mice or other mammals. Magpies have even been credited with possessing enough insight to search for traps around the bits of cotton placed by the trapper on the tips of shrubs to mark the sites. Years ago in Nevada and Utah, when coyotes were being systematically killed off, magpies attracted by the poisoned bait were also killed off in great numbers. The magpies' impact on short–circuiting coyote poisoning was severe enough that on occasions preliminary campaigns were instituted to deplete the magpie populations before going after the coyotes! Such a campaign occurred on Pyramid Lake Reservation in the winter of 1921–22. A population of magpies estimated at over 1000 birds was reduced to less than a dozen by the following winter."

Efforts to control magpie populations were made in an effort to protect game bird populations. A common method of control was to place strychnine on a small piece of pork rind that had about half an inch of fat on it. The pork rind was then nailed to a tree or the top of a fence post. This would attract the magpies and it was not uncommon to find 10 magpies dead at one place (Alcorn MS).

American Crow   *Corvus brachyrhynchos*

Distribution. From British Columbia to Newfoundland; south to the Gulf of Mexico and southern Florida.

Status in Nevada. Resident in the northern part of the state, but usually moves out of the colder northernmost areas in winter. Resident in winter in the southern part. Richards (1971) reported the eastern race of crow, *C. b. brachyrhynchos* occurs in Nevada and not the western race, *C. b. hesperis* as listed by Linsdale (1951).

The author has seen crows throughout the state, from Jarbidge in the north, to Moapa and Pahrump Valley in the south; from Lake Tahoe in the west, to Baker in the east. Crow populations in western Nevada are much higher in recent years (1980–88) than previously.

Richards (1971) reported, "The breeding range of the crow in the Great Basin was restricted almost entirely to riparian habitats. The Humboldt River, Reese River, Marys River in Nevada are stream drainages supporting some of the largest populations of breeding crows."

Records of Occurrence. Sheldon NWR (1982) reported the American Crow as an occasional spring and summer transient visitor.

LaRochelle (MS, 1967) reported at Ruby Lake NWR, the crow is generally found year round on the area with exception of mid–and late–winter. "Flocks of 100–200 in summer and fall. Many nest."

Kingery (1972) reported during the 1971–72 winter season, there were 300 crows at Ruby Lake.

In Paradise Valley and Winnemucca areas of Humboldt County, crows regularly nest in the willow thickets. Twenty or more active crow nests were seen in May, 1931 in the willows along the Humboldt River about three miles upstream from Winnemucca (Alcorn MS).

In the Lovelock area crows were seen in all months of the year. A person could see 10 or more each day in December, 1963. However, crows were seen only once in the nine trips into that area in January, 1965 (Alcorn MS).

About a mile southwest of Lovelock on April 26, 1961, two crow nests were seen high in cottonwood trees. An adult crow flew from each nest. Homer Ford unsuccessfully attempted to climb the tree to examine the contents of the nests. He could, however, see partially into one nest that contained three eggs (Alcorn MS).

Ryser (1970) reported on May 6, 1970 at the University of Nevada, Reno, a crow was sitting on a nest and in early July a second nest was built on top of a building on this campus.

In the Fallon area, these birds were seen 147 days from March 6, 1960 to August 29, 1967. A total of 835 were seen for an average of 5.7 per day. They were seen twice in January and February; 16 times in March; 29 in April; 27 in May; 12 in June; 9 in July; 13 in August; 16 in September; 10 in October; and once in November and December (Alcorn MS).

On April 4, 1939 a nest was examined five miles southwest of Fallon that contained six green–splotched eggs. On May 17, 1939 the same nest contained four partially feathered young that were quite large. The nest was about 20 feet above ground level and was built in a willow tree (Alcorn MS).

About nine miles west of Austin on April 27, 1961 this author, in company with Homer Ford, saw a crow nest about eight feet above ground level in a willow thicket. The nest contained one egg which Ford removed at 8:30 a.m. We returned to the nest on this same day at 4:30 p.m. and saw a crow fly from the nest. Examination revealed another egg in the nest (Alcorn MS).

Biale (MS) recorded these birds in the Eureka area 66 times from August 30, 1964 to August 9, 1973. He reported a very few nest in willows on the various ranches in that area. During this 9–year period he saw them twice in January; none in February; 12 times in March and April; 7 in May; 9 in June; twice in July; 8 in August; 6 in September; 5 in October; and once in November and December.

Gullion et al., (1959) reported for southern Nevada, "Common winter visitor in at least the Moapa and Virgin Valleys, arriving as early as October 10, (1951) and remaining as late as February 28 (1952) with one summer record from Pahrump Valley on August 25, 1951."

J. R. and A. A. Alcorn (MS) recorded crows in the Moapa Valley from December 15, 1959 to December 12, 1966. They were seen 6 times in January; 7 in February; 6 in March; once in April; twice in May; none in June; once on July 9, 1963; none in August or September; 5 times in October; 4 in November; and 5 times in December. On November 3, 1961 about 75 crows were seen in a field at the north end of the valley and November 26, 1963 over 300 were seen in the valley.

Austin (1971) for southern Nevada reported, "Winter resident in desert scrub and riparian areas; transient in montane forests. Usually occurs from October 10 to March 14 with an unseasonal occurrence on April 29."

Common Raven   *Corvus corax*

Distribution. The northern parts of both hemispheres. In America, from subarctic Alaska south to Central America. In the Eastern Hemisphere, south to Africa, India and China.

Status in Nevada. Widespread over entire state.

Records of Occurrence. Ravens are common in all months over most of Nevada. No attempt is made to list all the records which are available for all sections of the state.

Sheldon NWR (1982) reported the Common Raven as a common permanent resident spring, summer and fall; uncommon in winter.

At Wildhorse Reservoir, Elko County, a family of six young ravens were seen June 26, 1971 that appeared to have recently left the nest. Near Carlin, two Brewer's Blackbirds were chasing a lone raven June 28, 1971, and near Jiggs a raven nest was seen April 28, 1970 built at the cross–arm of a telephone pole (Alcorn MS).

LaRochelle (MS, 1967) reported at Ruby Lake NWR, "Common year–round with up to 65 using the area. A serious predator on waterfowl nests."

Kingery (1985) reported, "Ruby Lake attracted a cacophony of 220 Common Ravens November 23."

This author saw a nest at Lamance Creek, Santa Rosa Mountains, Humboldt County June 10, 1970. It was built in a tall Lombardi Poplar tree and contained young, with adults nearby.

In the lower valley south of Lovelock, three raven nests, built in cottonwood trees, were seen by the author April 25, 1961. Two of these nests had adults sitting on them. The third

Nest of young Common Ravens. Photo by A. A. Alcorn.

nest, about 30 feet above ground was built with sticks and lined with sheep's wool. This bulky nest contained one egg that was pipping along with five newly hatched ravens. Two adults flew excitedly around nearby as the nest was examined (Alcorn MS).

In the Fallon area this author examined three raven nests May 1, 1970. One contained six eggs plus a newly hatched young, another nest contained three nestlings about one–fourth grown, and the third was a newly made nest which contained three eggs (Alcorn MS).

Napier (MS, 1974) reported ravens at Stillwater WMA, Lahontan Valley, "A common year–round bird, most abundant from July to September when young take to wing and join together. They nest in cottonwood trees around the marsh. Major waterfowl nest predator, probably major limitation to nesting on Stillwater. Since ravens have been given protection status, their population has increased in the marsh."

Napier (MS, 1974) reported he examined a nest containing five eggs April 27, 1967 that had been built on a lookout pole. In June he returned to this nest and took one of the young ravens. The Napiers kept this raven as a pet until October, 1967. It took its first sustained flight when eight weeks old. They report it as showing great intelligence and being very friendly. "Its diet was canned dog food. Often it would share the food with a wild magpie, both eating at the same time."

Kingery (1986) reported, "A shut–off of irrigation water in the Carson River, Nevada combined lack of fresh water with ice and low oxygen and killed thousands of fish. One stretch of river, solid with dying and dead fish for 1/2 mile, attracted over 600 Common Ravens."

Biale (MS) reported seeing these birds in the Eureka area in all months. From May 1, 1964 to December 3, 1967, he recorded them 329 times. He reported a nest in a windmill contained four young about 10 days old June 6, 1965. A nest in a pinon tree contained four young partly feathered out June 8, 1966.

Biale (MS) also reported, "In northeast Antelope Valley, on March 28, 1971 a nest approximately 11 feet above the ground in a dead juniper tree was found. It was made nearly entirely of greasewood twigs (freshly snapped off), lined mostly with cow hair, with some three scraps of paper, and one piece of white cloth. The nest was a deep cup approximately 24 inches across and 18 inches deep, containing seven greenish eggs with brown spots. The elevation was approximately 6300 feet."

At Bartines Ranch on May 23, 1971, Biale (MS) reported, "A nest in an old barn, on roof support about 10 feet above ground made of large greasewood and sage twigs, with three partly grown young (approx. 6000 feet elevation)."

Biale (PC, 1984) reports a large increase in the population of ravens within the past three or four years in the Eureka area.

Near Mesquite, Clark County, a Western Kingbird was seen in flight chasing a raven June 9, 1961. The kingbird flew above the raven and appeared to alight on the raven and peck at its back as it flew (Alcorn MS).

Gullion et al., (1959) reported for southern Nevada, "Uncommon but widely distributed, permanent resident on the desert."

Austin (1971) reported for southern Nevada, "A permanent resident in riparian areas, winter resident in desert scrub, and transient in woodlands and montane forests."

In the Moapa Valley from April 25, 1957 to July 22, 1965 it was seen in all months (Alcorn MS). A. A. Alcorn (MS) recorded it 104 times in this valley during this same time period.

Food Habits. Ravens are usually considered to be scavengers and they are frequently seen as they patrol the major highways looking for road–killed rabbits, rodents, reptiles, birds and other carrion. However, they are opportunists and frequently kill small animals, especially those caught in unusual situations. An example of this is in the Lovelock area where they kill and eat numerous Meadow Mice *(Microtus montanus)* flooded out of their burrows as the fields are flood irrigated.

Wildlife managers and biologists at various wildlife areas in the state report ravens are destructive to waterfowl and other birds' nests. Napier (MS, 1970) reported nest destruction at Stillwater WMA was commonly the leading cause of duck nesting failure. Ravens account for most of the 44.8% of the duck nests destroyed.

Other observations by Alcorn relative to food habits of ravens in Nevada are given in the following accounts: On April 21, 1942 about three miles north of Lovelock a raven was seen flying with what was thought to be a whole pheasant egg in its beak.

On April 20, 1942, at seven miles northeast of Fernley in a marsh area, a raven was seen carrying an egg in its beak as it flew across in front of this author. It alighted on the ground about 200 yards away at which time this author fired a shot from a rifle. The raven flew away without the egg and examination revealed it was a pintail duck egg. The egg was not broken, but it was cracked on the side.

At the north end of Humboldt Sink, Mills shot a raven June 17, 1943. Its stomach contained one Gridiron–tailed Lizard *(Callisaurus dracoides)* that had been freshly killed.

From near Pelican Island in the Carson Sink, six ravens shot December 23, 1943 were examined. Three of their stomachs contained duck feathers and one contained duck feathers plus two catkins from a cottonwood tree. One contained mouse hair and the sixth stomach was empty.

In the Austin area on September 22, 1960, six ravens were eating on a road–killed Black–tailed Jackrabbit. In the same area the author saw over 30 ravens October 1, 1963. They were thought to be migrating birds. Many were eating road–killed jackrabbits scattered over a 5–mile stretch of the highway west of Austin.

A. A. Alcorn (MS) saw a raven down a sandgrouse in flight at Overton WMA on May 13, 1961. "It took the raven three tries to down the grouse and by the time I got there the raven had eaten the skin from around the neck, along with the crop of the sandgrouse."

Family **PARIDAE:** Titmice

Black–capped Chickadee   *Parus atricapillus*

Distribution. From Alaska to Newfoundland; south to California and east to western North Carolina.

Status in Nevada. Uncommon in extreme northeastern Nevada and accidental in western Nevada.

Records of Occurrence. Sheldon NWR (1982) reported the Black–capped Chickadee as seen fall, winter and spring; occasional winter resident.

Linsdale (1936) reported, "Resident along streams in the Snake River drainage system in northeastern part of the state. Nine skins were taken, September 16, 1934, on the Salmon River at Shoshone Creek, 5000 feet, Elko County."

Gullion (1957) reported, "A single bird was seen foraging in the willows and lower branches of cottonwoods along the south fork of the Humboldt River, 16 miles south of Elko, on April 11, 1954."

Kingery (1982) reported, "A Black–capped Chickadee which wintered at Jarbidge, Nevada, stayed until about March 1; the species rarely occurs in Nevada."

Kingery (1979) reported two Black–capped Chickadees at Carson City September 28.

Evenden (1952) reported, "At noon on October 10, 1950, several of these chickadees were watched at close range about picnic tables in the municipal park at Fallon, Churchill County. Since I was then unaware of the unusualness of this observation, no attempt was made to count them."

Mountain Chickadee   *Parus gambeli*

Distribution. From British Columbia and Alberta; south to Baja California and Texas.

Status in Nevada. Resident in most of the mountain ranges statewide; transient in valleys. Linsdale (1951) reported two races from Nevada: *P. g. abbreviatus* and *P. g. inyoensis*.

Linsdale (1936) reported this chickadee from many sections of the state.

The author has recorded them near Jarbidge, Lye Creek, Santa Rosa Mountains, Virginia City, Mount Rose, Lake Tahoe, Pine Grove, Freeman Basin, Stillwater Mountains, Clan Alpine Mountains, Toiyabe Mountains south of Austin, along Reese River in Nye County, Greenmonster Canyon in Toquima Range, east side of Snake Range, and in southern Nevada.

Records of Occurrence. Sheldon NWR (1982) reported the Mountain Chickadee as an occasional spring and fall migrant.

The Mountain Chickadee is reported by Ruby Lake NWR as common year round. Sara Brown (MS, 1986), biologist at the refuge, commented the Mountain Chickadee is, "Seen frequently in winter on the slopes bordering the [Ruby Lake] refuge."

Specimens in the USNM were taken in all months except June and September between 1967 and 1968 by Burleigh from the Reno, Galena Creek and Lake Tahoe areas.

The author recorded the Mountain Chickadee in the Fallon area 39 times between January, 1960 and August, 1968 (Alcorn MS).

Biale (MS) saw these birds 56 times in the Eureka area from August 2, 1964 to November 22, 1967. They were seen in all months and he reported a nest on June 10, 1965 in a willow tree trunk. The nest was in a flicker hole approximately 75 feet away from a Goshawk nest. The nest contained eight young about 10 days old, which Biale banded. On June 28, 1965, a nest in a hole near the base of a nearly dead big sage contained young. On July 7, 1965, this same nest contained six young (banded) ready to leave the nest.

Johnson (1965) reported, "Common in the Sheep Range between 7000 feet and 8200 feet in montane forest; less numerous outside this formation, both below it in pinon–juniper down to 6500 feet and above it in subalpine forest to at least 9000 feet. Common in coniferous growth between 7700 feet and 9600 feet in the Spring Range; however, during our period of field work this species was probably not the most common permanent resident, as stated by van Rossem (1936). On June 21, a nest with young found at 8100 feet in Macks Canyon."

Austin (1971) reported, "Present all year in southern Nevada. Permanent resident in woodland and montane forests and winter resident in desert scrub and riparian areas."

A. A. Alcorn (MS) saw two birds at Lee Canyon, Charleston Mountains, March 4, 1960; five birds at Kyle Canyon September 15, 1960; two birds at Lee Canyon July 18, 1963; one bird at Trough Spring July 19 and one August 7, 1963.

Gullion et al., (1959) reported, "Recorded twice from the pinon–juniper covered McCullough Range (at about 5500 feet elevation), about 16 miles northwest of Searchlight, on August 21, 1951 and March 29, 1952."

Kingery (1977) reported one Mountain Chickadee at Boulder City on November 1 to 7 and he (1979) reported three at Las Vegas October 13.

Plain Titmouse   *Parus inornatus*

Distribution. From southern Oregon to western Oklahoma; south to Baja California and western Texas.

Status in Nevada. Occurs statewide mostly in pinon pine and juniper forests. Linsdale (1951) reported two races in Nevada: *P. i. zaleptus* and *P. i. ridgwayi.*

Records of Occurrence. Specimens reported by Linsdale (1936) are from northwest Reno, Washoe County; northeast of Wells and at Hastings Pass in the Ruby Mountains, Elko County; Lehman Creek, White Pine County; northeast of Sharp, Nye County; Irish Mountain in Lincoln County; and Charleston Mountains, Clark County.

Sheldon NWR (1982) reported the Plain Titmouse as an occasional transient visitor summer and fall.

Specimens in the USNM collected by Burleigh include a male taken August 25, 1960 from

Ely; a male taken April 9, 1967 from Virginia City; a male taken September 6, 1968 from the Reno area; and a female taken March 13, 1969 from the Reno area.

Gabrielson (1949) reported them from, "Pine Nut Mountain, Douglas County, August 22, 1933."

Biale (MS) recorded these birds in all months in the Eureka area except August, September, October and December. He observed a nest below a Goshawk nest at Hunters Ranch, five miles east of Eureka June 10, 1965. He (PC, 1986) commented, "They are permanent residents in the Eureka area."

Johnson (1974) reported for the Grapevine Mountains it occurred locally and in very small numbers in 1939 and 1940.

Gullion et al., (1959) reported, "Recorded on four occasions from the higher (4300 feet to 4500 feet elevation) pinon–juniper covered parts of the Gold Butte area, on October 14, 1959, June 19 and August 31, 1953, and September 26, 1954."

Johnson (1965) reported, "In the Sheep Range we met this species on June 14 in pinon–juniper woodland in Hidden Forest Canyon at 6500 feet and again at 6900 feet. An adult female and three fully grown juvenile males were collected. This species was found twice in the Spring Range in mixed pinon–juniper–mahogany half–mile east of Macks Canyon, 7700 feet, on June 25; an adult and a juvenile of indeterminable sex were taken. Van Rossem (MS) recorded this titmouse only at Cold Creek on November 24 and 25, 1932, and expressed surprise at its rarity."

Austin (1971) reported for southern Nevada, "Present all year. Permanent resident in woodland areas and transient in montane forests."

Family **REMIZIDAE:** Penduline Tits and Verdins

Verdin   *Auriparus flaviceps*

Distribution. From southern California to south–central Texas; south in Mexico to Jalisco and San Luis Potosi.

Status in Nevada. Resident in the southern part of the state.

Records of Occurrence. This author recorded these birds in the Hiko area of Pahranagat Valley January 8, 1965 and April 14, 1965.

Austin (1971) reported for Clark County, "Species is a common permanent resident in areas dominated by woody legumes."

Austin and Bradley (1971) reported for southern Nevada, "Occurs all year. A permanent resident in riparian areas and winter resident in desert scrub."

J. R. and A. A. Alcorn (MS) observed Verdins in the Overton area 23 times from 1960 to 1965. They were recorded once in January and February; 6 times in March; 8 in April; twice in May; and 5 times in July.

This writer saw an adult feeding a young out of the nest near Overton on July 11, 1963 (Alcorn MS).

Gullion et al., (1959) reported, "Although reported to be restricted to the Colorado and Virgin Valleys by Linsdale (1951), this species is as widely distributed as the honey mesquite *(Prosopis juliflora)* in southern Nevada. Johnson and Richardson (1952) first recorded this bird away from the Colorado drainage, finding it in the Pahrump Valley and Ash Meadows in June,

1951. We had additional records for Pahrump Valley on May 19, 1951 and August 6, 1953 and 1954 and for Steward Valley (2400 feet elevation, Nye County) on May 20, 1951. On the upland desert areas the catclaw *(Acacia greggii)* suffices as habitat for Verdins, and in this environment they have been observed in the McCullough Range on April 16, 1951; Eldorado Mountains near Nelson on May 3, 1951; Searchlight area on February 6, 1953, and August 5, 1954; Muddy Mountains on February 11 and 28, 1954; and on numerous occasions in the Gold Butte area. Fledglings were noted in Las Vegas Valley on May 24, 1952.''

Family **AEGITHALIDAE:** Long–tailed Tits and Bushtits

Bushtit   *Psaltriparus minumus*

Distribution. From extreme southwestern British Columbia to western Oklahoma; south to central Guatemala.

Status in Nevada. Resident statewide on lower portions of the highest mountains and on desert ranges. Found most frequently in thickets. Linsdale (1951) reported two races occur in Nevada: *P. m. plumbeus* and *P.m. providentialis*.

Records of Occurrence. Sheldon NWR (1982) reported the Bushtit as an uncommon permanent resident spring, summer and fall; occasional in winter.

Two of these birds were observed at Lamance Creek on June 10, 1970 and about 10 were seen at Stonehouse Creek November 19, 1963 on the east side of Santa Rosa Mountains (Alcorn MS).

LaRochelle (MS, 1967) reported at Ruby Lake NWR Bushtits are seen in loose flocks or groups, generally in pinon, sagebrush edge.

One flock of over 10 was seen in Jersey Valley, (southeastern Pershing County) November 22, 1963 (Alcorn MS).

Specimens in the USNM collected by Burleigh include two females taken near Reno February 13, 1968 and a male taken from Galena Creek January 18, 1969.

Alcorn (1946) reported Bushtits as resident but not abundant in the Fallon area. Their presence in the Fallon area has diminished since that time. Observations from 1959 to 1966 reveal they were only seen May 20, 1959 (2 birds); November 4, 1962 (10 birds); and October 12, 1966 (less than 10 birds). The last observation occurred when about 12 were seen in a Russian olive tree at this author's ranch six miles west–southwest of Fallon on November 11, 1984.

Thompson (PC, 1988) reported a small flock of about 10 Bushtits seen at the sewage ponds southeast of Fallon December 20, 1987.

A flock of about 30 were seen in the Freeman Basin, Stillwater Mountains November 18, 1959. In the Austin area two were seen October 25, 1961 and about 10 were seen November 28, 1964 (Alcorn MS).

At Wilson Creek, Toquima Range, a loose flock of about 10 birds was seen July 4, 1969 (Alcorn MS).

Biale (MS) saw these birds in the Eureka area in all months except March. On April 25, 1965 he saw Bushtits building a nest in a large black sagebrush. He reported them over a wide area. From 1964 to 1975 he recorded them 78 times.

In southern Nevada, Gullion et al., (1959) reported, ''Common Bushtit is uncommon resident in brushy areas on the desert ranges. Recorded during most months of the year from the Mormon, Clover and Virgin Mountains, the Tule Desert and the Gold Butte area, these being all of the ranges on the eastern edge of Nevada's southern desert. Also noted on June 28, 1952 at Potosi Spring (5700 feet elevation) on the north slopes of Potosi Mountain.''

A flock of about 10 were seen on the north side of the Virgin Mountains, Clark County December 14, 1959 (Alcorn MS).

Austin and Bradley (1971) reported, ''In woodlands and montane forest it occurs all year. In desert scrub and riparian areas it usually occurs from December 21 to February 29 with an unseasonal occurrence on September 15. Winter resident in desert scrub and riparian areas, permanent resident in woodlands. Transient in montane forests.''

Family **SITTIDAE:** Nuthatches

Red–breasted Nuthatch　　*Sitta canadensis*

Distribution. From southeastern Alaska to Newfoundland; south to southern California and Florida.

Status in Nevada. Uncommon resident in some mountain ranges; transient statewide.

Records of Occurrence. Linsdale (1936) reported records of this nuthatch from Wheeler Creek, Quinn River Crossing, Big Creek and the Santa Rosa Mountains, Humboldt County. From Elko County, he reported them from Humboldt Valley, Jarbidge, Clover and Ruby Mountains. From central Nevada he reported them from Birch and Peterson Creek, Lander County. From eastern Nevada he reported specimens from Baker Creek, Snake Mountains and in southern Nevada from Saint Thomas, Clark County.

Sheldon NWR (1982) reported the Red–breasted Nuthatch as an occasional winter resident; seen from fall through spring.

This author saw them at Abel Creek, Santa Rosa Mountains, Humboldt County on October 5, 1963 (Alcorn MS).

LaRochelle (MS, 1967) reported, ''At Ruby Lake NWR, though reportedly a common summer resident, I have only one observation.''

Others were seen at Lake Tahoe 4–H Camp, July 20–22, 1966 (Alcorn MS).

In the Fallon area, one or two birds were infrequently seen in April, May and August from 1959 to 1975 (Alcorn MS).

In the Desatoya Mountains, two were observed at Carroll Summit on September 27, 1963 and about 10 were seen in Buffalo Canyon on November 10, 1963 (Alcorn MS).

In Monitor Valley, at the site of the old forest ranger station, one bird was seen August 26, 1966 (Alcorn MS).

Biale (MS) reported these birds in the Eureka area on January 18, 1967; August 31, 1969; September 1, 1969; October 27, 1963 and 1965; October 29, 1960; October 12, 1969; October 14 and 22, 1973; November 7, 1965; November 6, 1966; and November 2 and 7, 1969. He (PC, 1986) reported, ''They are mostly seen in the fall.'' He noted that in October of 1983, at a small spring on the south end of the Pancake Range numerous Red–breasted Nuthatches were seen in heavy pinon–juniper.

Biale (MS) reported these birds in White Pine County from Duck Creek Basin October 25, 1972 and on September 16, 1973 and from the White Pine Mountains October 15, 1972.

Red-breasted Nuthatch. Photo by Art Biale.

Johnson (1973) in his report on avifaunas of southeastern Nevada says, "June 20, 1953, Scofield Canyon (8800 feet elevation), several were seen in white fir and limber pine. In 1972 the species was scarce in the same canyon. June 19, 1972, one was heard (9000 feet, subalpine timber. May 21, 1970 West Virgin Peak (7700 feet elevation) one male was taken in clumps of firs."

Miller and Russell (1956) reported, "Apparently summer resident in small numbers. A male in breeding condition was taken in Trail Canyon, Esmeralda County, on May 26."

For southern Nevada in the Moapa Valley, two birds were seen September 22, 1963 and A. A. Alcorn reported two birds from Gold Butte September 15, 1963 (Alcorn MS).

Gullion et al., (1959) reported, "Two records away from Boulder City (where there are numerous fall records): one bird in a catclaw thicket near Catclaw Spring, in the Gold Butte area on September 28, 1954; and a small flock near Indian Springs in the Virgin Mountains, on November 27, 1953."

Austin and Bradley (1971) reported, "In woodland and montane forest it usually occurs from March 20 to April 2, and from July 30 to October 8. Unseasonal occurrences on February 14, in June and on November 27. In riparian areas it usually occurs from September 2 to November 5. Transient in riparian and woodland areas. Summer resident in montane forests."

White–breasted Nuthatch   *Sitta carolinensis*

Distribution. From southern British Columbia to Nova Scotia; south in Mexico to Veracruz.

Status in Nevada. An uncommon resident in some mountain ranges, statewide. Linsdale (1951) reported two races in Nevada: *S. c. nelsoni* and *S. c. tenuissima*.

Records of Occurrence. Linsdale (1936) reported records from the Carson Range, the Pine Grove Hills and the White Mountains. From central Nevada he reported them from Toiyabe and Quinn Canyon Mountains. From eastern Nevada he reported specimens taken from the Snake Range and Silver Canyon Mountains.

Six specimens for the USNM were collected by Burleigh at Reno, Galena Creek and Mount Rose during January, March, April, July and August of 1968 and 1969.

Ryser (1972) reported this bird from Sierra Valley (Carson Range) November 12, 1972.

Biale (MS) saw this bird in the Eureka area on October 10, 1964 and November 4, 1978. He (MS, 1986) reported additional single birds seen November 4, 1978 at Eureka and November 12, 1978 in the west Diamond Mountains.

Kingery (1979) reported one in Eureka and one in the mountains nearby were the first there in 10 years.

Biale (PC, 1984) reported one seen at Mt. Hamilton, White Pine County October 14, 1972.

A. A. Alcorn (MS) reported lone birds from the Charleston Mountains at Kyle Canyon September 15, 1960, at Deer Creek February 26, 1963 and at Lee Canyon July 18, 1963.

Monson (1951) reported one seen at Boulder City September 13, 1950.

Austin and Bradley (1971) reported, "This bird occurs all year. Transient in riparian areas. Summer resident in woodland areas. Permanent resident in montane forests."

Pygmy Nuthatch   *Sitta pygmaea*

Distribution. Mountainous areas from British Columbia to South Dakota; south to Baja California and through the Mexican highlands to Jalisco and Veracruz.

Status in Nevada. Resident in mountain ranges. Linsdale (1951) reported two races in Nevada: *S. p. melanotis* and *S. p. canescens*.

Records of Occurrence. Linsdale (1936) reported specimens from Carson City and Lake Tahoe region of western Nevada; Baker Creek, Snake Mountains of eastern Nevada; and from the Sheep and Spring Mountains of southern Nevada.

Heilbrun and Rosenburg (1981) reported 15 in the Truckee Meadows (Reno area) December 20, 1980 during the CBC.

Six specimens in the USNM were taken by Burleigh at Galena Creek, Carson Range. A male was taken June 12, 1968; an immature female July 23, 1968; an immature male September 4, 1968; a male March 19, 1969; a male April 10, 1969; and a male June 23, 1970.

Johnson (1973) reported at Irish Mountain, Lincoln County, "This nuthatch proved to be resident in small numbers between 7200 feet and 8200 feet elevation, in yellow pines on Mount Irish where previously it had not been recorded. May 24, 1972 a male was taken at 7600 feet on Mount Irish."

Gullion et al., (1959) reported seeing several birds in the pinon pines near Big Pine Spring, McCullough Range March 29, 1952.

In southern Nevada, Austin and Bradley (1971) reported, "Occur all year as a permanent resident in montane forests, winter resident in woodland and a visitant in riparian areas."

Family **CERTHIIDAE:** Creepers

Brown Creeper   *Certhia americana*

Distribution. From central Alaska, eastward to Newfoundland; south through Mexico to Nicaragua. Also occurs in Eurasia.

Status in Nevada. Resident in mountain ranges statewide. Linsdale (1951) reported three races in Nevada: *C. a. montana, C. a. zelotes* and *C. a. leucosticta.*

Records of Occurrence. Linsdale (1936) reported these birds from Jarbidge Mountains and Ruby Lake, Elko County; from Galena Creek and Glenbrook in the Carson Range; from the Lower Truckee and Carson Rivers in western Nevada; from Lehman and Baker Creeks in the Snake Range, White Pine County; and from the Hidden Forest of the Sheep Mountains.

Burleigh collected seven specimens for the USNM from Galena Creek, Mount Rose and North Lake Tahoe, all in the Carson Range in April, May, August, September and October, 1968.

Five were reported on the Truckee Meadows CBC in December of 1977.

Biale (PC, 1984) reported one at Faulkner Canyon, Monitor Range on July 11, 1984.

One was observed at the head of Clear Creek, west side of the Monitor Mountains, Nye County in July, 1961 (Alcorn MS).

Biale (MS) reported a Brown Creeper at Success Summit, Schell Creek Range, White Pine County on October 8, 1973.

Johnson (1973) reported, "New records of the Brown Creeper were obtained in Scofield Canyon, Quinn Canyon Mountains, where a specimen was taken at 8800 feet on June 20, 1953 and one was seen at 9700 feet in mixed bristlecone pine–limber pine forest on June 19, 1972; a male in Water Canyon, 7500 feet, and a probable female in Anderson Canyon, 8600 feet. This species was not found in the other mountain ranges under discussion, but it probably will be discovered in the fir forest on Mount Wilson, where the habitat seems ideal."

Johnson (1974) stated, "Miller (1945) did not report the creeper and commented that the forested areas on Potosi were probably too small and too scattered and heavily isolated for this species. However, I found a single singing male in a dense clump of white fir at 7300 feet elevation, suggesting probable residence."

A. A. Alcorn (MS) saw one at Trough Spring, Charleston Mountains, August 7, 1963.

Austin and Bradley (1971) reported, "Summer resident in montane forests and transient in riparian and woodland areas. Usually occurring from February 5 to October 30 with unseasonal occurrences on December 21 and 23."

Family **TROGLODYTIDAE:** Wrens

Cactus Wren *Campylorhynchus brunneicapillus*

Distribution. Southwestern United States from southern California to central Texas; south to Mexico (Michoacan).

Status in Nevada. A permanent resident in southern Nevada.

Records of Occurrence. Linsdale (1936) reported, "From a few localities in the southern part of the state. Recorded by the Death Valley Expedition as common about the ranch in Vegas Valley, and 'still more numerous among the mesquite in Vegas Wash near the Colorado River, where the birds were in full song, March 10.' On May 5, two nests were found at Bitter Springs in the Muddy Mountains. 'The species was common on the high mesa between the Muddy and Virgin Rivers May 7, where nearly every branching cactus contained the remnants of a nest, but all the young had hatched and flown away' (Fisher, 1893). A specimen in the Museum of Vertebrate Zoology was taken January 31, 1934 came from the Dead Mountains in southern Clark County (Benson)."

Gullion et al., (1959) reported, "A fairly common permanent resident, normally found in the Joshua tree forests on the desert ranges, but also regularly recorded from the mesquite bosques in Las Vegas Valley."

J. R. and A. A. Alcorn (MS) saw this wren on Mormon Flat, Lincoln County April 4, 1960; at Sheep Springs, Virgin Mountains (2 birds) June 13, 1961; Gold Butte (2 birds) June 22, 1962; Mormon Mountains (2 birds) June 25, 1962; Gold Butte (2 birds) January 26, 1963; Virgin Mountains (one bird) July 9, 1963; Virgin Valley (3 birds) July 24, 1963.

Ryser (1967) reported seeing one at Searchlight February 15–19, 1967.

Austin and Bradley (1971) reported, "Occurs all year in southern Nevada; permanent resident in desert scrub and riparian areas. Transient in woodland."

Rock Wren   *Salpinctes obsoletus*

Distribution. From British Columbia to northern Manitoba and Minnesota; south to southern Baja California, and through Mexico to Costa Rica.

Status in Nevada. Summer resident statewide; winter resident in southern Nevada.

Records of Occurrence. Linsdale (1936) reported records for many sections of the state and says, "Summer resident; found nearly everywhere in the state."

Sheldon NWR (1982) reported the Rock Wren as a common summer resident; uncommon spring and fall.

Ryser (1973) reported in early August, 1973, a number of Rock Wrens were seen in the Pine Forest Mountains, Humboldt County.

The author observed this bird in Elko County at Wildhorse Reservoir April 12, 1971; near Tuscarora October 3, 1964; near Elko May 30, 1971; and near Jiggs April 28, 1970 (Alcorn MS).

Ryser (1967) reported Rock Wrens in Lamoille Canyon from July 1 to 4, 1967.

LaRochelle (MS, 1967) reported for Ruby Lake NWR, "Are seen, though not often, along the east and west edges of refuge in rock situations. No nesting observed."

These birds have frequently been seen in the Pyramid Lake area in all spring and summer months (Alcorn MS).

Several were seen almost daily through the summer of 1981–82 at the McGowan residence near Yerington. Three juveniles were seen there July 21, 1981 (Alcorn MS).

Marshall and Giles (1953) saw a Rock Wren on the west side of Nutgrass Dike, Stillwater Marsh September 12, 1950.

Alcorn (1946) had no personal sightings of this bird in Lahontan Valley. Since that time, however, it has been seen by him near Fallon April 20 and 25, 1950 and six miles southwest of Fallon September 9, 1966 and May 19, 1968. One was collected near Fallon October 12, 1960.

On October 19, 1984, at six miles west–southwest of Fallon, this author wrote, "I was surprised to see a Rock Wren on the south step of the ranch house this evening. It was searching for food and worked its way around the corner of the house." This wren was seen again October 22, 1984 at this same location.

One was seen near Austin May 16, 1961 (Alcorn MS).

Biale (MS) recorded these birds in the Eureka area 21 days between May 2, 1964 and September 4, 1967. He reported them 3 times in May; 4 in July; 5 in August; 8 in September; and once in October. Since that time he has seen them on October 5, 1965; August 31, 1969; September 7, 1969; June 25, 1972; July 30, 1972; August 16, 1973; September 9, 1973; October 7, 1973; and March 31, 1974.

Biale (PC, 1984) reported seeing adults carrying food into a crevice in a rock at Antelope Valley on June 22, 1984. He (PC, 1986) reported a nest with six white eggs under a large flat rock in a rock wall in the Ruby Hill Mine area May 28, 1985.

Medin (1987) reported the Rock Wren as breeding in restricted habitats in the alpine environment of Bald Mountain, Snake Range.

In Lincoln County, one was seen near Hiko December 23, 1964 and another was observed on the Mormon Mesa April 4, 1960 (Alcorn MS).

Gullion et al., (1959) reported, "This wren is a common fall, winter and spring visitor, but seldom recorded on the desert in summer."

Austin and Bradley (1971) for southern Nevada reported, "A permanent resident in desert scrub and riparian areas and a summer resident in woodland and montane forests."

Canyon Wren *Catherpes mexicanus*

Distribution. From southern British Columbia, eastward to western Oklahoma; south to the Isthmus of Tehuantepec, southern Mexico.

Status in Nevada. Resident in selected areas.

Records of Occurrence. Sheldon NWR (1982) reported the Canyon Wren as uncommon spring, summer and fall; resident in summer.

Kingery (1978) reported one at Jarbidge January 4 for the first winter report for northern Nevada and he (1979) reported two wintering in summer homes closed for the season at Jarbidge.

Gabrielson (1949) noted Canyon Wrens at Thousand Springs Canyon, Humboldt County May 31, 1948; Wells, May 29, 1932; on the Little Humboldt River August 20, 1933, Elko County; and near Ely May 28, 1932, White Pine County.

Heilbrun and Kaufman (1977) reported one near Unionville in December, 1976, during the CBC.

Linsdale (1951) reported, "On July 23, 1948, F. Ruth (MS) saw one at Cave Rock, east shore of Lake Tahoe, Douglas County."

Orr and Moffitt (1971) listed this bird as a rare or possible irregular resident at Lake Tahoe.

Ridgway (1877) reported the capture of an adult male near Fort Churchill, Lyon County December 7, 1867.

Linsdale (1936) reported specimens from the Toiyabe Range, from near Candelaria, from Pahroc and Grapevine Mountains; and others were seen by Hall near Cave Spring, Esmeralda County.

One was seen on the rock outcrops at the west side of Walker Lake January 25, 1941 (Alcorn MS).

Biale (MS) reported seeing a Canyon Wren feeding three young in the Diamond Mountains June 23, 1968. One was also seen at Phillipsburg Canyon, West Diamond Mountains on July

11, 1975 and May 31, 1982. He (PC, 1984) saw one Canyon Wren in the north Pancake Range on December 14, 1980, January 3, 1982 and November 17, 1983. He estimates he has seen less than 10 birds in the Eureka area over a 20–year period. He (MS, 1986) reported additional sightings in the Eureka area November 14 and 15, 1984.

Johnson (1974) reported, "Canyon Wrens *(Catherpes mexicanus)* occurred locally and in very small numbers in the Grapevine Mountains in 1939 and 1940."

Austin and Bradley (1971) reported for southern Nevada, "Permanent resident in desert scrub and riparian areas. Summer resident in woodland and montane forests."

Castetter and Hill (1970) reported for southern Nevada, "Fairly common at higher elevations, likely a summer resident, but records only for March through May."

Bewick's Wren   *Thryomanes bewickii*

Distribution. From British Columbia to Ontario; south into Mexico to Jalisco and Puebla.

Status in Nevada. Resident in western and southern Nevada; transient. Linsdale (1951) reported two races from Nevada: *T. b. eremophilus* and *T. b. drymoecus*.

Records of Occurrence. Sheldon NWR (1982) reported the Bewick's Wren as an occasional transient visitor in spring.

Burleigh collected eight specimens for the USNM from the Reno–Galena Creek area. They were taken in 1967 and 1968 in January, March, August, October, November and December.

Observations by this writer in Idlewild Park, Reno revealed this wren from November, 1941 to August, 1942 as present in all seasons with records for February, March, April, May, August, November and December.

Ryser (1965) reported males in song along the Truckee River near the Thomas Ranch July 23, 1965.

The author has seen these wrens in the timber and thickets along the lower portions of the Truckee, Carson and Walker Rivers on numerous occasions in all seasons of the year.

Alcorn (MS) reported on the Truckee River about five miles upstream from Wadsworth, Washoe County on the evening of December 4, 1959, "About 9:30 p.m. Mills and I were looking at the swallow nests with the aid of a flashlight. These nests were located in a highway culvert that was about 40 yards long, eight feet wide and eight feet high. We saw something move in one of the nests, then we saw that it was a Bewick's Wren. It was apparently roosting in this nest and it came to the nest opening and perched there for about 30 seconds then flew out of the culvert into the darkness."

One Bewick's Wren was seen by the author at the McGowan residence in Yerington January 17, 1983 (Alcorn MS).

In the Fallon area from January 2, 1960 to November 16, 1967, these wrens were recorded on 126 days. They were seen frequently in all months (Alcorn MS).

In the city of Fallon on December 29, 1958, one Bewick's Wren was seen feeding on some insects by picking them from the bark and from crevices in the bark of fresh cut (green) cottonwoods. Examination revealed that the insects were leaf hoppers (Alcorn MS).

Nests, located in the walls of a barn, were observed near Fallon July 7, 1945. They contained adults feeding young. On May 17, 1960, a nest reported by Wendell Wheat contained two adults feeding three young. The nest was located in a loose pile of bee boxes against a shed. He reported a gopher snake got into the nest and the young left the nest. He killed the snake and put the young wrens back into the nest, but they would not stay. They could fly about two feet. When the author examined the nest area May 18, the three young were perched on the bark about 10 feet up on a leaning cottonwood tree. One adult gathered insects from the bark of this tree and was feeding all three young. Sometimes the adult fed half a beak of insects to one young and half to another young. On May 13, 1961 in the Fallon area another nest contained three eggs (Alcorn MS).

On June 15, 1964 at six miles west–southwest of Fallon, a gopher snake was found in a shed in a Bewick Wren nest. Examination revealed the snake had five Bewick Wren eggs in its stomach.

On September 14, 1966, at six miles west–southwest of Fallon, a nest was found in a cardboard box that was half full of clothes. The box lid was open a few inches, and the nest appeared to have been built the previous summer (Alcorn MS).

On April 7, 1988 at six miles west–southwest of Fallon, a feather–lined nest (mostly quail feathers) contained five eggs. The embryos were partially developed (Alcorn MS).

Biale (MS) reported one seen in the Eureka area May 5, 1968, April 28, 1982 and December 14, 1984. He considers them rare in the area.

Johnson (1965) reported, "In the Sheep Range at least three pair occurred in the mixed brush on the floor of Hidden Forest Canyon, from 6000 feet to 6400 feet. A male taken here on June 14 had testes in breeding condition. In the Spring Range, two Bewick Wrens were singing in the pinon–juniper at 7700 feet, half mile east of Macks Canyon on June 25. van Rossem found this wren only at Indian Springs 3300 feet, on the desert at the north base of the Spring Range."

In southern Nevada one was seen nine miles northeast of Beatty on December 17, 1959 and two were seen at Las Vegas April 2, 1960 and another at Las Vegas April 3, 1965 (Alcorn MS).

Gabrielson (1949) reported, "Taken at Nelson, Clark County, November 19, 1934."

Austin and Bradley (1971) reported these wrens occur all year in southern Nevada. "Permanent resident in desert scrub, riparian, and woodland areas and transient in montane forests."

House Wren   *Troglodytes aedon*

Distribution. From British Columbia to New Brunswick; south to Florida and in Mexico to Oaxaca. Also throughout South America.

Status in Nevada. Summer resident, mainly along streams in mountain ranges. Winter resident in the southern part of the state.

Records of Occurrence. Ridgway (1877) reported the range of the House Wren in Nevada as coextensive with the distribution of timber, and is governed by the presence or absence of trees, without special regard to their kind.

Sheldon NWR (1982) reported the House Wren as uncommon spring, summer and fall; summer resident and nester.

One was observed at Jarbidge May 30, 1968 (Alcorn MS).

Gabrielson (1949) reported the House Wren in Tuscarora, Secret Pass and Ruby Lake NWR, all of Elko County.

LaRochelle (MS, 1967) reported at Ruby Lake NWR, "Arrives in mid–spring and nests. One usually nested on top of an old gas pump beneath the cover."

Heilbrun and Rosenburg (1981) reported one at Ruby Lakes December 20, 1980 during the CBC.

Ryser (1967) reported House Wrens were present in Lamoille Canyon May 28 and 29, 1967.

In the Santa Rosa Mountains, Humboldt County, House Wrens were seen at their nests at Lye Creek June 11, 1968; however, on July 4, 1968 only one was observed in this nest area. Two nesting House Wrens were seen at Lamance Creek on the east side of these same mountains June 10, 1970 (Alcorn MS).

Orr and Moffitt (1971) reported for Lake Tahoe the House Wren is a fairly common summer resident.

Ryser (1965) reported, "males in song" were observed along the Truckee River near the Thomas Ranch July 23, 1965.

Kingery (1980) reported an out of season record of one at Reno January 23.

Burleigh collected a male specimen for the USNM at Galena Creek May 6, 1968.

Kingery (1977) reported them singing in Reno by February 10 and he (1980) reported, "In Reno, House Wrens nested in Alves' yard."

Along the Truckee River near Wadsworth, three were seen July 4, 1960 and on that same day about 10 were seen along the Carson River, east of Fort Churchill (Alcorn MS).

Two adults were feeding young at Wilson Canyon, Walker River July 4, 1969 (Alcorn MS).

Alcorn (1946) reported the House Wren as rare in the Lahontan Valley (Fallon area). Several reports of nesting House Wrens in this area were investigated and proved to be Bewick Wrens, which are common in this valley. However, a House Wren was seen along the Carson River 14 miles west of Fallon May 14, 1971 and a pair nested at Velma Watkins yard in June, 1977. A gopher snake ate the eggs and the adults were not seen again (Alcorn MS).

In Watkins' yard on June 15, 1986 an adult was seen carrying sticks to a nest box. On June 26 there were two adults at the nest box and on July 12 an adult carrying a beak full of worms entered the box. On July 29 examination revealed four feathered–out young in the nest box (Alcorn MS).

At this same location on June 13, 1987 two adults were carrying food to young in a nest in a gourd (Alcorn MS).

Linsdale (1936) reported House Wrens from all regions of the state. In the Toiyabe Range the greatest nesting activity is found in the month of June, with both eggs and young found during the last half of June.

Two adults were feeding young in a nest at Wilson Creek, Toquima Range July 4, 1969. The nest was in a hole about six feet above ground level in an aspen tree (Alcorn MS).

Biale (MS) reported House Wrens nest in the Eureka area and reported observations from May through September. He saw seven nests in nest boxes on June 7, 1984 containing from three to seven eggs. All the nests were located within a 4–acre area (elev. 7800 feet) about five miles south of Eureka.

A. A. Alcorn (MS) reported House Wrens in southern Nevada from Lee Canyon July 18, 1963 and from Trough Springs July 19, 1963, both localities in the Charleston Mountains.

Austin and Bradley (1971) reported for southern Nevada, "Summer resident in montane forests and winter resident in riparian areas. In woodland and montane forest it usually occurs from June 3 to October 6. Unseasonal occurrence on May 11. In desert scrub in riparian areas, it usually occurs from March 30 to May 27, and from September 5 to October 2. Unseasonal occurrences in January, February and December. Transient in desert scrub and woodland."

Heilbrun and Rosenburg (1982) reported two House Wrens on the Henderson CBC December 19, 1981.

Winter Wren   *Troglodytes troglodytes*

Distribution. Holarctic region from the Aleutians in Alaska eastward to Newfoundland and Iceland; south to California. Across Europe, Asia, India, China and Japan.

Status in Nevada. Uncommon visitor to Nevada.

Records of Occurrence. In that so few records are available for Nevada, all available records are listed.

Sheldon NWR (1982) reported the Winter Wren as an occasional transient visitor in the fall.

Linsdale (1936) reported Streator obtained one from the Cottonwood Range, Humboldt County September 19, 1896.

One was seen in a willow thicket at Wolf Creek, Elko County September 29, 1969 (Alcorn MS).

Kingery (1981) reported, "These wrens were seen at Ruby Lake and Eureka."

One was seen at Nightingale, Pershing County October 11, 1961 (Alcorn MS).

Ridgway (1877) reported one collected in the Truckee Bottoms near Pyramid Lake December 25, 1867.

Ryser (1966) reported one Winter Wren on October 16, 1966 at the Thomas Ranch on the Truckee River.

Ryser (1972) reported one at Mason Valley December 30, 1971.

Alcorn (1946) collected one feeding in an elm tree four miles west of Fallon April 5, 1939.

Biale (PC, 1984) reported one Winter Wren seen on the West Diamond Mountains October 6, 1974 and one bird five miles south of Eureka October 2, 1977.

Gullion (1957) reported, "On September 13, 1953, a single bird of this species was found living among the moss–covered stone ruins of the Hamilton water–pumping station, at 7600 feet elevation in Harris Canyon, three miles east of Hamilton, White Pine County.

Lawson and Mowbray (MS) saw one December 18, 1971 at Paradise Valley near Las Vegas.

Austin and Bradley (1968) reported, "Individuals observed December 22 and 30, 1963 at Gilcrease Ranch northwest of Las Vegas are the first records for Clark County."

Kingery (1980) reported, "Winter Wrens were discovered in Nye, White Pine and Lincoln Counties October 25, November 12; they probably are regular in the north although scarce in southern Nevada."

Kingery (1984) reported, "Winter Wrens reached Mesquite Nevada by October 12 and Boulder City, Nevada by October 20."

Kingery (1984 and 1985) reported Winter Wrens in the Las Vegas area April 6, September 21, October 6 and November 6."

Marsh Wren   *Cistothorus palustris*

Distribution. Formerly the Long–billed Marsh Wren. From British Columbia to New Brunswick; south to central Mexico.

Status in Nevada. A common resident of marsh areas, statewide. Linsdale (1951) reported two races from Nevada: *C. p. plesius* and *C. p. aestuarinus.*

Records of Occurrence. Sheldon NWR (1982) reported the Marsh Wren as an uncommon breeding summer resident, occasionally seen in winter.

LaRochelle (MS, 1967) reported for Ruby Lake NWR, "Found most of the year except in severe winters. Nests in large numbers in bulrushes."

Ridgway (1877) reported finding two nests with five eggs each on May 18, 1868 in the Truckee Bottoms near Pyramid Lake. According to Henshaw (1877) thousands of these wrens wintered in the sedgy margins of Washoe Lake.

Burleigh collected 12 specimens (4 females, 8 males) for the USNM from the Reno–Galena Creek area from 1967 to 1969, in the months of March, September, October, November and December.

In the marsh areas near Minden and Genoa, these wrens were frequently seen from March to November (Alcorn MS).

On May 26, 1960, "Two Marsh Wren nests were seen in the marsh areas about three miles northeast of Fallon. Both nests appeared to be newly constructed, but neither contained any eggs (Alcorn MS).

Napier (MS, 1974) reported them in the Stillwater Marsh, east of Fallon, as common all year. "Even heard singing during cold winter months. Nests in abundant numbers. No change in population status from 1967 to 1974."

In the Stillwater Marsh at the Canvasback Gun Club on May 12, 1941, this author waded in water about three feet deep and noted a number of Marsh Wren nests in the tules. Walking about 100 yards, 17 nests were seen. Inserting a finger into each nest I found all were empty, except one containing three young wrens. They were disturbed by my activity and left the nest. They were able to fly a little. Two of the nests were not completed and those nests being constructed were wet. Apparently, the material used in making the nest is wet so it can be bent without breaking. All materials seen in the new nests were dead, dry tules. No green tules were used in the partly–constructed nests (Alcorn MS).

Biale (MS) saw Marsh Wrens at the J. D. Ranch and Fish Creek, Eureka area, March 28, 1967 and has seen them in the months of March, April, May, August, November and December.

About nine miles northeast of Beatty one was seen December 17, 1959 and three were seen April 2, 1960 (Alcorn MS).

One bird was seen near Hiko March 18, 1965 and farther south, these wrens were seen each day this author visited Moapa Valley on March 10, 13, 14, and 15, 1962 (Alcorn MS).

Austin and Bradley (1971) reported for southern Nevada, "Occur all year and are a permanent resident in riparian areas."

Family **CINCLIDAE:** Dippers

American Dipper   *Cinclus mexicanus*

Distribution. Sometimes called the Water Ouzel. From the Aleutian Islands to central Yukon; south to Panama.

Status in Nevada. Resident statewide in mountain ranges along permanent streams.

Records of Occurrence. Linsdale (1936) reported these birds from the Pine Forest Mountains, the East Humboldt Mountains, West Humboldt Mountains, Truckee River, Chiatovich Creek in White Mountains, Toiyabe Mountains, and Snake Mountains.

Voget (MS, 1986) reported for Sheldon NWR the dipper is seen spring, summer and fall, with nesting observed in June of 1986.

In Elko County, along the Jarbidge River near Jarbidge, lone birds were seen September 22 and 24, 1959 (Alcorn MS).

Heilbrun and Rosenburg (1981) reported one near Jarbidge December 21, 1980.

LaRochelle (MS, 1967) reported for Ruby Lake NWR, "Found year round on Cave Creek and in the hatchery area. Sings (year round) like a clear–throated, high–pitched Robin or Mockingbird."

Orr and Moffitt (1971) reported in the Lake Tahoe area dippers may be found in the summer months, "Along nearly any of the swifter watercourses and many of the lakes."

They were observed in Idlewild Park, Reno on the Truckee River February 8, 12 and 13, 1942 and one was observed at Bronco Creek, Carson Range, August 1, 1962 (Alcorn MS).

Ryser (1985) commented, "Sometimes the dipper will forego the sound and spray of falling water and build above a smooth–flowing stream on the underbeam of a bridge. This they have been doing at a bridge over the Truckee River at Verdi, Nevada. The dipper builds a large, domed, Dutch oven–shaped nest. The entrance hole grows progressively larger along with the nestlings."

Johnson (1953) reported, "I identified the partly digested remains of a fledgling bird of this species from the stomach of a 10–inch, male brook trout caught by W. V. Woodbury at Hunter Creek, Washoe County, Nevada, in July, 1945. The Dipper was tightly compressed into the stomach, and the approximate size of the bird at the time of the capture by the fish could not be judged."

Burleigh collected four specimens in the USNM from the Reno area. Three females were collected October 10, 1967 and a male on September 17, 1968.

At Wilson Canyon, 12 miles southwest of Yerington, Lyon County, on April 16, 1939 a nest with two half–grown young was seen. The nest was about two feet above the water at the base of a cliff and two adults were carrying food to the young (Alcorn MS).

Mills (MS) on November 26, 1939 in the Fallon area wrote, "This afternoon I killed a Water Ouzel on Lewis' drop in the canal. Mama thinks we better keep it for our collection, but if the record is worth anything to you we might send it down for examination." Mills also reported seeing one three miles east–northeast of Fallon in mid–November, 1973.

In the Toiyabe Range, lone individuals were seen at Kingston Canyon June 20 and 21, 1963 and at the headwaters of the Reese River July 11, 1962 (Alcorn MS).

Biale (MS) reported he has never seen dippers in the Eureka area.

On the east side of the Snake Range one was seen August 8, 1968 at Baker Creek (Alcorn MS), and A. A. Alcorn (MS) reported them from Spring Creek October 6 and December 17, 1966.

One Dipper was seen at Birch Creek, Smoky Valley, Lincoln County, September 25, 1972 (Alcorn MS).

Jaeger (1927) reported the dipper nesting at Williams Ranch, Charleston Mountains, Clark County on June 20, 1926.

These birds are not common to southern Nevada. Lawson (PC, 1984) reported occasional wanderers are observed in southern Nevada. "They are usually juveniles."

Family **MUSCICAPIDAE:** Muscicapids

Golden–crowned Kinglet   *Regulus satrapa*

Distribution. From the Kenai Peninsula, Alaska to Newfoundland; south to Central America (Guatemala).

Status in Nevada. Reported as resident in the Snake Range in eastern Nevada and probably resident in some of the other mountains ranges. An irregular visitor and transient over the entire state. Linsdale (1951) reported two races from Nevada: *R. s. satrapa* and *R. s. apache.*

Records of Occurrence. Linsdale (1936) reported these birds on Bear Creek, Jarbidge Mountains September 13, 1934.

LaRochelle (MS, 1967) reported for Ruby Lake NWR, "Not often seen and less often identified as such." He reported two sightings for the area.

Kingery (1980) reported Golden–crowned Kinglets (16) at South Lake Tahoe December 16, 1979.

Johnson and Richardson (1952) reported, "A single bird was seen in Reno in October, 1944, and several were noted on February 29, 1948, along Hunter Creek at 5000 feet near Reno (Johnson). On March 20, 1948, H. I. Fisher and V. K. Johnson saw one at Galena Creek Ranger Station, also in Washoe County. This species is to be looked for as a resident in the Carson Range near Lake Tahoe."

Burleigh collected four specimens for the USNM from the Reno area. A male and a female were collected December 16, 1967 and a male January 1, 1968. A female was taken October 6, 1968 from the Mount Rose area.

Johnson (1956) reported, "A significant sight record is that of a fledgling being fed by an adult in a dense pocket of mature white fir at 7200 feet, half mile southwest of Daggett's Pass, Douglas County, July 29, 1953."

Ryser (1966) reported these birds from the Thomas Ranch, Truckee River October 16, 1966.

Specimens were taken from along the Carson River two miles west of Fort Churchill on October 30, 1959 and on the same date four miles east of Fort Churchill (Alcorn MS).

One was seen and taken from a growth of *sarcobatus* four miles north–northwest of Fallon on October 29, 1959 (Alcorn MS).

Biale reported one in Water Canyon (Eureka area), Diamond Range on November 9, 1975; and one five miles south of Eureka October 2, 1977.

Alden H. Miller and Ward C. Russell (1956) reported, "A male in breeding condition was taken on May 26 in Trail Canyon, Esmeralda County, 8900 feet, in limber pines. Probably the species nests in small numbers in the better stands of conifers in the range."

Warren M. Pulich and Allan R. Phillips (1951) for the Charleston Mountains reported, "A flock of Golden–crowned Kinglets was discovered almost immediately on our arrival at the upper picnic area at the upper edge of the yellow pine belt. It was estimated that there were about a dozen in the loose flock."

Gullion et al., (1959) reported, "Two additional records for the Spring Mountains. Imhof reported birds from Lee Canyon (7500 feet to 9000 feet elevation) on October 6, 1951 and Pulich recorded this species in the same area on November 17, 1951."

Austin and Bradley (1971) for Clark County reported, "Usually occurs from October 6 to January 28. Unseasonal occurrence on September 9. Visitant in riparian areas."

Heilbrun and Rosenburg (1982) reported two Golden–crowned Kinglets on the Henderson CBC December 19, 1981.

Austin and Rea (1976) reported, "Austin saw two males at Overton on 8 December 1968. A female was taken on 8 November 1968 near Henderson . . . . A male taken 26 October 1966 in Lee Canyon."

Ruby–crowned Kinglet   *Regulus calendula*

Distribution. From Alaska to Newfoundland; south to Guatemala.

Status in Nevada. Resident in major mountain ranges statewide; transient and winter visitor in valleys and deserts.

Records of Occurrence. Specimens in various museums represent all areas of Nevada.

Linsdale (1936) reported the Ruby–crowned Kinglet from eight of the 17 Nevada counties. Recent records are available for 13 counties.

Sheldon NWR (1982) reported the Ruby–crowned Kinglet as an occasional permanent resident summer, fall and winter; uncommon spring. It is thought to nest in the area.

Two specimens in the USNM were taken by Lewis at Ruby Lake NWR October 15 and 27, 1966.

Burleigh collected 23 specimens from 1966 to 1969 from the Carson Range, Washoe Lake and Reno areas. Five were taken in January; two in February; one in March; two in April; one in May and September; nine in October; and two in November.

In Idlewild Park, Reno, Ruby–crowned Kinglets were seen in 1941 and 1942 in September, October, November and December (Alcorn MS).

Ryser (1965) reported them as abundant and widespread in the Reno area.

Alcorn (1946) reported them in the Fallon area as a winter resident, seen in all months except June, July and August.

Apparently a severe population decline, beginning in 1961, occurred, continuing until 1967. Since that time the population appears stabilized.

Thompson (PC, 1988) reported seeing at least 10 at Stillwater WMA October 7, 1987.

Biale (MS) reported Ruby–crowned Kinglets in the Eureka area 13 times from 1966 to 1973. On October 6, 1969 he reported them abundant in the area. He saw a pair in Allison Creek, 8100 feet, Monitor Range on June 16, 1974.

Kingery (1980) reported them at Eureka in May.

In the Baker area, they were observed in 1967 on April 29 and October 24 and 27 (Alcorn MS).

Johnson (1973) reported, "New records verify the presence of this species as a common summer resident in the forest zones in Quinn Canyon Mountains between 7500 feet and 10,200 feet, Mount Irish between 7900 feet and 8500 feet (male taken at 8200 feet), Highland Range 7500 feet to 8800 feet and Mount Wilson 8200 feet to 9100 feet."

Johnson (1974) reported, "This species was not recorded from the Grapevine Mountains during 1939 to 1940, nor was it expected to occur. Surprisingly, at least two singing males were encountered on May 20, 1971 in medium–sized pinon at 7800 feet and 8000 feet elevations on the cool north slope of Wahguyhe Peak. The individual at the latter elevation was carrying nesting material and approached within five feet in response to my imitated owl calls. I know of no other breeding occurrences in the pinon association."

In Clark County they were seen December 14, 1959 on the north side of Virgin Mountains; on April 2, 1960 in Las Vegas; and on October 30, 1960 in the Gold Butte area.

Gullion et al., (1959) reported, "An uncommon winter visitor in brushy areas, arriving on the desert as early as September 26 (1954, Gold Butte area) and remaining as late as April 28 (1951, Las Vegas Wash)."

Austin and Bradley (1971) reported for Clark County, "In woodlands and montane forests it occurs all year. In desert scrub and riparian areas it usually occurs from September 10 to May 18, winter resident in desert scrub, riparian, woodland and montane forest areas."

Blue–gray Gnatcatcher   *Polioptila caerulea*

Distribution. From southern Oregon to southern Quebec and New Jersey; south through Mexico to Guatemala.

Status in Nevada. Summer resident.

Records of Occurrence. Sheldon NWR (1982) reported the Blue–gray Gnatcatcher as an uncommon breeding summer resident, with summer and fall records.

Gordon W. Gullion and Leonard W. Hoskins (1956) reported, "On June 7, 1953, Hoskins found a pair nesting in a riparian association along the south fork of the Humboldt River, about 16 miles south of Elko . . . . A second record for Elko County is that of a single bird seen by Gullion at Cherry Spring, about 11 miles southwest of Elko July 22, 1955."

Burleigh collected a male specimen for the USNM at Virginia City June 8, 1967.

Alcorn (1946) obtained a specimen April 27, 1942 from the Fallon area.

Linsdale (1936) reported a nesting pair with four eggs taken June 20, 1930 near the mouth of Kingston Creek, Toiyabe Range.

Biale (MS) reported the Blue–gray Gnatcatcher in the Eureka area May 13, 1966 and May 31, 1971. He (PC, 1984) considers these birds uncommon in the Eureka area. He (PC, 1986) reported a female, accompanied by a male, feeding a young cowbird in a nest in the Ruby Hill Mine area June 19, 1986. The nest also contained one unhatched egg.

Gullion et al., (1959) reported, "A permanent resident on the intermediate (3500 feet to 5000 feet elevation) desert ranges, being most frequently found in close proximity to springs and waterholes. This species normally occurs in willow and cottonwood thickets in oak–chaparral and pinon–juniper areas."

Austin and Bradley (1971) reported for Clark County, "A permanent resident in desert scrub. Winter resident in riparian areas. Summer resident in woodland areas. Transient in montane forests."

Heilbrun and Rosenburg (1981) reported six in the Henderson area December 20, 1980 during the CBC.

Black–tailed Gnatcatcher   *Polioptila melanura*

Distribution. From southwestern California to the lower Rio Grande Valley in Texas; south into Mexico to Tamaulipas.

Status in Nevada. Resident in southern Nevada.

Records of Occurrence. Kingery (1981) reported, "The description of a Black–tailed Gnatcatcher at Ruby Lake on May 2, relied on the tail and calls for identification."

Gabrielson (1949) reported, "Two at Nelson, Clark County, November 19, 1934."

Gullion et al., (1959) reported, "Recorded from Las Vegas Valley throughout the year, spreading to brushy washes in the lower desert ranges in April and remaining as late as August. This species is normally closely restricted to mesquite and catclaw thickets below an elevation of about 3000 feet."

Austin and Bradley (1971) reported, "It occurs all year. Permanent resident in riparian areas. Summer resident in montane forests."

The Audubon CBC usually lists one or two of these birds on the Desert Game Range or Henderson routes.

Western Bluebird *Sialia mexicana*

Distribution. From southern British Columbia and central Montana; south through the mountains to Mexico (Michoacan).

Status in Nevada. Resident, mainly in the western and southern parts of the state. Linsdale (1951) reported two races from Nevada: *S. m. bairdi* and *S. m. occidentalis*.

Records of Occurrence. Sheldon NWR (1982) reported the Western Bluebird as an occasional spring and fall migrant.

In Elko County one was seen near Elko April 5, 1961 (Alcorn MS).

LaRochelle (MS, 1967) had never seen the Western Bluebird at Ruby Lake NWR.

In western Nevada, they were reported from along the Truckee, Carson and Walker Rivers, where they often nest in abandoned woodpecker holes (Alcorn MS).

A serious decline in the population of these birds has occurred, at least in western Nevada, within the past 15 years. It has been suggested the starling has been responsible for the decline in the breeding population of these birds along the Carson River (Alcorn MS).

In the Reno area, they were reported during all seasons in 1942 (Alcorn MS).

The Audubon CBC for the Truckee Meadows (Reno area) from 1964 to 1974 revealed their presence only once, on December 23, 1967.

Kingery (1982) reported, "By early February, Western Bluebirds arrived at Reno."

On June 14, 1961, four adults were feeding young near Wadsworth (Alcorn MS).

Ryser (1966) reported Western Bluebirds at the Thomas Ranch, Truckee River, October 16, 1966 and he (1971) reported, "A small migratory party" of Western Bluebirds at the Paiute Reservation on the Truckee River February 17, 1971.

Ryser (1964) reported them in the Genoa area May 23, 1964.

One was seen at Minden February 19, 1942 (Alcorn MS).

In Mason Valley (Yerington area) these bluebirds were seen in all seasons from 1944 to 1962. Since 1962, this author has recorded them in Mason Valley on March 13, 1965 and a solitary bird was seen at the McGowan residence, near Yerington, April 6, 1982 (Alcorn MS).

Adults were feeding young near Fort Churchill, Carson River, May 15, 1960 (Alcorn MS).

Alcorn (1946) reported, "Resident but not numerous in the Lahontan Valley."

Since 1946, they have been recorded in the Fallon area only nine times in 1961; four in 1962; twice in 1963; and once in 1964 and 1967. Napier (MS, 1974) saw one February 8, 1969. The most recent record is of 10 seen five mile west of Fallon December 19, 1987 (Alcorn MS).

In White Pine County, Western Bluebirds have been reported as follows: August 16, 1961, one seen by J. R. Alcorn (MS) near Hamilton; April 20, 1967, seen by A. A. Alcorn (MS) near Baker; and March 23, 1985, an adult male seen by Biale (MS) in the northern Pancake Range.

J. R. and A. A. Alcorn (MS) reported seeing one near Hiko, Pahranagat Valley October 31, 1960 and about 10 on April 14, 1965.

Johnson (1973) reported on May 21, 1970, "Two pair of this species were found nesting in the firs at 7700 feet on the north face of West Virgin Peak. (One pair was entering an abandoned woodpecker hole in a 50 foot fir snag)."

In Moapa Valley, three were seen January 26, 1962 and three on February 18, 1963 (Alcorn MS).

One was recorded from Gold Butte area October 30, 1960 (Alcorn MS).

Gullion et al., (1959) reported, "Six records: small flocks in mesquite thickets (feeding on fruit of mistletoe) *(Phoradendron californicum)* in Las Vegas on January 12 and 21, 1952 and March 15, 1953; and in the Mohave Valley, January 25 and 31, 1952. One summer record from Hidden Forest (8000 feet elevation), Sheep Range, Clark County, June 27, 1951."

Austin and Bradley (1971) for Clark County reported, "Permanent resident in woodlands and summer resident in montane forests, winter resident in desert scrub and riparian areas. In woodlands and montane forests, it occurs all year. In desert scrub and riparian areas, it usually occurs from November 9 to April 20. An unseasonal occurrence on October 6."

Kingery (1982) reported, "They moved to the southern portion of the region [Mountain West] more than in most years to Las Vegas (one January 14), the first southern Nevada record in three years."

Mountain Bluebird   *Sialia currucoides*

Distribution. From Alaska to Manitoba; south to Arizona and western Oklahoma. Winters south in Mexico to Guanajuato.

Status in Nevada. The Mountain Bluebird is Nevada's state bird. Resident statewide. More common in the north in summer and south in winter.

Records of Occurrence. Sheldon NWR (1982) reported the Mountain Bluebird as a breeding summer resident; uncommon spring, common summer and fall.

Ryser (1973) reported adult and juveniles were seen in the Pine Forest Mountains of Humboldt County on September 21, 1973.

In the Santa Rosa Range, Lye Creek, a female was seen carrying food to a nest of young on July 4, 1968. About 10 were seen at Hinkey Summit in this same range November 1, 1963 (Alcorn MS).

In the Jarbidge area of Elko County, these bluebirds were present from May through September. Others were seen in the Midas, Tuscarora and Elko areas in June, August, September and October (Alcorn MS).

LaRochelle (MS, 1967) reported Mountain Bluebirds for Ruby Lake NWR, "Very common from mid–spring through summer."

A loose flock of about 10 were seen March 23, 1966 in the Ruby Valley (Alcorn MS).

Over 30 Mountain Bluebirds were seen at Elkhorn Canyon, Lander County, March 2, 1977 (Alcorn MS).

In the Fallon area, a sharp decline in the population of these bluebirds has occurred within

Mountain Bluebird. Photo by Art Biale.

the past 20 years. Alcorn (1946) for the Lahontan Valley (Fallon area), reported, "Repeatedly seen in November, December and January. Less often recorded in February and March." Since that time, this author observed them a few times each winter from 1958 until 1965. Since 1965, they have been seen only four times: on February 26, 1966 (5 birds); on January 25, 1967 (single bird); on March 6, 1976 (single bird); and February 26, 1987 (three seen) (Alcorn MS).

Napier (MS, 1974) reported two in the Fallon area February 27, 1969.

Thompson (MS, 1988) reported 10 Mountain Bluebirds at Stillwater WMA March 19, 1987. He noted, "They always show up when a storm comes through."

At the south end of the Reese River Valley, Nye County, they were seen each month from March to November (Alcorn MS).

The Mountain Bluebird's abundance in summer in the Eureka area is shown by Biale's (MS) recording them 298 times between May 3, 1964 and November 19, 1967. They are most abundant from March through September. From 1964 to 1974, he did not see them in December and had only two records for January: single birds on January 15, 1967 and January 24, 1971. He saw 105 on April 27, 1975 at Fish Creek Ranch, 6300 feet elevation, about 20 miles south of Eureka.

Biale (PC, 1988) reported, "There seems to be no real change in their population over the past 20 years."

J. R. and A. A. Alcorn (MS) in the Hiko and Pahranagat Valley area reported Mountain Bluebirds seen February 4, 1960 (2 birds); October 31, 1960 (12 birds); March 16, 1962 (one bird); March 17 and 18, 1965 (10 birds); April 7, 1965 (2 birds); and April 14, 1965 (9 birds).

J. R. and A. A. Alcorn (MS) recorded Mountain Bluebirds in the Moapa Valley area 3 times in February; 6 in March; 3 in April; on May 24, 1962; twice in October (October 20 and 31, 1960); and on December 18, 1959.

Gullion et al., (1959) reported, "Mountain Bluebird, sparingly recorded from low desert ranges in winter, from as early as September 2 (1951, Virgin Mountains) to April 26, (1951, Mormon Mountains). Recorded at one time or another in winter from all of the desert valleys and ranges in southern Nevada."

Johnson (1974) reported for southern Nevada, "This species was found only in one place by the earlier survey (Miller, 1946). In contrast, I found Mountain Bluebirds in five locations in 1971; at least two of these sites had active nests. In 1973, four territories were found. An increase in population size since the earlier fieldwork is strongly suggested."

Austin and Bradley (1971) for Clark County reported, "Transient in montane forests and winter resident in desert scrub, riparian and woodland areas. Usually occurs from August 26 to April 30. Unseasonal occurrence on June 18 and July 7."

Food. Information on the Mountain Bluebird's feeding habits in the Fallon area is taken from this author's notes. "On December 30, 1941, five Mountain Bluebirds were seen feeding on the (small berries) on the branches of the Russian olive tree located 20+ feet away. With difficulty, the birds were seen to swallow the berries whole. They usually picked the berries in flight then perched on a limb and swallowed the berry."

"On January 1, 1942 a lone Mountain Bluebird was seen feeding on the Russian olive berries that are clinging to the tree not over 20 feet from my house near Fallon. This tree has many small gray berries on it. Several of these bluebirds were seen yesterday feeding from this same tree. Laura E. Mills advises me that these birds have been feeding on the red asparagus berries that are located on their ranch about one mile south–southwest of my house (six miles west–southwest of Fallon). It has been snowing intermittently for the past five days and there is six inches of snow on the ground. These birds were swallowing the berries whole."

On January 3, 1942, "Three Mountain Bluebirds were seen eating small gray Russian olive berries from a tree located 20+ feet southeast of my house. These berries were swallowed whole."

Nest. J. R. and A. A. Alcorn (MS) reported at San Juan Canyon on May 3, 1986, three pair of Mountain Bluebirds were seen near abandoned woodpecker holes of aspen trees.

Biale (MS) reported nesting on June 26, 1966 in a box on a boxelder tree 14 feet above ground. One egg and a young (approx. eight days old) were in the nest.

On June 26, 1969, a nest was in a birdhouse on a post, five feet above the ground. The nest was made of juniper bark and contained two young birds two or three days old (6450 feet elev.) (Biale MS).

Biale (MS) reported on June 7, 1970 at Tonkin Summit, a Mountain Bluebird nest in a hole in a dirt bank four feet above the bottom of a wash. It was made of juniper bark and contained four blue–green eggs (6600 feet elev.).

On June 5, 1972, in Eureka, a nest was located in a gravel spreader. The nest was made of weed stems and juniper bark, lined with hair and grass. It contained one blue egg and two just–hatched young (6420 feet elev.).

Townsend's Solitaire   *Myadestes townsendi*

Distribution. From Alaska to west–central Mackenzie, Alberta and South Dakota; south to the mainland of Mexico.

Status in Nevada. Statewide summer resident in the higher mountain ranges. Winter visitor to the lower elevations.

Records of Occurrence. Sheldon NWR (1982) reported the Townsend's Solitaire as an uncommon permanent resident, occasionally seen in winter.

In the Jarbidge area of Elko County, three were seen September 22 and 23, 1959 and one was seen September 7, 1961 (Alcorn MS).

Heilbrun and Rosenburg (1981) reported 23 in the Jarbidge area December 21, 1980.

One was seen in the Tuscarora area September 30, 1965 (Alcorn MS) and A. A. Alcorn saw one at Ruby Lake NWR March 23, 1966.

Burleigh collected a female at this refuge May 23, 1968.

The scarcity of this bird in the Ruby Lakes area is demonstrated by a narrative report which stated one was observed on September 24, 1969 by Marshall and Dubbert. They thought, incorrectly, that it was a new addition to their refuge bird list.

One was seen October 7, 1961 at Solid Silver Creek, Santa Rosa Range, Humboldt County and another was seen April 26, 1961 in the Lovelock area, Pershing County (Alcorn MS).

In Idlewild Park, Reno, lone birds were seen 4 times in December, 1941; 3 in February, 1942; and 3 in March, 1942 (Alcorn MS).

At Fort Churchill, Carson River, one was observed May 15, 1960 (Alcorn MS).

Alcorn (1946) reported Townsend's Solitaires as a winter visitant in the Lahontan Valley (Fallon area), with records for January, February, September, October and November.

More recent records for this area are of solitary birds (unless otherwise noted) seen on May 9, 1960; October 16, 1960; October 20, 1961; November 26, 1961; December 8 and 18, 1961; September 22, 1963; September 16, 1965; August 28, 1968 (2 seen); April 28, 1969; September 23, 1971; and March 2, 1975 (Alcorn MS).

Solitary birds were observed at Cherry Creek, Clan Alpine Mountains, October 20, 1960; Porter Canyon, Toiyabe Mountains October 12, 1964; in the Toquima Range at Wilson Creek on July 4, 1969; and Northumberland Cave, October 4, 1966 (Alcorn MS).

Biale (MS) reported Townsend's Solitaires only nine times in the Eureka area from 1964 to 1967, inclusive. He saw them once in January; 5 times in October; and 3 times in November. Since 1967, he has reported them in the Eureka area on February 16, 1974; March 13, 1971; May 9, 1971; August 15, 1972; September 25, 1971; and 6 times in October. From September, 1975 through January, 1976, they were more numerous, probably due to an abundant supply of chokecherry berries.

Kingery (1980) reported a Townsend's Solitaire at Eureka in May.

Biale (PC, 1984) suspects they nest in the Eureka area, even though he has seen no nests. He observed one on July 4, 1978, two on August 4, 1984, an immature on August 12, 1985, and one on August 14, 1985.

Townsend's Solitaires were observed in the Kern Mountains October 26, 1967; at Baker Creek, Snake Range on May 12, 1967 and October 30, 1961; at Hiko, Pahranagat Valley April 23, 1964, and at Corn Creek, Clark County April 5, 1960 (Alcorn MS).

Gullion et al., (1959) reported, "A widespread and fairly frequent winter visitor to the desert, having been recorded as early as September 10, 1951 at Boulder Beach by Mrs. Poyser to as late as June 5, 1955 at Boulder City, also by Mrs. Poyser (Monson, 1955). Two summer records: Hidden Forest, Sheep Range, June 27, 1951 and reported by Mrs. Poyser at Lake Mead, August 16 and 23, 1952 (Monson, 1953)."

Johnson (1974) reported, "This species was unreported previously from Potosi Mountain.

I shot a female at 7100 feet elevation in ponderosa pines and white firs. The ova were not enlarged and the bird was slightly fat. Prior to collection, however, this individual had taken part in a chase with another solitaire, and subsequent breeding may have ensued.''

Austin and Bradley (1971) reported for Clark County, ''Woodland and montane forests all year. Riparian occurrence from August 16 to May 9. Unseasonal occurrence on June 5. Winter resident in riparian and woodland areas. Permanent resident in montane forests.''

Kingery (1984) reported Townsend's Solitaires at Las Vegas October 3, 1983. He (1987) reported, ''Las Vegas enjoyed an unusual influx of Townsend's Solitaires, with 20–30 feeding on Russian olives with robins and waxwings September 30 to October 19.''

Veery   *Catharus fuscescens*

Distribution. From British Columbia to Newfoundland; south to Brazil.

Status in Nevada. Rare transient.

Records of Occurrence. Ridgway (1907) reported a specimen, now in the USNM, taken at Mountain City, Elko County June 13, 1898 by W. K. Fisher.

Snider (1967) reported a Veery on the Desert Game Range on April 30 to May 16, 1967 and she (1970) reported, ''On May 23, 1970 a banded Veery seen near Las Vegas may have been the same individual banded by Dr. Charles G. Hansen two years earlier.''

Monson (1972) reported one at Corn Creek May 21, 1972. It was the Desert Game Range's third record for the locality in five years.

Kingery (1980) reported Veeries seen May 25 and 27 at Las Vegas by Vince Mowbray.

Gray–cheeked Thrush   *Catharus minimus*

Distribution. From Siberia and Alaska to Newfoundland; south to Peru and Brazil.

Status in Nevada. Hypothetical.

Records of Occurrence. Monson (1972) reported one was identified at Tule Springs Park, Las Vegas by Lawson May 12–13, 1972.

Swainson's Thrush   *Catharus ustulatus*

Distribution. Formerly *Hylocichla ustulatus*. From Alaska to Newfoundland; south to Peru.

Status in Nevada. Uncommon summer resident. Linsdale (1951) reported two races from Nevada: *H. u. swainsoni* and *H. u. almae*.

Records of Occurrence. Linsdale (1936) reported specimens from the Pine Forest and Santa Rosa Mountains of Humboldt County; from Mountain City and East Humboldt Mountains in Elko County; Lake Tahoe and Truckee Reservation in Washoe County; and the Toiyabe Mountains of Lander County.

A female specimen in the USNM was taken on May 24, 1967 at Ruby Lake NWR.

LaRochelle (MS, 1967) reported banding several at this locality. He reported, "Our only thrush and quite common in riparian situations."

Gullion et al., (1959) reported one record, a bird trapped and banded near Turkey Spring, in the Gold Butte area, September 17, 1954.

Austin and Bradley (1971) reported, "Transient in the riparian area of Clark County. Usually occurs from April 22 to May 25 and on September 27."

Kingery (1980) reported the Swainson's Thrush at Las Vegas October 13.

Hermit Thrush   *Catharus guttatus*

Distribution. Formerly *Hylocichla guttatus*. From Alaska to Newfoundland; south to Guatemala.

Status in Nevada. Statewide summer resident. Linsdale (1951) reported five races from Nevada: *H. g. guttatus, H. g. nanus, H. g. sequoiensis, H. g. polionotus* and *H. g. auduboni*.

Records of Occurrence. Sheldon NWR (1982) reported the Hermit Thrush as an uncommon to occasional migrant spring and summer.

In the Santa Rosa Range of Humboldt County, one was seen at Solid Silver Creek, October 7, 1961 and another at Lye Creek July 4, 1968 (Alcorn MS).

Four were seen near Jarbidge on August 22, 1963 and one was seen near Elko on June 7, 1967 (Alcorn MS).

A female specimen in the USNM was taken by Lewis at Ruby Lake NWR on June 1, 1967.

Ryser (1967) reported on July 1–4, 1967 in Lamoille Canyon, the Hermit Thrushes were still singing.

One was seen at Jersey Valley, Pershing County, November 22, 1963 (Alcorn MS).

On the east side of Lake Tahoe, these birds were seen on July 23–25, 1963 (Alcorn MS).

Orr and Moffitt (1971) listed them as a summer resident at Lake Tahoe.

On January 6, 1943 in the Fallon area, "Mills shot a thrush and Mr. J. S. Mills mounted it for display purposes. With their permission this author took the bird to Berkeley and Dr. Alden H. Miller (Museum of Vertebrate Zoology), identified this bird as *Hylocichla guttatus nanus*. The bird was a male." Since then, Hermit Thrushes have been observed in this same area May 24, 1965, December 10 1987 (one) and January 17, 1988 (one) (Alcorn MS).

Biale (MS) reported these birds 22 different days in the Eureka area from June, 1964 to October, 1973. He reported them in June, July, August, September and October. He (MS) saw two nests at the north fork of Allison Creek, Monitor Range June 16, 1974. One nest, 10 feet above ground in an aspen tree, was made of grass and stems and lined with fine grass. It contained four blue–green eggs (elev 8100 feet). Another nest was three feet above ground in a dead aspen tree. It was wedged between loose bark and the tree trunk. The nest contained three eggs. Another nest was at Faulkner Creek, Monitor Range (8300 feet elev.) on June 23, 1974. The nest was 30 feet above ground in a limber pine tree growing in a stand of aspen. An adult bird was in the nest.

A. A. Alcorn (MS) reported one from the Baker area August 26, 1957.

Bond (1940b) reported this thrush common in Lincoln County and, "A specimen taken May 25, 1939 on Wilson Peak and present on Table Mountain."

Johnson (1974) reported his records positively establish breeding status for the Hermit

Thrush in the Grapevine Mountains and he reported three singing males at Potosi Mountain May 25, 1971.

One was seen at Juanita Springs Ranch about seven miles south of Riverside, Clark County June 1, 1963 (Alcorn MS).

Austin and Bradley (1971) reported for Clark County, "Summer resident in montane forests, transient in riparian and woodland areas."

Kingery (1978) reported, "A Hermit Thrush wintered at Corn Creek and others occurred at Boulder City January 12 and at Verdi January 29 to give Nevada its first winter records in several years."

Kingery (1982) reported, "A Hermit Thrush wintered in a Las Vegas backyard."

Wood Thrush *Hylocichla mustelina*

Distribution. From southeastern North Dakota eastward to Quebec and Maine; south to northern Florida. Winters from southern Texas south through eastern Mexico to Panama.

Status in Nevada. Accidental.

Records of Occurrence. Kingery (1977) reported a Wood Thrush at Pahranagat NWR October 6 and 7, 1976.

Lawson (MS) reported a sight record at Tule Springs May 19, 1970.

Snider (1970) reported, "A Wood Thrush was observed on May 18 and 24 near Las Vegas and was photographed; this is the first record for Nevada."

American Robin *Turdus migratorius*

Distribution. From the limit of trees in northern Alaska to Newfoundland; south to Guatemala.

Status in Nevada. Statewide resident in summer; visitant to resident in winter. In summer, most abundant in the mountains and higher valleys of Nevada.

Records of Occurrence. The robin's abundance is indicated by the numerous records available for all 17 Nevada counties. In the summer months, it is generally more abundant in the middle and northern portions of the state. The majority of the robins leave the coldest northern parts in winter, but a few remain in less severe winters in the northern areas.

Sheldon NWR (1982) reported the American Robin as a breeding permanent resident; common spring, summer and fall and uncommon in winter.

LaRochelle (MS, 1967) reported many robins nest at Ruby Lake NWR, Elko County and some stay through the winter when the snow does not persist too long.

On February 22, 1970 five groups of one to five robins were judged to have recently arrived in the Ruby Valley (Alcorn MS).

In and near the city of Elko, they were seen 66 different days from April of 1965 to May of 1968. None were seen from November through February (Alcorn MS).

In 1942, in Idlewild Park, Reno, up to 30 robins were seen January 5 and others (mostly one to three) were seen from January 7 to 23. Twenty or more were observed on eight different days in June. From November of 1941 to December of 1942, robins were recorded 278 times in the Reno area (Alcorn MS).

Ryser (1966) reported robins were "very abundant" during the summer in Little Valley, Carson Range. In 1967 in Reno, he, "Witnessed the largest fall concentration of robins ever seen in Nevada. Over 300 were feeding on a bit of lawn at the University of Nevada (at Reno) during a light, but cold rainstorm."

Ryser (1985) reported, "One winter roost was located on the University of Nevada Campus in Reno. Several thousand robins would gather nightly to roost in the trees bordering the campus quadrangle. During the early morning, flocks would fan out to forage, returning at dusk to roost."

The abundance of robins in Lahontan Valley is indicated by their being recorded 1595 times from January, 1960 to December, 1967. They were reported many times in all months, with lesser numbers present in winter. Over 25 were seen at one time feeding on Russian olive berries on January 17, 1982 when the ground was covered with snow and the temperature was near zero degrees Fahrenheit. Their numbers increase noticeably beginning the last of February and the first of March (Alcorn MS).

Biale (MS) recorded these birds 518 times in the Eureka area from April 23, 1964 to October 15, 1967. He did not see robins during this period in December or January. However, he later reported them in these months (December 18, 1972 and 14 days in January of 1974). He reported robins stay in winter when feed is available.

J. R. and A. A. Alcorn reported seeing robins in Pahranagat Valley from 1960 to 1965 in March, April, May, June, October and November (Alcorn MS).

J. R. and A. A. Alcorn (MS) sighted robins in Moapa Valley from January of 1960 to June of 1965 a total of, 5 times in January; 3 in February; 12 in March; 9 in April; twice in May; 3 times in June and July; twice in October; and once in December.

Gullion et al., (1959) reported for southern Nevada, "Most common on the desert as a winter visitor from mid–October to early April, having been noted in the Gold Butte area from October 13 (1951) to April 6 (1953). Mrs. Poyser reported robins nesting in Boulder City during 1953 and 1954 (Monson, 1953; 1954)."

Austin and Bradley (1971) reported, "Permanent resident in woodlands and montane forests, and winter resident in desert scrub and riparian areas in Clark County."

Nest. Information on nesting and young, unless otherwise noted, is by this author and for the Fallon area. On April 17, 1959 the author wrote, "A pale–colored robin was seen in our yard this morning. It had a bunch of sticks in its mouth that it held for several minutes then dropped. It hopped 20 inches or more and started gathering up more sticks in its mouth. Later it dropped them. I judged this robin was picking up nest material. I suppose that this was a preliminary or warm–up for the real thing to come later because this nest material was later discarded and the robin left the area."

On April 27, 1959 at two separate places in Fallon the author saw robins carrying nest material. I wrote, "Two robins, one dark and one light–colored, gathered nest material (sticks) from our yard today."

On June 4, 1959, "A young robin was perched on our yard wire fence, it apparently was 'fresh out' of the nest. I slowly walked to it and grabbed it with my hand and held it for about 30 seconds. It cried for help and three adult robins responded and came to its rescue. It flew into a tree when I released it."

On May 8, 1941, a nest in a pear tree contained two eggs. A gopher snake 30 inches long was also up in the tree near the nest. Two adult robins were snapping their beaks which created a popping sound. The same nest on May 28, 1941 contained a dead young robin and two dead robins were on the ground under the nest.

On April 12, 1942. "Robins building a nest in Virginia Creeper vine under the eve of the house. The robins started today to line the nest with mud. On April 28, a broken robin egg was found on the ground under the nest. No eggs were in the nest."

"April 26, 1942, a recently finished nest on a rafter of an open shed, no eggs. May 16, 1942, nest built on pole fence, contained four eggs, two of which are pipped. On May 17 at 9:30 a.m. the nest contains two baby robins and two eggs. At 6:30 p.m. the nest contained three baby robins and one egg. On May 18, at 6:00 a.m. an adult flew from the nest which contained four baby robins. Their eyes are unopened. On May 23, at 7:15 a.m. I placed my finger in the nest and the baby robins opened their eyes and their mouths wide in anticipation of being fed."

"May 21, 1967, nest with four eggs. June 11, 1967 a robin has just now finished building its nest in a mulberry tree. June 11, 1967 an adult robin is feeding its large young in our yard. The young appear as large as their mother. May 31, 1970 an adult feeding young in nest. May 31, 1970 adult feeding a spotted large young out of the nest."

In the Eureka area, Biale reported nesting June 11, 1967. "Nest in an apple tree eight feet above the ground, four eggs (pipped) July 4, 1967, nest in mountain mahogany tree, eight feet above the ground, nest material: grass, small twigs, and mud. Two eggs (bluish–green) and one just hatched young elevation approximately 6500 feet."

Biale (PC, 1984) reported considerable nesting in the area. One nest was in a limber pine on Prospect Peak, five miles south of Eureka July 4, 1978 and contained four very young birds (8900 feet elev.).

Food. This author on January 28, 1941 wrote, "There are many Russian olive trees here and the robins are attracted by the small gray Russian olive berries. One robin was shot, in its stomach were 13 Russian olive seeds. In its throat, one Russian olive berry was noted."

"On September 6, 1942 and likewise for the past several days, I have noted numerous robins feeding on the new crop of (small variety) Russian olive berries. On one occasion over 12 robins were seen perched in one small Russian olive tree. They were all swallowing the whole berries."

"On December 28, 1958 an adult robin was seen feeding on the ground 20 feet from me this morning. Using its bill it leaned forward, pecked the ground covered with leaves and gave the leaves a flip to the side. For eight pecks it alternated flipping the leaves first left, then right. Then it found a Russian olive berry which it ate. Then it flipped the leaves right four times and found another berry which it ate. Then it flipped the leaves three times left and ate another berry. At one peck it seemed to hook its bill on a Virginia Creeper vine, which flipped up and the robin jumped back, then continued pecking. A White–crowned Sparrow approached within three feet, but kept its distance. This sparrow seemed interested in what the robin was doing or eating, but it was apparently afraid of the robin."

"November 11, 1941 about 15 robins in one Russian olive tree all were eating berries from the tree. November 11, 1941, about 10 robins in a Russian olive tree feeding on the larger red berries, one robin was seen to pick from the tree a large red berry and only after the second attempt, due to the large size of the fruit, was it able to swallow it."

"January 3, 1942, about 10 robins were feeding on Russian olive berries and they were seen to swallow whole both the small grey–colored and the large red–colored varieties of fruit."

"On July 13, 1957, one (adult female) robin was seen to pick and eat 10 honeysuckle (bush) berries from the plant in our front yard."

"On July 15, 1957, one robin was seen to pick 14 honeysuckle bush berries. These red berries were eaten as the robin picked them. Another robin ate 10 berries and another robin picked and ate seven berries. Some robins jumped up from the ground and picked the berries, other robins alighted in the bush to pick the red berries. One full–grown spotted juvenile robin was also seen eating the berries."

"On January 20, 1958, half of a fresh cut apple (delicious variety) was placed in the Russian olive tree in our yard. One robin perched on the limb and ate most of the meat of the apple, leaving only the skin."

"On January 26, 1958, two White–crowned Sparrows were eating an apple and a robin chased them away. On January 29, 1958, each day since January 20 I have cut an apple in the middle and placed half of it, face up, on a limb. It is secured there by a match–sized limb which protrudes up from the larger three inches limb. This match–sized limb runs through the apple. One robin had taken charge and kept all the other birds away as it eats the meat of the apple leaving only the skin."

"On June 12, 1959, an adult robin was seen to eat four red berries from the honeysuckle bush in our yard. The berries are just now starting to get their red color."

Behavior. Territorial behavior is evidenced by their fighting images in automobile rear view mirrors and house windows. Watkins (PC) reported at four miles west of Fallon on March 29, 1970 a robin was fighting her house window. From March 20 to 28, 1982, a robin was seen fighting its reflection in a window in the city of Fallon. This author wrote, "What appears to be the same robin has been fighting the window on a daily basis. This same bird also has selected the outside rear view mirrors on five different nearby trucks and cars, perches on the holding arms in front of the mirrors, defecates on the vehicle doors, and appears to be charmed at what it sees; nevertheless, it fights the mirrors, but not with the same vigor that it has for the window."

On April 25, 1973 the author was watching a robin at Fallon. It was on dry ground, on a nice sunny day, and was seen to sit down on the dry dirt. It then spread its wings out on the ground and laid there for about 10 seconds, as if it were soaking up the heat from the ground. It then proceeded to look for food.

Varied Thrush   *Ixoreus naevius*

Distribution. From Alaska and Mackenzie; south to Nevada, Arizona and Baja California. Casual throughout central and northeastern North America.

Status in Nevada. An irregular visitor.

Records of Occurrence. Sheldon NWR (1982) reported the Varied Thrush as an occasional transient visitor in spring and fall.

In Idlewild Park, Reno, one was observed May 7 and 8, 1942 (Alcorn MS).

Ryser (1965) reported these birds sighted in four different localities in western Nevada.

Alcorn (1946) reported these birds seen in Lahontan Valley in October and November with specimens taken November 3 and December 29, 1940.

Mills (PC) reported two were seen in the Fallon area November 13, 1966.

The author reported one at Smith Creek, Lander County October 10, 1987 (Alcorn MS).

Biale (MS) reported only one sighting for the Eureka area on October 6, 1974. Kingery (1982) reported one appeared at Eureka October 14, 1981.

Ryser (1967) reported one at Springdale, north of Beatty on February 15 and 19.

Austin (1969) reported for southern Nevada, "A male was found dead by Hansen at the mouth of Hidden Forest Canyon, Sheep Mountains, on November 4, 1965. This supplements the record by Cottam (1954) for the same area and sight record for Boulder City (Monson, 1952)."

Lawson (MS) reported an adult female collected at Corn Creek January 31, 1973.

Recent records are of two sightings at Corn Creek, (Kingery, 1978) and one at Las Vegas (Kingery, 1980).

Kingery has reported Varied Thrushes in the Las Vegas area in February, October and November nearly every year since 1980.

Family MIMIDAE: Mockingbirds, Thrashers and Allies

Gray Catbird   *Dumetella carolinensis*

Distribution. From southern British Columbia to Nova Scotia; south to the Canal Zone, Panama, Islands of the western Caribbean and Lesser Antilles.

Status in Nevada. An uncommon visitor.

Records of Occurrence. Sheldon NWR (1982) reported the Gray Catbird as accidental—one observed by a research crew in 1978.

Ruby Lake NWR manager Lewis collected a male at this refuge September 23, 1966.

Kingery (1980) reported, "The Gray Catbird singing in the Ruby Mountains of Elko County on June 17, apparently on territory, but also with no apparent mate, had picked a habitat similar to that of pair seen five years ago near Baker, 100 miles south. The remote canyons of eastern Nevada may support a small population of catbirds."

Kingery (1981) reported, "Gray Catbirds visited Ely, Nevada, May 17." He (1983) reported, "A Gray Catbird appeared at Baker, Nevada, June 2."

Cottam (1936) saw one at Alamo May 1, 1924.

Linsdale (1936) reported a female was taken at Cave Spring (6248 feet) Esmeralda County June 18, 1928 by Russell.

One was seen at the east side of Patterson Pass, Ely Range, Lincoln County October 2, 1974 (Alcorn MS).

Ryser (1985) reported, "There are a few sight records and specimens for northeastern Nevada and a nesting record by Al Knorr from the Toiyabe Range in central Nevada."

Kingery (1984) reported, "Gray Catbirds were found at Dyer, Nevada."

Lawson (MS) reported a specimen from Corn Creek October 22, 1974.

Kingery (1978) reported a catbird wintered at a Las Vegas golf course and another occurred in North Las Vegas January 29. He (1982) reported catbirds at Las Vegas October 19, 1981.

Northern Mockingbird   *Mimus polyglottos*

Distribution. From California, eastward to Maryland; south to the Isthmus of Tehuantepec and the Virgin Islands. Has spread north in recent years, now found casually to the southern part of Canada.

Status in Nevada. Summer resident over most of the state except for the extreme northernmost part. Present in winter in southern Nevada with limited numbers as far north in winter as Fallon and Carson City.

Records of Occurrence. Linsdale (1936) reported, "The farthest north occurrence of the Mockingbird for the state was reported by Dawson (1919) who saw two individuals July 28, 1918 on Duck Flat at a point some miles northwest of Sunkist (formerly Duck Lake) Washoe County. This locality is well above the 41st parallel of latitude and is within six miles of the California line."

In Elko County an adult mockingbird was seen on the east side of Huntington Valley June 16, 1972. It alighted in a juniper tree (Alcorn MS).

One was seen at Ruby Lake NWR by refuge manager Howard.

Dan Delany (PC, 1984) reported one seen at Gerlach July 29, 1984.

An unusually large number of sightings of mockingbirds was reported by Ryser (1967) who stated in the middle of July, 1967, these birds were seen in eight to 10 different places in the Reno–Sparks area. Also they were seen at Six Mile Canyon in the Virginia Range and throughout the Carson Valley with one immature in Carson Valley.

Heilbrun and Rosenburg (1981) reported one in the Truckee Meadows area (Reno) December 20, 1980.

Wright (PC, 1984) reported three sightings of the mockingbird in the Reno area. It was seen twice in 1982, in February and in mid–summer (stayed in neighborhood for three weeks). The third sighting was November 14, 1983.

Ryser (1985) commented, "In the Reno area the Northern Mockingbird may be more numerous during the winter than during the summer." He also believes Mockingbirds nest in the Reno area.

One was seen on Anaho Island, Pyramid Lake April 21, 1961 (Alcorn MS).

Kingery (1982) reported, "Stray mockingbirds were found at Carson City January 19."

Delany (PC, 1984) reported seeing a mockingbird that spring as he jogged in the Carson City area.

Alcorn (MS) reported along the Carson River from Lahontan Reservoir to the Fort Churchill area, an occasional mockingbird has been seen in May, June, July and August.

Two were seen in an elm tree in the town of Fallon November 22, 1957. In the same area one was seen January 20, 1945 and it was eating frost. One was seen at the same place January 26, 1945 eating snow (Alcorn MS).

A. A. Alcorn (MS) reported a mockingbird seen all winter (1973–74) at his place four miles west of Fallon. It was frequently seen throughout the winter feeding in his apple orchard on the frozen apples scattered over the ground. He reported a mockingbird, possibly the same one mentioned above, spent the winter of 1974–75 at the same orchard even though there was no fruit (due to complete crop failure in the 1974 season).

Saxton (PC) reported a mockingbird seen almost daily in Fallon from November, 1974 to

March, 1975 at a feeder in her yard. This bird was very possessive and chased all other birds away, including robins, sparrows, flickers and starlings. It was seen eating apple and suet.

Thompson (MS, 1988) reported a mockingbird at his home in Fallon January 5, 1986.

One Northern Mockingbird was observed at Capitola Alcorn's residence five miles west of Fallon on December 24, 1987 (Alcorn MS).

One was seen three miles southeast of Schurz May 14, 1963 and two were seen at Mina May 6, 1962 (Alcorn MS).

Delany (PC, 1984) reported mockingbirds were also seen regularly in Hawthorne and Babbitt in 1983 and 1984 and he reported seeing a male displaying territorial behavior. "I looked for a nest, but he was in a dense locust tree. I would be surprised if they weren't nesting there."

One was observed near Austin October 25, 1961 (Alcorn MS).

Biale (MS) reported them on 14 different occasions in the Eureka area. On July 9, 1967, one young was out of the nest, but unable to fly. He (PC, 1984) reported an increase of these birds into the Eureka area where he has seen two nests, both in juniper trees, containing three and four eggs. They were seen June 3 and 5, 1984.

Kingery (1985) reported, "A pair nested for the second year at Eureka."

One was seen to alight in a juniper tree in Spring Valley east of Conners Pass May 23, 1967 (Alcorn MS).

These birds were frequently seen in the Pahranagat Valley from 1960 to 1965. They were recorded 6 times in April; 13 in May; 4 in June; once in July; twice in August; and once in October (Alcorn MS).

In Moapa Valley these birds were recorded by J. R. and A. A. Alcorn (MS) 157 different days from April 25, 1957 to July 22, 1965. They were seen in all months, but most often seen from mid–April through June. One nest with two young was seen June 16, 1962 (Alcorn MS).

Gullion et al., (1959) reported, "At least sometimes a winter resident in the Mohave Valley, having been noted there on January 25 and 31, 1952. Mockingbirds move into the higher and more northerly areas in late March (March 31, 1952, Blue Diamond, 3200 feet elevation, Clark County) and remain at least to mid–August (August 21, 1951, McCullough Range). A pair with fledglings recently out of the nest was seen at 4000 feet elevation in the Mormon Mountains on August 10, 1954. The parent birds were having some difficulties preventing a Loggerhead Shrike *(Lanius ludovicianus)* from preying upon their fledglings."

Austin and Bradley (1971) reported, "Permanent resident in desert scrub and riparian areas and a summer resident in woodlands."

Sage Thrasher   *Oreoscoptes montanus*

Distribution. From extreme southern British Columbia and Saskatchewan; south to Mexico (Tamaulipas).

Status in Nevada. Common summer resident over entire state, especially in areas where tall sagebrush is abundant. Winter resident in southern Nevada.

Records of Occurrence. Linsdale (1936) reported Sage Thrashers from 12 of 17 Nevada counties. It has since been observed in all 17 counties. It is found in greatest abundance in the sagebrush *(Artemisia)* that covers a large part of the state.

Sheldon NWR (1982) reported the Sage Thrasher as a common breeding summer resident, seen spring through fall.

Ryser (1973) reported in the Pine Forest Mountains, Humboldt County, many were seen in early August, 1973.

In the area from Wildhorse Reservoir to Jarbidge, they were numerous from April into September. In the Elko area these birds were seen on 32 different days from April through August (Alcorn MS).

LaRochelle (MS, 1967) reported at Ruby Lake NWR he saw these birds on six occasions between June 26, 1962 and May 19, 1966, with one sighting on April 28, 1970.

This author saw an adult fly from a nest north of this refuge June 26, 1962 (Alcorn MS).

In the Fallon area in some years, they visited the valley in late August, September and October. A few were seen in this valley in the autumn of 1960, 1962, 1963, 1966 and 1968. About 10 birds were repeatedly seen six miles west–southwest of Fallon from August 27 to October 19, 1966 and were often seen feeding on the Russian olive berries (Alcorn MS).

Sage Thrashers were frequently seen from May into September of each year in the Reese River Valley of Lander and Nye Counties (Alcorn MS).

Biale (MS) indicated their abundance in the Eureka area by recording them 179 times from May 1, 1964 to September 17, 1967. They were mostly seen from April to September; however, he saw one February 28, 1965 and another March 16, 1966. Late fall records are for October 11 and November 30, 1964 and October 10, 1965. They were seen from 6000 feet in the valley to 8500 feet elevation in the mountains.

Biale reported nests as follows: "June 9, 1965, nest built in large greasewood, four young approximately 10 days old. May 20, 1969, Hay Ranch, a nest in a thick greasewood bush about 12 inches above ground was made of twigs and lined with fine grasses and horse hair. It contained three very young birds and one bluish–brown spotted egg (elev. approx. 6000 feet). On May 23, 1969, the fourth egg in the nest had hatched. May 24, 1970 in Willow Wash area, Sage Thrasher's nest in black sage 12 inches above ground, made of dry sage twigs, and lined with sage bark and a few small feathers (probably from the female bird). Three blue–green eggs with brown spots, elevation about 6250 feet."

A Sage Thrasher was seen at Comins Lake near Ely June 5, 1962; one was seen near Baker June 5, 1962; two were seen on Delamar Flat east of Hiko May 4, 1962; and one was seen at Hiko March 17, 1965 (Alcorn MS).

J. R. and A. A. Alcorn (MS) for the Overton area from April 6, 1960 to April 21, 1964 recorded these thrashers 21 times. They were seen twice in January; 4 times in March and April; 3 in May; once in June and July; and twice in September, October and November.

Austin and Bradley (1971) reported, "Summer resident in woodlands and a winter resident in desert scrub in Clark County."

Brown Thrasher   *Toxostoma rufum*

Distribution. From southeastern Alberta, eastward to southwestern Quebec and New Brunswick; south to Texas and Florida.

Status in Nevada. A straggler to central Nevada and an uncommon visitor to the southern part of the state.

Records of Occurrence. Richard C. Banks and Charles G. Hansen (1968) reported, "Gullion

(1957) presented the first record of this species in Nevada, a bird banded but not collected, in Eureka County. Hansen collected the first specimen from Nevada, providing the second record, on 3 October 1963. The bird, a male, had been observed for several days before it was captured in the net."

Biale (MS) reported one near Eureka May 24, 1977.

One was seen by the author near Northumberland Cave, Nye County October 4, 1966 (Alcorn MS).

Kingery (1986) reported a Brown Thrasher, "Feeding on a small lawn among shrubs at a desert highway rest stop at Tonopah, Nevada June 16."

Kingery (1985) reported, "Brown Thrashers strayed far west into Nevada; one October 26 at Beatty and one November 11 at Lund."

Kingery (1987) reported, "Brown Thrashers also strayed to Kirch WMA July 6–20."

Castetter and Hill (1979) reported a Brown Thrasher at Mercury, Clark County on September 28, 1975.

Hansen collected one specimen, now in the USNM, from Corn Creek October 3, 1963.

Monson (1964) reported a Brown Thrasher at Corn Creek Station, Desert Game Range, October 3.

Snider (1970) reported a Brown Thrasher was seen on occasion at the Desert Game Range from April 4 to May 5, 1970 and again from October 10 to November 15.

Snider (1971) reported a Brown Thrasher wintered on the Desert Game Range and was still present April 6, 1971.

Lawson (MS) reported a specimen taken at Corn Creek January 31, 1973.

Kingery (1984) reported, "Brown Thrasher at Las Vegas November 20, the first in southern Nevada since 1979." He (1986, 1987) reported one at Las Vegas June 1–2 and November 9–11, 1986.

Bendire's Thrasher   *Toxostoma bendirei*

Distribution. From southeastern California to New Mexico; south to Sinaloa, Mexico.

Status in Nevada. Uncommon, possibly a few are resident in the southern part of the state.

Records of Occurrence. Biale (PC, 1984) reported two birds and a nest containing four eggs seen in a black sage in Little Smoky Valley, Nye County on May 20, 1984.

Jewett (1940a) reported he collected one pair of this species from three miles north of Delamar, Lincoln County May 16, 1939.

Gullion et al., (1959) reported,"Two records: on June 26, 1954, an adult male was collected by Gullion from an open stand of Mohave Yucca *(Yucca schidigera)* on the south edge of the Tule Desert (3000 feet elevation), 10 miles east–southeast of Carp. On July 1, 1954 at least five more birds of this species were seen in the dense Joshua tree forest between the south end of the Multichrome Range and the McCullough Range, 11 miles northwest of Searchlight."

Austin and Bradley (1965) reported, "On June 21, 1961 a juvenile female was collected by Gerald Perske at 8400 feet in Clark Canyon, Charleston Mountains, Clark County. This specimen was collected in a fir–pine forest with an undergrowth of small shrubs. This is noteworthy in that the Bendire Thrasher is commonly associated with desert scrub at the lower elevations. This juvenile may have wandered into the higher elevations after being reared in the low desert."

Kingery (1981) reported, "The Bendire's Thrasher at Las Vegas on May 11 was the first in six years."

Curve-billed Thrasher   *Toxostoma curvirostre*

Distribution. From central and southeastern Arizona to extreme western Oklahoma and western and central Texas; south to central Tamaulipas, Mexico.

Status in Nevada. Uncommon in southern Nevada and accidental in northwestern Nevada.

Records of Occurrence. Voget (MS, 1986) reported the Curve-billed Thrasher as accidental at Sheldon NWR with only two sightings. He reported, "One observation by a research crew in June, 1978 and one observation by refuge personnel in May, 1984; both near Thousand Creek."

Gullion et al., (1959) reported, "One record: in the course of a five and one–half hour count at a quail waterhole at Maynard Spring (2200 feet elevation), Gold Butte area, on August 10, 1953, a single individual of this species was seen foraging among the litter under large catclaws for more than an hour, often coming to within 20 feet of the truck in which Gullion was sitting."

Snider (1970) reported, "May 29, 1970 one was found in Red Rock Canyon National Recreation area near Las Vegas.

Lawson (TT, 1976) reported, "Present in Red Rock Canyon, Spring Mountain Range, and nesting."

Crissal Thrasher   *Toxostoma crissale*

Distribution. Formerly *Toxostoma dorsale*. From southeastern California to western Texas; south to south–central Mexico.

Status in Nevada. Permanent resident in the southern part of the state; seen most often in mesquite thickets.

Records of Occurrence. A female was collected by Burleigh for the USNM from North Lake Tahoe April 12, 1964.

Linsdale (1936) reported, "Resident in southern end of the state, in Clark County, north to Charleston Mountains."

Fisher (1893) reported a nest containing three eggs at Cottonwood Springs, east base of the Charleston Mountains May 8, 1891.

J. R. and A. A. Alcorn (MS) for Moapa Valley from 1962–64 observed this bird once in January and March; twice in April; 4 times in May and June; 5 in July; 3 in August; and once in September.

Ryser (1967) reported this thrasher as present at Paradise Valley by Las Vegas February 15–19, 1967. Snider and Kingery, in 1965, 1968, 1969, 1970, 1972 and 1973, regularly listed this thrasher in the Overton, Desert Game Range, Henderson and Ash Meadows areas.

Gullion et al., (1959) reported, "These birds are normally closely restricted to the mesquite thickets in the Las Vegas, Pahrump and Moapa Valleys, being common year–round residents in these areas. However, they occasionally are seen in situations outside of this habitat."

Austin and Bradley (1971) reported, "Permanent resident in riparian areas of Clark County."

LeConte's Thrasher   *Toxostoma lecontei*

Distribution. Desert regions from California, Utah and Arizona; south to central Baja California and northwestern Sonora.

Status in Nevada. An uncommon resident in the southern part of the state.

Records of Occurrence. Linsdale (1936) reported, "Resident. Reported in summer from several places south of the parallel of 37 degrees. Localities given by Fisher (1893) are as follows: Table Mountain, Amargosa Desert, male, May 6, 1891. Pahrump Valley, Clark County, several on February 11, 1891. Ash Meadows, Nye County, three males and full grown young shot on April 29, among yuccas. Vegas Valley, Clark County, one killed, May 1, 1891. Virgin and Lower Muddy Valleys, common; a nest found in a branching cactus on a mesa between those rivers."

Gullion et al., (1959) reported, "Not common, but recorded from Eldorado Mountains south to the California state line, Bird Springs area (4300 feet elevation, Clark County), Ash Meadows, Mormon Mountains, Tule Desert and Kane Springs Wash. The Kane Springs Wash record, for August 7, 1954, apparently is the most northerly record for this species in Nevada. Records extend from February 6 (1953, Searchlight) to as late as August 27 (1952, Eldorado Mountains near Nelson)."

"This thrasher is most commonly seen in areas of Mohave yucca and Joshua trees, but on July 1, 1954 one bird was seen in the creosote–bush–bur—sage type of desert at the south end of Eldorado Valley. Phillips reported encountering an adult with six young south of Searchlight on August 16, 1952. A bird collected by Gullion, at 2800 feet elevation, eight miles southeast of Searchlight, on June 4, 1952 had been identified at the Museum of Vertebrate Zoology as *T. l. lecontei*."

Austin and Bradley (1971) for southern Nevada reported, "A permanent resident in riparian areas, a winter resident in desert scrub areas and transient in woodlands."

Kingery (1983) reported, "Near Las Vegas, LeConte's Thrashers met with some breeding success despite increased people activity."

Family **MOTACILLIDAE:** Wagtails and Pipits

Olive Tree–Pipit   *Anthus hodgsoni*

Distribution. Formerly Indian Tree–Pipit. From northern and eastern Asia, India, China and Japan.

Status in Nevada. Accidental. Only one record for the United States.

Records of Occurrence. Burleigh (1968) collected the only Olive Tree–Pipit reported in the United States 10 miles south of Reno May 16, 1967. The specimen is now in the USNM.

Water Pipit   *Anthus spinoletta*

Distribution. Of irregular distribution. From Alaska to Newfoundland and Greenland; south to Guatemala. Also, throughout Europe, Siberia, south in winter to Africa, India and Japan.

Status in Nevada. Resident.

Records of Occurrence. Gabrielson (1949) reported, "Numerous at Charles Sheldon Refuge, Washoe County, September 4, 1931."

In Elko County, Water Pipits were seen in the Tuscarora and Midas areas in October, 1964, 1966 and 1970 (Alcorn MS).

In the Santa Rosa Mountains, Humboldt County, they were observed in October and November of 1960, 1962 and 1963 (Alcorn MS).

Kingery (1977) reported, "Near Reno the Alves came across a fantastic flock of 2000 Water Pipits."

In western Nevada, Burleigh collected 20 of these from the Reno area in 1967–69 in January, February, April, October, November and December.

This author has recorded them in western Nevada from the vicinity of Reno, Minden, Yerington and Fallon.

Alcorn (1946) reported them as winter residents in the Fallon area with frequent records from October to April. From January, 1960 to December, 1975, small numbers of these birds were seen in winters in this area from October through April, with one May 11, 1971 record of a lone bird at Carson Lake.

Two were seen near Austin October 15, 1964 (Alcorn MS).

Biale (MS) recorded these birds 19 times in the Eureka area from 1964 to 1973. He observed them in April, May, September, October, November and December. He (PC, 1984) does not consider the Water Pipit common to the Eureka area.

Two were seen in the Eureka area May 16, 1960 (Alcorn MS).

Medin (1987) reported the Water Pipit as a common breeder in the alpine environment on Bald Mountain, Snake Range. He found two nests, one on June 27, containing five eggs and the second, on July 6, containing four eggs.

A. A. Alcorn (MS) saw one near Baker, White Pine County, April 29 and two on April 30, 1967.

Five were seen nine miles north–northeast of Beatty October 7, 1958 (Alcorn MS).

In Pahranagat Valley near Hiko, up to 30 birds were seen from October, 1960 to December, 1964. They were seen in March, May, October, November and December (Alcorn MS).

In Moapa Valley from 1958 to 1967, they were seen in winter from October through April, with one record of May 4, 1962 (Alcorn MS).

Gullion et al., (1959) reported, "Recorded in desert areas from as early as October 8 (1951) by Imhof at Camp Three (4200 feet elevation, 39 miles east–southeast of Beatty, Nye County) to as late as March 31 (1951) in the Pahranagat Valley."

Lawson (PC, 1979) reported Al Knorr observed courtship flights at Wheeler Peak, Arc Dome and Mount Rose. Lawson (MS, 1981) reported on July 16 to 23, 1978, he observed courtship flight and feeding of young at Mount Charleston.

Austin and Bradley (1971) reported for Clark County, "Usually occurs from September 20 to May 29. Winter resident in desert scrub and riparian areas. Transient in woodlands and montane forests.''

Sprague's Pipit   *Anthus spragueii*

Distribution. In Canada from Alberta to Manitoba; south to Veracruz, Mexico.

Status in Nevada. Hypothetical.

Records of Occurrence. Lawson (MS, 1981) reported that he and his son Karl saw one at Pahranagat Valley NWR, Lincoln County, December 22, 1978.

Family **BOMBYCILLIDAE:** Waxwings

Bohemian Waxwing   *Bombycilla garrulus*

Distribution. From Alaska to Nova Scotia; south to California and Texas. Also, across northern Europe and northern Asia; south to Italy and Japan.

Status in Nevada. A regular winter visitor; transient.

Records of Occurrence. Gullion (1957) reported, ''A flock numbering as high as 28 birds was present in and around Elko . . . from January 3 to March 5, 1955. At least six of these waxwings were in the Elko area from March 5 to 7, 1956. While in the Elko area, these waxwings fed on the berries of ornamental shrubs around residences, especially on the fruit of the snowberry.''

Kingery (1980) reported three at Jarbidge January 1 and later reported 25 at Jarbidge March 2.

A flock of 14 was observed by Ruby Lake NWR personnel at a bird feeder, near refuge headquarters in the 1971–72 winter.

Kingery (1979) reported Bohemian Waxwings at Ruby Lakes in late November. He (1985) reported, ''A flock December 13–20 at Ruby Lake peaked December 14 at 55.''

Kingery (1987) reported a flock of 70 Bohemian Waxwings spent the entire winter at Ruby Valley.

Heilbrun and Kaufmann (1977) reported 20 at Unionville during the CBC.

Ryser (1964) reported a large number of Bohemian Waxwings in the Reno area during the winter and all spring of 1964. He (1968) reported in Truckee Meadows, several small flocks were present from late November to December 9, 1968.

One flock of 12 was observed on the University of Nevada, Reno campus. Ryser (1969) reported waxwings were still in the area March 1, 1969.

Two male specimens in the USNM, collected by Burleigh, are from the Reno area. One was taken November 25, 1968 and the other January 18, 1969.

Kingery (1985) reported Bohemian Waxwings in Reno September 24, 1984.

Alcorn (1946) reported one taken from a small flock on January 29, 1942 from near Fallon, Lahontan Valley.

Saxton (PC) reported one at her feed station in Fallon January 2, 1966.

Mills (PC) saw a flock of five near Fallon November 18, 1968 and this writer saw a flock of about 10 on November 4, 1969.

Biale (MS) reported seeing these birds eating wild rose hips in the Eureka area. He saw these birds December 5, 1965; January 19, 23 and 26 of 1966; February 2 and 11 of 1966; March 27 and December 8 of 1966; and May 15, 1967. He saw 12 in the Fish Creek Range, south of Eureka, January 11, 1976 and he saw 162 in the town of Eureka January 13, 1976. Lesser numbers were seen from January 14 through January 20, 1976.

Grater (1939) reported, "Transient visitant. One collected on April 30, 1938 at Willow Beach where it had flown from the Nevada shore on the opposite side of the Colorado River. First record for southern Nevada."

Snider (1967) reported about 5000 or more spent a month in Las Vegas. Snider (1969) reported 12,000 seen from December 12, 1968 through March 31, 1969 at Las Vegas.

Kingery (1977) reported Bohemian Waxwings remained in numbers throughout Nevada with counts like 100 at Las Vegas March 20. Kingery (1979) reported 51 Bohemian Waxwings March 31 at Las Vegas.

Austin and Bradley (1971) reported for southern Nevada, "Usually occurs from November 2 to May 4. Winter resident in riparian areas. Transient in montane forests."

Cedar Waxwing   *Bombycilla cedrorum*

Distribution. From southeastern Alaska to Newfoundland; south to the Gulf Coast of the United States, casually to Cuba and Columbia.

Status in Nevada. An irregular visitor to most sections of the state, with nesting reported from Reno.

Records of Occurrence. Sheldon NWR (1982) reported the Cedar Waxwing as an occasional migrant spring, fall and winter.

LaRochelle (MS) reported Cedar Waxwings at Ruby Lake NWR as a regular winter and spring visitor. He banded several in 1963.

The largest number seen at one place in Nevada by the author was a Stonehouse Creek, Santa Rosa Range on November 13, 1963, when about 100 were seen (Alcorn MS).

In Reno, up to 30 birds were seen on three days in September of 1942 (Alcorn MS).

Ryser (1967) reported, "In the northwest sections of Reno, several large flocks, numbering in the 100s have been foraging during the past two months (October and November)."

Ryser (1969) reported December 31, 1969 a large flock was present since fall in Reno. He (1985) reported nesting on the UNR campus "off and on over the years."

Ryser (1968) reported on December 9, 1968, "A few days ago in Carson City more than 700 birds in flocks were seen; the next day, one couldn't even be found." He (1973) again reported numerous flocks in Carson City October 29, 1973 with over 1200 in the city.

Kingery (1984) reported six to 40 Cedar Waxwings at Carson City from March through April.

Alcorn (1946) noted in the Fallon area these birds were, "Seen most frequently in September, October, November and December. Less often recorded in January, May and June."

From 1960 to 1975, these birds continued to be of irregular occurrence in the Lahontan Valley, but they usually were seen in some winter months of each year. In this area on February

Cedar Waxwing. Photo by A. A. Alcorn.

13, 1962 one lone bird, in company with starlings, was eating an apple on the ground (Alcorn MS).

Patricia Lott (PC) reported Cedar Waxwings in her yard in Fallon, February 26, 1975. They were seen regurgitating partially digested Pyrocantha berries.

In Fallon on December 9, 1957 this author wrote, "One Cedar Waxwing was seen perched on a tree limb of a Russian olive tree. It flew out and alighted near the tip of the limb where it reached a Russian olive berry, which it swallowed whole. Then it flew back to alight again on the limb near the trunk of the tree" (Alcorn MS).

At the University of Nevada Experiment Farm, located about one mile south of Fallon on June 9, 1971, John McCormick phoned to say there were four dead birds under the trees in their yard. This author identified them as Cedar Waxwings. A fresh bird was skinned then cut open for examination. It was a female and its gizzard contained two elm beetles and about seven flower heads that looked like locust blossoms. (The local locust trees were in full bloom.) The city of Fallon was spraying Seven and Diazanon for the control of elm beetles and this author had seen their spray crews spraying all the trees including locust, elm, ash, mulberry, poplar and Russian olive. It is assumed this was the cause of the waxwings' deaths.

Biale (MS) reported these birds 24 times in the Eureka area from 1965 to 1973. They were seen once in February; twice in March; 3 times in April; 6 in May; 3 in June; once in August and September; and 7 times in October. He saw them eating chokecherries on October 6, 1969. In January, 1976 Biale reported over 100 Cedar Waxwings on four different days, with 162 on January 13.

Lehman Caves National Monument park personnel (MS) reported Cedar Waxwings from the Lehman–Baker area of the Snake Range once in August and October, and six times in November.

Gullion et al., (1959) stated, "The Cedar Waxwing is an erratic visitor to the desert. Recorded in three localities where mistletoe berries were available in mesquite thickets."

Austin and Bradley (1971) reported for southern Nevada, "This bird is winter resident in riparian areas. Transient in woodland and montane forests. In woodlands and montane forest, it occurred on February 21, March 7 and from October 7 to December 19. In desert scrub and riparian areas, it usually occurred from September 3 to June 3. Unseasonal occurrences are August 8 and June 16."

Kingery (1979) reported Cedar Waxwings in small groups at Las Vegas in early June. He (1984) reported, "200 at Las Vegas in May feeding on ripening mulberries."

Family PTILOGONATIDAE: Silky Flycatchers

Phainopepla   *Phainopepla nitens*

Distribution. From central California, southern Nevada, to western Texas; south in Mexico to Puebla and Veracruz.

Status in Nevada. Resident in southern part of the state. Stragglers reported as far north as Fallon, Storey County and Eureka.

Records of Occurrence. Linsdale (1936) reported, "Resident in southern Nevada, occurring farther north in summer, but the northern limit has not been definitely determined."

Ryser (1985) reports a record for Storey County, but no details were given.

Alcorn (1946) reported in the Fallon area, "A lone individual was seen August, 1943. One was shot on October 11, 1943 by Mills three and a half miles west–southwest of Fallon."

Biale (MS) reported one seen in Eureka October 26, 1973 that appeared to be a female or juvenile.

Behle (1976) reported, "Cottam (1936) found it regularly at Alamo in Pahranagat Valley in summer from May to September. Gabrielson (1949) took a male at Glendale on 20 November 1934. A nest with two young was discovered at Overton on 1 June 1954."

Two pair were seen near Hiko, Pahranagat Valley June 11, 1965 (Alcorn MS).

J. R. and A. A. Alcorn (MS) reported these birds 98 times in Moapa Valley from December 15, 1959 to June 11, 1965.

One was seen in Pahrump Valley March 16, 1962 (Alcorn MS).

Austin and Bradley (1971) reported, "A permanent resident in desert scrub and riparian areas. Transient in woodlands."

Family LANIIDAE: Shrikes

Northern Shrike   *Lanius excubitor*

Distribution. From Alaska to southern Labrador; south from California to Maryland. Also in Eurasia and Japan.

Status in Nevada. Winter visitor and transient in limited numbers to all sections of the state.

Records of Occurrence. Sheldon NWR (1982) reported the Northern Shrike as an occasional winter resident.

Gullion and Hoskins (1956) reported five records and two specimens, from Elko County in 1955.

Two specimens in the USNM were taken by Lewis from Ruby Lake NWR. One, a male was taken November 10, 1966. The other, a female, was collected January 29, 1968.

Burleigh collected a female specimen at Reno December 15, 1967.

Rubega and Stejskal (1984) reported one Northern Shrike in Carson City December 30, 1983 during the CBC.

Alcorn (1946) reported these birds in the Fallon area as a winter visitant in limited numbers with records for December and January. One was taken January 13, 1941.

At this author's residence six miles west of Fallon, lone birds, possibly often the same bird, were seen on six different days in December, 1973 and three days in January, 1974. A Northern Shrike was flying in pursuit of a White–crowned Sparrow February 23, 1974 and a lone Northern Shrike was seen March 7, 1974 (Alcorn MS).

Linsdale (1936) reported Bailey took a female on the Reese River, Lander County November 22, 1890.

Biale (MS) recorded these birds 41 times in the Eureka area from January 4, 1965 to December 25, 1967. They were seen 6 times in January; 4 in February; twice in March; 3 in October; 11 in November; and 15 times in December. From 1967 until 1975, he repeatedly reported them in these same winter months.

Austin and Bradley (1968) reported, "An adult observed at close range (breast barring very evident) near Tule Springs on September 29, 1962 is the first record for Clark County."

Austin (1969) reported, "A female collected at Henderson Slough (near Henderson) on October 18, 1966."

Loggerhead Shrike   *Lanius ludovicianus*

Distribution. From southern British Columbia to Quebec; south to Baja California and Nevada.

Status in Nevada. Common resident statewide. Most of them move out of the colder northern parts of the state in mid–winter. Linsdale (1951) reported three races from Nevada: *L. l. gambeli, L. l. nevadensis* and *L. l. sonoriensis.*

Records of Occurrence. Sheldon NWR (1982) reported the Loggerhead Shrike as a common permanent resident spring, summer and fall; uncommon in winter.

Ryser (1973) reported in northern Nevada, "All through the foothill canyons out of Denio Junction from the junction at Cedarville, California and then on to Gerlach, a stretch of over 200 miles, I sighted shrike after shrike, flying through the shrublands in early August."

In Humboldt County they were seen from Hinkey Summit on November 1, 1963, from Paradise Valley on April 28, 1969 and on May 12, 1966 (Alcorn MS).

LaRochelle (MS, 1967) reported these birds at Ruby Lake NWR as commonly seen except in mid–winter. No nesting observed. This writer recorded these shrikes on numerous occasions in Elko County in the vicinity of Wildhorse Reservoir, Jarbidge, Owyhee Desert, Midas,

Tuscarora, Elko and Ruby Valley. Most of this author's records are for April through October (Alcorn MS).

Gullion and Hoskins (1956) reported, "On December 3, a Loggerhead Shrike was seen below the snow–line along the west side of the Great Salt Lake Desert, eight miles southwest of Wendover, 38 miles northeast of the collection site for the Boreal Shrike and 1500 feet lower."

In the Fallon area from 1960 to 1975, these birds were not seen as often as prior to 1960. However, records are available for all months (Alcorn MS).

Schwabenland (MS, 1966) reported them at Stillwater WMA in Lahontan Valley as, "Commonly seen at all times of the year. More abundant during summer. Nesting probably occurs, but none observed."

Biale (MS) recorded these birds 170 times in the Eureka area from April, 1964 to November, 1967. Most of his records are for April through September. He reported that, "May 24, 1970, Roberts Creek Ranch area, nest in black sage 30 inches above ground, of roots, dry sage twigs, lined with feathers, sagebrush and dry grass. It contained six eggs, off–white, with light brown spots, (elevation approximately 6500 feet)."

Biale (PC, 1984) reported another nest containing six eggs on June 3, 1984 in Antelope Valley in a juniper tree.

J. R. and A. A. Alcorn (MS) recorded these shrikes 136 times in southern Nevada from 1960 to 1965. They were seen in all months.

Gullion et al., (1959) reported, "A common permanent resident in all parts of the desert. This is one of the several species which summers on the desert independently of free water supplies."

Austin and Bradley (1971) for Clark County reported, "Occurs all year as a permanent resident in desert scrub and riparian areas and a summer resident in woodlands."

Family **STURNIDAE:** Starlings and Allies

European Starling   *Sturnis vulgarus*

Distribution. From Iceland, across northern Eurasia; south to northern Africa, southern India and China. Introduced in North America; has spread through southern Canada and the United States. Also introduced in the West Indies, South Africa and New Zealand.

Status in Nevada. Common, widespread resident over the entire state. Fewer in mid–winter in the northern part, as these birds migrate southward. The first record of these birds in Nevada was on August 13, 1938 at Las Vegas (Cottam, 1941b).

Records of Occurrence. Starlings have been seen in all 17 Nevada counties within the past 10 years. So many records are available, that no attempt will be made to list all of them.

Sheldon NWR (1982) reported the European Starling as a common permanent resident year round.

The abundance of the starling in the Fallon area is evidenced by their being seen 2402 times between January 1, 1960 and August 25, 1969. They were present in all months, with the greatest number seen were during October and November migration.

Approximately 30,140 starlings were gathered at a cattle feedlot southeast of Fallon on

European Starling. Photo by A. A. Alcorn.

January 4, 1964 (Alcorn MS). According to Gary Snow (PC, 1988) who now operates this feedlot, starlings continue to be a problem, with tens of thousands there each winter.

Biale (MS) reported starlings 691 times in the Eureka area from April 26, 1964 to December 30, 1967. They were seen in all months, but he reported them less often and saw fewer birds in mid–winter months. He (MS, 1984) thinks there has been a decrease in the number of starlings in the Eureka area in recent years.

In Pahranagat Valley, starlings were observed on 50 different days from December 18, 1959 to March 19, 1965. They were recorded in all months except July, August and September (Alcorn MS).

In the Overton area, J. R. and A. A. Alcorn (MS) observed starlings on 94 different days from December 13, 1959 to December 14, 1966. They were observed in all months, but more numerous in fall and winter.

A. A. Alcorn (MS) reported about 500 starlings, mostly in small flocks, flew past his house in Overton in about half an hour on October 19, 1961. Austin and Bradley (1971) reported, "Occur all year in Clark County. Permanent resident in riparian areas and transient in woodlands."

History. On March 6, 1890, 80 imported starlings were released in New York City's Central Park and 40 more were released there April 25, 1891. These and later importations were made by people who longed for the familiar sight of birds they remembered from their native European land.

In the west, the Portland Song Bird Club released 35 pair of starlings in 1889 and 1892, but they disappeared within a few years. The starlings became established in the east from these early introductions and expanded their range north, west and south.

The first appearance of this bird in Nevada was reported by Cottam (1941) who stated, "Because of the interest ornithologists have had in the movement and distribution of the European Starlings and also because of the unusual economic significance of this bird, it seems appropriate to record a field observation of this species at Las Vegas, Nevada. The following is quoted from a letter from Dr. M. M. Ellis of the Fish and Wildlife Service: 'On August 12, 1938, Dr. B. A. Westfall . . . and I saw three adult starlings in the trees in front of the Post Office at Las Vegas, Nevada. This was about 9:00 in the morning. Dr. Westfall is a trained ornithologist having considerable experience in that field, and I am well enough informed on the common birds to know starlings without a doubt. We were both surprised to see these birds so far west and followed them as they flew from bush to tree in the vicinity of the U.S. Post Office. Our observations lasted over 15 or 20 minutes, during which time we were very close to the birds. Business matters took us on, but we were so impressed with our find that we discussed it several times during the day and the next morning looked for the birds again. We did not see them'."

Several years prior to the sighting of three starlings at Las Vegas, this author and Mills had been alerted that the starling was moving west. We were constantly looking for them in Nevada, especially in the Fallon area, but to no avail. With Cottam's report we intensified our efforts. Finally, this author saw two birds at Springdale about 10 miles north of Beatty on December 14, 1945. From this day until February 21, 1946 on four occasions, from two to 27 starlings were seen at this same location.

After several years of searching in the Fallon area, Mills reported seeing starlings on February 2, 1947 in the Soda Lake District west of Fallon. Accompanied by this author we returned within an hour to the site and we saw 22 starlings feeding on Russian olive berries. We again returned to the area February 3, 1947 and collected two specimens, one of which is now in the University of California Museum of Vertebrate Zoology.

They were again seen in the Fallon area in the winter of 1947 and 1948. Their numbers increased each year until the winter of 1963–64, after which time their population declined for several years. The decline may have been due to control operations carried out by the U. S. Fish and Wildlife Service in Lovelock, Fallon and southern Nevada areas, in an attempt to reduce the number of starlings at various feedlots and dairies. A decline was also reported in other western and mid–western states. Some attribute this to the widespread use of experimental lethal baits, others to a natural leveling off of the population.

Nesting. Observations made in 1958 and 1959 indicate starling nesting was established over most of Nevada.

Mills (PC) reported a starling seen a few miles east of Lahontan Dam, "flying back and forth" to a nest site June 15, 1958. Mills also reported an adult starling four miles southwest of Fallon, had a beak full of bugs, and was flying toward a nest site May 16, 1959.

In this same year, Mills and the author visited various starling nesting sites and saw adults carrying beaks full of grasshoppers and other insects to nest sites along the Carson River in Lyon and Churchill Counties. All the nest sites seen were located in woodpecker and flicker holes in cottonwood trees.

Wheat (PC) reported two starling nests seen in the early summer of 1959. Both nests were in ''woodpecker'' holes in trees on the west side of Washoe Valley. He cut the trees down and found one nest with four young starlings and the second nest with three young.

On May 12, 1962 this author wrote, ''I think there may be five times more starling nests here in the Fallon area as compared with last year. Mills believes there may be 500 nests in this valley now.''

In this same year, the author saw starling nests in Lander County and other parts of the state. A. A. Alcorn (MS) reported on May 22, 1962 in Moapa Valley he saw, ''Two adult starlings carrying food to a nest in a dead limb in a cottonwood tree. Young birds were heard in the nest.'' On May 29, 1964 in the same area he saw two ''feathered out'' young starlings sticking their heads out of a nest hole in a cottonwood tree.

Observations to determine when starling nesting begins resulted in the following information: at Winnemucca on March 31, 1964 a starling with a beak full of nest material flew up under the eave of a house where it was building a nest behind some boards.

A nest box examined near Fallon on May 8, 1964 contained a partially constructed nest. When examined on May 19, 1964, it contained six blue–green starling eggs. Noisy, fuzzy young were seen in the nest on May 28 and were heard daily from May 28 until June 17 when the young left the nest.

Examination of over 100 nests over a 5–year period revealed no nests were found with more than six eggs.

After leaving the nest, an adult often accompanies the young for a day or two. The young join other juvenile birds and flocks of 10 to 40 juveniles may be seen associating together.

Mills (PC) reported at three miles west of Fallon on May 31, 1964 he saw a small flock of juvenile starlings. On June 4 he saw three flocks of juveniles. One contained 12 birds and the other two flocks contained about 20 birds each.

Frequently, dead young starlings are found in nests with their brains eaten. Kestrels were suspected of doing this and evidence in support of this activity follows. At this author's ranch west of Fallon on January 16, 1976, ''A kestrel was seen pecking at a live starling on the ground. The hawk's wings were spread so as to prevent the starling from escaping. As I approached, the hawk attempted to fly away with the starling, but dropped it as a shotgun was fired in the general direction. Examination revealed that the starling was a male, its stomach contained three seeds and considerable pulp of Russian olive berries. The starlings' left shoulder was punctured, probably by the hawks talons. The right side of the starlings' cranium was opened where the hawk was apparently eating the starlings' brains.''

Food. On January 23, 1961 beef hamburger, cracked corn, whole wheat kernels, kernels of barley and canned soft dog food were placed near the author's observation window. Within half an hour, starlings had hastily eaten the hamburger, prefering it to the other foods. They also ate the dog food and a little cracked corn, but did not eat any of the wheat or barley. However, on January 19, 1963 at about 10 miles west of Fallon, a starling was shot and its stomach contained 36 kernels of barley along with a considerable amount of chaff. Other starlings were shot at this same place January 25, 1963. One stomach contained 83 kernels of wheat, another contained 125 kernels of rye.

At a cattle feedlot near Fallon on March 23, 1965, Jerry Page put out 20 pounds of apples, cut into quarters and placed them in a trough. The next day all of the apples had been eaten by the starlings. He then put rolled barley in one trough, and rolled barley covered with melted lard in another. He reported the starlings were fighting for the lard–covered barley and ate all of it, while the plain rolled barley was left untouched.

At a feedlot southeast of Fallon on February 21, 1961 a company employee had put one ounce of strychnine in five pounds of lard cracklings in an effort to kill the starlings. Only 20 dead starlings were found under nearby trees. Four were females, 16 were males. Their stomachs were examined and most contained cracked corn and chaff. Some contained green alfalfa leaves or a green grass–like plant, one had two whole kernels of wheat and one had one whole kernel of barley.

On August 17, 1961 at the Charles H. "Sam" Currie Ranch, five miles west–southwest of Fallon, a field of 10 acres of corn was badly damaged by birds eating the kernels of corn in the milk stage. Currie stated blackbirds were doing the most damage, but about 200 starlings had come to the field two days before and started eating the corn. He shot at them and they left.

Large numbers of migrating starlings feed on the fruit of the Russian olive from October to November of each year in western Nevada. The starling's preference for the fruit of the Russian olive tree is indicated in this author's notes of October 28. "At my ranch (west of Fallon) this afternoon I saw about 3000 starlings. One Russian olive tree near the trail house is loaded with fruit." On October 29, 1971 I wrote, "There isn't any fruit on this tree today. What the starlings didn't eat they knocked to the ground."

At this same place on November 18, 1965 I wrote, "At present there are no Russian olive berries on the 100 + Russian olive trees on my 40–acre ranch. The starlings have cleaned them from the trees, mostly since the first of this month. The several hundred starlings that were hanging around these trees, left the area when the fruit was gone."

Various fruits are also eaten by starlings. Near Fallon on July 21, 1964, Mills saw about 200 starlings feeding on buffalo berries. Mills (PC) reported five miles southwest of Fallon on July 23, 1965 the starlings were flocking to the area to eat the fruit of the buffalo berries *(Shepherdia argenta)*. He shot three of these birds and examination by the author revealed all were juvenile birds. Two stomachs were full of buffalo berries and the third stomach was full of buffalo berries plus a grasshopper.

Near Hiko, A. A. Alcorn (MS) reported starlings ate the apples from several trees before he got them picked.

In the Fallon area on January 15, 1964 the stomach contents of a female was 20% green clover leaves. An occasional green alfalfa leaf was found in other stomachs examined (Alcorn MS).

The starlings' preference for insects as a food item is shown by the stomach contents of a female examined near Fallon on January 25, 1963. It contained about 40 lady bug beetles and a large number of black–colored beetles. Others examined contained spiders, gastropods, grasshoppers, earthworms and alfalfa weevils.

Near the Carson Lake, southeast of Fallon on March 27, 1963 a flock of over 100 starlings scattered over a wide area, were flight feeding on a fuzzy looking insect that was "everywhere" in the air" (Alcorn MS).

Population. In January, 1966, populations of starlings in the Elko and Winnemucca area were estimated at about 500 birds, most of which were concentrated around the city garbage dumps.

In the winter of 1963–64 the highest starling population in the Lahontan Valley (Fallon area) was estimated at over 100,000 with about 35,000 of these concentrated at the feedlot of Nevada Cattle Feeding Company southeast of Fallon. The following winter (1964–65) there was a big

European Starlings often flock to feedlots where they cause problems for feedlot operators. Photo by A. A. Alcorn.

decline in population. It was estimated population was 60,000 birds with about 12,000 birds at this feedlot. In the winter of 1965–66, another decline resulted in population estimates of 30,000 with a concentration of about 8000 at this feedlot. From 1965 to 1975, the starling population in the Lovelock–Fallon area never approached more than half the numbers seen in the winter of 1963–64.

In January, 1966, this author estimated that there were fewer than 30,000 starlings in all of Clark County, with concentrations of 7000 at Searles Dairy near Moapa; 6000 at LDS Dairy at Las Vegas; 2000 in the Mesquite area; and 2000 in the Overton area with a few groups widely scattered over the county.

Co–workers considered these estimates to be on the conservative side, and some wildlife biologists estimated twice that number. In that this author made all of the counts, however, they should be comparable. This population decline may have been due to control operations of the U.S. Fish and Wildlife Service, Division of Wildlife Services, whose control work resulted in killing over 100,000 starlings from 1963 to 1966. About 20,000 of the starlings killed were in southern Nevada and 70,000 or more in the Lovelock and Fallon areas. Efforts to control starlings at various dairies and feedlots were undertaken in some years. An example of the effectiveness of some of these efforts at a cattle feedlot southeast of Fallon on December 18, 1972 is indicated in the author's notes of that date that stated, "Starlicide put out yesterday to control starlings resulted in the death of about 20,000 starlings."

Family **VIREONIDAE:** Vireos

Bell's Vireo   *Vireo bellii*

Distribution. From California to Wisconsin; south to El Salvador and Nicaragua.

Status in Nevada. Summer resident in the southern part of the state. Linsdale (1951) reported two races from Nevada: *V. b. arizonae* and *V. b. pusillus*.

Records of Occurrence. Linsdale (1936) reported, "Three males from the Colorado River, opposite Fort Mohave, May 7 and 10, 1934. The Death Valley Expedition obtained a male of this species at Ash Meadows, Nye County May 30, 1891 by Bailey."

Gullion, et al., (1959) reported, "Recorded from the Dead Mountains and the Mohave Valley from March 17 (1952) to as late as August 16 (1951). A record of one bird from Las Vegas Valley, June 2, 1952."

Austin and Bradley (1971) reported, "Summer resident in riparian areas. Usually occurs from April 20 to September 20. Unseasonal occurrence in March."

Kingery (1984) reported, "Observers found a single Bell's Vireo in August, 1983 at Mesquite, Nevada." He (1986) reported, "A Bell's Vireo singing, visited Las Vegas August 22."

Gray Vireo   *Vireo vicinior*

Distribution. From southern California to Oklahoma; south in Mexico to southern Sonora.

Status in Nevada. Uncommon summer resident in the south. Also noted in central Nevada.

Records of Occurrence. Kingery (1980) reported, "A Gray Vireo in the Toiyabe Range near Round Mountain, Nevada, had set up territory in pinon–juniper habitat with a six to eight foot understory; this observation extends the range of this species to central Nevada."

Linsdale (1936) reported, "Summer resident in southern part of the state. On June 8, 1891 a male was shot by Nelson on Grapevine Mountain, Nye County. On June 10, he saw several among pinons in Wood Canyon and found one carrying nesting material (Fisher, 1893). A skin in the Museum of Vertebrate Zoology was taken (a male) May 27, 1931 by Russell, one and one–half miles southwest of Oak Spring, 5850 feet elevation, Nye County."

Gullion et al., (1959) reported, "Recorded from the desert ranges in summer, from as early as April 28 (1953, Boulder City) to as late as September 3 (1951, Gold Butte area). This species has been noted in the Dead, Mormon and Clover Mountains, on the Tule Desert, and in Las Vegas Valley."

Johnson (1965) reported, "We found this species in the Sheep Range among scattered pinon, juniper and brush on slopes near the bottom of Hidden Forest Canyon from 6100 feet to 6500 feet. No specimens were collected. Although van Rossem did not find the Gray Vireo in the Spring Range, a field party from the Museum of Vertebrate Zoology that visited the north side of Potosi Mountain in June, 1940 recorded it commonly and collected three males . . . two on June 12 and one June 14."

Austin and Bradley (1971) reported, "Summer resident in woodlands. Usually occurs from April 28 to September 3. Transient in riparian areas."

Solitary Vireo   *Vireo solitarius*

Distribution. From British Columbia to Newfoundland; south to Nicaragua.

Status in Nevada. Statewide summer resident in mountain ranges. Linsdale (1951) reported two races from Nevada: *V. s. plumbeus* and *V. s. cassinii*

Records of Occurrence. Lewis collected a male specimen now in the USNM at Ruby Lake NWR October 10, 1966.

Burleigh collected a male specimen for the USNM near Reno January 7, 1967.

Biale (MS) recorded these birds about 10 miles northwest of Eureka on August 8, 1965 and July 2, 1967.

Johnson (1973) reported on May 24, 1972 on Mount Irish (7800 feet to 8100 feet) two males of two pair were taken and on the next day one was singing at 6900 feet. He reported in Sawmill Canyon Mountains, a singing, mated male was taken at 7500 feet on May 26, 1972 and on this same date, one was singing in pinon–juniper and cottonwoods at Cherry Creek Campground at 6600 feet.

Johnson (1974) wrote, "In 1971 I found the species common in the pinon association (Grapevine Mountains) and collected a series of specimens in breeding condition between 6900 feet and 7500 feet elevation from May 18 to 20."

"In 1940, Miller collected a singing male of this species, but of the form *V. s. cassinii* at 8000 feet elevation on the northeast side of Potosi Mountains. Although the bird was moderately fat, it showed testes in breeding condition and Miller concluded the individual was established for breeding. In 1971, I collected a male *V.s. plumbeus* at 7000 feet elevation in open yellow pines. The bird was with an apparent mate. A second male Solitary Vireo sang from the opposite canyon, but was never seen."

Austin and Bradley (1971) reported for southern Nevada, "A summer resident in woodland and montane forests. Transient in riparian areas. In woodlands and montane forests it usually occurs from April 28 to September 16. In riparian areas it usually occurs from May 5 to May 23 and from September 5 to September 30. Unseasonal occurrences on March 18 and October 24."

Yellow–throated Vireo   *Vireo flavifrons*

Distribution. From southern Manitoba to Quebec; south to Panama; casually to Columbia and western Venezuela.

Status in Nevada. An accidental straggler to southern Nevada.

Records of Occurrence. Linsdale (1936) wrote, "A male Yellow–throated Vireo was collected May 29, 1932 by Baldwin, at Crystal Springs, 4000 feet, Pahranagat Valley, Lincoln County, Nevada. This specimen constitutes not only the first record of this species for Nevada, but the first one known to me for the western United States."

Lawson (MS, 1981) reported one singing male Yellow–throated Vireo was seen May 25 to 27, 1975 at Corn Creek Field Station. One was also seen at Corn Creek by Mowbray October 27, 1974.

Hutton's Vireo   *Vireo huttoni*

Distribution. From southwestern British Columbia through the western United States and western Texas; south to Guatemala.

Status in Nevada. Uncommon transient in Nevada.

Records of Occurrence. LaRochelle (MS, 1967) reported these birds were first observed at Ruby Lake NWR when mist netted in 1963. At this same refuge, Lewis collected two females and a male September 20, 1966.

Ryser (1985) reported the Hutton's Vireo was recorded in the West Humboldt Mountains of Nevada by R. E. Wallace.

Burleigh obtained a male near Reno January 26, 1968.

Herlan collected one at Corral Springs, Lyon County May 2, 1964.

Snider (1969) reported on May 4 and 8, 1969, "Several were reported on the Desert Game Range. One or more were seen near Las Vegas. One was found at Oak Creek, Nevada, near Las Vegas."

Kingery (1984) reported, "Later vireos included a Hutton's Vireo November 9 at Las Vegas."

Warbling Vireo   *Vireo gilvus*

Distribution. From southeastern Alaska to New Brunswick (casually to Newfoundland); south to El Salvador.

Status in Nevada. Common statewide summer resident in mountain ranges. Linsdale (1951) reported two races from Nevada: *V. g. leucopolius* and *V. g. swainsonii.*"

Records of Occurrence. Sheldon NWR (1982) reported the Warbling Vireo as an uncommon summer resident, seen spring through fall.

Ryser (1967) reported these birds at Lamoille Canyon, Ruby Mountains, from July 1 to 4, 1967.

Lewis took an immature female at the Ruby Lake NWR September 24, 1966. At this same refuge, LaRochelle (MS, 1967) reported the first records were of those he mist netted in 1963.

Burleigh collected six specimens from Lake Tahoe and Reno areas in 1966, 1967 and 1968. These were taken in the months of May, June and September.

Alcorn (1946) reported in the Lahontan Valley (Fallon area) one lone individual of the race *leucopolius* was taken on May 21, 1942 and a specimen of the race *swainsonii* was taken September 14, 1942.

One was seen four miles west of Fallon on September 21, 1971 (Alcorn MS).

At Kingston Canyon in the Toiyabe Range a nest with three eggs was seen June 20, 1963 (Alcorn MS).

One was seen at Wilson Creek, east side of Toquima Range, July 4, 1969 (Alcorn MS).

Biale (MS) reported Warbling Vireos 11 times in the Eureka area from 1965 to 1973. On August 2, 1967 he observed one feeding a young cowbird. He (PC, 1984) reported two nests in chokecherry trees not over five feet tall, five miles south of Eureka on July 15, 1984. One nest contained three young ready to leave the nest and two unhatched eggs. The other nest contained one well–feathered cowbird and one unhatched egg (7800 feet elev.).

Kingery (1980) reported a late vireo at Ely, Nevada on October 17 was probably a Warbling Vireo.

Johnson (1965) reported, "Despite the virtual absence of aspens and other broad–leafed trees, the Warbling Vireo was very common in the Sheep Range where it inhabited ponderosa pines and white firs generally between 6900 feet and 8400 feet . . . Van Rossem (1936) reported the Warbling Vireo as 'present in limited numbers' and with a surprisingly restricted distribution in the Spring Range; 'it was seemingly confined to aspen thickets from 8500 feet to 9000 feet'."

Austin and Bradley (1971) reported, "Summer resident in montane forests. Transient in riparian and woodland areas. In woodland and montane forests it usually occurs from May 16 to August 31. Unseasonal occurrence on October 3. In riparian areas it usually occurs from April 19 to May 30 and from August 14 to October 17. Unseasonal occurrence on June 18 and July 25 and 27."

Philadelphia Vireo   *Vireo philadelphicus*

Distribution. From British Columbia to Newfoundland; south to Panama and northwestern Columbia.

Status in Nevada. Hypothetical.

Records of Occurrence. Lawson (MS, 1981) reported three records: three birds at the Oasis Ranch, about 10 miles south of Dyer, Esmeralda County May 23–28, 1981; one in Lawson's back yard, bathing and preening, was observed by the Lawsons May 6, 1981; and another individual was reported by Helen Kittinger at Sunset Park, Las Vegas May 19, 1981.

Red–eyed Vireo   *Vireo olivaceus*

Distribution. From British Columbia to Nova Scotia; south to the Amazon Basin from Columbia to Peru and Brazil.

Status in Nevada. Accidental straggler to Nevada.

Records of Occurrence. Lewis collected a female specimen for the USNM at Ruby Lake NWR June 12, 1967.

Kingery (1978) reported one heard, but not seen at Virginia City May 13.

Kingery (1982) reported one at Dyer September 13, 1981.

Mowbray (MS, 1986) reported the Red–eyed Vireo as rare in northeastern Nevada.

Snider (1971) reported Red–eyed Vireos at the Desert Game Range on July 25, 1970 and September 5, 1970 (one seen each date). She (1971) also reported one September 20, 1970 (photographed). On October 4, 1970, two were seen. In 1971, Snider reported one was near Las Vegas June 5, 1971.

Kingery (1986) reported a Red–eyed Vireo at Las Vegas June 6.

Family **EMBERIZIDAE:** Emberizids

Tennessee Warbler   *Vermivora peregrina*

Distribution. From southeastern Alaska to Newfoundland; south to Columbia. Rarely to California.

Status in Nevada. Rare migrant in southern Nevada.

Records of Occurrence. Austin (1969) reported, "Two specimens were taken at Corn Creek on 26 May 1965. These are the first Nevada records."

Snider (1967) reported one was seen at the Desert Game Range (Corn Creek) May 7, 1967.

Kingery reported one at Las Vegas October 4, 1979, in May 1983 and May 4 and 9, 1986.

Lawson (MS, 1981) reported these birds as a rare migrant in southern Nevada.

Orange–crowned Warbler   *Vermivora celata*

Distribution. From Alaska to southern Labrador; south in winter to Florida and Guatemala.

Status in Nevada. Summer resident and winter visitant. Linsdale (1951) reported three races from Nevada: *V. c. celata, V. c. orestera* and *V. c. lutescens.*

Linsdale (1936) reported this warbler from eight Nevada counties, covering all sections of the state.

Records of Occurrence. Sheldon NWR (1982) reported the Orange–crowned Warbler as a breeding uncommon summer resident, seen spring through fall.

LaRochelle (MS, 1967) reported at Ruby Lake NWR several were mist netted and banded in 1963. Lewis collected a female, now in the USNM, at Ruby Lake NWR May 21, 1967.

In the Reno area and nearby Carson Range, these birds have been recorded in May, July, August, September and October (Alcorn MS).

In the Fallon area they have been seen or collected in January, February, May, September and November. This author observed an Orange–crowned Warbler in Fallon February 7, 1958 feeding in a Russian olive tree. It came upon a partially eaten apple that had been fastened to one of the tree limbs. This author wrote, "To my surprise it started eating the apple meat and continued to do so until it had swallowed five or six times. Then it continued its search for insects in this and nearby trees (Alcorn MS)."

Kendrick (PC) saw three in his yard in Fallon feeding on cut–up ripe tomatoes December 17, 1980. One was seen five miles west of Fallon December 19, 1987 (Alcorn MS).

Biale (MS) saw one about five miles north of Eureka September 24, 1974. It was sleeping on a haystack of baled alfalfa hay. Another was seen in the town of Eureka May 15, 1975 and another June 20, 1986.

Johnson (1973) reported, "A significant southward extension of known breeding range for this species in Nevada was provided by the collection of a male, singing on territory in an aspen grove, Scofield Canyon, Quinn Canyon Mountains."

Austin and Bradley (1971) reported for Clark County, "Transient in desert scrub, riparian, woodlands and montane forests. In woodlands and montane forest it usually occurs from June

21 to October and on May 12 and June 6. In desert scrub in riparian areas it usually occurs from August 6 to May 20. Unseasonal occurrences on June 12, July 3 and August 6."

Kingery (1977) reported two dozen Orange–crowned Warblers in southern Nevada December through January. He (1982) reported the peak of 200 Orange–crowned Warblers at Las Vegas September 22 to October 6, 1981.

Nashville Warbler   *Vermivora ruficapilla*

Distribution. From British Columbia to Nova Scotia; south to Guatemala.

Status in Nevada. Summer resident in the Carson Range of western Nevada; transient elsewhere.

Records of Occurrence. LaRochelle (MS, 1967) reported two mist netted and banded at the Ruby Lake NWR in 1963.

Kingery (1982) reported, "One Nashville Warbler at Ruby Lake." He (1983) noted, "One at Elko, Nevada August 5 and 16 and September 7."

Burleigh collected an immature male for the USNM at Galena Creek, Carson Range August 19, 1968.

Linsdale (1951) reported these birds as a summer resident at Lake Tahoe.

Ryser (1985) reported, "Uncommon breeding species" in the Warner Mountains, at Lake Tahoe Basin and in the Carson Range.

Austin (1969) wrote, "There appears to be only four records of this species for southern Nevada. (Linsdale, 1936; Gullion et al., and Hayward et al., 1963). I have 11 spring records between 16 April and 20 May, and six fall records between 16 September and 10 October."

Austin and Bradley (1971) reported for Clark County, "Transient in riparian and montane forests. Usually occurs from April 13 to May 20 and from August 13 to September 20. Unseasonal occurrence on October 10."

Kingery (1977) reported a Nashville Warbler at Las Vegas June 3 and he (1983) reported one at Las Vegas April 10.

Virginia's Warbler   *Vermivora virginiae*

Distribution. From east–central California and central Nevada to Colorado; south to Mexico.

Status in Nevada. Uncommon summer resident; seen most frequently in pinon–juniper belt on lower slopes of mountains in all sections of the state except for the western part.

Records of Occurrence. Linsdale (1936) reported these birds from the East Humboldt Mountains, Ruby Mountains, Toiyabe Mountains and on Irish Mountain.

LaRochelle (MS, 1967) reported at Ruby Lake NWR in 1963, five Virginia Warblers were mist netted. He reported them as often seen.

Biale (MS) reported an adult carrying nest material in the Diamond Mountains on June 23, 1968. He also reported at Bartines Ranch, about 20 miles west of Eureka on May 23, 1971, one of these warblers was found dead from a storm. He believes these birds nest in the Eureka area in the upper mahogany belt.

One was seen at the Lehman Caves National Monument near Baker September 24, 1959 (Alcorn MS).

Johnson (1974) reported in the Grapevine Mountains, "The Virginia's Warbler was not recorded during the earlier surveys. On May 19, 1971 a singing male was taken at 6900 feet in Purshia and small pinons in Phinney Canyon. No others were detected either in 1971 or 1973."

Austin and Bradley (1971) reported for southern Nevada, "Summer resident in woodlands. Usually occurs from May 1 to September 13. Transient in riparian and montane forest areas."

Kingery (1980) reported one Virginia Warbler as Las Vegas September 17.

Lucy's Warbler   *Vermivora luciae*

Distribution. From southeastern California and southern Nevada to Colorado; south to Mexico.

Status in Nevada. Uncommon summer resident in the southern part of the state.

Records of Occurrence. Linsdale (1936) reported, "Summer resident along the Colorado River, represented in the Museum of Vertebrate Zoology by a female taken May 8, 1934 on the Colorado River, opposite Fort Mohave."

Johnson (1956) reported records in southern Nevada for March, April, May and June. He also extended the breeding range as far north as Overton and Caliente.

Austin and Bradley (1971) reported for southern Nevada, "Summer resident in riparian areas. Usually occurs from March 21 to September 14."

Kingery (1987) reported, "A late Lucy's stopped at Las Vegas November 8."

Northern Parula   *Parula americana*

Distribution. From northern Canada south to central Florida; through Central America to Costa Rica.

Status in Nevada. A straggler to the southern part of Nevada.

Records of Occurrence. Kingery (1987) reported a Northern Parula at Dyer May 31.

Kingery (1977) reported a male May 16 and a female May 29 at Lida. He (1984) reported, "A Northern Parula responded to a Screech/Pygmy Owl imitation at Lida, Nevada June 1."

Johnson (1965) reported, "Christman shot a female in a thicket of squaw bush at the base of a north–facing cliff in Hidden Forest Canyon, 7000 feet, Sheep Range, on June 13, 1963. This is the first report of this species for Nevada, although there are scattered records of this "Eastern" species, some of which presumably represent stragglers for other western states."

Snider (1969) reported this warbler seen, "May 4, 1969 one at Desert Game Range." She (1971) reported one October 4 and 6, 1970, on the Desert Game Range. On May 27, 1972 at least one was seen at Corn Creek, Desert Game Range. Kingery (1979) reported a Northern Parula at Las Vegas June 5.

Kingery (1977) reported a singing male and female May 31 provided the first spring record at Las Vegas in five years.

Kingery (1978) reported a Northern Parula August 3 at Corn Creek and he (1981) reported Las Vegas had a Northern Parula September 29, the first fall record since 1977.

Kingery (1982) reported, "Las Vegas had its first in two years, a female Northern Parula June 6." He (1983) reported, "A Northern Parula was at Las Vegas May 22." He (1984) reported, "A singing Northern Parula at Las Vegas May 12–15."

Yellow Warbler   *Dendroica petechia*

Distribution. From Alaska and Canada; south to Peru and Brazil.

Status in Nevada. Common summer resident in riparian areas over the entire state. Linsdale (1951) reported four races from Nevada: *D. p. rubiginosa, D. p. morcomi, D. p. brewersteri* and *D. p. sonorana.*

Records of Occurrence. Linsdale (1936) reported Yellow Warblers from nine of the 17 Nevada counties. Recent observations of this warbler cover all 17 counties.

Sheldon NWR (1982) reported the Yellow Warbler as an uncommon breeding summer resident, present spring through fall.

In Elko County lone birds were observed at Wildhorse Reservoir May 29 and 30, 1971 (Alcorn MS).

Ryser (1967) reported seeing these birds in Lamoille Canyon May 28 and 29 and from July 1 to 4, 1967.

LaRochelle (MS, 1967) reported banding these birds at the Ruby Lake NWR in 1963. Lewis collected five specimens for the USNM at Ruby Lake NWR in May and June, 1968.

The author saw other Yellow Warblers in the Ruby Valley May 18, 1966 and at Harrison Pass August 22, 1965.

At Lamance Creek, Santa Rosa Range, what was thought to be a breeding Yellow Warbler was seen June 10, 1970 (Alcorn MS).

In the Reno area they were observed from April 30 and all months into September (Alcorn MS).

Burleigh collected nine specimens, now in the USNM, in May, July, August and September. Eight were taken in the Reno area and one was taken from Washoe Lake area.

A bright–colored male Yellow Warbler was seen near Yerington April 29, 1983 (Alcorn MS).

At four miles west of Fallon on June 27, 1943 a male and female Yellow Warbler were seen attending a nest containing young birds. The nest was situated high in a cottonwood tree. During this time, Yellow Warblers were a common summer resident, seen from May through August. Since then, their numbers have decreased significantly (Alcorn MS).

Biale (MS) recorded these birds 67 times in the Eureka area from May 10, 1964 to September 2, 1967. They were seen in all months from May through September.

Kingery (1980) reported a Yellow Warbler at Preston, White Pine County October 22.

A. A. Alcorn (MS) saw one in the Baker area May 11, 1967.

J. R. and A. A. Alcorn reported two Yellow Warblers at Panaca May 4, 1962 and in the Pahranagat Valley from one to six were observed May 5, 23 and 25, 1962. One was seen April 23, 1964 and in 1965, from one to five were seen April 4, 6 and 14 (Alcorn MS).

J. R. and A. A. Alcorn saw these birds at Gold Butte May 10, 1962. In Moapa Valley they were seen April 23; eight times in May; and June 6, all in 1962 (Alcorn MS).

In Clark County, Austin and Bradley (1971) reported, "Summer resident in riparian areas. Transient in woodlands and montane forests. Usually occurs from April 15 to October 3. Unseasonal occurrences on October 22 and 23."

Kingery (1981) reported, "At Las Vegas on May 11, 30–50 Yellow Warblers were seen bathing along small streams that run between ponds."

Chestnut–sided Warbler   *Dendroica pensylvanica*

Distribution. From east–central Alberta to Nova Scotia; south to Panama. Casual in Nevada and California.

Status in Nevada. Rare.

Records of Occurrence. Ryser (1969) reported one at Carlin, Elko County June 4, 1969.

Kingery (1980) reported one male singing at Dyer June 1 and a female there May 30, 1981.

Lawson (MS, 1981) reported from one to four were seen at Corn Creek on seven occasions in October, 1974 (one collected) and another one was at Corn Creek May 27, 1973.

Kingery (1984, 1985) reported Chestnut–sided Warblers at Las Vegas May 27 and October 11, 1984.

Magnolia Warbler   *Dendroica magnolia*

Distribution. From British Columbia to Newfoundland; south through Central America to Panama.

Status in Nevada. A straggler.

Records of Occurrence. Kelleher (1970) reported, "Evans (Raymond) banded one adult and one juvenile near the field station of the Whittell Forest and Wildlife Area in Little Valley, Nevada."

Miller collected one specimen in the Grapevine Mountains, Nye County June 7, 1940.

Kingery (1978) reported Mowbray saw one at Corn Creek October 10 and Lawson saw one at Las Vegas October 19.

Kingery (1980) reported one at Las Vegas from September 25 to October 8.

Lawson (MS, 1981) reported one at Corn Creek October 14, 1974. He reported a specimen taken from Vegas Wash October 20, 1974. Also, Lawson reported one in Oak Creek Canyon October 18, 1975 and one at Corn Creek October 10, 1977.

Kingery (1983, 1984) reported, "Magnolia Warblers at Las Vegas in May." He (1986) reported one at Las Vegas October 13.

Cape May Warbler   *Dendroica tigrina*

Distribution. Across central Canada; south into Central America.

Status in Nevada. Accidental.

Records of Occurrence. Lawson (MS, 1981) reported a male Cape May Warbler at the Corn Creek Field Station from July 26 through August 3, 1977. It was photographed July 28.

Black–throated Blue Warbler   *Dendroica caerulescens*

Distribution. From Alberta to Nova Scotia; south to Cuba, the Bahamas, the Virgin Islands and Bermuda.

Status in Nevada. An uncommon visitor.

Records of Occurrence. Kingery (1987) reported, "Two Black–throated Blue Warblers stopped at Stillwater October 7 [1986]."

Cruickshank (1964) reported at Unionville, Pershing County a male was observed for two days at close range in January, 1964.

A male specimen in the USNM was taken at the Desert Game Range at Corn Creek by C. G. Hansen October 11, 1964.

Snider (1971) reported a single bird seen at the Desert Game Range at Corn Creek October 4, 1970.

Kingery (1977) reported a Black–throated Blue Warbler at Las Vegas December 26 through January 15. He (1977) reported, "Five in southern Nevada with one still present on December 3; seen so often, no longer warrants rare status in Nevada."

Lawson (MS, 1981) reported seeing as many as 22 in one day during the fall (mid–September to mid–October) in the Las Vegas Valley. One of four males was taken as a specimen at Hansen Springs near Corn Creek on October 10, 1974.

Since 1983, Kingery has reported one or two Black–throated Blue Warblers each year in the Las Vegas area.

Yellow–rumped Warbler   *Dendroica coronata*

Distribution. The Audubon and Myrtle Warblers were combined into the Yellow–rumped Warbler. From Alaska and Canada; south through Mexico and Central America to Panama.

Status in Nevada. Summer resident in higher ranges, winter visitor to the valleys and southwestern and southern parts of Nevada. Linsdale (1951) reported three races from Nevada: *D. c. auduboni*, *D. c. memorabilis* and *D. c. coronata*.

Linsdale (1936) reported, "The population of Audubon Warblers in the mountains of western Nevada seems to be almost exactly intermediate between the races *auduboni* and *memorabilis*. In the absence of a detailed study of the whole species, I have considered all the summer birds in Nevada as one race. Some of the transient and winter birds, however, are obviously migrants from localities well within the summer range of *auduboni*."

After Linsdale's 1936 account, the Myrtle Warbler *(Dendroica coronata)* and Audubon's Warbler *(Dendroica auduboni)* were declared the same species and called the Yellow–rumped Warbler.

Records of Occurrence. Linsdale (1936) reported these birds from most areas in Nevada with records from 12 of the 17 Nevada counties. So many records are available for this warbler, no attempt will be made to list all of them.

Sheldon NWR (1982) reported the Yellow–rumped Warbler as an uncommon breeding summer resident, present spring through fall.

Ryser (1973) reported at Blue Lake in the Pine Forest Mountains of Humboldt County, many of these warblers were present in early August, 1973.

In the Santa Rosa Range, these birds were seen at Solid Silver Creek October 7, 1961 and at Lye Creek July 4, 1968 (Alcorn MS).

LaRochelle (MS, 1967) reported banding both the Myrtle and Audubon forms in 1963 at the Ruby Lake NWR.

Burleigh collected 27 specimens for the USNM from Reno, Mount Rose, North Lake Tahoe, Galena Creek and Virginia City in 1967, 1968 and 1969. Specimens were taken in all months except June and July.

In the Fallon area, Yellow–rumped Warblers have been recorded in all months except June and July. Most recent records are for April, May, October and November. Both the Myrtle and Audubon forms were seen May 12, 1971 in the area six miles west of Fallon (Alcorn MS).

Biale (MS) reported these warblers in the Eureka area 61 days from May 26, 1964 to October 22, 1967. They were most often seen in May, but all months from April to October are represented. Since that time, he has recorded an earlier spring record (March 28, 1971). In Faulkner Canyon, Monitor Range, on June 23, 1974, Biale saw a nest in a willow tree 20 feet above ground, with adults feeding fledglings. Biale also reported seeing the Myrtle form in April 1965 and 1966.

Records from the Baker and Lehman Caves National Monument vicinity represent April, May, August and November (Alcorn MS).

J. R. and A. A. Alcorn recorded Yellow–rumped Warblers in Moapa Valley from 1957 to 1965 once in January and February; 6 times in March; 8 in April; 6 in May; once in September; 3 times in October; and twice in November (Alcorn MS).

In Pahrump Valley, Nye County, two were seen October 31, 1961 (Alcorn MS).

The CBC for the Desert Game Range, Henderson and Ash Meadow routes, consistently reveal these birds in December of each year.

Austin and Bradley (1971) reported for Clark County, "Summer resident in montane forests. Winter resident in desert scrub and riparian areas. Usually occurs from October 3 to May 15. Unseasonal occurrence on July 26. Transient in riparian, woodlands and montane forest areas. In woodland and montane forests it usually occurs from March 16 to October 26. In riparian and desert scrub areas, it usually occurs from September 13 to May 31."

Black–throated Gray Warbler   *Dendroica nigrescens*

Distribution. From British Columbia to Colorado; south to Guatemala.

Status in Nevada. Summer resident in most of the mountain ranges.

Records of Occurrence. Linsdale (1936) reported Black–throated Gray Warblers from the East Humboldt and Ruby Mountains, near Wells and Ruby Lake, all in Elko County; from Galena Creek and Carson Range, Washoe County; from six localities in the Toiyabe Range of central Nevada; from Schellbourne Pass, Water Canyon and Lehman Creek, White Pine County; from Pine Grove, Mineral County; from the White Mountains, Silver Peak Range and

Mount McGruder, Esmeralda County; from the Quinn Canyon Mountains and Hot Creek and Belted Ranges, Esmeralda County; from Groom, Baldy and Irish Mountain, Lincoln County; and from the Juniper Mountains, Clark County.

Sheldon NWR (1982) reported the Black–throated Gray Warbler as an occasional to rare migrant spring and summer.

LaRochelle (MS, 1967) reported only one sightting at Ruby Lake NWR.

Burleigh collected a female specimen for the USNM near Reno October 8, 1967.

Gabrielson (1949) reported one near Eastgate, Churchill County, August 14, 1933.

Biale (MS) reported Black–throated Gray Warblers 20 times in the Eureka area from 1964 to 1970. They were seen 3 times in May; twice in June; 9 in July; and 3 times in August and September. The earliest record was May 9 and the latest record was September 5.

One was reported at Lehman Caves National Monument, Snake Range May 28, 1959 (Alcorn MS).

Ryser (1967) reported one at Hancock Summit February 15–19, 1967.

A. A. Alcorn (MS) saw one at Trough Springs, Charleston Mountains, July 19, 1963.

Austin and Bradley (1971) reported, ''Summer residents in woodlands and montane forests. Transient in riparian areas. In woodlands and montane forests it usually occurs from April 18 to October 3. Unseasonal occurrence April 1. In desert scrub and riparian areas it usually occurs from April 18 to May 8 and from September 5 to October 10. Unseasonal occurrences January 14, July 31 and November 30.''

Townsend's Warbler   *Dendroica townsendi*

Distribution. From Alaska and Yukon; south to Nicaragua. Other casual reports for various areas in North America.

Status in Nevada. An uncommon transient statewide.

Records of Occurrence. Linsdale's (1936) entire account is given, ''Transient; probably not very numerous over most of the state. In Washoe County, a male taken May 21, 1934, is from the Truckee River, 12 miles northwest of Wadsworth (Compton). Ridgway (1877) on September 8, 1868 saw a Townsend's Warbler in an alder thicket in the East Humboldt Mountains, Elko County and shot an adult male September 24, 1868 in Thousand Spring Valley, Elko County. An adult male was taken, September 18, 1934 on Lehman Creek, 8500 feet, Snake Mountains, White Pine County (Behle). In the White Mountains, Esmeralda County a female was taken, May 15, 1927 on Chiatovich Creek, 8200 feet. Several individuals were noted May 11 along this same stream feeding in willows at 7000 feet (Linsdale, MS).''

Sheldon NWR (1982) reported the Townsend's Warbler as an uncommon summer resident, seen spring through fall and thought to nest in the area.

Kingery (1980) reported Townsend's Warblers seen in Elko County in three mountain ranges August 10 and September 6.

Kingery (1982) reported Townsend's Warblers at Ruby Lake October 17, 1981.

Burleigh collected an immature male from North Lake Tahoe, 9000 feet, August 24, 1967 and a male from Reno April 29, 1969.

Napier (MS, 1974) reported seeing what he thought was a Townsend's Warbler May 11, 1970 on the lower Carson River, northeast of Fallon.

Thompson (MS, 1988) reported Townsend's Warblers seen at Stillwater WMA October 7, 1986.

Biale (MS) reported the following records: one, Eureka area May 8, 1965; one, dead from a storm, at Bartines Ranch, 20 miles west of Eureka May 23, 1971; one at Davis Canyon, Diamond Mountains, September 12, 1971; and one at Phillipsburg Canyon, Diamond Mountains September 9, 1973.

Gullion et al., (1959) reported, "Three spring records: one bird, Eldorado Mountains, east of Nelson, May 5, 1951; and one bird, Dead Mountains, May 7, 1951 and May 23, 1953. One fall record: one bird, Charleston Mountains, October 22, 1949. These spring records are the first for this species in the lower desert areas in Nevada."

Austin and Bradley (1971) for Clark County reported, "Transient in desert scrub, riparian, woodlands and montane forest areas. In woodland and montane forests, it usually occurs from April 29 to May 23 and on September 13 and October 23 and 30. Unseasonal occurrence on February 13."

Kingery (1982) reported Townsend's Warblers at Las Vegas in October, 1981.

Hermit Warbler   *Dendroica occidentalis*

Distribution. From southwestern Washington south through the coast ranges and the Sierra Nevada of California. Winters chiefly from Mexico to Nicaragua.

Status in Nevada. Uncommon. Reportedly breeds in extreme southern Nevada.

Records of Occurrence. Linsdale's (1936) entire account is given: "Transient; available records only in fall, but doubtless occurs in spring and along central western border as a summer resident."

"On August 29, 1868, a single individual of this species was seen in the lower portion of one of the eastern canyons of the East Humboldt Mountains, Elko County (Ridgway, 1877). An immature female was shot, September 13, 1934 on Bear Creek, 7000 feet, Jarbidge Mountains, Elko County (Linsdale)."

"At Birch Creek, 7500 feet, in the Toiyabe Mountains, Lander County, an immature male was taken August 31, 1931. Others were seen in the same mountain range on September 2 and 5, 1931. On Galena Creek, 7000 feet, Washoe County, three specimens, an adult male and two immature birds, were collected August 9 and 10, 1932. One was taken August 17, 1934 between Fallon and Hazen, Churchill County (Arnold)."

Since 1936, these birds have not been reported from Nevada except on rare occasions. Ryser (1985) listed this warbler as, "An uncommon breeding species in the mountainous western rim of the Great Basin in the Carson Range and in the Sierra Nevada."

Orr and Moffitt (1971) reported these warblers as a fairly common summer resident of the Lake Tahoe region. Of 10 specimens reported in the California Academy of Science collection from this area, however, none were taken in Nevada.

Grater (1939) collected a specimen September 13, 1938 at Hemenway Wash in southern Nevada.

Austin and Bradley (1971) reported for Clark County, "Transient in riparian, woodland and montane forest areas. In woodlands and montane forests it usually occurs on May 1 and 12 and from July 30 to September 16. Unseasonal occurrence June 25. In desert scrub and riparian areas it usually occurred on April 27 and 29 and on September 13."

Lawson (TT, 1976) reported the Hermit Warbler is an early fall migrant in mountains, late August and early September in southern Nevada.

Black–throated Green Warbler   *Dendroica virens*

Distribution. From Canada, south to Mexico, Central America and the West Indies.

Status in Nevada. Hypothetical.

Records of Occurrence. Lawson (MS, 1981) reported Mowbray saw one at Corn Creek October 15, 1978. Lawson saw one male at the same place May 26, 1980.

Kingery (1983) reported Nevada's third and fourth records were May 26, 1983 at Las Vegas and May 28, 1983 at Dyer.

Blackburnian Warbler   *Dendroica fusca*

Distribution. From central Alberta and the eastern United States; through Central America to Peru and Bolivia.

Status in Nevada. Rare visitor to southern Nevada.

Records of Occurrence. Kingery (1978) reported one at Beatty October 7, 1977.

Kingery (1977) reported one at Pahranagat NWR October 6, 1976.

Lawson (MS, 1981) reported lone individuals at Corn Creek Field Station September 19, 1973 and on six different days from October 4 to 20, 1974.

Kingery (1982) reported one at Las Vegas November 6, 1981 and he (1985) reported another at Las Vegas October 14, 1984.

Yellow–throated Warbler   *Dendroica dominica*

Distribution. From northern midwestern and eastern United States; south to Central America.

Status in Nevada. Hypothetical.

Records of Occurrence. Kingery (1977) reported Nevada's first record April 22, 16 miles southeast of Austin. "A bird foraging in low trees and providing good views."

Lawson (MS, 1981) commented that the bird reported by Kingery was seen by Bruce Sorrie and Barbara Flarris of the Point Reyes Bird Observatory. It was a male Yellow–throated Warbler seen at Kingston Canyon April 22, 1977.

Kingery (1980) reported Nevada's second record October 24, 1979 at Duckwater, Nye County.

Grace's Warbler   *Dendroica graciae*

Distribution. From southern Nevada, Utah and Colorado; south to Nicaragua.

Status in Nevada. Summer resident in southern Nevada.

Records of Occurrence. Johnson (1973) reported, "A major northwestward extension of known breeding range was recorded in 1972 when this species was discovered breeding on Mount Irish and in the Quinn Canyon Mountains, Sawmill Canyon, where a singing male accompanied by a very worn individual of indeterminate sex were collected in ponderosa pine and white fir on a steep slope."

Austin (1969) reported, "A single bird was seen near Deer Creek Canyon, 8000 feet, Spring Mountains June 10, 1966. This supplements the records of Jaeger (1927) for the same range and those of Johnson (1965) for the Sheep Range."

Johnson (1974) reported, "This species has not been reported previously from Potosi Mountains. On 25 May, 1971 I collected a singing male at 7100 feet in open yellow pines, indicating colonization of the mountain at some time in recent decades."

Austin and Bradley (1971) reported for Clark County, "A summer resident in montane forests. Usually occurs in June."

Pine Warbler   *Dendroica pinus*

Distribution. From Manitoba to Maine; south to eastern Texas.

Status in Nevada. Hypothetical.

Records of Occurrence. Mowbray (MS, 1986) reported the Pine Warbler as uncommon in southern Nevada. He (PC, 1988) reported two records for southern Nevada. One September 29, 1973 and another September 15, 1987.

Palm Warbler   *Dendroica palmarum*

Distribution. From Mackenzie to Newfoundland; south to Honduras and northeastern Nicaragua.

Status in Nevada. A straggler.

Records of Occurrence. Biale (MS) reported one Palm Warbler seen five miles south of Eureka September 4, 1977.

Lawson (MS, 1981) reported one at Tonopah October 6, 1974. He also reported collecting a female at Tule Springs Park, 14 miles northwest of Las Vegas, Clark County, October 16, 1974, and from one to two birds seen on eight different days at Corn Creek Field Station and Boulder Beach Sewage Ponds in October of 1974.

Kingery (1985, 1986) reported two Palm Warblers appeared at Las Vegas during separate visits.

Bay-breasted Warbler   *Dendroica castanea*

Distribution. From British Columbia to Newfoundland; south to South America (Venezuela).

Status in Nevada. Rare.

Records of Occurrence. Kingery (1980) reported, "Late Bay–breasted Warblers were seen at Dyer June 10."

Kingery (1983) reported, "Nevada's fourth Bay–breasted May 21 at Dyer, fed deliberately for 10 minutes within a 25 foot area."

Lawson (MS, 1981) reported a male at Corn Creek, May 27, 1975. He also reported one October 10 and 11, 1977 at Corn Creek.

Blackpoll Warbler   *Dendroica striata*

Distribution. From Alaska to Newfoundland; south in South America to Chile.

Status in Nevada. Rare to uncommon migrant.

Records of Occurrence. Lewis collected a male specimen for the USNM at Ruby Lake NWR, Elko County May 22, 1967.

Kingery (1980) reported, "Hawk watchers in the Goshutes found one Blackpoll October 7."

Kingery (1977) reported Blackpoll Warblers from Pahranagat NWR October 7, 1976.

A Blackpoll Warbler was collected by Lawson at Hansen's Spring, Desert Game Range October 2, 1974, and they were seen eight times by various observers in the autumn of 1973 and 1974 (Alcorn MS).

Lawson (MS, 1981) reported Mowbray saw four at Corn Creek September 16, 1979. Lawson saw one in Macks Canyon, Spring Mountains (9000 feet elev.) October 6, 1974.

Lawson (MS, 1981) reported one seen at his home in Las Vegas October 31 and November 1, 1979. He considers them a rare to uncommon fall migrant in southern Nevada.

Kingery (1982) reported blackpolls at Las Vegas September 22 and October 14, 1981.

Cerulean Warbler   *Dendroica cerulea*

Distribution. From Nebraska to Ontario; south to Bolivia. Casual to California.

Status in Nevada. Accidental.

Records of Occurrence. Phillips et al., (1964) reported, "One found dead at Boulder Beach on the Nevada side of Lake Mead, June 6, 1954."

Lawson (MS, 1981) reported, "This specimen is now at the University of Nevada, Las Vegas."

Black–and–White Warbler   *Mniotilta varia*

Distribution. From Mackenzie to Newfoundland; south to the West Indies and Ecuador.

Status in Nevada. Uncommon transient.

Records of Occurrence. Linsdale (1951) reported only one Black–and–White Warbler record in Boulder City, Nevada. Since that time they have been reported from several localities.

Kingery (1980) reported, "Nevada's first summer Black–and–White Warbler was foraging and singing near Wells July 6."

Ryser (1985) reported a summer specimen at the Eugene Mountains, Pershing County.

Johnson and Richardson (1952) reported, "H. I. Fisher and V. Mowbray positively identified an individual of this species at Galena Creek, September 13, 1947."

Ryser (1967) reported Black–and–White Warblers in May of 1967 from Verdi and Reno.

Kingery (1977) reported two different birds at Reno on March 21 and 22, and one singing at Lida May 23 and 24.

From 1965 to 1970, these birds were reported at the Desert Game Range, Corn Creek, Sheep Range and the Charleston Mountains, all in southern Nevada **(Audubon Notes).**

Austin and Bradley (1971) reported, "A visitant to montane forests, transient in riparian areas, usually occurs from May 4 to June 12 and on September 30 with periods of greatest abundance in December and February."

Kingery (1977) reported, "Las Vegas had a singing Black–and–White Warbler June 18 and also these warblers were at Las Vegas May 10 and 25."

Kingery (1982) reported a Black–and–White Warbler at Las Vegas September 12 and October 19, 1981. He (1983) reported, "Black–and–White Warbler at Baker, Nevada May 12 and one early in Las Vegas April 26." Kingery (1984) reported one in Las Vegas June 19, 1984.

American Redstart *Setophaga ruticilla*

Distribution. From Alaska to Newfoundland; south to Brazil.

Status in Nevada. A regular spring and fall transient in the southern part of the state. Irregular transient elsewhere.

Records of Occurrence. LaRochelle (MS, 1967) banded one female at the Ruby Lake NWR in 1963.

Castetter and Hill (1979) reported one at Cane Springs, Nevada Test Site, May 21, 1977.

John Sadowski and Wayne Elliott (MS, 1976) saw one at Porter Springs, southwest end of Seven Troughs Range, Pershing County, September 2, 1974.

Biale (MS) reported an individual bird seen daily in the Eureka area from May 12 to May 19, 1966.

A. A. Alcorn (MS) reported single birds at Hiko, Pahranagat Valley on May 15 and 21, 1965.

Austin (1971) reported, "Pulich and Phillips (1953) indicated a desert flight line for the American Redstart. There are now over 100 records in fall and 50 in spring for the area included (desert areas of southern Nevada, southeastern California and southwestern Arizona), substantiating this region as an important inland pathway for this species."

Snider (1969) reported, "September 2, 1968, there were usual scattered reports of American Redstarts in the region, single birds were reported near Las Vegas."

Snider (1970) reported from May 11 to 29, from one to seven was the largest influx ever for the Las Vegas area.

Austin and Bradley (1971) reported for Clark County, "Transient in riparian areas. Usually occurs from May 2 to 29 and from August 18 to September 28. Unseasonal occurrence on April 20 and October 21."

Mowbray (PC, 1988) reported American Redstarts are, "Seen in most months of the year.

It used to be unusual to see them, but now it's unusual not to see them. They are of regular occurrence in southern Nevada.''

Prothonotary Warbler   *Protonotaria citrea*

Distribution. Through Great Lakes region, south to South America (Columbia). Casual in western North America.

Status in Nevada. Accidental to western and southern Nevada.

Records of Occurrence. Ryser (1963) reported finding a dead male Prothonotary Warbler under a lone lodgepole pine close to a creek in the Carson Range about 18 miles south of Reno.

Kingery (1985) reported, ''A Prothonotary Warbler presented itself for close scrutiny November 9 at Reno.''

Lawson (MS, 1981) reported, ''Prothonotary Warbler, one on October 11, 1975 photographed at Cottonwood Cove, Lake Mohave, Clark County.''

Kingery (1986) reported a Prothonotary Warbler at Las Vegas September 21, 1985.

Worm–eating Warbler   *Helmitheros vermivorus*

Distribution. From northeastern Kansas to Maine; south to Cuba and Panama.

Status in Nevada. Accidental straggler to Clark County.

Records of Occurrence. Johnson (1965) reported, ''Christman took a male of this species in a shady thicket of squaw bush at the base of a north–facing cliff in Hidden Forest Canyon, 7000 feet, Sheep Range, on June 13.''

Lawson (MS, 1981) reported three records for southern Nevada: one at Boulder City on September 29, 1964; a specimen obtained in the Hidden Forest Canyon, Sheep Range on June 13, 1963; and Polly Long was with Nora Poyser in her yard at Boulder City when she observed a Worm–eating Warbler May 9, 1981.

Kingery (1983) reported, ''A Worm–eating Warbler at Las Vegas in May.''

Ovenbird   *Seiurus aurocapillus*

Distribution. From British Columbia to Newfoundland; south to Panama, Columbia, Venezuela and the Lesser Antilles.

Status in Nevada. Accidental.

Records of Occurrence. Kingery (1980) reported one appeared at Sutcliff, Pyramid Lake October 9, 1979.

Alcorn (1946) reported one found dead by W. H. Alcorn four miles west of Fallon June 12, 1941.

Kingery (1982) reported an Ovenbird at Dyer and he (1986) reported another one there May 16–17.

Monson (1972) reported one at Tule Springs Park, Las Vegas October 17, 1971.

Kingery (1971) reported one at Corn Creek May 19. He (1980) reported Mowbray saw individuals at Las Vegas October 4 and 22, and November 1. Kingery (1984) reported one Ovenbird at Las Vegas May 27–June 9 and he (1987) reported one September 21, 1986.

Phillips et al., (1964) reported one seen at Boulder City June 16, 1954.

Northern Waterthrush   *Seiurus noveboracensis*

Distribution. From Alaska to Labrador and Newfoundland; south to Venezuela.

Status in Nevada. Accidental transient.

Records of Occurrence. LaRochelle (MS, 1967) reported the banding of one of these birds at the Ruby Lake NWR, Elko County. At this same refuge an immature female was collected September 18, 1966 by Lewis.

Biale (PC, 1984) reported one seen in Diamond Valley on May 24, 1981.

Hansen collected one from Corn Creek September 11, 1963.

Snider (1971) reported unusual summer Northern Waterthrushes September 12, 1970 on the Desert Game Range and on August 14, 1971 at Corn Creek. Monson (1972) reported one at Corn Creek September 4, 1971.

Austin and Bradley (1971) reported for Clark County, "Transient in riparian areas. Usually occurs from April 29 to May 18 and from September 1 to October 1. Unseasonal occurrences in August and November 1."

Kingery (1977) reported one at Las Vegas May 10–11 and a specimen was taken from Hansen Springs, Desert Game Range May 26, 1976. He also reported, "One Northern Waterthrush at Las Vegas April 28–29." He (1984) reported, "Northern Waterthrushes at Boulder City, Nevada and Las Vegas, including one November 2 on an apartment complex lawn." Kingery (1987) reported one at Las Vegas August 31, 1986.

Kentucky Warbler   *Oporornis formosus*

Distribution. From Nebraska eastward; south to South America (Venezuela). Casual throughout western North America.

Status in Nevada. Rare.

Records of Occurrence. Lawson (MS, 1981) reported two records. First, Guy McCaskie, P. Lehman, Dr. J. M. Longham, et al., saw a female near Dyer, Esmeralda County from May 23 to 28, 1981.

Lawson and his wife reported a male at the Red Rock Canyon area, Clark County April 20, 1977, was Nevada's second record.

Kingery (1986) reported one at Las Vegas April 26, 1986. He (1987) reported a "surprising number" of Kentucky Warblers at Las Vegas May 23.

Connecticut Warbler   *Oporornis agilis*

Distribution. East–central Canada and the north–central United States; south to Brazil and Columbia in South America.

Status in Nevada. Accidental.

Records of Occurrence. Ryser (1976) reported Hansen trapped and banded one Connecticut Warbler at Corn Creek May 18, 1967.

MacGillivray's Warbler   *Oporornis tolmiei*

Distribution. From Alaska to Saskatchewan; south to Panama.

Status in Nevada. Summer resident in most of the mountain ranges.

Records of Occurrence. Linsdale (1936) reported this warbler in most of the major mountain ranges and from 11 of the 17 Nevada counties. "Abundant about moist ground about springs and streams in the valleys and in meadows and along streams in the mountains."

Sheldon NWR (1982) reported the MacGillivray's Warbler as uncommon in spring and fall; occasional in summer. It is thought to breed in the area.

Ryser (1967) reported these birds at Lamoille Canyon on May 28 and 29, 1967. He (1970) reported them from the Ruby Mountains on July 4, 1970.

LaRochelle (MS, 1967) reported for Ruby Lake NWR, "The commonest of our warblers. Banded 10 in 1963."

Lewis collected a male specimen, now in the USNM, at Ruby Lake NWR May 22, 1967. Another male was taken at this same place June 1, 1968.

Burleigh collected seven specimens in May, June, August and September of 1967 and 1968 in the Carson Range and nearby Reno area.

In the Fallon area MacGillivray's Warblers have been seen in May, August and September, with one unseasonal occurrence of one bird in Fallon June 7, 1971 (Alcorn MS).

Biale (MS) reported MacGillivray's Warblers in the Eureka area 13 times from 1965 to 1972. They were seen 4 times in May; twice in July; 3 times in August; and 4 times in September. Biale also reported they breed in the Eureka area in June.

Kingery (1980) reported MacGillivray's Warblers at Eureka in May.

Johnson (1973) reported, "This species was established for nesting in at least two places in mixed thickets of chokecherry, wild rose and cottonwoods along the stream in Sawmill Canyon, between 6600 feet and 6800 feet elevation, Quinn Canyon Mountains. This is the only known breeding station for this species in this part of Nevada."

Austin (1968) reported, "A nest of this species with four eggs found near Little Falls, Kyle Canyon, 8200 feet, Spring Mountains, 10 June 1965 appears to be the first breeding record for southern Nevada."

Austin (1971) reported, "Transient in riparian, woodland and montane forest areas. In woodlands and montane forest it usually occurs on May 18 and 23 and from August 10 to September 20. Unseasonal occurrences on June 10 and 14. In desert scrub and riparian areas it usually occurs from May 2 to June 8 and from August 10 to October 12. Unseasonal occurrence on November 12."

Heilbrun and Stotz (1980) reported four in the Henderson area December 15, 1979.

Kingery (1981) reported at Las Vegas 15 MacGillivray's Warblers were seen bathing with other warblers May 11.

Common Yellowthroat   *Geothlypis trichas*

Distribution. From southeastern Alaska to Newfoundland; south to Panama.

Status in Nevada. Summer resident statewide and an occasional winter visitor to southern Nevada. Linsdale (1951) reported two races from Nevada: *G. t. occidentalis* and *G. t. scirpicola.*

Linsdale (1936) reported these birds from nine of 17 Nevada counties. This author has recent records for seven Nevada counties.

Records of Occurrence. Sheldon NWR (1982) reported the Common Yellowthroat as common spring and fall; and an occasional summer breeder.

LaRochelle (MS, 1967) reported for Ruby Lake NWR the Common Yellowthroat is common in summer in the marsh areas where it nests.

Ryser (1967) reported for Ruby Lake NWR the Common Yellowthroat was heard singing throughout the marsh July 1 to 4, 1967.

Heilbrun and Stotz (1980) reported one yellowthroat at Ruby Lake NWR December 18, 1979.

This author saw Common Yellowthroats at Ruby Lake NWR June 20, 1966 and August 21 and 22, 1965. They were also seen in the Elko area June 16, 1965 and July 11, 1966 (Alcorn MS).

Common Yellowthroats were seen in summer months along the Truckee River near Wadsworth, in the Minden area and along the Walker River near Schurz (Alcorn MS).

In the Fallon and Stillwater Marsh areas of Lahontan Valley, they were recorded six times in 1960; four times in 1961 and 1962; once in 1963; three times in 1964; and only once from 1964 to 1970. On July 15, 1974 at Cattail Lake, Napier saw a male yellowthroat singing and heard several more (Alcorn MS).

Several were observed in May and June of 1962 along Reese River west of Austin (Alcorn MS).

In the Eureka area, this bird was seen May 7, 8 and 16, 1960 (Alcorn MS).

Biale (MS) reported them from the Eureka area three times in May, twice in June, and on September 14, 1969.

Austin and Bradley (1971) reported for Clark County, "Summer resident in riparian areas, transient in woodlands. Usually occurs from March 30 to October 11. Unseasonal occurrence on December 30."

Heilbrun and Kaufman (1977) reported one in the Henderson area during the CBC in December 1976.

Hooded Warbler   *Wilsonia citrina*

Distribution. From southeastern Nebraska to Ontario to Rhode Island; south to Costa Rica; rarely to central Panama. Casual in western North America.

Status in Nevada. An uncommon spring migrant.

Records of Occurrence. Austin collected specimens from Wassuk Range, Mineral County on June 6, 1981.

Kingery (1981) reported a female Hooded Warbler at Dyer May 27. He (1987) reported one at Dyer May 22 and 23.

Lawson (MS, 1981) reported others saw Hooded Warblers at the west end of Lida, Highway 6, Esmeralda County, May 27, 1981.

Kingery (1987) reported a Hooded Warbler at Lida May 16.

Lawson (MS, 1981) reported, "One October 20, 1961, specimen from Corn Creek Field Station by C. Hansen is in the USNM; one observed May 6, 1968 at Boulder City, by Pauline Long; one observed May 5, 1974 at Cottonwood Cove, Lake Mohave by Carol Reiglie; one male and female observed June 4, 1975 also at Cottonwood Cove, Lake Mohave by William Prange; and one male observed and photographed June 6, 1975 at this same place by Prange."

Wilson's Warbler *Wilsonia pusilla*

Distribution. From Alaska to Newfoundland; south to Panama.

Status in Nevada. Common summer resident in the higher valleys and mountain ranges. Linsdale (1951) reported three races from Nevada: *W. p. pusilla, W. p. pileolata* and *W. p. chryseola.*

Records of Occurrence. Sheldon NWR (1982) reported the Wilson's Warbler as an uncommon spring and fall migrant; occasional in summer.

Lewis collected three males for the USNM at Ruby Lake NWR. They were mist netted September 23, 1966; May 23, 1968 and June 2, 1968.

Burleigh collected 18 specimens for the USNM from the Reno area and the Carson Range from 1967 to 1970. One was taken in January; two in April; five in May; four in June; one in July; three in August; and one in September and October.

A Wilson's Warbler was seen at a feed and water station near Yerington May 10, 1983 (Alcorn MS).

Alcorn (1946) reported several seen in the Fallon area and one taken May 30, 1942. Since then they have been seen on May 27, 1959 and on May 14, 1971. They are probably a regular transient visitor to this area in spring and fall (Alcorn MS).

An adult Wilson's Warbler flew from a nest on the ground containing three eggs on July 4, 1969 at Wilson Creek, east side Toquima Range (Alcorn MS).

Biale (MS) reported these birds 14 times in the Eureka area from 1965 to 1971. They were seen nine times in May; once in June; and four times in September. He (PC, 1986) reported Wilson's Warblers are seen mostly during April, May and September of each year in the Eureka area.

A male was reported at Lehman Creek, Snake Range by park personnel on September 13, 1961.

One was seen two miles south of Hiko, Pahranagat Valley May 4, 1965 and one was observed in Moapa Valley April 23, 1964 (Alcorn MS).

Gullion et al., (1959) reported, "Spring and fall migrant in southern Nevada."

Austin and Bradley (1971) reported for Clark County, "This bird is transient in desert scrub, riparian, woodland and montane forest areas. In woodlands and montane forests, it usually occurs from April 29 to June 14 and from August 10 to September 20. In desert scrub and riparian areas it usually occurs from April 4 to June 10 and from August 7 to October 6. Unseasonal occurrences on October 23 and December 22."

Heilbrun and Stotz (1980) reported one Wilson's Warbler seen December 15, 1979 in the Henderson area during the CBC.

Kingery (1981) reported at Las Vegas on May 11, "Approximately 150 Wilson's Warblers were seen bathing along small streams that run between the ponds." He (1982) reported Wilson's Warblers at Las Vegas November 11, 1981.

Canada Warbler   *Wilsonia canadensis*

Distribution. From Alberta to Nova Scotia; south to South America (Peru).

Status in Nevada. Hypothetical.

Records of Occurrence. Lawson (MS, 1981) reported two seen at Coleville Bay, Lake Mead from September 6 to 13, 1972 and one was seen at Tule Springs Park October 13, 1977.
Kingery (1983) reported a Canada Warbler at Boulder City October 3, 1982.

Red–faced Warbler   *Cardellina rubrifrons*

Distribution. Southwestern United States; south to El Salvador and Honduras.

Status in Nevada. Rare.

Records of Occurrence. Lawson (MS, 1981) reported, "George Austin saw one during the first week of September, 1974 in Clark Canyon, Spring Mountains. Also one was seen on September 2 and 4, 1974 in Mack's Canyon, Spring Mountains by Lawson. Another one was seen September 6, 1974 at Scout Canyon, Spring Mountains by the Lawsons. There were also two to six sightings during September on the north side of the Mountain Range, so at least two to three birds are involved due to the distance between canyons."

Painted Redstart   *Myioborus pictus*

Distribution. From Arizona to Texas; south to Nicaragua.

Status in Nevada. In southern Nevada uncommon summer resident and transient.

Records of Occurrence. Kingery (1975) reported an immature bird seen at Macks Canyon, Charleston Mountains on August 20, 1975.
Austin and Bradley (1968) reported, "A sight record by Austin at the mouth of Eldorado Canyon on the Colorado River April 26, 1963 is verified by a specimen from the Hidden Forest, Sheep Mountains June 8, 1963 and a sight record from the Clover Mountains in Lincoln County on June 27, 1963 (Johnson, 1965). These are the only records for Nevada."
Since that time, Austin and Bradley (1971) reported the Painted Redstart as, "A summer resident in montane forest. Usually occurred on April 20, in June and on September 4. Visitant in riparian areas."
Lawson (MS, 1981) reported, "A rare to uncommon breeding species at Mount Charleston."
Kingery (1983) reported, "A Painted Redstart at Las Vegas April 17 to May 1 was the first in southern Nevada since 1976." Kingery (1984) reported, "A Painted Redstart turned up on Mt. Charleston, west of Las Vegas September 7–10 in an area where they may breed."

Kingery (1986) reported one Painted Redstart at Las Vegas May 10, 1986.

Lawson (MS, 1981) reported a male specimen from the Fort Mohave area below Davis Dam taken February 27, 1976.

Yellow–breasted Chat   *Icteria virens*

Distribution. From British Columbia to Ontario and New Hampshire; south to Panama.

Status in Nevada. Summer resident, most numerous in brushy thickets along rivers and streams.

Records of Occurrence. Voget (MS, 1986) reported the Yellow–breasted Chat as an occasional spring and summer resident at Sheldon NWR. He commented, "Individual banded 25 May 1982 at Badger Mountain. Several individuals observed in Virgin Valley and Thousand Creek Gorge during the summer of 1986—suspected nesting."

The author and Quiroz found the Yellow–breasted Chat abundant while traveling by boat on the south fork of the Owyhee River July 11 and 12, 1967.

Gabrielson (1949) reported Yellow–breasted Chats from Lamoille on August 18, 1933 and Harrison Pass in the Ruby Mountains June 1, 1948.

LaRochelle (MS, 1967) reported at Ruby Lake NWR the Yellow–breasted Chat was first seen in 1964 and has been seen every summer since that time. Two males in the USNM were taken from this refuge October 29, 1966 and May 16, 1968.

On the east side of the Santa Rosa Range at Lamance Creek several chats were heard and seen June 10, 1970 (Alcorn MS).

Burleigh collected a female specimen at Reno June 23, 1968.

One was seen along the Truckee River near Wadsworth July 4, 1960 (Alcorn MS).

Alcorn (1946) reported the Yellow–breasted Chat as a summer resident in the Lahontan Valley (Fallon area) seen frequently from May to August, inclusive. Since that time, this bird has almost vanished from this area and the author has only two records. One was May 18, 1960 and the other September 9, 1974 (Alcorn MS).

Along the Reese River, Nye County, these birds were seen July 12, 1965 and at Kingston Canyon, Toiyabe Range June 20–21, 1963 (Alcorn MS).

Biale (MS) reported chats in the town of Eureka September 12 and 22, 1966 and September 2, 1967. All were single birds and all were seen in a hop vine on a fence. Later dates reported by Biale include September 12, 1972 and May 19, 1975 when singles were observed.

Near Baker, White Pine County, one was seen and heard singing in the spring of 1967 (Alcorn MS).

One was observed at Springdale, north of Beatty, on May 23, 1962 (Alcorn MS).

Gullion et al., (1959) reported for southern Nevada, "It is a widespread spring migrant with earliest arrival date of May 3 (1951, Mohave Valley). Three summer records from hot, mesquite bosques; heard in Moapa Valley, June 28, 1952; heard in Mohave Valley July 19, 1951, and one bird seen there August 4, 1954. Our latest fall record is for September 17 (1952, Boulder City)."

Austin and Bradley (1971) reported for Clark County, "Summer resident in riparian areas. Transient in woodlands. Usually occurs from April 29 to September 17."

Hepatic Tanager   *Piranga flava*

Distribution. From southern California to Texas; south to Peru and southern Brazil.

Status in Nevada. A rare summer resident in the forests of southern Nevada; accidental to Eureka.

Records of Occurrence. Kingery (1977) reported a Hepatic Tanager at Eureka. Biale (MS) reported this tanager was a male seen six miles south of Eureka June 19, 1977.

Austin and Bradley (1971) reported the Hepatic Tanager in southern Nevada as, "A summer resident in montane forests. Visitant in woodlands. Occurs in June."

Lawson (MS, 1981) reported one record at Tule Springs Park December 30, 1973 to January 2, 1974.

Lawson (TT, 1976) commented, "This bird is an uncommon to rare breeding species in the mountains of southern Nevada."

Kingery (1981) reported a pair of Hepatic Tanagers stayed at Las Vegas May 4–9, the fourth record in 10 years. He (1982) reported Hepatic Tanagers at Las Vegas May 7.

Summer Tanager   *Piranga rubra*

Distribution. From southeastern California to Maryland; south through Mexico to Brazil.

Status in Nevada. Resident along the Colorado River in southernmost Nevada; transient elsewhere.

Records of Occurrence. Linsdale (1936) reported, "In May, 1905, Hollister (1908) found this tanager to be common in cottonwoods along the Colorado River in the extreme southern tip of Nevada, Clark County. A male and female were taken May 7 and 9, 1934 in the same locality."

Biale (MS) observed one in Eureka May 19, 1972.

A male specimen in the Nevada State Museum was collected by Funk from Meadow Valley Wash, 15 miles south of Caliente, Lincoln County May 8, 1968.

Gullion et al., (1959) reported, "Summer resident in Mohave Valley areas, where a single bird was seen at McCullough Spring, McCullough Range on August 22, 1951."

Kingery (1980) reported one at Dyer May 31.

Austin (1969) stated, "A breeding male was taken in Pine Creek Canyon at the eastern base of the Spring Mountains, about 20 miles west of Las Vegas. Especially noteworthy is one seen in my backyard in Las Vegas December 4, 1966."

Austin and Bradley (1971) reported, "Summer resident in riparian areas and usually occurs from April 29 to October 2 in Clark County."

Snider (1964) reported a specimen taken at Corn Creek Station on the Desert Game Range September 26, 1963.

Kingery (1977) reported Summer Tanagers summered at Corn Creek, but no nests were found. He (1978) reported these tanagers stayed as late as September 18.

Kingery (1980) reported two at Las Vegas May 2.

Scarlet Tanager   *Piranga olivacea*

Distribution. From eastern North Dakota to Quebec and New Brunswick; south in South America to Peru and Bolivia.

Status in Nevada. A straggler.

Records of Occurrence. Jerry Page (PC) reported Alan Sands saw a Scarlet Tanager in McGill Canyon, Jackson Mountains July 10, 1974.

Scott (1968) reported, "A Scarlet Tanager seen at Genoa in western Nevada on June 8 must have been an accidental straggler."

Ryser (1968) reported one seen at a canyon stream above Genoa, Douglas County July 1, 1968.

Lawson (MS, 1981) has a record of one from Grapevine Mountains September 13, 1978.

G. Thompson collected a specimen, now in the Nevada State Museum, at Meadow Valley Wash in Lincoln County on June 9, 1971.

Western Tanager   *Piranga ludoviciana*

Distribution. From Alaska to Saskatchewan; south through El Salvador to Panama.

Status in Nevada. Summer resident and transient. Found most commonly in mountain ranges in areas where mountain mahogany is growing.

Records of Occurrence. Sheldon NWR (1982) reported the Western Tanagers as uncommon in spring and an occasional summer breeder.

These birds were seen at Lamance Creek, Santa Rosa Range in June, 1967 (Alcorn MS).

In the Jarbidge area Western Tanagers were observed in September of 1960 and 1961. In the Elko area they were seen in May and June (Alcorn MS).

Ryser (1967) reported Western Tanagers at Lamoille Canyon July 1–4, 1967.

LaRochelle (MS, 1967) reported the Western Tanager at Ruby Lake NWR as very common in mid–spring, with many captured in mist nets and banded. This author (MS) saw one there May 27, 1965.

Six were seen on Mount Rose, Washoe County June 16, 1960.

Burleigh collected eight specimens (five in May, two in June, one in August) from Reno and the Carson Range from 1967–69.

Ryser (1964, 1967) reported Western Tanagers at Genoa May 23, 1964 and September 10, 1967.

In the Fallon area these tanagers are transient. From 1961 to 1967 they were seen 7 times in May; 3 in June; on July 25, 1966; and on August 4, 1963 (Alcorn MS).

Biale (MS) reported seeing these birds in the Eureka area from 1964 to 1973 a total of once in April; 16 times in May; 10 in June; 8 in July; 7 in August; 20 in September; and twice in October (October 2, 1969 and October 4, 1971). He (PC, 1984) reported a female sitting in a nest 20 feet above ground in an aspen tree in Faulkner Canyon, Monitor Range on July 4, 1981 (elev. 8000 feet).

Kingery (1987) reported 120 Western Tanagers present during a "cold spell" at Dyer May 22 and 23.

A. A. Alcorn (MS) saw four in the Virgin Valley of southern Nevada June 11, 1963. They were observed in the Moapa Valley in May and September and on Charleston Mountain in May and July.

Austin and Bradley (1971) reported, "Summer resident in woodlands and montane forests, transient in desert scrub and riparian areas. In montane forests and woodlands it usually occurs from May 16 to September 20. Unseasonal occurrence on September 30 and October 7. In desert scrub and riparian areas it usually occurs from April 29 to June 2 and from August 17 to September 22. Unseasonal occurrence on July 18."

Kingery (1982) reported Western Tanagers at Las Vegas October 19–30, 1981. He (1984) reported 35 at Las Vegas on May 15, 1984.

Northern Cardinal   *Cardinalis cardinalis*

Distribution. From South Dakota to Nova Scotia; south through Mexico to British Honduras.

Status in Nevada. A straggler to southern Nevada.

Records of Occurrence. Austin and Bradley (1968) reported, "A male was observed in a mesquite thicket along the road to Red Rock Canyon in the Spring Range February 4, 1962. This is the first record for Nevada."

Snider (1971) reported the second Nevada record, a male cardinal near Las Vegas May 10, 1971.

Lawson (MS, 1981) reported a female cardinal observed from late–December 1974 to mid–April 1975 in Las Vegas. Also, Lawson reported a female was seen August 3, 1975 at the Overton WMA by J. Blake.

Pyrrhuloxia   *Cardinalis sinuatus*

Distribution. Formerly *Pyrrhuloxia sinnatus*. Mainly from Baja California to Texas; south in Mexico to Tamaulipas.

Status in Nevada. Accidental.

Records of Occurrence. Lawson (MS, 1981) reported Nevada's only record is a photo taken by Frank Long of a male Pyrrhuloxia December 15, 1979 at Las Vegas Wash near the Boulder Highway.

Kingery (1980) reported on this same bird. "The Pyrrhuloxia on the Henderson CBC gave Nevada its first record."

Rose–breasted Grosbeak   *Pheucticus ludovicianus*

Distribution. From British Columbia to Nova Scotia; south to Ecuador and Venezuela.

Status in Nevada. Migrant.

Records of Occurrence. Voget (MS, 1986) reported the Rose–breasted Grosbeak as accidental with one observed at Badger Mountain, Washoe County June 5, 1983. He commented, "I suspect more frequent occurrences, but there has been a lack of adequate investigation."

Lewis collected a male specimen at Ruby Lake NWR November 15, 1966.

Kingery (1977) reported one at Ruby Lake NWR May 12–13. He (1980) reported one at Ruby Lake September 19, 1979.

Burleigh collected a male specimen for the USNM at Reno May 25, 1968.

Ryser (1964) reported one seen May 23, 1964 at Genoa and he (1973) reported one seen at Reno November 15, 1973.

Kingery (1985) reported, "Among the many Rose–breasted Grosbeaks, a notable one visited Stagecoach, Nevada May 16–18 for a first county [Lyon] record."

Biale (MS) reported one Rose–breasted Grosbeak at Eureka August 1, 1982.

Kingery (1982) reported, "A late migrant Rose–breasted Grosbeak occurred at Eureka July 3."

Grater (1939c) reported one observed June 20, 1938 at Saint Thomas.

Rickard (1960) reported, "An occurrence of the Rose–breasted Grosbeak in southern Nevada on May 17, 1959, a male and a female were seen at Indian Springs Ranch, Clark County, Nevada."

Kingery (1977) reported one at Lida, May 24–25.

Kingery (1981) says, "These birds are common in late May as a migrant in southern Nevada." He (1983) reported, "Rose–breasted Grosbeaks were at Las Vegas in June."

Black–headed Grosbeak   *Pheucticus melanocephalus*

Distribution. From southern British Columbia to southwestern Saskatchewan; south in Mexico to Oaxaca and Veracruz.

Status in Nevada. Summer resident over the entire state, but not abundant. Linsdale (1936) reported Black–headed Grosbeaks from 12 of the 17 Nevada counties. He (Linsdale, 1951) wrote, "Summer resident; common in shrubby vegetation bordering streams in the mountains. Sometimes occurs in mountain mahogany."

Records of Occurrence. Sheldon NWR (1982) reported the Black–headed Grosbeak as an uncommon spring and occasional summer migrant.

One was seen at Lye Creek, Santa Rosa Range July 4, 1968 and in the Elko area June 7, 1967 and May 25, 1968 (Alcorn MS).

LaRochelle (MS, 1967) reported banding one at Ruby Lake NWR in 1963.

In western Nevada, one was observed on Mount Rose June 16, 1960. In Reno they were observed from May 27 to August 14, 1942 (Alcorn MS).

Burleigh collected a male in Reno May 14, 1967.

Ryser (1967) reported one Black–headed Grosbeak at Genoa September 10, 1967; one in Mason Valley (Yerington area) June 7, 1945; two along the Carson River east of Fort Churchill on June 12 and July 4, 1960.

Black–headed Grosbeaks were repeatedly seen at a feed station located at Frances McGowan's residence near Yerington from May to August of 1981 and 1982 (Alcorn MS).

Alcorn (1946) reported this grosbeak in the Fallon area as a summer resident with records from May through August. A nest with four eggs was seen May 30, 1945. Since then, these birds have been seen only in these same months. Mills reported on June 28, 1961 he saw an adult Black–headed Grosbeak feeding two young grosbeaks and a young cowbird about three miles west of Fallon. All were out of the nest and Mills noted the young cowbird was larger than either of the grosbeaks.

A. A. Alcorn (MS) reported a nest containing five grosbeak and one cowbird egg nine feet above ground level in an apple tree four miles west of Fallon on June 17, 1973.

At Kingston Canyon, Toiyabe Range, four Black–headed Grosbeaks were seen June 20 and 21, 1963 (Alcorn MS).

In the Toquima Range, four grosbeaks were seen May 17, 1961 at Moor's Creek and two were seen in a grove of aspen trees at Wilson Creek September 6, 1970 (Alcorn MS).

Biale (MS) recorded this bird in the Eureka area 41 days from May 10, 1964 to September 19, 1967. It was seen 14 times in May; 11 in June; 4 in July; 7 in August; and 5 times in September. Since 1967, Biale has seen them repeatedly in these same months, with one unseasonal record on December 5, 1969.

In the Snake Range at Lehman Caves National Monument, park personnel reported Black–headed Grosbeaks August 24, 1957 (one male); September 17, 1959 (one female); and August 23, 1964 (one female).

Near Hiko, Pahranagat Valley, one was seen May 25, 1962 (Alcorn MS).

A. A. Alcorn (MS) reported one in Virgin Valley June 11, 1963; two at Trough Springs, Charleston Mountains July 19, 1963; four at the same place August 7, 1963; and one in the Moapa Valley August 16, 1963.

Cruickshank (1962) reported one observed December 31, 1961 at Boulder City during the CBC.

Monson (1972) reported one at Tule Springs Park, Las Vegas, November 12, 1971.

Austin and Bradley (1971) reported for Clark County, "A summer resident in woodlands and montane forests. Transient in riparian areas. In woodlands and montane forests it usually occurs from May 1 to September 28. In desert scrub and riparian areas it usually occurs from April 23 to June 5 and from August 4 to October 2. Unseasonal occurrences on April 4 and December 31."

Blue Grosbeak   *Guiraca caerulea*

Distribution. From southern California north to North Dakota and south to Panama.

Status in Nevada. Summer resident mainly in southern Nevada with limited numbers in western Nevada. Transient elsewhere. Linsdale (1951) reported two races from Nevada: *G. c. interfusa* and *G. c. salicarius*.

Records of Occurrence. Howard (MS, 1972) reported a male and a female at Ruby Lake NWR May 8 and 9, 1972.

Ryser (1969) reported the Blue Grosbeak as nesting in the Truckee Meadows in the summer of 1969.

In the Fallon area, one was observed six miles southwest of Fallon May 6 and 8, 1963. A pair was seen July 19 and a female July 24, 1969. On July 19, 1969, a fledgling Blue Grosbeak was seen. Other sightings include July 1 and August 7, 1971 (males) and June 18, 1978. Since that time, one or two have been seen most years (Alcorn MS).

A. A. Alcorn (MS) observed one in the Baker area at the Spring Creek Fish Rearing Station May 31, 1967. Another was seen by Frank Dodge August 8, 1968.

Near Hiko, Pahranagat Valley, they were observed in May and June, 1965 (Alcorn MS).

J. R. and A. A. Alcorn (MS) observed these birds in the Moapa Valley area from May 1960 to June 1965. They were seen once in March and April; 10 times in May; 9 in June; 10 in July; 5 in August; and once in September. The earliest record was March 18, 1964; the latest was September 5, 1963.

Blue Grosbeaks were observed at three different places in Pahrump Valley June 16, 1961 (Alcorn MS).

Austin and Bradley (1971) reported for Clark County, "Summer resident in riparian areas. Usually occurs from April 29 to September 13."

Lazuli Bunting   *Passerina amoena*

Distribution. From British Columbia to Saskatchewan; south in Mexico to central Veracruz.

Status in Nevada. Fairly common summer resident over most of the state. Seen frequently along streams.

Records of Occurrence. Sheldon NWR (1982) reported the Lazuli Bunting as an occasional spring migrant.

Lazuli Buntings were common at Lamance Creek, east side of the Santa Rosa Range, on June 10, 1970 (Alcorn MS).

Single birds were repeatedly seen on the south fork of the Owyhee River as the author and Quiroz traveled by boat down this river July 11 and 12, 1967 (Alcorn MS).

One was seen near Tuscarora August 23, 1965 and one near Elko June 20, 1967 (Alcorn MS).

Ryser (1967) reported them at Lamoille Canyon, Ruby Mountains July 1–4, 1967. He (1970) also reported them in the Ruby Mountains July 4, 1970.

Lewis collected a specimen for the USNM at Ruby Lake NWR May 22, 1968.

LaRochelle (MS, 1967) reported Lazuli Buntings were often seen at Ruby Lake NWR in loose groups feeding on dandelion seeds in the lawns. Five were banded in 1965.

Burleigh collected specimens at Reno May 26, July 19 and August 27, 1967.

Mills (PC) reported seeing one at Lahontan Reservoir September 11, 1960.

Alcorn (1946) reported Lazuli Buntings in the Fallon area in June, July and September. Since 1946, they have been seen in Fallon as early as May 7, 1966 and one young was seen July 26, 1968. They are uncommon in the area and, although one or more has been seen most years, they are not a consistent nester in the Fallon area (Alcorn MS).

About 20 were seen feeding in the Kendrick's back yard in Fallon on May 16, 1983. Colleen Ames saw one May 20, 1984 six miles west–southwest of Fallon (Alcorn MS).

Lazuli Buntings are common along the streams in the Toiyabe Range. They were seen south of Austin at Big Creek and Kingston Canyon in May 1961 and June of 1963 and 1965 (Alcorn MS).

Biale (MS) reported these buntings in the Eureka area from 1964 through 1972 a total of 33 times in May; 9 in June; 4 in July; 6 in August; 10 in September; and twice in October.

Kingery (1987) reported 60 Lazuli Buntings at Dyer during a "cold spell" May 22 and 23.

Johnson (1974) reported, "This species was not recorded in 1939 and 1940 (Miller, 1946) although Fisher (1893) earlier had reported it from the Grapevine Mountains. On May 29 and

30, 1973, singing males were noted in mixed shrubs and pinons in Phinney Canyon at 5600 feet and 6700 feet elevation.''

Gullion et al., (1959) reported, "Uncommon spring and fall migrant in southern Nevada."

Austin and Bradley (1971) reported for Clark County, "Summer resident in woodlands and montane forests. Transient in riparian areas. In woodlands and montane forests it usually occurs from April 18 to August 22. In desert scrub and riparian areas it usually occurs from April 25 to May 26 and August 17 to September 17. Unseasonal occurrences on October 29."

Kingery (1979) reported Las Vegas had 10 Lazuli Buntings October 1, 1978.

Indigo Bunting   *Passerina cyanea*

Distribution. From southeastern Saskatchewan to southern Quebec and New Brunswick; south to northwestern Columbia. Casual in the west to Oregon and California.

Status in Nevada. Possible summer resident in southern Nevada; transient.

Records of Occurrence. Kingery (1980) reported two near Jarbidge June 11.

Lewis collected two specimens, now in the USNM, at the Ruby Lake NWR. A male was taken June 1 and a female June 7, 1967.

Kingery (1980) reported an Indigo Bunting at Ruby Lake September 29, 1979.

Herlan collected a specimen, now in the Nevada State Museum, at Peavine Canyon, 6300 feet elevation, Toiyabe Range on July 15, 1966.

Biale (PC, 1986) reported an Indigo Bunting in the Eureka area from May 20–28, 1981.

Kingery (1983) reported, "Indigo Buntings have become widespread in the region [Mountain West]; the 45 reported included one at Eureka in early May."

Miller (1946) reported one taken June 2, 1940 from the Grapevine Mountains, Nye County.

Richardson (1952) reported, "On June 14, 1951, in a mesquite and atriplex thicket one mile southeast of Fairbanks Spring at the north end of Ash Meadows, Nye County, Nevada, an adult male Indigo Bunting was collected. The bird was shot by Ned K. Johnson after I had followed it for almost an hour. It sang repeatedly, never going more than 100 yards from where it was first observed. The condition of the specimen was excellent. It was the first seen in company with another bird of similar size, but no female bunting was found despite search on succeeding days."

Austin (1969) reported one taken at Corn Creek May 25, 1965 and one from Pine Creek June 21, 1966.

Kingery (1980) reported, "Territorial pairs graced Caliente, Nevada."

Snider (1970–71) reported Indigo Bunting at Corn Creek and the Desert Game Range April 24, 1970 (13); and once on July 12, August 16 and September 12 of 1970.

Kingery (1979) reported six or seven apparently pure pair as being suspected of nesting at Mount Charleston.

Kingery (1982) reported, "The Nevada Indigo Bunting was a male which sang May 7 to June 27 without a female; Mowbray feels they probably nested in past years though not this year."

Lawson (MS, 1981) reported, "These buntings are uncommon to common in southern Nevada."

Painted Bunting   *Passerina ciris*

Distribution. From New Mexico to North Carolina; south to Cuba and Panama.

Status in Nevada. Hypothetical.

Records of Occurrence. Lawson (MS, 1981) reported one immature bird seen at the Corn Creek Field Station, Clark County October 8–9, 1974 by the Lawsons and Mowbray.
Kingery (1982) reported a Painted Bunting stayed at Las Vegas October 27–31, 1981.

Dickcissel   *Spiza americana*

Distribution. From eastern Montana to Ontario; from Massachusetts south to Columbia and Venezuela. Casual in the west to California.

Status in Nevada. A straggler to southern Nevada.

Records of Occurrence. Pulich and Gullion (1953) reported, "Mrs. Nora Poyser first recorded a flock of about six Dickcissels at her residence near the Boulder Boat Dock on the shores of Lake Mead September 18, 1949. These birds remained around her house for several days. On September 6, 1951, she again reported the presence of six Dickcissels at her residence. She had first seen this flock the day before. One bird was trapped, banded and carefully examined in hand by Pulich. This small flock of birds remained around the banding station until September 8, but unfortunately, no others were captured. These two occurrences mark the first report of the Dickcissel in Nevada."

Austin and Bradley (1968) reported Austin had observed an adult male in his back yard in Las Vegas May 24, 1964.

Lawson (MS) reported Mowbray recorded Dickcissels at Corn Creek January 3, 1970 and September 17, 1977.

Cruickshank (1971) reported them from the Desert Game Range January 3, 1971 during the CBC.

Kingery (1984) reported a Dickcissel at Las Vegas May 25–29. He (1987) reported, "Las Vegas saw its first Dickcissel since 1984 on May 17."

Green–tailed Towhee   *Pipilo chlorurus*

Distribution. From Oregon to Montana; south in Mexico to Morelos.

Status in Nevada. Summer resident in most of the mountain ranges.

Records of Occurrence. Linsdale (1936) reported these birds from 13 of 17 Nevada counties. "A great majority of the Green–tailed Towhees in Nevada live in sagebrush, but where they are available, usually on the higher slopes, other types of shrubby vegetation are inhabited."

Sheldon NWR (1982) reported the Green–tailed Towhee as an uncommon to common summer resident from spring through fall.

Lewis collected one female at the Ruby Lake NWR May 20, 1967.

Burleigh collected four specimens for the USNM in 1967 and 1968 from the Reno and North Lake Tahoe areas in May, July, August and September.

Heilbrun and Kaufman (1978) reported one in the Truckee Meadows during the CBC December 17, 1977.

Ryser (1968) reported many Green–tailed Towhees present in western Nevada in the 1968 season.

They were observed in the Carson Range at Big Meadows August 1, 1962; at Hunters Lake, September 12, 1964; and at Virginia City May 8, 1962 (Alcorn MS).

Single birds were seen in the Fallon area May 10 and 12, 1963 and May 8, 1965 (Alcorn MS).

In the Desatoya Range, single Green–tailed Towhees were seen on Carroll Summit September 27, 1963 and at Campbell Creek, Lander County August 17, 1967 (Alcorn MS).

Green–tailed Towhees were observed in the Austin area in May and June from 1961 to 1966. South of Austin in the Toiyabe Mountains, on Big Creek one was seen September 6, 1970 and at Reese River several were seen September 22, 1963 (Alcorn MS).

One was seen in the Stoneburger Basin, Toquima Range September 8, 1963 and one in the Monitor Range at Dobbin Summit, Nye County August 28, 1960 (Alcorn MS).

Biale (MS) recorded these birds in the Eureka area between April 26, 1964 and September 17, 1967: on April 26, 1964; 22 times in May; 18 in June; 12 in July; 15 in August; and 17 times in September. They were not seen from October to March.

Biale (MS) reported a nest about 15 inches above ground level in a black sagebrush on July 16, 1967 (elev. 7600 feet). The nest contained four creamy white eggs with brown flecks. The nest was make of cheat grass and lined with horse hair. Biale reported another nest in a snowberry bush five miles south of Eureka on June 23, 1984. The nest was 12 inches above the ground and contained four eggs (elev. 7800 feet).

A Green–tailed Towhee was seen August 16, 1961 at the site of the old mining town of Hamilton, White Pine County (Alcorn MS).

Lehman Caves National Monument personnel reported one August 26, 1957.

A. A. Alcorn (MS) reported one at Trough Spring, Spring Mountains on July 10, 1963. Single birds were observed in the Moapa Valley only on three occasions on April 22 and 29, 1964 and on May 6, 1964.

Austin and Bradley (1971) reported for Clark County, "Summer resident in woodlands and montane forests. Transient in riparian areas. Unseasonal occurrences on November 27 and December 30 and 31. In woodlands and montane forest it usually occurs from April 18 to September 30. In riparian areas it usually occurs from April 1 to May 11 and from September 5 to the 30th."

Kingery (1979) reported the first spring arrival of a Green–tailed Towhee at Las Vegas April 21.

Rufous–sided Towhee   *Pipilo erythrophthalmus*

Distribution. From southern British Columbia to southwestern Maine; south to Guatemala.

Status in Nevada. Summer resident in most of the mountain ranges and higher valleys. Winter resident in western and southern Nevada. Linsdale (1951) reported two races from Nevada: *P. e. curtatus* and *P. e. montanus*.

Rufous-sided Towhee. Photo by A. A. Alcorn.

Records of Occurrence. Linsdale (1936) reported these birds from 12 of 17 Nevada counties. Recent records represent 14 counties.

Sheldon NWR (1982) reported the Rufous–sided Towhee as an uncommon spring and fall migrant.

In Humboldt County one Rufous–sided Towhee was seen at Lamance Creek, Santa Rosa Range on June 10, 1970 and three were seen at Quinn River Crossing April 3, 1962 (Alcorn MS).

The author has seen Rufous–sided Towhees over a wide area in Elko County, including Jarbidge, Midas, Tuscarora, Jiggs and Ruby Valley. They were seen in spring, summer and fall months, with usually only one or two birds seen at a time.

LaRochelle (MS, 1967) reported just one observation at Ruby Lake NWR—a lone bird seen in May 1964.

Orr and Moffitt (1971) for Lake Tahoe reported, "Migrant in small numbers and occasional winter visitant."

In Reno single birds were seen November 13 and 24, 1941 and on March 8, 1942 (Alcorn MS).

Fourteen specimens in the USNM were taken in all months except April, May, September and November from Reno and the nearby Carson Range area between 1967 and 1969.

In the Fallon area, single birds have been seen in the months of February, March, April, September, October and December (Alcorn MS).

In the Austin area one was seen February 27, 1962; several seen November 6, 1964; over

10 seen at Porter Canyon on October 12, 1964; one at Kingston Canyon April 27, 1961; and one at the Testolin Ranch on the Reese River May 17, 1961 (Alcorn MS).

Biale (MS) observed Rufous–sided Towhees in the Eureka area 70 times in all months between July 12, 1964 and October 8, 1967. They were seen 14 times in January; 18 in February; twice in March; 3 in April; 6 in May; 3 in June; 6 in July; 4 in August; 10 in September; and 4 in October, November and December. He (MS) reported a nest with four eggs on the ground under a black sage in the Diamond Mountains June 23, 1968. The nest material was sage bark, lined with cheat grass (elev. 8300 feet).

Gabrielson (1949) reported, "A nest containing four nearly fledged young was found on Lehman Creek, White Pine County, May 27, 1932."

At the Lehman Caves and Baker areas, from one to three were seen in January, March, April, May, June and July (Alcorn MS).

J. R. and A. A. Alcorn (MS) reported the following records in Clark County: one at Gold Butte October 30, 1960; one at Cold Creek Field Station, Spring Mountains July 24, 1962; and two at Rosebud Springs, Spring Mountains July 16, 1963.

Austin and Bradley (1971) reported, "Permanent resident in woodlands. Transient in montane forests. Occurs all year. Winter resident in desert scrub and riparian areas."

Abert's Towhee   *Pipilo aberti*

Distribution. From southeastern California to New Mexico; south in Mexico to Sonora.

Status in Nevada. Resident in the southern part of Nevada.

Records of Occurrence. J. R. and A. A. Alcorn (MS) recorded these birds 81 times in the Moapa Valley from October 1960 to January 3, 1967. They were seen 9 times in January; 8 in February; 13 in March; 5 in April; 11 in May; 8 in June; 9 in July; 4 in August; none in September; 4 in October and November; and 3 times in December.

Ryser (1967) reported one at Paradise Valley near Las Vegas February 15–19, 1967.

Austin and Bradley (1971) for Clark County reported, "A permanent resident in riparian areas."

Cassin's Sparrow   *Aimophila cassinii*

Distribution. From southeastern Arizona to Kansas; south in Mexico to Guanajuato.

Status in Nevada. Accidental. Funk (1969) reported, "The Cassin's is found occasionally on dry flats among low brush and short grass in eastern Nevada."

Records of Occurrence. Linsdale (1936) reported, "An adult male shot on May 26, 1891, by Merriam in Timpahute Valley, Lincoln County was, 'in worn breeding plumage' and attracted his attention by flying up from the desert brush and singing in the air (Fisher, 1893)."

Rufous–crowned Sparrow   *Aimophila ruficeps*

Distribution. From central California to southwestern New Mexico, Colorado and Oklahoma. In Mexico to Oaxaca and Veracruz.

Status in Nevada. Rare.

Records of Occurrence. Single birds were reported by Lehman Caves National Monument personnel on May 10, 1958; June 1, 1959; and July 13, 1961 (Alcorn MS).

Lawson (MS, 1981) reported five records for Nevada: On May 30, 1966, in the Spring Mountains with George Austin, Lawson observed one Rufous–crowned Sparrow; on April 8, 1973 at Tule Springs Park he saw one bird; on April 10, 1973 at Fort Mohave below Davis Dam, two were seen; and on May 3 and 4, 1973, at Corn Creek Field Station, Lawson saw one each day.

Kingery (1981) reported, "A Rufous–crowned Sparrow at Las Vegas on May 6 was the first record since 1973."

American Tree Sparrow   *Spizella arborea*

Distribution. From Alaska to Nova Scotia; south to California, Texas and North Carolina.

Status in Nevada. An uncommon winter visitor. Funk (1969) reported, "A rare bird in Nevada, the Tree Sparrow is a winter visitor to willow thickets, fence rows and open weed fields."

Records of Occurrence. Linsdale (1936) reported, "A winter visitant; probably of more regular occurrence than the few records indicate. At Truckee Meadows, Ridgway (1877) collected an adult male on November 19, 1867. At Pahrump Ranch, Clark County, E. W. Nelson saw 'quite a number' of tree sparrows the latter part of February, 1891. The birds appeared quite suddenly one morning before a storm which filled the valley with rain and the mountains with snow, (Fisher, 1893)."

Lewis collected a female specimen at Ruby Lake NWR November 6, 1967.

Heilbrun and Rosenburg (1983) reported eight tree sparrows during the Unionville CBC January 1, 1983.

Alcorn (1940) reported a specimen collected by Earl J. Alcorn, four miles west of Fallon November 25, 1939. None have been seen since then in the Fallon area.

Biale (MS) recorded one at Eureka on November 17, 1971, January 30, 1979 and one in Diamond Valley November 25, 1984.

Heilbrun and Rosenburg (1982) reported six American Tree Sparrows in the Ely area December 19, 1981 during the CBC.

Gullion et al., (1959) reported, "One additional record: 'a number' of birds in the Eldorado Mountains, five miles northwest of Nelson (3000 feet elev.), on March 13, 1951."

Austin and Bradley (1971) reported, "Transient in riparian areas and usually occurred from September 20 to April 18."

Kingery (1979) reported one Tree Sparrow at Las Vegas January 22.

Lawson (MS, 1981) reported the Tree Sparrow as rare in southern Nevada. Usually seen every fall in late October and November.

Chipping Sparrow   *Spizella passerina*

Distribution. From Alaska to Newfoundland; south to Central America (Nicaragua).

Status in Nevada. Summer resident in most of the mountain ranges in the state. Winter resident in the southern part. Funk (1969) reported, "Distributed throughout the state in the summer, this common sparrow is often seen under scattered coniferous trees."

Records of Occurrence. Linsdale (1936) reported Chipping Sparrows from all sections of the state with records from 11 Nevada counties.

Sheldon NWR (1982) reported the Chipping Sparrow as an uncommon spring, summer and fall migrant.

Ryser (1973) reported many of these birds in the Pine Forest Mountains, Humboldt County in early August of 1973.

One was observed April 14, 1971 near Jiggs, Elko County (Alcorn MS).

LaRochelle (MS, 1967) reported only nine sightings at Ruby Lake NWR and commented that the Chipping Sparrow is not often seen.

Lewis collected a male specimen at Ruby Lake NWR September 20, 1966.

Burleigh collected 11 specimens for the USNM from the Reno area and nearby Carson Range from 1966 to 1969. The earliest specimen was taken April 26 and others were obtained in May, June, July and August.

What was suspected to be a mated pair was seen on Mount Rose June 16, 1960 (Alcorn MS).

Ryser (1966) reported these sparrows as very abundant in Little Valley in the summer of 1966 and he (1970) reported large flocks seen on May 11, 1970 along Franktown Road in Washoe Valley.

At a feed station near Yerington, Chipping Sparrows were seen April 29, May 10 and May 14, 1983 (Alcorn MS).

Alcorn (1946) in the Fallon area reported specimens taken July 30 and September 21, 1941. About 10 were seen in a loose flock in this area May 8, 1961. Two were seen in Fallon April 29, 1982 (Alcorn MS).

Biale (MS) recorded this bird 142 times in the Eureka area from 1964 to 1972, inclusive. The earliest and only April record was April 27, 1969. They were seen 33 times in May; 20 in June; 18 in July; 42 in August; 26 in September; and twice in October. Nesting in the Eureka area was observed June 18, 1967 at Mahogany Hills. The nest was in a sagebrush two and one–half feet above ground and was made of grass fibers and lined with horse hair. The nest contained two blue, brown–spotted eggs (elev. approx. 7200 feet). On July 16, 1967 a nest 30 inches above ground was found in a black sage. The nest, made of plant fibers and lined with horse hair, contained two eggs that were blue with brownish spots (elev. approx. 7700 feet).

A. A. Alcorn (MS) reported one seen near Baker May 19, 1963.

Biale (MS) reported Chipping Sparrows from White Pine Mountains October 15, 1972.

In the Hiko area, 10 or more of these sparrows were seen April 23, 1965; about 10 were seen May 7, 1965; and a loose flock of about 25 was seen May 14, 1965. All were thought to be migrating (Alcorn MS).

J. R. and A. A. Alcorn (MS) reported two in the Moapa Valley, southern Nevada April 29, 1963 and 10 or more on April 22, 1964.

A. A. Alcorn (MS) recorded one at Rosebud Springs July 16, 1963; one at Lee Canyon July 18, 1963; and two at Trough Spring July 19, 1963—all in the Spring Mountains.

Gullion et al., (1959) reported, "Recorded in the spring from April 25 (1951, Mormon Mountains) to May 19 (1951, Pahrump Valley) and in the fall from August 22 (1951, McCullough Range) to November 3 (1951, Las Vegas Wash). One winter record: one bird, Boulder City, January 13, 1952."

Austin and Bradley (1971) for Clark County reported, "A summer resident in woodlands and montane forests. Winter resident in riparian areas. Transient in desert scrub. In woodlands and montane forests it usually occurs from April 13 to October 24. In desert scrub and riparian areas it usually occurs from August 12 to May 17."

Clay–colored Sparrow   *Spizella pallida*

Distribution. From the Mackenzie to New York, including parts of Illinois and Michigan; south in winter to Mexico.

Status in Nevada. Rare.

Records of Occurrence. Kingery (1978) reported one at Pahranagat September 23, 1977 and he (1984) reported one at Las Vegas May 13.

Lawson (MS, 1981) reported in southern Nevada in recent years individuals have been seen each fall at Pahranagat and Corn Creek, but he considers them rare. He reported one seen by Mowbray August 27, 1972 at Tule Springs Park and Lawson saw one October 9 and November 1, 1972 at the Corn Creek Field Station.

Brewer's Sparrow   *Spizella breweri*

Distribution. From Yukon to Saskatchewan; south in Mexico to Jalisco.

Status in Nevada. Widespread summer resident over the entire state. Funk (1969) reported, "A common little sparrow in the sagebrush and scrubs, the Brewer's Sparrow can be seen in the summertime throughout the state."

Linsdale (1951) reported, "One of the most abundant birds in the state. Lives in sagebrush at all altitudes both in valleys and on the mountains."

Records of Occurrence. Sheldon NWR (1982) reported the Brewer's Sparrow as an abundant breeding summer resident; common in spring and fall.

Ryser (1973) reported many were seen in early August, 1973 at Blue Lake in the Pine Forest Mountains of Humboldt County.

One was seen near Jarbidge September 4, 1960 (Alcorn MS).

LaRochelle (MS, 1967) saw one at Ruby Lake NWR April 9, 1966 and Lewis collected a male at this same refuge June 24, 1967.

One was seen at Wrights Canyon, Pershing Range on May 13, 1970 (Alcorn MS).

Burleigh collected a female June 8, 1967 at Virginia City; a female at Reno August 1, 1967; an immature male north of Lake Tahoe (9000 feet elev.), September 3, 1967; and a male at Galena Creek August 4, 1968.

Alcorn (1946) reported specimens taken in the Fallon area August 22 and September 4, 1941. Nine were seen in this area August 19, 1960; one was seen each day on May 7 and 24, 1965; and one was singing in an elm tree on May 13, 1970. It apparently does not nest in this area and was only seen in spring and late summer migrations. In the eastern part of Churchill County about 10 of these birds were seen in Buffalo Canyon May 13, 1963 and one was seen at the Alpine Ranch May 2, 1971 (Alcorn MS).

One was seen at Big Creek south of Austin September 6, 1970 and one was seen at Clear Creek on the east side of the Monitor Range June 6, 1968 (Alcorn MS).

Biale (MS) reported seeing these birds in the Eureka area from 1964 to 1973 inclusive, 5 times in April (earliest April 9); 19 in May; 13 in June; 7 in July; 19 in August; and 15 times in September (latest September 24).

Biale (MS) reported a nest June 14, 1964 contained four eggs. Another nest, partially constructed June 13, 1965, contained three eggs on June 21. It was located two feet above ground in a sagebrush. Another nest seen July 6, 1969 at Cottonwood Ranch on the west slope of the Diamond Range was in black sage three feet above the ground and contained one egg. It was made of fine grass and lined with horse hair (elev. approx. 6500 feet). On June 7, 1970 at Three Bar area, a nest in a black sage, one foot above ground was made of dry grass and stems, lined with horse hair. It contained three blue–green, brown spotted eggs (elev. 6500 feet).

Heilbrun and Rosenburg (1982) reported Brewer's Sparrows in the Ely area during the CBC December 19, 1981.

In the Pahranagat Valley a group of 10 or more Brewer's Sparrows was seen April 23, 1964 and another group was seen May 7, 1965 (Alcorn MS).

Gullion et al., (1959) reported, "The Brewer Sparrow is a regular and common spring and fall migrant. This species is abundant in the desert in spring from about March 24 (1952, Boulder City) to May 26 (1951, Meadow Valley) and in the fall from about the end of August to mid–October. Mrs. Poyser reported this species at Boulder Beach on February 27, 1951 and Pulich saw fledglings (and collected one) just out of the nest at Potosi Spring on June 28, 1952."

Austin and Bradley (1971) reported, "A summer resident in woodlands. Winter resident in riparian areas. Transient in desert scrub. In woodlands it usually occurs from April 18 to September 30. In desert scrub and riparian areas it usually occurs from February 1 to May 20 and from August 20 to October 10. Unseasonal occurrences on November 19 and December 22 to 30."

Field Sparrow   *Spizella pusilla*

Distribution. From southeastern Montana to southern New Brunswick; south to the Gulf States to Tamaulipas.

Status in Nevada. Hypothetical.

Records of Occurrence. Lawson (MS, 1981) reported one immature bird was seen at Tule Springs Park September 14, 1968 by him with Mowbray and Dr. G. Meade.

Black–chinned Sparrow   *Spizella atrogularis*

Distribution. From central California to Texas; south in Mexico to Oaxaca.

Status in Nevada. A scarce summer resident in southern Nevada. Funk (1969) reported, "This bird may be seen in southern Nevada in the summer."

Records of Occurrence. Linsdale (1936) reported, "Summer resident in southern part of the state; scarce. A single specimen was collected on June 15, 1932 by Sheldon on Trout Creek, one mile above Williams Ranch, Clark County."

Austin and Bradley (1971) reported for Clark County, "Summer resident in woodlands. Transient in riparian areas. Usually occurs from April 13 to August 30."

Lawson (MS, 1981) reported for southern Nevada, "Well established breeding colonies in

Cabin Canyon, Virgin Mountains; Lower Kyle Canyon and vicinity of Lovell Canyon Road and Red Rock Summit Road, Spring Range.''

Vesper Sparrow   *Pooecetes gramineus*

Distribution. From British Columbia to Nova Scotia; south into Mexico (Oaxaca).

Status in Nevada. Uncommon summer resident and transient. Funk (1969) reported, ''Uncommon statewide, the bird can be seen along roadsides, in fields, and in surrounding sagebrush near agricultural areas.''

Linsdale (1936) reported this bird from eight of 17 Nevada counties, covering all parts of the state.

Records of Occurrence. Sheldon NWR (1982) reported the Vesper Sparrow as a common breeding summer resident; uncommon spring and fall.

Ryser (1973) reported many were seen in the Pine Forest Mountains, Humboldt County in early August of 1973.

In the area between Midas and Tuscarora, Elko County, one was observed July 11, 1961 and two others were seen September 25, 1966 (Alcorn MS).

In the Elko area, these birds were seen from May through August (Alcorn MS).

Near Halleck, Page (PC) reported a nest with four eggs seen in June of 1965.

LaRochelle (MS, 1967) reported for Ruby Lake NWR, ''A very common sparrow generally seen flitting about in sagebrush.''

A male specimen in the USNM was taken at Ruby Lake NWR on May 14, 1967.

In the Ruby Valley, about 10 Vesper Sparrows were seen March 23, 1966. The latest fall record for this area was of one seen September 29, 1965 (Alcorn MS).

Burleigh collected specimens from Reno and north of Lake Tahoe in May, June and July of 1967–69.

Ryser (1985) noted, ''I first detected it nesting in the subalpine meadows in the Carson Range when I inadvertently caught fledglings in my small–mammal snap–trap lines.''

Abut 10 Vesper Sparrows were seen at Barrett Creek, Shoshone Mountains on September 22, 1965 and over 10 were seen in one group at Potts in Monitor Valley August 26, 1966 (Alcorn MS).

Biale (MS) recorded these birds in the Eureka area in summer months from 1964 to 1973. He saw them twice in March; 10 times in April; 12 in May; 11 in June; 5 in July; 10 in August; 16 in September; and once in October. The earliest spring date was March 27 and the latest fall date was October 4.

Biale (MS) reported a nest with three gray–green spotted eggs seen in the west side of the Diamond Range north of Eureka (elev. 7100 feet). The nest was on the ground under a sage on July 12, 1975.

Biale (MS) also reported this bird in the Duck Creek Basin, White Pine County September 16, 1973.

A. A. Alcorn (MS) saw two of these sparrows in the Virgin Valley March 26, 1963 and in the Moapa Valley he saw one January 28, 1963 and two on September 26, 1963.

Austin and Bradley (1971) reported for Clark County, ''Transient in riparian areas. Usually occurs from March 7 to May 1 and from September 11 to October 11. Unseasonal occurrence on November 18 and on December 22 and 28.''

Kingery (1976) reported two at Corn Creek January 2 through February 12. He (1980) also reported two Vesper Sparrows arrived at Las Vegas February 24.

Lark Sparrow   *Chondestes grammacus*

Distribution. From southern British Columbia to Ontario; south to El Salvador.

Status in Nevada. Summer resident and transient. Limited occurrence in winter in southern Nevada. Funk (1969) reported, "This distinct sparrow is uncommon, but may be seen in the summer in sagebrush areas throughout the state."

Linsdale (1936) reported these birds from nine of 17 Nevada counties. Recent records are available for 12 counties.

Linsdale (1951) reported, "Occurs over most of the state, but not continuously distributed; lingers late in fall and possibly remains through the winter in some sections."

Records of Occurrence. Sheldon NWR (1982) reported the Lark Sparrow as an occasional spring and summer migrant.

In the Tuscarora area from one to six Lark Sparrows were observed at five places July 11, 1966 and about 10 were seen each day on August 22 and 23, 1965 in this same area (Alcorn MS).

In the Elko–Halleck area, Lark Sparrows were observed in April, May and June from 1965 to 1968 (Alcorn MS).

LaRochelle (MS, 1967) reported for Ruby Lake NWR, "A not uncommon sparrow generally in grassy meadows and sage."

Lewis collected seven specimens, now in the USNM, at this refuge from May 14, 1967 to May 7, 1969.

In Ruby Valley on May 27, 1965 about 10 were seen and at this same location others were seen May 19, 1966 (Alcorn MS).

Lark Sparrows are seldom seen in the Fallon area. Alcorn (1946) reported two from the Fallon area May 3, 1941. Records for the Lahontan Valley from 1960 to 1967 reveal five seen April 3, 1962; from one to 10 were recorded in May of five different years; and one was seen June 24, 1965 and August 3, 1960.

Biale (MS) recorded this sparrow 131 times in the Eureka area between April of 1964 and September 3, 1967. From 1964 to September 1973, the earliest record was April 28 (1969) and the latest was October 4 (1972). During this 1964 to 1973 period, Biale saw them 8 times in April; 104 in May; 22 in June; 23 in July; 26 in August; 15 in September; and once in October. More Lark Sparrows were seen in May than in all other months combined, indicating a major migration through the Eureka area during this month.

On July 5, 1964 Biale saw three young in a nest on the ground near a small sagebrush in an old burn area. He (PC, 1986) reported another nest on the ground contained four feathered young on June 21, 1985.

In the Hiko area of Pahranagat Valley, about 10 were seen April 22, 1961; about 10 on May 6, 1965; about 30 on May 14, 1965; and about 10 on May 15, 1965. These were probably all migratory birds (Alcorn MS).

In the Moapa Valley from 1961 to 1965, Lark Sparrows were seen 5 times in April; 6 in May; and once in June, July, August and September (Alcorn MS).

Gullion et al., (1959) reported, "A spring and fall migrant, being noted from about April 30

(1951, Las Vegas Valley) to May 24 (1951, Pahranagat and Meadow Valleys) in the spring and from August 8 (1953, Meadow Valley) to about September 10 (1951, Boulder Beach) in the fall.''

Austin and Bradley (1971) reported for Clark County, ''This bird is transient in desert scrub, riparian, woodland and montane forest areas. Usually occurs from April 21 to June 3 and from August 13 to October 11. Unseasonal occurrences on December 22.''

Black–throated Sparrow   *Amphispiza bilineata*

Distribution. From Oregon to Wyoming; south to Jalisco, Mexico.

Status in Nevada. Summer resident over entire state, winter resident in southern portion. Funk (1969) reported, ''This common sparrow can be seen in the sagebrush areas of the state as well as in cactus and mesquite.''

Linsdale (1936) reported the Black–throated Sparrow generally occurs over the lower valleys of the state, but is not recorded above 5000 feet elevation in the northern part. Their widespread occurrence is indicated by Linsdale's recording them from 12 of the 17 Nevada counties.

The author's and others records available for the Black–throated Sparrow for the past 30 years either do not adequately reveal the occurrence of this bird in Nevada, or a major decline in the population has occurred.

Records of Occurrence. Sheldon NWR (1982) reported the Black–throated Sparrow as a common breeding summer resident; uncommon spring, fall and winter.

LaRochelle (MS, 1967) reported at Ruby Lake NWR Black–throated Sparrows are regularly seen amongst sparse vegetation. ''I have seen several in the Sand Dunes area east side of the Refuge.''

One was seen at the mouth of Wrights Canyon, Humboldt Range on May 13, 1970 (Alcorn MS).

In the Fallon area one was seen May 9, 1962 and another June 12, 1971 (Alcorn MS).

Biale (PC, 1986) reported these birds are becoming more common in the Eureka area during summer months. His earliest record is of one seen April 27, 1970 when eight inches of snow was on the ground. On May 23, 1971 Biale found three dead from a storm. In 1972 he reported them on May 7 and 21. Biale also has four June and two July records.

On July 9, 1967 Biale (MS) observed a nest containing four young. The nest was in a greasewood *(Sarcobatus)* bush about 12 inches above ground. It was made of grass fibers and lined with sheep wool. Both the male and female were feeding young. Elevation was about 6350 feet, considerably higher than that quoted by Linsdale.

Biale (MS) reported another nest at Ruby Hill Mine area on June 24, 1984. The nest was located in a small sage about six inches above the ground and contained three white eggs (elev. 6900 feet).

A. A. Alcorn (MS) reported seeing single birds near Baker April 19 and 29, 1967.

One was observed near Tonopah May 6, 1962 (Alcorn MS).

Gullion et al., (1959) reported for southern Nevada, ''A common spring, summer and fall resident, being absent from Nevada's deserts for not over two months in mid–winter, if that long . . . .Pulich found a nest containing three eggs in a sagebrush near Potosi Spring June 28, 1952.''

On Mormon Mesa, six of these birds were seen April 4, 1960. Two were carrying nest materials; one to a nearly complete nest (Alcorn MS).

A. A. Alcorn (MS) reported a Black–throated Sparrow on the northwest side of the Virgin Mountains, April 4, 1960 and two at Sheep Springs, Virgin Mountains June 13, 1961.

Austin and Bradley (1971) reported for Clark County, "A permanent resident in desert scrub. Summer resident in woodlands. Winter resident in riparian areas. Usually occurs from February 6 to December 31."

Kingery (1977) reported Black–throated Sparrows began migrating in southern Nevada in February. Heilbrun and Rosenburg (1981) reported three in the Henderson area December 20, 1980.

Sage Sparrow   *Amphispiza belli*

Distribution. From central Washington to Colorado; south in Mexico to Chihuahua.

Status in Nevada. Summer resident in valleys over the entire state and winter resident mainly in the southern part. Linsdale (1951) reported two races from Nevada: *A. b. nevadensis* and *A. b. canescens*.

Funk (1969) reported, "Found throughout the state in sagebrush areas."

Linsdale (1936) reported Sage Sparrows from 12 of 17 Nevada counties.

Records of Occurrence. Sheldon NWR (1982) reported the Sage Sparrow as a common breeding summer resident; uncommon spring, fall and winter.

LaRochelle (MS, 1967) reported only one positive record for the Ruby Lake NWR—July 9, 1965.

Heilbrun and Kaufman (1979) reported one Sage Sparrow was seen at Ruby Lake NWR during the CBC.

In Pershing County near Lovelock, three were seen February 17, 1965. Two were observed near Nightingale October 11, 1961 (Alcorn MS).

Kingery (1977) reported Sage Sparrows had reached Lovelock February 5 and 30 were in Reno February 25.

Burleigh collected five specimens for the USNM near Reno in April, May, July, August and September from 1967 to 1969.

In Mason Valley (Yerington area) two were seen each day on January 7, 9 and 11, 1961 (Alcorn MS).

From 1960 through 1968 in the Fallon area, these birds were seen a total of 4 times in January; once in February, March, April and June; twice in August; once in October; and 4 times in December. In 1971, Sage Sparrows were observed west of Fallon on February 8, July 10, September 8 and December 26 (Alcorn MS).

In Churchill County, two were seen near Eastgate January 8, 1970 and four were seen in Buffalo Canyon April 27, 1962 (Alcorn MS).

In the Austin area, one was observed August 29, 1961 and near the Reese River, Nye County, one was seen June 12, 1963 (Alcorn MS).

Biale (MS) reported these birds in the Eureka area from 1965 through 1967 a total of twice in February; 6 times in March; 3 in May; once in June; twice in July; and once in September and December. From 1970 through 1973, Biale saw them 5 times in February; 7 in March; 9

in April; 11 in May; 10 in June; twice in July; 5 in August; 6 in September; 3 in October; and once in November and December.

Biale (PC, 1986) reported at Kobeh Valley, 12 miles west of Eureka, an unfinished Sage Sparrow nest was on the ground under a black sage May 12, 1985.

In the Pahranagat Valley one was observed March 6, 1961 and another on July 9, 1963 (Alcorn MS).

One was recorded from near Springdale north of Beatty December 13, 1959 (Alcorn MS).

In the Moapa Valley one was recorded March 14, 1962 and one February 6, 1963 (Alcorn MS).

Gullion et al., (1959) reported, "A winter visitant on the desert, being recorded from as early in the fall as August 5 (1954, Pahrump Valley) and remaining in the spring as late as June 28 (1954, Mormon Mountains). However, this species was recorded from the Indian Spring area northwest of Las Vegas by van Rossem (1936) in July, so perhaps it is best considered a permanent resident, at least on the higher desert areas. A bird collected along the Davis Dam Highway (Dead Mountains) by Pulich on November 30, 1951, was identified as *A. b. nevadensis* by A. R. Phillips."

Austin and Bradley (1971) reported for Clark County, "Occurs all year. Summer resident in woodlands. Winter resident in desert scrub and riparian areas. Most abundant from September 16 to March 31."

Lark Bunting   *Calamospiza melanocorys*

Distribution. From southern Alberta to Manitoba; south in Mexico to Hidalgo.

Status in Nevada. Irregular visitor mostly to southern Nevada.

Records of Occurrence. Linsdale (1951) reported, "Irregular winter visitant; recorded in April and May from Humboldt, White Pine and Clark Counties. Jewett (1940) found this bird common in southern Clark County in early February, 1940. On December 25, 1948, W. Pulich saw about 40 in a flock on the shore of Lake Mead (Monson, 1949)."

Ryser (1985) reported the Lark Bunting from Washoe County, but gives no specifics.

Kingery (1983) reported a Lark Bunting at Gardnerville February 27.

Billy Meck (PC) saw one May 20, 1967 in the Stoneburger Basin near the head of Wilson Creek, Toquima Range, Nye County.

A male was seen 20 miles west of Eureka May 14, 1964 (Alcorn MS). One was observed by A. A. Alcorn (MS) in the Eureka area June 13, 1967.

Gullion et al., (1959) reported, "Erratic occurrence in winter and spring. From about May 2 to 14, 1952, flocks were recorded on the desert from Searchlight area, Las Vegas Valley, Coyote Wells (2300 feet elev., Lincoln County), Pahranagat Valley, Delamar Flat (5000 feet elev., Lincoln County), and Meadow Valley. Gullion collected an adult male at 3000 feet elev., 16 miles north of Searchlight on May 3, 1952. This species has been previously recorded on Nevada's southern deserts in April, 1891 by Fisher (1893), in February, 1940 by Jewett (1940b) and in December, 1948 by Pulich (Monson, 1949)."

Monson (1972) reported one at Mormon Farm, Las Vegas September 6, 1971.

Austin and Bradley (1971) reported for Clark County, "Transient in desert scrub in riparian areas. Occurs in February, April, May and December."

Kingery (1983) reported a Lark Bunting at Las Vegas December 4.

Savannah Sparrow   *Passerculus sandwichensis*

Distribution. From Alaska to Labrador; south to El Salvador.

Status in Nevada. Linsdale (1936) reported this sparrow from 12 of 17 Nevada counties.

The Savannah Sparrow is a common resident of meadows and grassy marsh areas, mostly in the valleys. Most move out of the colder northeastern part of Nevada in winter and those in southern Nevada move out in the summer.

Linsdale (1936) reported two races from Nevada: *P. s. alaudinus* and *P. s. nevadensis*.

Funk (1969) reported, "This bird of the field is common throughout the state in summer as well as winter."

Records of Occurrence. Sheldon NWR (1982) reported the Savannah Sparrow as an occasional spring and summer resident, nesting in the area.

LaRochelle (MS, 1967) reported for Ruby Lake NWR, "Found year round. Several winter along water inlet and collection ditch."

This author's own records for this bird in the Ruby Valley and Elko area indicate its presence in summer months from April through August (Alcorn MS).

Records for the Lovelock, Reno, Minden, Yerington and Fallon areas indicate its abundance (especially in summer months) in western Nevada (Alcorn MS).

Burleigh collected specimens in the Reno areas from 1967 to 1969. Savannah Sparrows were collected in January, March, April, September, October and November, indicating the year round occurrence of this sparrow in the Reno area.

In the Fallon area from April 4, 1960 to October 3, 1967, this sparrow was seen once in January, February and March; 3 times in April; 5 in May; 4 in June and July; and twice in October. These records do not indicate the abundance of this bird in this area as suitable habitat was not regularly censused. However, it does indicate the presence of Savannah Sparrows in all seasons (Alcorn MS).

Schwabenland (MS, 1966) reported at Stillwater WMA the Savannah Sparrow is found in good numbers throughout the year.

In the Eureka area from 1964 to 1973, the earliest spring record was March 27. The latest fall record was October 28. During this period, Biale recorded them twice in March; 6 times in April; 12 in May; twice in June; 10 in July; and once in August, September and October (Alcorn MS).

Kingery (1981) reported a late Savannah Sparrow at Eureka November 7, 1980.

Biale (MS) reported a Savannah Sparrow nest at the Hay Ranch west of Eureka May 16, 1969. The nest was a cup on the ground, partly under a cow chip and was made of grass, lined with very fine grass. Five bluish–white, brown–spotted eggs were in the nest (elev. 6000 feet). At the Hay Ranch on May 18, 1969, another nest was on the ground slightly under the south side of a cow chip. The nest was made of grass and contained five bluish–white eggs splotched with reddish–brown (elev. 6000 feet).

At the same location Biale found another nest May 29, 1969 in a moist meadow under a cow chip. It was made of grass and lined with fine grass. It contained two bluish–white eggs that were lightly splotched brown (elev. 6000 feet).

A Savannah Sparrow was seen at Comin's Lake southeast of Ely June 5, 1962 (Alcorn MS).

In the Pahranagat Valley these sparrows were observed in March and May of 1965 (Alcorn MS).

In the Moapa Valley about 10 were seen March 5, 1962; two on March 12, 1962; three on May 4, 1962; two on May 5, 1962; two on February 23 and 24, 1963; and two on April 22 and 23, 1964 (Alcorn MS).

One Savannah Sparrow was seen in the Pahrump Valley October 31, 1961 (Alcorn MS).

Gullion et al., (1959) reported, "Six records scattered through the year as follows: several birds, Pahrump Valley, August 25, 1951; one bird, Boulder Beach, March 1, 1952; Camp Three, September 28 and October 3 and 9, 1951, reported by Imhof; and one bird collected in Las Vegas Wash on April 5, 1952, by Pulich (identified as *P. s. nevadensis* by A. R. Phillips)."

Austin and Bradley (1971) reported for Clark County, "A winter resident in riparian areas, usually occurring from September 8 to May 18. Transient in desert scrub."

Grasshopper Sparrow   *Ammodramus savannarum*

Distribution. From British Columbia to Maine; south to southern California and Florida; also from southern Mexico to Panama and Ecuador.

Status in Nevada. Rare summer resident in northeastern Nevada and transient. Funk (1969) reported, "This sparrow is uncommon but can be seen around agricultural areas in northeastern Nevada."

Ryser (1970) reported, "An uncommon summer resident, transient visitant: meadows and pastures grassy areas; statewide."

Records of Occurrence. Linsdale (1936) reported, "In the Ruby Valley, Elko County, Ridgway (1877) collected one on July 22, 1868, a juvenile male. From August 9–14, 1872, Nelson (1875) observed the species in the meadows along the Humboldt River, but found it not uncommon."

LaRochelle (MS, 1967) reported for Ruby Lake NWR, "I have only three sightings from mid–summer."

Ellis and Nelson (1952) list this sparrow as uncommon in the Carson River Basin. This author has visited the area many times in the past 40 years and has never seen them.

Hoffman (1881) records the Grasshopper Sparrow as breeding near Eureka, Eureka County. In recent years Biale has not seen this bird in the Eureka area.

Lawson (MS, 1981) reported three from Comins Lake near Ely June 5, 1970.

Kingery (1980) reported, "An alfalfa field at Dyer held several singing Grasshopper Sparrows; mated pairs apparently on territory. He (1984) reported one at Las Vegas.

Linsdale (1936) reported, "On May 7, 1934 a female was taken on the Colorado River, opposite Fort Mohave."

LeConte's Sparrow   *Ammodramus leconteii*

Distribution. From British Columbia to Quebec; south through central United States to Gulf Coast.

Status in Nevada. Hypothetical in southern Nevada.

Records of Occurrence. Mowbray (PC, 1988) reported a LeConte's Sparrow was seen October 2, 1975 at Overton WMA. "The next day Lawson and I looked for it, but we couldn't find it."

Fox Sparrow   *Passerella iliaca*

Distribution. From Alaska to Newfoundland; south to California and Florida.

Status in Nevada. Summer resident in the mountain ranges, except for southern Nevada where it is transient. Linsdale (1951) listed six races for Nevada: *P. i. fulva, P. i. schistacea, P. i. olivacea, P. i. megarhynchus, P. i. monoensis* and *P. i. canescens.*

Funk (1969) reported, "This is a very common bird in the summer throughout Nevada."

Records of Occurrence. Linsdale (1936) reported this sparrow from the Pine Forest and Santa Rosa Mountains, Humboldt County; from the Jarbidge Mountains, Elko County; the Tahoe area Carson Range; the Walker River Range; the Wassuk Range; White Mountains; Shoshone Range; Toiyabe Range; Toquima Range; Monitor Range; Schell Creek Range; and the Snake Range.

Sheldon NWR (1982) reported the Fox Sparrow as an occasional spring and fall migrant.

On July 4, 1968 these birds were noted at Lye Creek, Santa Rosa Range and one adult was seen setting on a nest (Alcorn MS).

The absence of this bird from some valleys is noteworthy. It was not seen or collected by LaRochelle or Lewis at Ruby Lake NWR.

Ryser (1967) reported at Lamoille Canyon from July 1 to 4, 1967 male Fox Sparrows were abundant and still engaged in territorial singing.

Five specimens in the USNM were taken by Burleigh from the Carson Range in the months of April, June and September of 1968–1970. One specimen was also taken by Burleigh from Reno September 30, 1968.

Heilbrun and Kaufman (1979) reported 18 Fox Sparrows seen in the Truckee Meadows area in December of 1978 and another (1981) was reported December 20, 1980.

Ryser (1967) reported in early July 1967, males were singing vigorously in territory in western Nevada.

Gabrielson (1949) reported a specimen taken August 12, 1933 west of Carson City.

Alcorn (1946) in the Lahontan Valley (Fallon area) reported a specimen taken April 23, 1941 and one seen by Mills March 29, 1942. The author has not seen them in this valley since then. However, Thompson (MS, 1988) reported Fox Sparrows seen along the East Canal in Fallon October 7, 1986.

One was observed in the Freeman Basin of the Stillwater Range, Churchill County on November 18, 1959 (Alcorn MS).

Biale (MS) recorded this bird in the Eureka area six days in April, 1967, on October 8, 1967 and October 5, 1974. He reported them three days in April, 1972 and on June 18, 1972.

Johnson (1973) reported for the Quinn Canyon Mountains, "An isolated colony of this species at the southern boundary of its range in the Great Basin was found in willows, wild rose and cottonwoods along a small stream at Cherry Creek Campground (Quinn Canyon Mountains). At least two males were singing on territories on May 27 and June 18 (1972). No specimen could be obtained."

Austin and Bradley (1971) reported for Clark County, "Transient in riparian, woodlands and montane forest areas. Usually occurs from August 29 to October 29."

Lawson (MS, 1981) reported in southern Nevada this bird is a rare winter visitor, with birds seen in Vegas Wash and Corn Creek.

Song Sparrow   *Melospiza melodia*

Distribution. From the Aleutian Islands to Newfoundland; south in Mexico to Puebla.

Status in Nevada. Summer resident, visitant and transient in all areas of the state. Most commonly found in thickets along streams and marsh areas. Linsdale (1951) reported five races from Nevada: *M. m. montana, M. m. fallax, M. m. fisherella, M. m. merrilli* and *M. m. saltonis*.

Records of Occurrence. Sheldon NWR (1982) reported the Song Sparrow as a common breeding summer resident; uncommon spring and fall.

An immature female specimen in the USNM was taken at the Ruby Lake NWR by Lewis on October 25, 1966.

Other Song Sparrows were observed in this area March 24, 1966; June 20, 1966; August 21, 1965; and December 17 and 22, 1965 (Alcorn MS).

Burleigh collected a male at Winnemucca February 14, 1958.

Twenty–three specimens in the USNM were taken by Burleigh from the Reno area from 1966 to 1969. All months are represented except May, June, July and August.

The author, however, in the Reno area in 1942, noted Song Sparrows 6 times in May and June; 9 in July; and once in August, as well as in all other months.

In the Lahontan Valley (Fallon area) these sparrows are present in all months. Joe T. Marshall in November, 1942 examined 17 specimens taken by this author from this area and identified three races: *merrilli, fisherella* and *montana* indicating migratory birds from Oregon, Idaho and probably other areas. In the Stillwater Marsh area, Napier (MS, 1974) reported nesting in small numbers and wrote, "Seems that in eight years (1967–74) their numbers have decreased."

Biale (PC, 1984) reported a Song Sparrow nest at Kingston Canyon (Groves Lake), Lander County on July 14, 1979. It was located in black sage about 18 inches above ground and was made of grass and weed stems and lined with fine roots. It contained four pale blue eggs with brownish spots.

At Willow Creek, seven miles east–southeast of the Potts Ranger Station (west side Monitor Range) this author saw an adult Song Sparrow fly from a nest containing five eggs on June 16, 1945 (Alcorn MS).

Biale (MS) recorded these sparrows in the Eureka area from October 4, 1964 through 1973 a total of 23 times in January; 22 in February; 27 in March; 13 in April; once in May; none in June or July; 4 in August; 6 in September; 17 in October; 22 in November; and 24 times in December.

These birds were not abundant in the Moapa Valley and between 1959 and 1963 were observed twice in February; 4 times in March; twice in October; and once in December (Alcorn MS).

Gullion et al., (1959) reported, "A permanent resident, recorded from the Mohave and Las Vegas Valleys, the Flat Nose Ranch east of Pioche, and Roger's Spring (1500 feet elev.) in the

Song Sparrow. Photo by Art Biale.

Muddy Mountains south of Overton, all within the range given by Linsdale (1951). Also noted in the Pahrump Valley on May 19 and 20, 1951, a locality from which this species has not been previously recorded.''

Austin and Bradley (1971) reported for Clark County, ''A permanent resident in riparian areas, transient in woodlands and montane forests, occurs all year.''

Lincoln's Sparrow   *Melospiza lincolnii*

Distribution. From Alaska to Newfoundland; south to Panama.

Status in Nevada. An uncommon resident in limited localities and transient over most of the state.

Linsdale (1951) reported two races from Nevada: *M. l. lincolnii* and *M. l. alticola*.''

Funk (1969) reported, ''This bird lives in the marshes throughout the state, but is present only in small numbers.''

Records of Occurrence. Sheldon NWR (1982) reported the Lincoln's Sparrow as an occasional spring, summer and fall resident, with nesting thought to occur in the area.

A male was caught in a mouse trap at Favre Lake, Ruby Mountains on August 26, 1970 (Alcorn MS).

Lewis collected one female specimen for the USNM at Ruby Lake NWR on October 12, 1967.

Burleigh collected 11 specimens from Reno, Washoe Lake and the nearby Carson Range in July, September, October and November of 1967 and 1968.

Alcorn (1946) reported specimens taken in the Fallon area on March 25 and September 4 of 1941 and on September 12, 1942.

On April 18, 1971 one was caught in a mouse trap in a grassy situation six miles west–southwest of Fallon (Alcorn MS).

Biale (MS) recorded these birds in the Eureka area from 1964 to 1973 a total of 6 times in March; 54 in April; 11 in May; and twice in September and October.

This sparrow was observed at Lehman Caves by park personnel August 26, 1957 and A. A. Alcorn (MS) saw one near Baker October 5, 1966.

Austin and Bradley (1971) reported for Clark County, "A winter resident in riparian areas, transient in montane forests and usually occurs from August 22 to May 15."

Lawson (MS, 1981) reported, "Lincoln's Sparrow is a rare fall and common spring migrant in southern Nevada."

Kingery (1983) reported, "A Lincoln's Sparrow wintered at Las Vegas."

Swamp Sparrow   *Melospiza georgiana*

Distribution. From Mackenzie to Newfoundland; south to Texas, the Gulf Coast and Florida to central Mexico.

Status in Nevada. A straggler. Funk (1969) reported, "This sparrow is occasionally found in northeastern and western Nevada."

Records of Occurrence. Ellis (1935) reported a female taken from the west side of Ruby Lake, Elko County December 17, 1927.

LaRochelle (MS, 1967) reported this bird at Ruby Lake NWR as not abundant, but seen spring and summer.

Monson (1972) reported one was seen at Las Vegas February 16, 1972.

Kingery (1975) reported, "Swamp Sparrows, singles, were reported at Overton April 19 and Las Vegas May 3." He (1977) reported Swamp Sparrows that wintered at Las Vegas remained until April 25.

Lawson (MS, 1981) reported, "This sparrow is rare in fall and a winter visitor in southern Nevada. One seen on October 14, 1974 at Corn Creek and another at Tule Springs on the same date."

White–throated Sparrow   *Zonotrichia albicollis*

Distribution. From Yukon to Newfoundland; south from California to Florida.

Status in Nevada. A rare visitor to Nevada. Funk (1969) reported, "This bird is occasionally found in southern Nevada."

Records of Occurrence. Linsdale (1936) reported only one record of this sparrow in Nevada. It was a male, collected by D. A. Feathers on Cottonwood Creek, 7400 feet Mount Grant, Mineral County July 15, 1934.

Lewis collected a female at the Ruby Lake NWR November 15, 1966.

The author observed a White–throated Sparrow six miles west–southwest of Fallon on January 12, 1985. "The white throat and orange–colored part of the side head stripes was very plain to see." It was observed almost daily until the end of March.

Thompson (PC, 1988) reported for the Fallon area, "A few White–throated Sparrows at the feeder at my house. We pick up a couple every year during the CBC. They're definitely around."

Ryser (1970) reported a record from White Pine County.

Snider (1969) reported three birds near Overton November 20, 1968 and Snider (1971) reported one was on the Desert Game Range November 8 and 15, 1970.

Austin (1969) reported records from Las Vegas on December 17, 1967 and January 7 and 9, 1968 (immature in back yard); Tule Springs Park, April 22, 1966; Las Vegas, May 2, 1967; Lee Canyon (elev. 8000 feet), October 12, 1966; and Corn Creek October 17 and 21, 1960.

Kingery (1977) reported a White–throated Sparrow stayed the winter at Las Vegas. He (1978) reported one at Las Vegas May 9.

Lawson (TT, 1976) commented, "This sparrow is a regular but rare fall and winter visitor in southern Nevada."

Golden–crowned Sparrow   *Zonotrichia atricapilla*

Distribution. From Alaska and Yukon; south through the western United States to Baja California. Casual in the eastern United States.

Status in Nevada. Rare winter visitor. Funk (1969) reported, "In fall and winter this bird can occasionally be seen in dense thickets around the western and southern parts of the state."

Records of Occurrence. Only a few records are available for this bird in Nevada. Linsdale (1936) has only three records, one reported by Ridgway (1877) was collected in the West Humboldt Mountains, Churchill County on October 7, 1867 and two were taken by Henshaw at Lake Tahoe October 13, 1876.

Voget (MS, 1986) reported at Sheldon NWR the Golden–crowned Sparrow is a transient visitor and migrant spring and fall. He commented, "One individual banded 8 October 1985 at Thousand Creek. Generally observed with migrating flocks of White–crowned Sparrows."

Lewis collected one male specimen for the USNM at the Ruby Lake NWR October 26, 1967.

Three specimens in the USNM were taken by Burleigh from the Reno area. One, an immature male, was collected October 28, 1967; a female was taken November 19, 1967; and a female was taken October 2, 1968.

Ryser (1970) reported on December 6, 1970 at least six were in the Truckee Meadows and he (1972) reported them in the Truckee Meadows December 16, 1972.

Scott (1971) reported, "These birds were seen on several occasions in Reno. This is the third year that this bird has been seen there, either in fall or winter and may represent a real expansion of their range."

Ryser (1970) reported on October 2, 1970 one of these sparrows was seen below the big bend of the Truckee River on the Paiute Reservation.

Drennan and Rosenburg (1986) reported 16 Golden–crowned Sparrows in Carson City December 18, 1985 during the CBC.

Alcorn (1940) reported one collected four miles west of Fallon April 30, 1936. Mills and Lemburg (PC) reported one at a feeder in Fallon in March 1975.

Biale (MS) trapped and banded one in Eureka on November 16, 1964 and recaptured the same bird in the same trap November 18, 1964. He (PC, 1984) reported one seen in Eureka May 7, 1984.

Gullion et al., (1959) reported for southern Nevada, "A single bird was taken from a quail trap in Willow Basin (3550 feet elev.), Gold Butte area, 26 miles southeast of Overton on March 31, 1953."

Austin (1969) reported, "Additional records for January 11, 1962 at Las Vegas (two birds); April 22, 1966, Kyle Canyon, 7100 feet; October 14, 1964, Sheep Spring, Sheep Mountains (Hansen); October 17, 1966, Kyle Canyon, 5800 feet, (immature dead on road); and November 9, 1966, Kyle Canyon 7100 feet suggest that this species is regular visitant in southern Nevada."

Kingery (1979) reported one to three Golden–crowned Sparrows wintered at Las Vegas.

Lawson (MS, 1981) reported one collected at Tule Springs January 4, 1978.

White–crowned Sparrow   *Zonotrichia leucophrys*

Distribution. From Alaska to Newfoundland; south in Mexico to Michoacan.

Status in Nevada. Summer resident in mountain ranges and winter resident in the valleys. Linsdale (1951) reported two races from Nevada: *Z. l. gambelii* and *Z. l. oriantha*.

Funk (1969) reported, "Year–round one can see this common bird in thickets and low brush throughout Nevada."

Linsdale (1936) reported these birds from eight of the 17 Nevada counties, but he listed them as numerous over most of the state.

In the past 10 years these sparrows have been seen in all counties in Nevada. No attempt will be made to list the many localities from which they were observed.

Records of Occurrence. Sheldon NWR (1982) reported the White–crowned Sparrow as an occasional summer resident with breeding thought to occur in the area. It is uncommon spring and fall.

LaRochelle (MS, 1967) reported at Ruby Lake NWR he has seen them in all months of the year. However, they were mostly seen from late fall until early winter and again in the spring.

In the Lovelock area these birds were abundant in winter from October through April (Alcorn MS).

In the Reno area these birds were observed from September into April (Alcorn MS).

Heilbrun and Rosenburg (1982) reported 509 White–crowned Sparrows on the Truckee Meadows CBC on December 19, 1981.

Burleigh collected specimens in the Carson Range, southwest of Reno from 1967 to 1969, representing the months of April, May, June, July, September and October.

Ryser (1968) reported innumerable flocks have accumulated in the valleys in western Nevada.

They are common in winter in the Yerington area. The first ones to arrive at a feed station

White-crowned Sparrow. Photo by A. A. Alcorn.

near Yerington were on September 28, 1981. They were abundant until May, 1982 (Alcorn MS).

In the Fallon area from 1960 through 1965, these birds were recorded 123 times in January; 109 in February; 114 in March; 80 in April; 18 in May; none in June, July or August; 40 in September; 98 in October; 112 in November; and 119 times in December. The earliest September arrivals were September 3, 1961; two were seen September 10, 1960 and about 30 were seen September 17, 1965 (Alcorn MS).

Biale (MS) reported this species in the Eureka area 280 times between April 1964 and December 1967, inclusive. It was not seen in January and the only February record during this time is February 17, 1966. It was reported 8 times in March; 70 in April; 66 in May; 7 in June; not seen in July; twice in August; 32 in September; 44 in October; 19 in November; and on December 3, 1966.

Biale (MS) reported a nest in the southeast Diamond Mountains in a snowberry bush contained three young about a week old on June 23, 1968 (elev. 7600 feet).

J. R. and A. A. Alcorn (MS) saw these birds in the Baker area in April, May, September, October and December.

In the Pahranagat Valley these sparrows were observed from 1960 to 1965 4 times in January; twice in February; 10 in March; 6 in April and May; none in June, July, August or September; on October 18, 1963; 3 times in November; and 6 times in December (Alcorn MS).

They were observed in the Pahrump Valley of southern Nye County October 30 and 31, 1961 and March 16, 1962 (Alcorn MS).

J. R. and A. A. Alcorn (MS) recorded these sparrows in the Moapa Valley and nearby areas

133 times from April 25, 1957 to January 5, 1967. They were observed 27 times in January; 17 in February; 27 in March; 24 in April; 11 in May; once in June; none in July, August or September; 8 in October; and 9 times in November and December.

Austin and Bradley (1971) reported for Clark County, "This sparrow is a winter resident in desert scrub and riparian areas. Transient in woodlands and montane forests. In woodlands and montane forests it usually occurs from March 9 to May 4 and from September 7 to November 28. Unseasonal occurrence on December 23. In riparian and desert scrub areas it usually occurs from September 10 to May 24."

Food. In the Fallon area on December 31, 1941 the author saw many of these sparrows feeding on the Russian olive berries that were hanging on the trees. These sparrows were seen to perch on a limb and peck at the berries; apparently eating the pulp from around the hard seed. On December 31, 1941, clearing away snow from a small area about 12 feet east of my house, I placed a small amount of wheat on the ground. Many were seen feeding on this wheat. Several apples were placed in this clearing and were readily eaten by these birds.

On May 5, 1959, three White–crowned Sparrows and one House Sparrow were feeding near our window. They were picking up Chinese elm seeds and eating the kernel. The House Sparrow was very efficient and worked the kernel out rapidly. The White–crowned Sparrows were not efficient and I was never sure that they were eating the kernel or some object stuck to the seeds.

On March 18, 1972 three White–crowned Sparrows were feeding in a cottonwood tree on the catkins and on March 21, 1972 seven of these sparrows were feeding on catkins in the same tree.

Harris' Sparrow *Zonotrichia querula*

Distribution. From Mackenzie to Nova Scotia; south to southern California and Florida.

Status in Nevada. An uncommon winter visitor. Funk (1969) reported, "This rare bird visits western and southern Nevada in the winter. Staying in thickets near rivers and streams, the bird goes unnoticed, and thus very few have been recorded."

Records of Occurrence. Gullion (1957) reported collecting an adult female at the Cottonwood Ranch, 19 miles west of Contact, Elko County on November 5, 1955.

Lewis collected a male specimen for the USNM at Ruby Lake NWR October 29, 1966.

Ryser (1968) reported one observed the previous two winters at a feeder in the southwest section of the Truckee Meadows (Reno area). He (1970) reported one at this same feeder.

Ryser (1972) reported these sparrows present at several different sites in the Truckee Meadows December 16, 1972.

Kingery (1981) reported, "The one which wintered through March 29 at Gardnerville, Nevada was as far west in our region [Mountain West] as one could get."

Kingery (1984) reported, "Harris' Sparrows—one at Carson City in November."

Drennan and Rosenburg (1986) reported a Harris' Sparrow at Carson City during the CBC December 28, 1985.

At a feed station near Yerington, one was seen December 10, 1982 until April 4, 1983 at which time it was joined by another. These two were seen at this feed station until April 29 (Alcorn MS).

A Harris' Sparrow was collected six miles west of Fallon December 23, 1974 and another November 28, 1975 (Alcorn MS).

Biale (MS) recorded one in Eureka December 14, 1965. He reported one at his feeder in Eureka from December 3, 1983 to May 7, 1984. "I saw him nearly every day." He (MS) reported another Harris' Sparrow at the Eureka sewage pond November 4, 1986.

Kingery (1977) reported Harris' Sparrows lingered, with late records on May 17 and May 22–25 at Lida.

Austin (1969) reported a male was collected near Hiko, Lincoln County November 12, 1966.

In the Las Vegas area one or more of these birds were reported by **Audubon Notes** as present each winter from 1967 to 1973, with a male specimen from Corn Creek January 31, 1973.

Kingery (1982) reported one at Las Vegas November 17, 1981. He (1987) reported, "Harris' Sparrows wintered throughout the Region [Mountain West], in all four states; farthest west were three wintering at Kirch WMA and one February 21–22 at Las Vegas.

Lawson (TT, 1976) commented this bird is a rare to uncommon winter resident in southern Nevada.

Dark–eyed Junco   *Junco hyemalis*

Distribution. Transcontinental in boreal forests from Alaska to Labrador; south into Mexico.

Status in Nevada. Many have previously listed two species of juncos in Nevada, the Gray–headed Junco or Slate–colored Junco, *J. caniceps*, and the Oregon Junco *J. oreganus*. However, they have been consolidated into one, the Dark–eyed Junco.

The author recorded them 845 times (8578 birds) statewide between January 1960 and March 1974.

Records of Occurrence. Sheldon NWR (1982) reported the Dark–eyed Junco as an uncommon permanent resident spring through fall; occasional in winter and thought to breed in the area.

A specimen was collected by Lewis at Ruby Lake NWR October 27, 1967.

LaRochelle (MS, 1967) reported them abundant at Ruby Lake NWR from early spring until mid–winter.

Heilbrun and Stotz (1980) reported two Slate–colored and 38 Oregon Juncos seen at Ruby Lakes during the CBC December 18, 1979.

Johnson (1950) recorded both races of this junco at his residence in Reno from January 19 to February 13, 1949.

Heilbrun and Rosenburg (1982) reported three Slate–colored and 88 Oregon Juncos in the Truckee Meadows during the CBC December 19, 1981.

In the Fallon area, Alcorn (1946) reported taking a specimen in the Lahontan Valley of *J. h. cismontanus* December 26, 1940; two specimens of *J. o. montanus* November 20, 1942; and a specimen of *J. o. shufeladti* March 16, 1945. Two juncos were observed by this author six miles west–southwest of Fallon October 5, 1984.

In the Fallon area these birds are usually seen from late September until mid–March.

Biale (MS) recorded this bird in the Eureka area 222 times from October 16, 1964 to December 31, 1967. He saw them mostly in winter months and had no records for June through

Dark-eyed Junco. Photo by Art Biale.

September. In the same area, Biale recorded them 416 times between October of 1964 and March of 1974.

Biale (MS) recorded *J. c. caniceps* 25 times in the Eureka area from April 2, 1964 through October 8, 1967. He saw them 10 times in April; twice in July; 3 in August; 8 in September; and once in October. Biale saw a nest with five young June 23, 1974 and two nests on July 28, 1974 contained three young each. He also recorded this bird 42 times between April, 1964 and November, 1973.

Cottam (1942) reported a female specimen collected by Luther J. Goldman near the summit of Silver Peak Mountains, Esmeralda County on October 6, 1915.

A. A. Alcorn (MS) reported up to 15 birds seen in the Pahranagat Valley in January, February, April and November. He also reported them in the Moapa Valley in December and January; in the Gold Butte area in October; and in the Pahrump Valley in October.

Austin (1971) for Clark County reported *J. h. caniceps*, "In montane forests usually occur from March 21 to November 13. In riparian areas, it usually occurs from September 30 to February 14. Winter resident in riparian areas. Transient in woodlands. Summer resident in montane forests."

A. A. Alcorn (MS) saw an adult *J. h. caniceps* on a nest with five eggs on Mount Charleston June 18, 1962.

Heilbrun and Stotz (1980) reported one Slate–colored Junco seen at Mount Charleston December 28, 1979 during the CBC.

McCown's Longspur   *Calcarius mccownii*

Distribution. From southern British Columbia to Manitoba; south into Mexico to Durango.

Status in Nevada. Rare.

Records of Occurrence. Lawson (MS, 1981) reported three to five McCown's Longspurs were seen at Jacks Creek, Elko County October 14, 1978; four to six in the Baker area, White Pine County October 14, 1978; seven to 10 in the farm area of Panaca, Lincoln County on October 13, 1978; nine to 13 near Elgin October 15, 1978; and three to four at Tule Springs Park February 21, 1973.

Lapland Longspur   *Calcarius lapponicus*

Distribution. From Alaska to Greenland to Siberia; south to California and northern Florida. In Eurasia south to France, Italy, eastern China and northern Japan.

Status in Nevada. Rare.

Records of Occurrence. Ridgway (1877) wrote, "During the more severe portion of winter, individuals of this species were frequently detected among the large flocks of Horned Larks . . . around Carson City. They were recognized by their peculiar and unmistakable notes."
    Ryser (1970) reported this bird for Nye, Ormsby and Washoe Counties.
    Kingery (1977) reported two Lapland Longspurs at Diamond Valley, Nevada on March 2.
    Snider (1969) reported one near Las Vegas January 12 and February 6, 1969.
    Kingery (1978) reported one reached Corn Creek October 6.
    Kingery (1984) reported, "Lapland Longspurs spread south to Pahranagat November 11 [1983]." He (1984) also reported, "Lapland Longspurs straggled southwest to Las Vegas January 14."

Chestnut–collared Longspur   *Calcarius ornatus*

Distribution. From Alberta to Manitoba; south into Mexico to Veracruz. Casual throughout Canada and the Atlantic coast to Virginia.

Status in Nevada. Rare visitor to southern Nevada.

Records of Occurrence. Lawson (1974) reported the first Nevada record of the Chestnut–collared Longspur on January 28, 1973 when they collected one, "In a pasture at Tule Springs Park, 14 miles northwest of Las Vegas, Clark County. One of two or three present, the bird was prepared as a study skin by Donald Baepler, University of Nevada Las Vegas. The specimen, an adult male, was moderately fat. The feathers on the stomach and nape of the neck were beginning to wear into breeding plumage. The stomach was almost black and the nape rufous–brown in color. The coloration of the toes of the left foot was unusual; the left and middle toes were white, right toe brownish white, and the spur was brown. This first specimen record for Nevada supports a sight record from the same location, November 30, 1971 (Monson, 1972)."

Lawson (MS, 1981) reported an immature male Chestnut–collared Longspur taken at the north intersection of U.S. 95 and Corn Creek Road October 9, 1974. It was one of 16 birds present.

Kingery (1984) reported at Chestnut–collared Longspur at Las Vegas April 15.

Snow Bunting   *Plectrophenax nivalis*

Distribution. Circumpolar arctic areas. South in America from Oregon to Florida. Casually to Bermuda and northern Africa.

Status in Nevada. Rare in winter.

Records of Occurrence. Sheldon NWR (1982) reported the Snow Bunting as an occasional winter resident.

Cruickshank (1958) reported one at Ruby Lake NWR December 28, 1957.

Tsukamoto (MS) reported, "Hoskins sighted a single bird above the Knoll Creek Field Station, near White Peaks, Granite Mountains, Elko County on December 3, 1958. N. J. Papez and the author observed a lone bird on Little Island, above Dave Creek, Jarbidge Mountains, Elko County, on December 18, 1963. It is interesting to note that, although this species is characteristically a flocking bird, on both occasions only singles were seen which may indicate accidental status."

Kingery (1976) reported, "A Snow Bunting perched on a fence outside Eyre's window at Diamond Valley to permit a good leisurely view."

Biale (MS) reported one five miles south of Eureka November 1, 1981.

Kingery (1986) reported a Snow Bunting at Kirch WMA December 27, 1985.

Jim and Marian Cressman reported one at the Lake Mead Fish Hatchery, Clark County on January 1, 1984. Kingery (1984) on this same bird commented, "Nevada had its first photographed Snow Bunting, found January 1–2 and immortalized by photographs."

Bobolink   *Dolichonyx oryzivorus*

Distribution. From British Columbia to Nova Scotia; south to South America (Brazil and Paraguay).

Status in Nevada. Present in summer in northeastern Nevada; transient elsewhere.

Records of Occurrence. Linsdale (1936) reported records from Quinn River Crossing, Humboldt County June 3, 1909.

Tsukamoto (MS) reported, "I have observed this bird in Lamoille Valley in the wild hay fields on many occasions during the early summer. A large flock (100 plus) of these birds was observed here on 27 May 1966. I have also observed them on the Salmon Falls Creek at the confluence of Shoshone Creek near Jackpot."

Linsdale (1936) reported Ridgway collected an adult female in Ruby Valley, Elko County on August 28, 1868 and Oberholser shot two males at Franklin Lake on June 25 and 26, 1898.

Two specimens in the Nevada State Museum were collected three miles north of Lamoille on May 20, 1965.

Several males were seen near Lamoille June 21, 1967 (Alcorn MS).

Quiroz (MS) reported seeing two males June 5, 1970 at this same locality.

Kingery (1981) reported, "A colony of Bobolinks at Lamoille, Nevada, sported 37 males and three females June 13."

LaRochelle (MS, 1967) reported at Ruby Lake NWR the Bobolink is not often seen in the area.

Kingery (1980) reported 10 to 15 Bobolinks at Ruby Lake, May 29; and one at Independence Valley.

Kingery (1982) reported 50 to 100 in grain fields at Ruby Lake in mid–September, 1981.

Kingery (1983) reported, "New nesting sites were found at six locations near Elko, probably Nevada's only nesting site."

Kingery (1986) reported, "Northeastern Nevada has scattered colonies of breeding Bobolinks: they were reported this summer from Lamoille, Elko/Wells and White Pine County."

Biale (MS) saw one in Eureka May 19, 1974.

Linsdale (1936) reported, "Holt (MS) observed the species May 26, 1915 at Peavine Ranch, Toiyabe Mountains, Nye County."

Kingery (1980) reported a singing male at Dyer May 31.

Lawson (TT, 1976) reported, "Dr. Baepler has seen these birds regularly at Panaca in late summer."

Tsukamoto (MS) reported, "Two birds were seen adjacent to Tule Reservoir on the Kirch WMA in Nye County."

Monson (1972) reported one of these birds September 6, 1971, "At Corn Creek, Desert Game Range; two at Mormon Farm near Vegas, October 10, 1971; and one at Tule Springs Park, October 12, 1971."

Lawson (MS, 1981) reported them at Corn Creek June 2, 1973 and two adult males May 28, 1975.

Red–winged Blackbird   *Agelaius phoeniceus*

Distribution. From British Columbia to southwestern Newfoundland; south to Costa Rica and the northern Bahamas.

Status in Nevada. Common resident in summer over entire state; restricted to valleys in winter. They usually leave the colder northeastern part of the state in winter.

The Red–winged Blackbird's abundance in Nevada is indicated by their being recorded 2411 times statewide from January of 1960 to March of 1974. No attempt is made to list all the localities where these blackbirds have been seen. They have been observed in all 17 Nevada counties.

Records of Occurrence. Sheldon NWR (1982) reported the Red–winged Blackbird as a common summer resident; uncommon spring and fall.

LaRochelle (MS, 1967) reported these blackbirds arrived at Ruby Lake NWR in early spring, from February 10 to 20, and nest throughout the marsh areas. He reported banding several hundred from this area.

On May 30, 1960 along the Carson River about five miles east of Fort Churchill, "Mills and this author saw over a dozen Red–winged Blackbird nests. Most were near the tops of young cottonwood trees that were eight to 12 feet high. One tree was bent over and we looked into a nest containing one egg. All these nests were made of inner bark fiber of the cottonwood tree (Alcorn MS).''

The abundance of Red–winged Blackbird in the Fallon area is evidenced by their being recorded 1240 times between January 4, 1960 and October 9, 1967. They were frequently seen in all months (Alcorn MS).

Napier (MS, 1974) reported for Stillwater WMA, some of these birds stay all year, but most abundant from March to September or October. He also reported they nest widely in the marsh. In spring and summer they are widely dispersed and in fall they frequently assemble into large flocks. On March 3, 1970, males, scattered one at a place, were seen showing territorial and display behavior. On this same date Napier reported in Stillwater Marsh these birds had been singing on terriorty for the past several weeks.

By April they are dispersed over all of the riparian area in this valley. In winter they frequent only selected areas and are in flocks from 10 individuals to thousands (Alcorn MS).

On November 15, 1962 a flock of over 2000 was seen; on November 18, 1962 a flock of over 3000 was seen; and on January 25, 1965 a flock of over 2000 was seen. Many of these blackbirds have a common roosting place during the winter months in the Mahala–Massey section of Lahontan Valley. In late afternoon they leave from the various feeding grounds, mostly from the livestock feedlots in the south and southeastern part of the valley, and fly over 20 miles to their roost. They are frequently joined by starlings and other blackbirds. As one group after another meets in flight, they form a long "cola" or chain extending for miles in length. Their path in some sections runs parallel to the Fallon to Hazen highway and people driving in the evening over this route are amazed at the miles of long, continuous flight of birds. From a distance, the flight sometimes appears to be a long horizontal column of smoke. In excess of 100,000 birds roost in this common roost (Alcorn MS).

On June 1, 1941 about four miles west of Fallon, a nest built in wire grass about 10 inches above the edge of the water in a drain ditch was seen. The nest contained three young blackbirds and one egg. The young blackbirds appeared to be only a day or two old. On June 4, 1941 the three young were larger and the egg was nowhere to be seen (Alcorn MS).

Biale (MS) recorded these birds in the Eureka area 209 times between April 27, 1964 and October 18, 1967. Red–winged Blackbirds were abundant in May, June and July and were seen on November 26, 1964. They were not recorded in December or January during this three–year period.

Since that time Biale has recorded them on December 14, 1969, January 15, 1972, December 3, 1972 and December 31, 1973. He reported nesting June 28, 1964 when 15 nests were examined. One nest contained three young and the other 14 nests contained from two to five eggs for an average of 3.14 eggs per nest. On June 19, 1966 Biale examined 26 nests of which 16 contained a total of 52 eggs, nine nests contained a total of 28 young and one nest contained three eggs and one newly hatched young.

J. R. and A. A. Alcorn (MS) reported these birds in the Pahranagat Valley 63 times from February 4, 1960 to March 19, 1965. They were seen twice in January and February; 12 times in March; 11 in April; 12 in May; 6 in June; 4 in July; 3 in August; twice in October and November; and 5 times in December.

J. R. and A. A. Alcorn (MS) observed these blackbirds in the Moapa Valley a total of 183 times from April 25, 1957 to January 4, 1967. They were seen in all months.

Austin and Bradley (1971) reported these birds are, ''A permanent resident of Clark County in riparian areas.''

Female Red-winged Blackbird. Photo by A. A. Alcorn.

Food. The many tons of insects and pest worms eaten by these birds each year contribute considerably to the natural balance and check effect on farmland pests. On some occasions, however, these birds are a problem. In the Fallon area, problems in the corn fields and at the livestock feedlots sometimes arise. Every corn grower in this area experiences some damage to their crop from Red–winged Blackbirds in August (Alcorn MS).

In some instances, 90% of the kernels were eaten from the cob in the fields where the corn was in the milkstage (the period just prior to hardening of the kernels). The blackbirds especially like milo and related grains and eat them before harvest time. These birds eat the mature kernels of corn, wheat, barley and other grains. The kernels are chewed up into small pieces before being swallowed, as examination has revealed only small pieces in blackbird stomachs (Alcorn MS).

Red–winged Blackbirds are especially fond of sunflower seeds and are seen from August through September of each year feeding on the seeds. On December 27, 1970 several of these birds were feeding on the large domesticated sunflower heads still standing in the author's yard. The birds were reaching over the edge in an attempt to get the seeds. Two were seen attempting to get the seeds in flight from underneath the heads (Alcorn MS).

Tricolored Blackbird   *Agelaius tricolor*

Distribution. From southern Oregon south through California to northwestern Baja California.

Status in Nevada. Rare transient.

Records of Occurrence. Ryser (1970) reported these birds as a rare transient and visitant in marshes, meadows and pastures in Washoe County.

Lawson (1972) reported seeing 10 to 12 Tricolored Blackbirds at Reno March 9, 1972.

Ryser collected three specimens in the Truckee Meadows. Two males were taken April 12, 1969 and one female was collected April 12, 1974.

Ryser (1985) comments, "There is a limited but regular migratory flow in the spring through Truckee Meadows."

Western Meadowlark   *Sturnella neglecta*

Distribution. From British Columbia to Ontario into Mexico.

Status in Nevada. Summer resident over entire state; winter resident in western and southern Nevada. These birds leave the colder northeastern part of Nevada during the winter.

Linsdale (1936) reported these birds from 14 of 17 Nevada counties. Recent records are available for all 17 counties.

Records of Occurrence. No attempt is made to list all of the records available. Their widespread abundance is indicated through the author's recording them 2608 times statewide between January 1, 1960 and December of 1971.

Sheldon NWR (1982) reported the Western Meadowlark as an abundant breeding summer resident; common spring and fall.

Records for the Elko area from 1962 to 1968 are from March through October of each year (Alcorn MS).

LaRochelle (MS, 1967) reported for Ruby Lake NWR, "Very common; arrive at this refuge in early spring and nest on the area."

Western Meadowlarks are common in all months in the Lovelock, Reno, Minden and Yerington areas (Alcorn MS).

Their abundance in the Fallon area is indicated by their being recorded 1909 times between January 2, 1960 and January 3, 1968. A total of 7026 birds were seen for an average of 3.7 birds per census covering all months during this time (Alcorn MS).

Biale (MS) reported these birds 213 times in the Eureka area from May 3, 1964 to December 25, 1967. He reported them most often from March through October with one January record (January 11, 1967); twice in February; 5 times in November; and 3 times in December. He (MS) reported them in 1973 a total of 3 times in March; 4 in April; 16 in May; 15 in June; 5 in July; 6 in August; 9 in September; and 6 times in October. None were seen in January, February or December of this year.

Kingery (1984) reported, "At Eureka, Nevada 150 fed by a highway during a six–inch snow."

In the Pahranagat Valley, Western Meadowlarks were seen in all months and were reported 48 times from 1960 to 1965 (Alcorn MS).

In Moapa Valley, Western Meadowlarks were seen in all months and were reported 177 times between December 15, 1959 and May 1965. They were seen 20 times in January; 15 in February; 28 in March; 26 in April; 20 in May; 12 in June; 15 in July; 7 in August and September; 11 in October; 8 in November; and 7 times in December (Alcorn MS).

Austin and Bradley (1971) reported for Clark County, "Permanent resident in riparian areas.

Transient in desert scrub and woodlands. Occurs all year. Most abundant from August 30 to May 8."

Nest. A meadowlark nest was found on the ground four miles west of Fallon on May 11, 1941. It had a roof and lining made of fine grass and contained five meadowlark and three California Quail eggs. This nest was again examined May 16 and the meadowlark eggs were nowhere to be found. The quail eggs were broken, apparently by some predator (Alcorn MS).

At this same location on April 27, 1942 a meadowlark nest was located in grass on the ground and contained five eggs. Another nest containing five meadowlark and two California quail eggs was seen April 25, 1943. This nest was in an alfalfa field flooded with four inches of water. The quail eggs were in the bottom of the nest, but the meadowlark eggs were floating in the nest. The nest was examined daily until May 3 when an adult meadowlark was flushed from the nest. The nest contained one meadowlark egg, four baby meadowlarks and the two quail eggs. On May 8 there were two baby meadowlarks and two quail eggs. On May 14 there was nothing in the nest except the two quail eggs. Examination of these eggs revealed one was rotten and the other contained a well developed, but dead chick (Alcorn MS).

Biale (MS) reported a nest in Little Smoky Valley, Nye County on June 12, 1984. The nest was located on the ground, roofed over with grass. It contained three, white with reddish–brown, spotted eggs (elev. approx. 6600 feet).

Yellow–headed Blackbird   *Xanthocephalus xanthocephalus*

Distribution. From northern Alaska to Nova Scotia; south into Mexico.

Status in Nevada. A common summer resident in marsh areas over the entire state. The abundance of this bird in summer in Nevada is indicated by Linsdale (1936) who recorded these birds from 11 of the 17 Nevada counties. Usually flocks containing less than 50 birds are seen.

Records of Occurrence. Sheldon NWR (1982) reported the Yellow–headed Blackbird as a common breeding summer resident; uncommon spring and fall.

At Ruby Lake NWR from 1964 to 1968, refuge personnel banded 2503 of these blackbirds. On June 7, 1969 a total of 212 nestlings were banded. Howard (MS, 1972) reported nesting at this refuge occurs over a wide area.

In the Elko area from 1961 through 1968, these birds were recorded each month from April into September. The earliest date was April 11, 1966 when one was seen and the latest date was one observed September 26, 1962 (Alcorn MS).

Kingery (1980) reported two seen at Ruby Lake NWR December 18, 1979. They usually are seen in riparian areas in groups of up to 15 birds.

Hope M. Gladding (1941) reported nests near Minden, Douglas County in 1941 and for "several seasons" previous to 1941. She reports, "I have no idea of the size of the colony, but it seems to me it must number in the hundreds."

In the Fallon area these birds were observed 546 times from January 4, 1960 to May 10, 1968. In 1965, this bird was observed once in January, February and March; 8 times in April; 17 in May; 14 in June; 20 in July; 12 in August; 5 times in September; and none were seen in October, November or December. In recent years, however, a few have been observed in these months (Alcorn MS).

Male Yellow-headed Blackbird. Photo by A. A. Alcorn.

On January 31, 1940, Mills (MS) wrote, "Out at Ed Wade's near Fallon today, there was a flock of about a 1000 blackbirds, which is nothing out of the ordinary this winter, but the thing that took my eye right off the bat was the fact that at least 200–300 of them were male Yellowheads. I know that we have seen them in winter before, but that is a rather positive record and in about the wintriest month we have. I remember seeing a big flock of them in December several years ago down at Carson Sink going to roost in the tules."

Mills (PC) also reported a flock of mixed blackbirds, numbering in the thousands at his ranch west of Fallon December 24, 1959. He said the drain ditch bank was black with birds for a distance of 200 yards and there were an additional several thousand in the nearby field and trees. He saw at least 200 Yellow–headed Blackbirds in the bunch; the others were Red–winged and Brewer's.

Near Fallon on April 14, 1961, two flocks were seen with over 900 birds in one and over 800 in the other (Alcorn MS).

Nesting was observed in Lahontan Valley at Carson Lake, Mahala, Stillwater Marsh and many other small areas where a patch of tules was growing (Alcorn MS).

Napier (MS, 1974) reported at Stillwater Marsh, "A few spend the winter with Red–winged Blackbirds. Most arrive in March and many nest here."

Biale (MS) recorded these birds in the Eureka area 112 times from April 26, 1964 to July 2, 1967. They were seen 11 times in April; 73 in May; 19 in June; 8 in July; and once in August. His records from 1964 to 1974 show these birds arrive in the area in the latter part of April and leave the area in the late summer. In most years he has not seen them after July. Biale reported only three September records during this 10–year period, all in 1971.

To this, the author adds a September 4, 1964 record of over 10 seen in the Eureka area.

Biale (MS) found 11 nests made mostly of coarse reeds July 2, 1967. These nests were located in tules and grass and three nests contained two eggs each, six nests contained four eggs each, one nest contained three young and one nest contained two eggs and two young.

In the Pahranagat Valley they were observed from March 29 to August 9 with most records for April and May (Alcorn MS).

J. R. and A. A. Alcorn (MS) reported these birds 103 times in the Moapa Valley from April 9, 1960 to January 4, 1967. Their records reveal these birds were seen in all months except November. They were seen 6 times in January; twice in February; 9 in March; 23 in April; 28 in May; 17 in June; 7 in July; 4 in August; 3 in September; and twice in October and December.

Austin and Bradley (1971) reported, "Summer resident in riparian areas, usually occurring from March 17 to November 30 with unseasonal occurrence on December 30."

Rusty Blackbird   *Euphagus carolinus*

Distribution. From Alaska to Newfoundland; south to the Gulf Coast and Florida.

Status in Nevada. Rare.

Records of Occurrence. Ryser (1985) reported, "We acquired a sight record for Reno on an Audubon CBC [December 15, 1973], and I observed an individual in a flock of Brewer's Blackbirds on the University of Nevada campus in Reno during the fall of 1983."

Biale (PC, 1986) reported a Rusty Blackbird at the Eureka Sewage Pond November 8, 1986.

Kingery (1978) reported one at Beatty December 10, 1977.

Snider (1970) reported three seen at Tule Springs Park on November 9 and 10, 1969.

Kingery (1978) reported one at Las Vegas March 25, 1978.

Lawson (MS, 1981) reported specimens collected by him at Tule Springs October 22 and 29, 1974; and two specimens from Vegas Wash March 21, 1978. He considers them a rare fall and spring migrant and winter resident.

Kingery (1981) reported, "Rusty Blackbirds appeared at Las Vegas October 18, 1980." He (1986) reported, "On October 13, Las Vegas had its first Rusty Blackbird in five years."

Brewer's Blackbird   *Euphagus cyanocephalus*

Distribution. From British Columbia to southern Ontario; south into Mexico.

Status in Nevada. Resident. Linsdale (1936) listed this blackbird as the most numerous and widely distributed of the blackbirds. No attempt will be made to list the many recent locality records available. Their abundance is indicated by their being seen 1654 times statewide between 1964 and 1974.

Records of Occurrence. Sheldon NWR (1982) reported the Brewer's Blackbird as an abundant breeding summer resident; common spring and fall.

In Elko County, these blackbirds were seen in the Elko area from April through September of each year (Alcorn MS).

LaRochelle (MS, 1967) reported at Ruby Lake NWR these birds are a common summer resident first appearing in mid–spring.

A. A. Alcorn (MS) saw about 25 at Ruby Lake NWR October 5, 1965.

On the south fork of the Owyhee River, a nest containing four eggs was seen July 11, 1967. Near the nest a female Brewer's Blackbird was attacking a gopher snake (Alcorn MS).

A nest with four eggs was seen at Wildhorse Reservoir May 29, 1971. In the Elko area, a female flew from a nest containing four young June 12, 1967 (Alcorn MS).

In the Lovelock area they were observed in all months and were present at the cattle feedlots throughout the winter (Alcorn MS).

In Idlewild Park, Reno, these blackbirds were seen 129 times from March 1942 to June 1965. They were observed in all months except January and were reported only once in February. However, they do occur in these months at nearby ranches (Alcorn MS).

Near the Carson River, five miles east of Fort Churchill, on June 12, 1960, an adult flew from a nest on the ground containing five eggs. The nest was on a grass–covered bank (Alcorn MS).

In the Yerington area, Mason Valley, these blackbirds were seen in all months. In summer, lone individuals were seen. This is in contrast to those seen in winter months when they sometimes gather in large flocks. On February 14, 1944 a flock of about 500 of these birds was seen near Yerington and on January 7, 1961 a flock of about 200 was seen (Alcorn MS).

In the Fallon area they were recorded 1048 times between January 4, 1960 and October 18, 1967. In 1965, a sample year, they were seen 12 times in January; 13 in February; 8 in March and April; 22 in May; 14 in June; 20 in July; 18 in August; 19 in September; 16 in October; 9 in November; and 10 times in December. The number seen each day varied from two to more than 500, with larger numbers seen in winter months. Although it nests in this area it is not abundant during the nesting season. On June 1, 1964, a nest on a horizonal limb of a cottonwood tree contained four young Brewer's Blackbirds. The nest was about 10 feet above ground (Alcorn MS).

Biale (MS) reported these blackbirds in the Eureka area on 507 different days between April 26, 1964 and December 8, 1967. During this time, he saw them November 23, 1964 and November 3, 1967; on December 8, 1967; and January 31, 1966 and January 18, 1967. He reported them most often from April through September.

In the Baker and Lehman Creek areas of White Pine County, these birds were seen in all seasons, but none were seen in January or February (Alcorn MS).

Fred Packard (MS) observed nesting in May and June at Lehman Caves National Monument.

J. R. and A. A. Alcorn (MS) reported these birds in the Hiko area 55 times between January 13, 1960 and June 28, 1965. They were seen 3 times in January; once in February; 8 in March; 10 in April; 9 in May; 5 in June; twice in July, August, September and October; 3 in November; and 7 times in December.

J. R. and A. A. Alcorn (MS) reported these birds in the Moapa Valley area 128 times from April, 1957 through December, 1965. They were observed in all months with flocks over 5000 birds seen in some winter months. They were less often seen in July and August than in any other month.

Gullion et al., (1959) reported for southern Nevada, "A permanent resident in valley areas, occasionally wandering onto deserts."

Austin and Bradley (1971) reported, "Winter resident in riparian areas and transient in woodlands. Usually occurs from August 9 to May 18."

Nest. Biale (MS) provided extensive nesting records for the Eureka area. Selected reports follow: A nest containing five eggs was seen June 2, 1964 in a greasewood brush. June 11,

1965, two nests on the ground with very little cover, contained five and six eggs. A nest with five eggs in a greasewood brush were seen on a bank on the same day. On June 3, 1965 Biale saw three nests in a greasewood bush *(Sarcobatus)* and three in a rabbitbrush bush *(Chrysothamnus)*. Of these, one contained three eggs, four contained four eggs and one contained five eggs. On June 13, 1965 Biale saw three nests in a big sage *(Artemisia)* that contained four, five and four eggs. On July 5, 1965, two nests built between bales of hay in a haystack contained four and two young. On June 6, 1966, a nest in a rabbit brush contained five very young blackbirds and a nest in a greasewood bush contained three young that were well–feathered out with two dead young in the same nest. On June 19, 1966 a nest on a horizontal apple tree limb contained one egg. This nest was eight feet above ground.

Great–tailed Grackle   *Quiscalus mexicanus*

Distribution. Formerly Boat–tailed Grackle. From the southwestern United States to Peru.

Status in Nevada. The Great–tailed Grackle is a newcomer to Nevada, first appearing in the southern portion of the state in the 1970s and gradually expanding their range to as far north as Ruby Lakes.

Records of Occurrence. Kingery (1978) reported, "Great–tailed Grackles strayed to Ruby Lakes May 15 for the first report for northern Nevada." He (1982) reported, "Great–tailed Grackles have moved north to Ruby Lakes (three males during May in three different places)."

Kingery (1985) reported, "Great–tailed Grackles continued to expand . . . reports from Ruby Lakes."

Alberson (PC) reported five miles southeast of Fallon on May 21, 1984 he saw one of these birds.

Biale (PC, 1984) reported an adult male in Eureka May 12, 1984. He (PC, 1986) reported that at Cold Creek Ranch in Newark Valley, about 10 Great–tailed Grackles were seen July 6, 1986. He also found three nests in willows. Two of the nests were empty and one contained two unfeathered grackles. Adults were seen feeding fledglings.

Snider (1971) reported one seen near Las Vegas May 17–29, 1970. It was photographed as the first Nevada record.

Snider (1971) reported one seen near Las Vegas August 30, 1970 and July 5 and 24, 1971.

Ryser (1972) reported during August of 1972, five Great–tailed Grackles were seen in Las Vegas Wash and Mormon Farm on two occasions; in August, eight were in Las Vegas Wash and Mormon Farm on one occasion.

Ryser (1973) reported, "At Tule Springs Park this spring the courtship of four was observed."

Kingery (1977) reported, "The Las Vegas colony of Great–tailed Grackles, 25 birds, were seen at Davis Dam on April 19 by Mowbray and three were seen May 12 at Mesquite by the Lawsons."

Kingery (1977) reported, "Las Vegas reported the largest flock yet of Great–tailed Grackles with 50 on October 21 and smaller flocks at Lake Mead, Logandale, Lake Mohave and Davis Dam."

Kingery (1980) reported the Great–tailed Grackle nested at Beatty, was seen at Mesquite and, "All over the place at Crystal Springs near Hiko."

Heilbrun and Kaufman (1977) reported one Great–tailed Grackle seen in the Henderson area December 19, 1976 during the CBC. Two were reported in 1980 and 1981.

Lawson (MS, 1981) reported, "A male was seen 17 and 18 of May, 1970 at Tule Springs Park for the first record. Now they are very common in the Las Vegas Valley."

Kingery (1984) reported, "Lawson described the colonization of Nevada by Great–tailed Grackles. Overton hosted the first major colony north of Las Vegas. They moved northeast to Hiko and Pahranagat and northwest to Pahrump on the California border and from there north to Death Valley and Beatty. This summer they were found in eastern Nevada at Panaca and Ruby Valley and in central Nevada between Tonopah and Austin. They have not spread so far in the rest of the Mountain West."

Common Grackle   *Quiscalus quiscula*

Distribution. From northeastern British Columbia to southwestern Newfoundland; south and east of the Rockies to the Gulf Coast.

Status in Nevada. Transient.

Records of Occurrence. Linsdale (1936) reported, "On May 22, 1932, Robert H. Poultney shot a male Common Grackle near Crystal Spring, 4000 feet, Pahranagat Valley, Lincoln County. The specimen was prepared by Hall and is now in the Museum of Vertebrate Zoology. This record is the first known to me for this species in the region west of the Rocky Mountains in the United States."

Kingery (1982) reported a Common Grackle at Jarbidge November 11 and 12, 1981.

Ryser (1985) reported this grackle has been seen at Ruby Lakes.

Kingery (1986) reported, "Several Common Grackles spent late July in Elko."

Alcorn (1940) reported, "One was found dead four miles west of Fallon, Churchill County on April 14, 1938. This bird was given to Mrs. Anna Bailey Mills who agreed with the identification of the species. It was found dead along with about nine blackbirds that had apparently been killed with poisoned rolled oats that were distributed in the area for ground squirrels."

Kingery (1987) reported, "Stillwater produced a second Nevada breeding record for Common Grackles, with the discovery of two adults and three fledglings there August 19."

Biale (MS) saw one of these birds in the Eureka area January 13 and 28, 1968. He banded this bird February 22 and saw it again March 8, 1968.

Kingery and Lawson (1985) reported, "A Common Grackle appeared May 14 at Eureka, Nevada."

Mowbray (PC, 1987) reported Common Grackles nesting at Dyer at the Circle L. Ranch during Memorial Day weekend of 1987.

Kingery (1977) reported Mowbray saw a male at Las Vegas May 11 and he (1980) reported a Common Grackle at Lund October 24.

Lawson (MS, 1981) reported this grackle as an occasional migrant in southern Nevada with only three or four records.

Kingery (1983) reported, "Two Common Grackles appeared May 25 in a Las Vegas yard."

Brown–headed Cowbird   *Molothrus ater*

Distribution. From southeastern Alaska to southern Newfoundland; south from Florida and in Mexico to Veracruz.

Status in Nevada. Common summer resident statewide. Winter visitor in limited numbers in the western part and resident in southern Nevada. Linsdale (1951) reported two races from Nevada: *M. a. artemisiae* and *M. a. obscurus*.

Records of Occurrence. Sheldon NWR (1982) reported the Brown–headed Cowbird as a common breeding summer resident, uncommon spring and fall.

Two males were found dead at Elko on April 28, 1970. They had recently eaten poisoned grain being used for Pocket Gopher control. Others were seen in this area in June and July (Alcorn MS).

Gabrielson (1949) reported six individuals at Harrison Pass, Ruby Mountains in Elko County on June 1, 1948.

Lewis collected a female at the Ruby Lake NWR May 27, 1968 and a male October 18, 1966.

LaRochelle (MS, 1967) reported these birds arrive at this refuge in mid–spring. At this refuge Alcorn (MS) saw six in one group June 26, 1962 and three in another.

Heilbrun and Rosenburg (1982) reported one Brown–headed Cowbird at Ruby Lake NWR on December 19, 1981.

At a cattle feedlot near Lovelock one was seen January 20, 1965; about 10 were seen at the same locality from February 4 to 13, 1965. About 20 cowbirds were seen in the Lovelock area March 21, 1966 and others were seen in April, May, June, July and August. An unusually large number of cowbirds were noted in this area at one feedlot when about 50 were counted August 15, 1965. This is the only record for this month in this area (Alcorn MS).

Burleigh collected one female specimen now in the USNM at Reno on January 26, 1969.

These birds were seen in Reno from April through July (Alcorn MS). They are usually reported in this area during the December CBC.

Two males were seen in Mason Valley (Yerington area) January 5, 1961. In recent years (1979–1984) others were observed in this valley in May, June, July and August (Alcorn MS).

In the Fallon area these birds were recorded in all months. They are more abundant from April through July. Combined figures for 1961 through 1965 indicate cowbirds were seen 5 times in January; twice in February; once in March; 18 in April; 88 in May; 56 in June; 39 in July; 6 in August; twice in September; and once in December. They were not seen in October or November during this time (Alcorn MS).

Biale (MS) reported seeing cowbirds 159 times in the Eureka area between April 23, 1964 and September 5, 1967. During this time they were most frequently seen in May, June, July and August. They were not seen in November, January, February or March and they were seen once in December (December 13, 1964). Biale saw a Warbling Vireo feeding a young cowbird on August 2, 1967.

In the Moapa Valley area from October 30, 1960 to March 23, 1967, cowbirds were seen 8 times in January; twice in February; 3 in March; 4 in April; 6 in May; 5 in June; 3 in July; once in August and September (about 150 seen September 13, 1965); 3 in October; none in November; and twice in December (about 100 seen December 14, 1966) (Alcorn MS).

Johnson and Richardson (1952) reported three specimens in breeding condition taken at Ash Meadows in June of 1951.

Austin and Bradley (1971) reported for Clark County, "Permanent resident in desert scrub and riparian areas. Summer resident in woodlands. Transient in montane forests. In woodlands and montane forests it usually occurs from April 29 to June 29. Unseasonal occurrence on August 29. In desert scrub and riparian areas it occurs all year."

Orchard Oriole   *Icterus spurius*

Distribution. From southeastern Saskatchewan; south to Columbia and Venezuela. Rarely to California.

Status in Nevada. Accidental in Clark County.

Records of Occurrence. C. G. Hansen collected an immature male at Corn Creek September 4, 1965.

A. A. Alcorn (MS) reported one at Brier Spring July 12, 1962.

Lawson (MS, 1981) reported a first year male seen by Mowbray and himself at Corn Creek June 15 to July 17, 1977, singing until late June.

Hooded Oriole   *Icterus cucullatus*

Distribution. From California and southern Nevada to southern Texas; south through the Yucatan Peninsula to the British Honduras.

Status in Nevada. Uncommon summer resident in southern Nevada.

Records of Occurrence. Kingery (1982) reported, "A Hooded Oriole feeding from a hummingbird feeder at Ruby Lake provided the region's most northerly record."

Kingery (1984) reported, "A Hooded Oriole came to a Reno home for the second year."

Snider (1966) reported, "Two Hooded Oriole nests were at Littlefield, Nevada on June 14 (1966)."

Kingery (1975) reported a Hooded Oriole visited the Pickslay feeder in Las Vegas May 20-23.

Linsdale (1936) reported, "A female was taken June 12, 1929 by Sheldon at Pahrump, Nye County."

Johnson and Richardson (1962) reported a male taken April 3, 1950 at the north end of Ash Meadows. Another male was collected at Pahrump June 21, 1951. "This bird was actively foraging and then making repeated trips into a tall ornamental palm, probably indicating the first nesting record for Nevada."

One was collected on the Desert Game Range May 12, 1962.

Snider (1971) reported a pair spent the summer in Las Vegas in 1971.

Kingery (1975) reported 16 Hooded Orioles recorded in the Las Vegas area during May and one at Pioche. He (1984) reported, "A male Hooded Oriole wintered at Las Vegas, feeding from a hummingbird feeder."

Lawson (MS, 1981) reported this oriole is an uncommon summer resident primarily where there are cottonwood trees.

Northern Oriole   *Icterus galbula*

Distribution. From British Columbia to Nova Scotia; south to Costa Rica.

Status in Nevada. Summer resident in the valleys and along streams in lower parts of mountain ranges. The author has seen Northern Orioles in all sections of Nevada. They are most commonly seen among tall trees in riparian areas of the valleys and lower parts of mountains.

Records of Occurrence. Sheldon NWR (1982) reported the Northern Oriole as an uncommon breeding summer resident, uncommon spring and fall.

A male was seen at Lamance Creek, east side of Santa Rosa Range, Humboldt County June 10, 1970. They were repeatedly seen near Elko in May and June (Alcorn MS).

The earliest records for spring arrivals in western Nevada was Ryser's (1967) report of one seen in Reno on March 17, 1967.

Burleigh collected one male at Reno April 21, 1968.

This author has records of Northern Orioles in Reno only for the months of May, June and July. These orioles were also seen in these months near Wadsworth, Genoa, Dayton, Fort Churchill, Yerington and Schurz (Alcorn MS).

At Lahontan Reservoir two nests containing young were seen July 7, 1959; one nest with young July 4, 1960; two young left a nest June 30, 1974. These birds are common nesters around the shores of this reservoir and nests examined there were constructed mostly of monofilament fish line, some with hooks and sinkers still attached (Alcorn MS).

In the Fallon area, these birds were seen only in the months of May, June, July, August and one September record (September 11, 1964) when two were seen (Alcorn MS).

Napier (MS, 1974) reported at Stillwater WMA the only oriole identified from 1967 to 1974 was of this species. He reported them as commonly nesting in the Indian Lakes area northeast of Fallon.

Baumann (MS, 1971) reported an abandoned nest in the Indian Lakes area built near the end of a Russian olive tree limb. The nest was made almost entirely of monofilament type of fishing line with several hooks still attached.

Biale (MS) recorded these orioles 60 times in the Eureka area from May 1964 to September 1967. They were seen 18 times in May; 31 in June; 8 in July; twice in August; and on September 1, 1967. From 1967 to 1974 he reported them frequently in May, June and July with an occasional August sighting, but Biale had no records for other months. He (MS, 1986) reported a nest in a juniper two miles north of Eureka contained two eggs June 25, 1986.

A. A. Alcorn (MS) reported these birds in the Baker area of White Pine County on May 12 and June 10, 1967.

In the Hiko area these orioles were seen three times in May of 1962 and twice in May of 1965 (Alcorn MS).

Kingery (1980) reported a Northern Oriole at Kirch WMA October 23. He (1984) reported a Northern Oriole stopped at Kirch April 30.

J. R. and A. A. Alcorn (MS) in the Moapa Valley saw this bird twice in April; 11 times in May; 7 in June; and twice in July.

Gullion et al., (1959) reported, "Present in the desert valleys and in the cottonwoods about springs in the desert ranges from as early as April 16 (1951, McCullough Range) to as late as September 7 (1951, when a bird was banded at Boulder Beach)."

Austin and Bradley (1971) reported, "A summer resident in riparian and woodland areas of Clark County. It is transient in montane forests and usually occurs from April 3 to October 1 with one unseasonal occurrence on October 23."

Scott's Oriole   *Icterus parisorum*

Distribution. From southern California, central Nevada to western Colorado; south through Texas to Mexico.

Status in Nevada. Summer resident in central and southern Nevada.

Records of Occurrence. Kingery (1976) reported, "Scott's Orioles nested near Unionville, Nevada 200 miles north of any other Nevada breeding record and 300 miles west of the recent Utah nesting records."

Linsdale (1936) reported, "Recorded from as far north as 10 miles east of Stillwater, where on May 11, 1898, a mated male and female were collected among junipers in a little valley (Oberholser, 1918)."

None have been reported by Ryser, Napier or Alcorn in western Nevada in the Reno, Fallon, Stillwater and Eastgate areas.

Biale (MS) reported these birds in the Eureka area July 9, 1967; June 1, 1969; May 31, 1970; June 6, 1971; and July 4, 1972.

Scott (1959) reported on August 9, 1959, "This bird was reported feeding young on Ruby Hill at Eureka in central Nevada. This is a remarkably far north record."

Biale (PC, 1984) reports a nest seen in the Pancake Range, White Pine County on June 19, 1980. It was located in a juniper tree about seven feet above ground and contained four greenish–white with brown eggs (elevation 6800 feet).

Kingery (1980) reported three Scott's Orioles at Caliente August 10, 1979.

J. R. and A. A. Alcorn (MS) reported from the Moapa Valley as seen April 4, 1960; on July 1, 1960; July 13 and 14, 1961; June 29, 1962; July 12, 1962; July 23, 1963; and April 16, 1964.

Austin and Bradley (1971) reported, "A summer resident in desert scrub and woodlands. Transient in riparian and montane forest areas. It usually occurs from April 21 to August 30."

Family **FRINGILLIDAE:**
Fringilline and Cardueline Finches and Allies

Brambling   *Fringilla montifringilla*

Distribution. From tree limits in Eurasia south to the Mediterranean, south China and Japan.

Status in Nevada. Hypothetical.

Records of Occurrence. Kingery (1978) reported an "amazing" Brambling appeared in a trailer court at Sutcliff (west side of Pyramid Lake), Nevada from October 31 to November 1, 1978. It fed on pyrancanthus and on the ground with House Sparrows. Presumably wild, it allowed no approach closer than 20 feet on the ground and 10 feet when in a tree.

Rosy Finch   *Leucosticte arctoa*

Distribution. The former Black Rosy Finch *(L. atrata)* was combined with the Gray–crowned Rosy Finch *(L. arctoa)*. From the Aleutian Islands and Alaska to western Mackenzie; south to California, New Mexico and Nebraska.

Status in Nevada. Summer resident on higher mountain ranges and winter resident mainly in the northern half of the state. They are known to breed on most mountain ranges in the northern two–thirds of the state.

Records of Occurrence. Voget (MS, 1986) reported at Sheldon NWR, "Approximately 50 birds spent two weeks in the vicinity of Thousand Creek in December 1985."

Lawson (MS, 1981) reported the race *atrata* from Hinkey Summit, Santa Rosa Mountains, Humboldt County December 18, 1975.

In the Santa Rosa Range of Humboldt County about 30 were seen in Solid Silver Canyon October 21, 1958; about 25 were seen at Stonehouse Creek Canyon October 22, 1963; and over 1000 were seen in this same canyon November 13, 1963 (Alcorn MS).

Over 20 were seen near Deeth March 14, 1969 on the highway from Dinner Station to Northfork; on November 27, 1970 flocks of 12, 40 and 80 finches were seen (Alcorn MS).

LaRochelle (MS, 1967) reported seven were observed near Gardner Peak about 14 miles north of Ruby Lake NWR on October 26, 1963 (elev. 9300 feet).

Ryser (1967) reported a flock of *L. atrata* west of Reno March 10, 1967 and Cruickshank (1970) recorded them in the Truckee Meadows December 20, 1969 during the CBC.

Burleigh collected seven specimens, three males and four females, on Mount Rose in Washoe County in November 1968.

Alcorn (1943) reported collecting two specimens from a flock of over 50 near Ramsey, Lyon County November 15, 1941.

A flock of approximately 500 birds were seen feeding alongside the highway in Douglas County, two miles west of Wellington on March 3, 1961. The birds were of various color variations and probably represented more than one race (Alcorn MS).

In Wilson Canyon near the Walker River in Lyon County about 10 were seen January 11, 1961 (Alcorn MS).

Thompson (MS, 1988) reported about 40 Rosy Finches feeding on a south slope of the Sand Springs Mountains on January 18, 1987.

In Clan Alpine Mountains of Churchill County, about 100 were seen near the head of Cherry Creek October 20, 1960 and over 50 were seen at the site of the old mining town of Wonder, Churchill County, November 7, 1961 (Alcorn MS).

On March 1, 1962 a flock of about 1500 were seen at Elkhorn Pass, Lander County. Some were no more than 20 feet away as this author sat in an auto watching them feed on what appeared to be Russian thistle seeds. At this same location on February 12, 1965, a flock of about 200 were seen (Alcorn MS).

On February 27, 1962 about 1500 were eating Russian thistle plants on Dr. Helming's Ranch in Reese River Valley (Alcorn MS).

Biale (MS) reported flocks of these birds in the Eureka area between 1964 and February 23, 1974 a total of 8 times in January; 11 in February and March; 3 in October; 5 in November; and 24 times in December.

In the Lehman Caves area, park personnel reported two of these finches August 26, 1957 and July 14, 1961.

Medin (1987) reported Rosy Finches were frequent visitors to his study plot on Bald Mountain and they, "Apparently nested off the area in nearby cliff habitats."

Ned Kendrick (PC) reported seeing three large flocks of these birds on Montgomery Pass, southern Mineral County on December 13, 1959.

Ryser (1971) reported a flock of about 30 seen feeding close to a gas station in Tonopah on March 20, 1971.

Gullion (1957) reported, "On February 12, 1954 a flock of 100 to 150 Rosy Finches was watched feeding on the shoulders of the highway east of Dutch John Mountains, about 36 miles north of Pioche, Lincoln County. These finches had apparently found a preferred food since they returned to feed at the same site immediately following the passage of each of five automobiles."

Lawson (MS, 1981) reported four *L. atrata* at Pioche December 21, 1975.

Pine Grosbeak *Pinicola enucleator*

Distribution. Forests of northern Hemisphere, from northern Scandinavia, Siberia, Alaska and Canada. In America it winters south from California to Virginia.

Status in Nevada. Uncommon.

Records of Occurrence. Tsukamoto (MS) reported, "This species is a new record for Elko County. Hoskins observed a pair on February 22, 1954, on a pinon pine on the west side of Spruce Mountain, Pequop Mountains, about two miles below the Sprucemont Mine, Elko County. The author observed a flock of seven birds feeding on pinon pine buds on the summit above Victoria Mine, in the Dolly Varden Mountains, Elko County. Observation was made for approximately five minutes at a distance of only 25 feet."

A specimen was taken by Tsukamoto 10 miles southeast of Wells, Elko County November 4, 1966.

Ryser (1970) reported several in the west Humboldt Mountains and Ruby Mountains.

Cruickshank (1964) reported nine Pine Grosbeaks (two males) at Unionville on January 1, 1964 during the CBC.

A male specimen was taken by Burleigh on Mount Rose on August 24, 1968.

Lawson (MS, 1981) reported six Pine Grosbeaks seen at Slide Mountain on June 17, 1969.

Ryser (1968) reported, "To date, there are only a few records of the Pine Grosbeak in Nevada and all are of recent vintage. July, 1958, three Pine Grosbeaks along the west fork of Gray Creek in the Carson Range in a mature subalpine forest of white pines and red firs. Several summers prior to 1958, Pine Grosbeaks were seen repeatedly along the wet southeast shore of Marlette Lake. Also have been encountered in the Hobart Reservoir region in the Carson Range. This summer, encountered Pine Grosbeaks near Incline Lake."

Ryser later collected a juvenile female specimen at Hobart Reservoir on August 13, 1968.

Ryser (1965) reported one pair about three miles south of Little Valley.

Johnson and Banks (1952) reported, "On the West Fork of Gray Creek, one mile west and one and three–quarter miles north of Rose Nob in the Carson Range of Washoe County, three individuals of this species were seen in a mature subalpine forest of white pines and red firs on July 10, 1958. The grosbeaks were apparently attracted to a commotion started by others birds as a result of imitated Pygmy Owl calls. The birds warbled repeatedly from perches in the uppermost boughs of the firs, but remained together and gave no indication of being established on territories. Two grayish individuals were present in addition to the single red male obtained."

Kingery (1984) reported Pine Grosbeaks in Reno February 14.

Drennan and Rosenburg (1986) reported the Pine Grosbeak seen in the Carson City area December 28, 1985 during the CBC.

Biale (MS) reported three seen in the Cottonwood area south of Diamond Mountain, north of Eureka on October 29 and 31, 1972. Kingery (1980) reported Pine Grosbeaks were at Eureka in late October and November.

Biale (MS) reported two in the Duck Creek Basin, east of McGill on October 25, 1972.

Purple Finch   *Carpodacus purpureus*

Distribution. From British Columbia to Newfoundland; south to Baja California, the Gulf Coast and Florida.

Status in Nevada. Rare.

Records of Occurrence. Voget (MS, 1986) reported the Purple Finch as an occasional migrant spring and fall at Sheldon NWR.

Rubega and Stejskal (1984) reported eight Purple Finches on the CBC at Ruby Lake December 20, 1983.

Cruickshank (1959, 1960) reported Purple Finches on the Reno CBC December 29, 1958 and December 25, 1959. He (1971) reported 12 on the Truckee Meadows CBC December 26, 1970.

Regarding the preceding records for the Reno area, Ryser (1985) commented, "Purple Finches have been reported on several Audubon CBC's from Reno. However, I strongly believe these records have all been based on misidentified House or Cassin's Finches. As yet, all the reported sightings in Reno that I have checked out have been misidentifications. As far as I know, documentation does not yet exist for the presence of this finch in western Nevada."

Ryser (1970) reported this bird as a rare transient in Nye County.

Snider (1969) reported this bird at the Desert Game Range September 14, 1968. Snider (1971) reported one November 16, 1970.

For southern Nevada, Mowbray reported one to six were seen in April, May and October from 1971 to 1974.

Kingery (1985) reported, "Las Vegas had six Purple Finches February 10–16, birds that came when Nevada was situated between a high pressure system to the west and a low to the east, a condition which produced frigid arctic winds."

Cassin's Finch *Carpodacus cassinii*

Distribution. From British Columbia to southwestern Manitoba; south in the western United States to Baja California and the highlands of Mexico.

Status in Nevada. Summer resident in most of the higher mountain ranges in the state; winter resident and visitant at lower elevations.

Records of Occurrence. Linsdale (1936) reported, "Common from 7000 feet elev. in the Pine Forest Mountains. Taylor (1912) reported a nest in limber pine containing five young on June 26, 1909."

Sheldon NWR (1982) reported the Cassin's Finch as an uncommon summer resident, present spring through fall; thought to breed in the area.

LaRochelle (MS, 1967) reported mist netting and banding these finches at the Ruby Lake NWR in 1963. Lewis collected two specimens at this refuge on September 20, 1966 and May 29, 1968.

Twenty–nine specimens in the USNM were collected by Burleigh between 1966 and 1970. They were taken from the Carson Range and nearby Reno–Virginia City areas with all months represented.

Alcorn (1946) had no record of these birds in the Fallon area. Since then, two were seen feeding in Fallon March 20, 1962. They were eating barley that had been thrown out in the yard to attract birds. On March 24, 1962, there were eight of these birds feeding at the same place. From one to 10 were seen in the area September 8 and 9, 1965 and on April 29, 1967; about 10 were seen feeding in an elm tree in Fallon on April 18, 1971 (Alcorn MS).

Six were observed in the Stillwater Range May 10, 1961; two in the Clan Alpine Mountains October 23, 1960; from two to 10 were seen in the Desatoya Mountains October 23, 1960, and September 27, 1963; about 10 on March 18, 1966 at Ellsworth in the Paradise Range; and about 10 at Barrett Creek in the Shoshone Mountains, Nye County. One was seen at Austin on May 16, 1961 and two February 19, 1963 (Alcorn MS).

Biale (MS) reported these birds 209 times in the Eureka area between May 1964 and August 1967. They were seen in all months except December and January with only one November record. He found a dead male Cassin's Finch in Windfall Canyon August 26, 1969. In this same area, the totals for three years from 1970 through 1972 were: twice in January; 10 in February; 46 in March; 45 in April; 66 in May; 57 in June; 55 in July; 6 in August; 3 in October; and twice in November. They were not seen in September or December during this 3–year period. Nest building was observed May 17, 19 and 24, 1970 and two young were seen June 29, 1970.

Biale (PC, 1984) reported a nest five miles south of Eureka on June 14, 1970 in a mountain mahogany tree about nine feet above the ground. It was made of dry grass and lined with horse hair. The nest contained four blue eggs marked with a few spots (elev. approx. 7800 feet).

Kingery (1981) reported, "Cassin's Finches were also scarce, although they wintered at Eureka for the first time."

A. A. Alcorn (MS) reported from four to about 50 finches in the Baker and Lehman Caves area February 15, 1958; May 18, 1961; June 18, 1961; October 6, 1966; and May 11, 1967.

Miller (1946) reported this finch as occurring only sparingly in the Grapevine Range.

Johnson (1974) reported in 1971 on Potosi Mountain, "The species was numerous and conspicuous on 18 and 20 of May between 6900 feet and 8000 feet elevation in the pinon association. On the latter date, at least 15 were seen; many males were in full song and flight song display was noted, indicating active breeding. In late May, 1973, Cassin's Finches were numerous between 6700 feet and 8200 feet. An increase in numbers from low population levels in the 1940's is indicated."

A. A. Alcorn (MS) reported two at Lee Canyon July 18, 1963 and one at Trough Spring July 19, 1963, both in the Spring Mountains.

Heilbrun and Rosenburg (1981) reported a total of 10 in the Mount Charleston area December 30, 1980.

Miller (1945) reported one individual from Potosi Mountain and Johnson (1974) reported in 1971 on this mountain, "Seven singing males were found in pine–fir in two canyons between 7000 feet and 7500 feet elevation, suggest a pronounced increase in the population level of this species since the work of 1940."

Austin and Bradley (1971) reported for Clark County, "A permanent resident in montane forests and a winter resident in riparian and woodlands."

House Finch   *Carpodacus mexicanus*

Distribution. From British Columbia to Nebraska; south in Mexico to the state of Oaxaca.

Status in Nevada. Summer resident in the northeastern part of the state and resident elsewhere. Linsdale (1936) reported these birds are generally distributed through a wide range of conditions and sometimes occur in flocks for a large part of the year.

Linsdale (1951) reported, "The House Finches in Nevada have been separated as a race *C. m. solitudinis* by Moore with the type from Fallon." The other race occurring in Nevada is *C. m. frontalis*.

No attempt will be made to list all of the records available for Nevada.

Records of Occurrence. Sheldon NWR (1982) reported the House Finch as an uncommon breeding summer resident; present spring through fall.

The author saw one House Finch eating wild currants on the south fork of the Owyhee River July 12, 1967.

These finches were seen in flocks of 10 to 50 in the Jarbidge area in September of 1960 and 1961. They were commonly heard or seen each summer in the city of Elko. The earliest spring record was April 9, 1968 and the latest fall record was September 27, 1962 (Alcorn MS).

LaRochelle (MS, 1967) reported these birds at Ruby Lake NWR as common from early spring through fall.

Records for Reno for 1942 through 1965 indicated these birds were recorded once in March; 4 times in April; 15 in May; 12 in June; 8 in July; 5 in August; 7 in September; 4 in October; and once in December (Alcorn MS).

Burleigh collected specimens from Reno in January, February, March, April, September and December.

Ryser (1970) reported in late May and early June, large winter flocks were still intact in Reno.

A flock of over 100 birds was seen at Virginia City January 1, 1962 (Alcorn MS).

House Finches were seen year round at the McGowan residence near Yerington from 1981 to 1983 (Alcorn MS).

House Finch. Photo by A. A. Alcorn.

Their abundance in the Fallon area in indicated by their being seen 437 times from February of 1960 to August 21, 1967. From 1961–63 they were recorded 5 times in January; 7 in February; 20 in March; 33 in April; 37 in May; 20 in June; 16 in July; 3 in August; 9 in September; 8 in October; 7 in November; and once in December (Alcorn MS).

In most instances only one or two birds were observed at one place in April, May and June. In fall and winter months small flocks contained 10 to 20 birds. Five of these birds were seen eating apples in a tree four miles west of Fallon November 10, 1941 (Alcorn MS).

Biale (MS) recorded this bird 347 times in the Eureka area between May, 1964 and December, 1967. Most commonly observed from April through September. In this period they were not observed in January, November or December except January 9, 1966; November 22, 1967; and December 1, 1967.

More recently, Biale (MS) in 1971 and 1972 saw these birds 4 times in April; 49 in May; 56 in June; 32 in July; 33 in August; 31 in September; and 4 times in October. They were not seen in any other month during this period.

One flock of over 100 birds were seen near Hiko November 25, 1961. Normally only one or two birds were observed April through July in this area (Alcorn MS).

J. R. and A. A. Alcorn (MS) recorded these birds in the Moapa Valley area 90 times between April 1957 and January 1967. They were seen twice in January; 3 times in February; 5 in March; 15 in April; 19 in May; 14 in June; 10 in July; twice in August; 4 in September and October; and twice in November and December.

Austin and Bradley (1971) reported for Clark County, "A permanent resident in desert scrub and riparian areas. Summer resident in woodlands. Transient in montane forests. In woodlands and montane forests it usually occurs from March 16 to November 9. In desert scrub and riparian areas it occurs all year."

Heilbrun and Rosenburg (1981) reported a total of 133 in the Henderson area December 20, 1980.

Nesting. At Lahontan Reservoir on May 24, 1959 the author saw a female House Finch fly from a nest located in part of an old tire casing. This casing was nailed to the loading pier located at the Lahontan Boat landing. The casing was nailed horizontally to the end of the boards, acting as a bumper to prevent the boats from bashing against the ends of these boards. I wrote, "I could feel some eggs in the nest. On June 4, 1959, with the aid of a mirror I saw six eggs in this nest. On June 6, 1959 I saw two young House Finches and some eggs in the nest. On June 13, 1959 there were four young finches in the nest and no eggs. On June 21, 1959 two young House Finches were seen perched on the edge of the nest. I did not look into the nest. On June 26, 1959 the nest had no birds in it. The nest was intact, but empty."

On July 12, 1959 at this same reservoir a House Finch was seen to fly from this nest. Examination revealed three eggs in the nest. This nest was located in an old tire casing situated about 50 feet east of the nest found May 24, 1959. On July 18, 1959 a House Finch flew from this nest. The nest contained three eggs. On August 1, 1959, examination of the nest revealed droppings on the edge in two places. No birds were in or near the nest (Alcorn MS).

The following year at this same place, a House Finch nest in an old tire at the boat dock contained four eggs on June 5, 1960. In the town of Fallon on June 30, 1959 an adult House Finch was seen feeding its young. The young House Finch was perched on an elm tree limb and appeared feathered out with hard feathers except it had fuzzy feathers on its head (Alcorn MS).

On June 13, 1971 an adult was feeding a fledgling at Eastgate, Churchill County (Alcorn MS).

Biale (MS) recorded nesting in the Eureka area May 15, 1965 and May 8, 1966 with four eggs found in hop vines. A nest containing one egg was built on a ceiling timber about eight feet above the ground in an old shed on June 11, 1967. On July 18, 1969 a nest 20 feet up in a boxelder tree was made of weed stems and lined with fine grasses. It contained three young nearly ready to leave the nest (elev. 6500 feet). A House Finch nest was located 15 feet above ground under the eaves of a building at Eureka on May 29, 1970. The nest was made of weed stems and dry grass, lined with fine dry grass. It contained five blue–green eggs with dark spots and flecks (elev. 6450 feet). On June 7, 1972 a nest was found in a paper towel dispenser at a service station. The nest was made of grass, weeds and wool and contained five blue–green eggs with dark spots and lines (elev. 6450 feet).

Red Crossbill   *Loxia curvirostra*

Distribution. Conifer forests of northern hemisphere. In America, from Alaska to Newfoundland; south to Nicaragua in Central America.

Status in Nevada. Summer resident and transient. Linsdale (1951) reported four races from Nevada: *L. c. bendirei, L. c. grinnelli, L. c. benti* and *L. c. stricklandi*.

Records of Occurrence. Linsdale (1951) reported Red Crossbills from the East Humboldt Mountains, the Shell Creek Range, the Quinn Canyon Mountains, the Grapevine Mountains, the Snake Range, the Charleston Mountains and Lake Mead.

Gullion (1957) reported, "A flock seen in the subalpine firs at 7500 feet elevation on the east side of Merritt Mountain, east of Mountain City, Elko County, on July 26, 1955."

Johnson (1958) reported specimens taken by Miller and Russell in Elko County June 22 and 23, 1955 and June 26, 1956. He reported the Red Crossbill was recorded from Washoe, Humboldt, Elko, Douglas, Lyon, Esmeralda, Lincoln and Clark Counties.

A male specimen was taken by Burleigh on November 1, 1968 north of Lake Tahoe.

Ryser (1966) reported these birds were present in Little Valley July 18, 1966 for the second consecutive summer.

Rubega and Stejskal (1984) reported 42 present during the Truckee Meadows CBC December 15, 1983.

Alcorn (1946) reported, "These birds were seen and specimens taken by Anna Bailey Mills July 18, 1919."

Biale (PC, 1984) reported three birds seen at Eureka August 25, 1969; two birds seen five miles south of Eureka July 28, 1979; and four birds at Squaw Wells, Nye County August 17, 1980.

Biale (PC, 1986) reported two additional records: one at Fish Creek Range March 30, 1985 and another five miles south of Eureka June 5, 1985.

Kingery (1978) reported 10 at Ely November 21, 1977.

At Lehman Caves, park personnel observed two males and one female March 22, 1963.

Miller and Russell (1956) reported, "Family groups containing streaked juveniles were taken on Indian Creek at 7400 feet, Esmeralda County on May 31 and June 1."

Johnson (1973) reported, "Crossbills commonly seen in the Quinn Canyon Mountains in 1972. Flocks of up to 15 individuals were seen in mixed subalpine forest in Scofield Canyon. No specimens were obtained. At Mount Irish, the species was numerous. On May 24, 1972 many family groups were seen indicating breeding in the range in the recent past. An adult male and a streaked juvenile were collected. On May 18–20, 1972 at Sawmill Peak, Clover Mountains, small flocks passed overhead repeatedly on these dates. Two males were taken."

Johnson (1974) reporting on this bird in the Grapevine Range said, "They may have been absent in 1971 and 1973, when persistent hunting over nine days failed to disclose any in the most promising habitat of luxuriant pinon. However, to conclude local extinction even for the crossbill would be improper. This nomadic species is unpredictable in breeding occurrences anywhere in the forests of the western United States, being dependent upon the irregular production of conifer seeds. It is likely that individual crossbills resident on one mountain range in one year may be the same birds found breeding later in another range when the conifers of the first range are producing insufficient seeds."

Kingery (1981) reported, "Several were seen at Lida, Nevada, May 27."

Johnson (1965) reported "family groups and other small flocks" of Red Crossbills in the Sheep Range. In the Spring Range, a small flock was, "Heard flying over the mixed conifers near camp at 8900 feet on June 17, one was shot from a group of three feeding June 18, and one was collected on June 23."

Austin and Bradley (1971) reported for Clark County, "A permanent resident in montane forests. Transient in riparian areas. Occurs all year."

Lawson (1977) reported two Red Crossbills at Corn Creek May 16, 1967.

Kingery and Lawson (1985) reported, "Red Crossbills continued to roam from their mountain conifers, with small flocks seen in Las Vegas."

White–winged Crossbill   *Loxia leucoptera*

Distribution. Throughout most of North America and the Old World.

Status in Nevada. Hypothetical.

Records of Occurrence. Ridgway (1877) reported encountering this bird among cedars on the eastern slope of the Ruby Mountains August 12, 1868.

Kingery (1984) reported, "A White–winged Crossbill seen drinking from a stream with reds in 'scarlet alternate plumage' in the Sheep Range 20 miles north of Las Vegas, gave Nevada its second record."

Common Redpoll   *Carduelis flammea*

Distribution. Circumpolar arctic and subarctic areas extending south to England, the Alps, Russia, Siberia and Alaska to Newfoundland; south to France, Italy and Japan. In North America from Alaska to Newfoundland; south to Texas.

Status in Nevada. Rare winter visitor to northern and western Nevada.

Records of Occurrence. Voget (MS, 1986) reported at Sheldon NWR the Common Redpoll is an, "Occasional transient visitor. Last observation December, 1984 at Thousand Creek."

Ellis (1935) reported a male and a female collected from the west side of Ruby Lake three miles south of the White Pine County line November 2, 1929.

Gullion et al., (1959) reported, "On November 16, 1955, one of a pair was taken at 7000 feet on Sun Creek, in the O'Neil Basin, 46 miles north of Wells, Elko County. This specimen proved to be a fat, adult female. Both birds were feeding on the seeds of a big sagebrush close to willow–aspen association in the stream bottom."

Tsukamoto (1966) reported, "Two sightings were made in Elko County in the winter of 1964. A flock of about 200 birds was observed by the author on January 26, 1964, feeding on the south–facing slope of Crittenden Reservoir Dam, 17 miles north of Montello. A flock of approximately 10 birds was seen by the author and Hoskins near the Nevada Fish and Game Department Owyhee District Headquarters in Elko."

Ryser (1966) reported one December 14, 1966 in the Como region of the Pinenut Range.

Scott (1967) reported one seen on a peak near Reno November 26, 1966.

The Baumann's (PC) reported seeing a male from less than 30 feet in their yard in Carson City February 25, 1970.

Kingery (1984) reported, "The first real snow at Carson City November 20 (1983) brought two Common Redpolls."

Hoary Redpoll   *Carduelis hornemanni*

Distribution. From Alaska to Greenland; south to the northern United States. Also from northern Scandanavia to northern and eastern Siberia.

Status in Nevada. Hypothetical.

Records of Occurrence. Ryser (1985) reported, "There are no documented records for the Hoary Redpoll, *Carduelis hornemanni*, in the Great Basin. But, during late winter on a university field trip, my ornithology class and I saw a flock feeding on the ground in a park along the Truckee River at Verdi, Nevada. We were able to closely approach the flock in our cars for careful study."

Pine Siskin   *Carduelis pinus*

Distribution. From Alaska to Newfoundland; south to Veracruz.

Status in Nevada. Summer resident and transient over the state.

Records of Occurrence. Linsdale (1936) listed the Pine Siskin in summer from the Pine Forest Mountains, East Humboldt Mountains, Carson Range, Toiyabe Range, Snake Range and the Charleston Mountains.

Sheldon NWR (1982) reported the Pine Siskin as an occasional fall and uncommon winter migrant.

LaRochelle (MS, 1967) reported several Pine Siskins mist netted and banded at Ruby Lake NWR in 1963.

Ryser (1967) reported these birds observed at Lamoille Canyon June 1–4, 1967.

About 10 were seen in Reno November 19, 1941 (Alcorn MS).

Seven specimens in the USNM are from North Lake Tahoe, the Carson Range and nearby Reno. They were collected by Burleigh in May, June, August, October and November in 1967 and 1968.

Five Pine Siskins were seen near Fallon June 3, 1964 and on the following day two were seen at the same location. Two were seen six miles west–southwest of Fallon January 16, 1982. The temperature was zero degrees Fahrenheit. Six were seen in the Kendrick's yard in Fallon on June 30 and 31, 1982 (Alcorn MS).

Thompson (MS, 1988) reported 10 Pine Siskins seen along the East Canal near Fallon on October 7, 1987.

Biale (MS) in the Eureka area reported from three to 10 of these birds seen 101 times between May 14, 1964 and October 4, 1967. He saw them once in April; 13 times in May; 4 in June; none in July; 15 in August; 41 in September; 17 in October; 4 in November; and once in December. He (PC, 1986) comments they are irregular in numbers from year to year.

Kingery (1977) reported Pine Siskins at Ely April 25.

Lehman Caves National Monument park personnel reported one Pine Siskin observed May 28 and 29, 1958.

A. A. Alcorn (MS) saw about 20 in the Baker area May 31, 1967.

Kingery (1987) reported 200 Pine Siskins during a "cold spell" May 22–23.

A. A. Alcorn (MS) reported one at Trough Springs, Charleston Mountains July 19, 1963.

Heilbrun and Rosenburg (1982) reported 12 Pine Siskins in the Mt. Charleston area December 19, 1981 during the CBC.

Snider (1970) reported two Pine Siskins found and a female collected above Davis Dam, Clark County June 3, 1970 and on June 5, 1970 five breeding pair were found on Mount Charleston and two females were collected.

Austin and Bradley (1971) reported, "A permanent resident in montane forests. Winter resident in riparian areas. Transient in woodlands. In woodlands and montane forests it occurs all year. In riparian areas it usually occurs from October 21 to May 25."

Lesser Goldfinch   *Carduelis psaltria*

Distribution. From Washington to Colorado; south to South America.

Status in Nevada. Uncommon widespread summer resident, less common in winter.

Linsdale (1936) reported these goldfinches from Humboldt, Lyon, Lander, Esmeralda, Nye, Lincoln and Clark Counties and comments, "Widespread but not numerous in Nevada."

Records of Occurrence. Sheldon NWR (1982) reported the Lesser Goldfinch as an occasional spring and fall migrant.

Gabrielson (1949) reported Lesser Goldfinches seen at Star Valley and Lamoille, Elko County May 30, 1932; north of Elko on the edge of the Humboldt National Forest on May 31, 1932."

Heilbrun and Stotz (1982) reported 68 Lesser Goldfinches seen during the CBC at Truckee Meadows December 19, 1981.

Kingery (1984) reported, "40 Lesser Goldfinches wintered at a Carson City feeder."

Alcorn (1946) reported one taken in the Lahontan Valley February 5, 1944 and two of these goldfinches were collected four miles west of Fallon April 20, 1944.

Biale (MS) reported one or two of these birds on five occasions in the Eureka area: on September 15, 1964, September 10, 1968, August 26, 1972, August 22, 1974 and June 25, 1984.

One female was reported at Lehman Caves on April 10, 1959 (Alcorn MS).

Gullion et al., (1959) reported, "A winter and breeding resident, but with no mid–summer records available. Dates of occurrence extend from August 14 (1951) to June 28 (1952, both at Boulder City). Pulich observed fledglings in Boulder City on May 20, 1952."

Austin and Bradley (1971) reported for Clark County, "Permanent resident in riparian areas. Occurs all year. Transient in woodlands and montane forests."

Heilbrun and Rosenburg (1980) reported 33 Lesser Goldfinches at the Desert Game Range December 22, 1979.

Lawrence's Goldfinch   *Carduelis lawrencei*

Distribution. From central California west of the Sierra Nevadas, northern Baja California, southern New Mexico, Arizona, northern Sonora and western Texas.

Status in Nevada. A rare transient in southern Nevada.

Records of Occurrence. Johnson and Banks (1959) reported, "On April 4, 1958, a singing male was taken on the Colorado River flood plain in a thicket of small Fremont cottonwoods, tamarisks and arrowweed one–half mile west and one–half mile south of the Snyder Ranch in extreme southern Clark County. One–half hour later a second bird was heard singing in dense arrowweed and tamarisk growth near the site of the first collection, but it could not be located."

Lawson (MS, 1981) reported three seen September 11, 1963 at Corn Creek Field Station; one September 10, 1968 at Red Rock Canyon; one October 1, 1972 at Tule Springs Park by Mowbray; two males seen October 1, 1972 at Corn Creek; two males and a female seen from October 7 to November 4, 1974 at Corn Creek; and two males October 7, 1974 at Tule Springs Park.

American Goldfinch   *Carduelis tristis*

Distribution. From southern British Columbia to Newfoundland; south in Mexico to Veracruz.

Status in Nevada. Summer resident in limited areas; transient statewide. Linsdale (1951) reported these goldfinches as summer residents with a small number of records indicating a small population that stays close to the valleys.

Records of Occurrence. Gabrielson (1949) reported observations or specimens from Elko, Churchill, Nye and Clark Counties.

Gullion (1953) reported, "On November 18, 1955 a flock of about 20–30 goldfinches observed at the mouth of Bishop Creek Canyon, nine miles north of Wells, Elko County."

About 40 were seen in the Elko area May 18, 1966 (Alcorn MS).

LaRochelle (MS, 1967) reported at Ruby Lake NWR five were seen April 17, 1963. Lewis collected a female at this refuge October 4, 1966 and a male November 10, 1967.

Burleigh collected nine specimens at Reno from 1966 through 1969. They were taken in January, March, April, October, November and December, which is not indicative of nesting in this area. However, Alcorn (MS) in 1942, observed these birds in Reno once in April; twice in May; once in June; 4 times in July and September; and 5 times in October. In that year they were thought to be a breeding summer resident in the area.

In winter these birds were normally seen in small groups of about 10 to 15 birds. A group like this was seen at Virginia City on January 1, 1962 (Alcorn MS). Sometimes larger numbers are seen as reported by Mills who saw about 200 at Fernley January 19, 1961.

Kingery (1984) reported 10 American Goldfinches wintered at a Carson City feeder.

The author recorded over 100 American Goldfinches near Yerington April 20, 1983. Some were brightly colored and were feeding on dandelion seeds growing in a lawn at the McGowan residence (Alcorn MS).

In the Fallon area from 1960 through 1966, they were seen 9 times in January; 14 in February; 7 in March; 19 in April; 13 in May; once on June 1, 1964; none in July or August; once in September and October; 9 in November; and 11 times in December (Alcorn MS).

Napier (MS, 1974) reported these birds as abundant in migration through Fallon in time to coincide with seed pods on the elm trees, but before leaves erupt. On April 25, 1967 Napier reported them abundant in Fallon.

Biale (MS) reported the American Goldfinch 59 times in the Eureka area between April 29, 1964 and November 10, 1967. In this period Biale saw them once in January; none in February and March; 4 in April; 24 in May; once in June; none in July; 6 in August; 11 in September; 5 in October; 7 in November; and none in December. Since 1967 he has seen them occasionally.

Kingery (1982) reported, "At Hiko on January 9, an American Goldfinch was caught in mid–air, providing the first southern Nevada record in nine years."

A. A. Alcorn (MS) saw one in the Virgin Valley on January 25, 1963.

In Moapa Valley a group of about 10 of these birds were seen April 22, 1964 (Alcorn MS).

Austin and Bradley (1971) reported for Clark County, "Winter resident in riparian areas. Transient in montane forests. Usually occurs from September 19 to May 14."

Evening Grosbeak   *Coccothraustes vespertinus*

Distribution. From British Columbia to Nova Scotia; south to central California and in Mexico to Oaxaca.

Status in Nevada. Resident in some mountain ranges in western and southern Nevada.

Records of Occurrence. Linsdale (1936) listed the Evening Grosbeak in Nevada as, "Resident, at least on the higher slopes of the mountains along the western border of the state." He reported specimens or sightings from Galena Creek, Granite Mountains, Washoe Valley and Chiatovich Creek.

Gullion (1957) reported, "This species seems to be a common and regular fall and spring visitant in the Elko area."

Lewis collected a male specimen at the Ruby Lake NWR October 4, 1966.

From one to 25 Evening Grosbeaks were seen in Reno on five occasions in March 1942. One was also seen there April 19, 1967 and three were seen May 21, 1962 (Alcorn MS).

Burleigh collected a female specimen in Reno March 13, 1968.

Ryser (1965) reported numbers of breeding grosbeaks in the Carson Range in the summer of 1965 and he (1966) reported large numbers nesting in Little Valley (near Reno) during the summer of 1966. He (1970) reported in winter and early spring, large numbers in juniper forests at the eastern base of the Carson Range just above Reno. On May 11, 1970 he reported a flock of over 100 showed up at the UNR campus and remained for several days.

In Yerington, up to 20 birds were seen April 2, 1982 and in this same area a flock of 12 was seen at the McGowan residence near Yerington on May 15, 1982 (Alcorn MS).

These birds are an irregular visitor to the Fallon area, being reported at irregular intervals in most years. In 1963, they were seen only once in this area, when eight were recorded on October 21, 1963. In contrast to 1963, with only one sighting, there were numerous sightings for the spring of 1964. About 20 were seen March 20, 1964 and they were observed a total of seven times in March; 12 in April; 16 in May; 7 in June; and one was seen July 4, 1964. No others were reported until May 1, 1966 when one was seen and the next report was of three seen by Napier (MS, 1974) on May 21, 1967. Watkins (PC) observed over 100 of these birds four miles west of Fallon May 14, 1971 (Alcorn MS).

From 200 to 300 were seen along the Carson River in Jack Humphrey's yard west of Fallon May 22, 1971; and about 20 were seen by Lloyd Alcorn at his place west of Fallon May 27, 1971. On June 10, 1971, Napier (MS, 1974) reported two in his yard in Fallon (Alcorn MS).

A. A. Alcorn (MS) saw 15 in the Fallon area June 11, 1984 and 10 were seen in the Fallon area February 19, 1988.

Biale (MS) reported seeing these birds in the Eureka area on 152 different days from 1964 to 1973. In this period they were seen 3 times in January; 7 in February; 13 in March; 32 in April; 49 in May; 5 in June; none in July, August or September; 25 in October; 12 in November; and 6 times in December. In this area they are an irregular visitor. An example of this is their being seen only once in 1965 and on 41 different days in only four months in 1967.

In the Baker area and Lehman Caves National Monument vicinity, they were observed in February, October and November.

Johnson (1973) reported, "On Mount Wilson at 8100 feet, a mated pair of Evening Grosbeaks were collected in Douglas firs and aspen. Their bills were crammed with one and a half–inch long, light green caterpillars which had been gathered from foliage in the aspen tops; presumably these larvae were to be carried to young . . . . At 8000 feet, in dense firs and aspen in the same canyon, another male in breeding condition was taken. These are the only positive breeding records for this species in Nevada, despite the report of Linsdale (1936). In the Virgin Mountains, an Evening Grosbeak was present in the fir clumps at 7700 feet on the north face of West Virgin Peak May 21, 1970; however, no solid evidence of breeding was obtained."

Seven were observed at Panaca on April 10 and four were at the same place May 4, 1962. One was seen near Hiko, Pahranagat Valley on November 30, 1964 (Alcorn MS).

Kingery (1985) reported Evening Grosbeaks arrived a month early in Las Vegas on October 23.

Kingery and Lawson (1985) reported, "Evening Grosbeaks stayed in the lowlands late; at Las Vegas to May 20."

A. A. Alcorn (MS) saw eight or more at Overton in Moapa Valley feeding in a Chinese pistachio tree November 2, 1961 and he saw another bird in this same valley May 12, 1964.

Food. Mrs. Lois Saxton (PC) reported on December 17, 1972 about 12 Evening Grosbeaks had been seen almost daily since October of 1972. They visited her bird feeder and ate the sunflower seeds. The Napiers reported up to 50 birds eating sunflower seeds at their yard bird feeder also.

On February 28, 1958, the author wrote, "Evening Grosbeaks were seen feeding on the Russian olive seeds in our back yard today. Five birds were in one tree. These seeds are attached to the tree and have been left there by other birds. The English Sparrows last autumn ate most of the meat from the seeds. I cannot tell whether they are eating these seeds or the remaining pulp from the outside of the seed."

On March 2, 1958, "About 15 Evening Grosbeaks were seen feeding in my yard this morning. One adult was seen to pick up kernels of wheat, mouth them one at a time, then discard them. From 12 feet away I couldn't tell exactly what it was doing. It may have been cracking the wheat open eating the heart of the grain, and discarding the rest. The other grosbeaks appeared to be eating the Russian olive seeds."

On March 30, 1958, the author wrote, "Evening Grosbeaks were seen each and every day in this month in the vicinity of our house here in Fallon. They were feeding on the Russian olive seeds which they pick up or dig from the ground. They then crack open the seed and eat the meat."

Family **PASSERIDAE:** Old World Sparrows

House Sparrow   *Passer domesticus*

Distribution. Formerly English Sparrow. Introduced in North America and many other countries including South Africa, Australia and New Zealand. Native in the British Isles, most of Europe and across central Siberia, south to Morocco, Egypt, India, Ceylon and southern

Burma. Established in North America from the southern Yukon to Newfoundland south through South America.

Status in Nevada. Common resident in cities, towns and ranches statewide. No attempt is made to list all of the localities for which records are available. They have been seen in every town in the state from Jarbidge to Searchlight and from Reno to Baker.

Records of Occurrence. Sheldon NWR (1982) reported the House Sparrow as a common permanent resident, present in all seasons.

LaRochelle (MS, 1967) reported, "Quite common from mid–spring until early winter."

On February 4, 1982 this author wrote, "I don't know what happened to the House Sparrows. Two males were seen here (Yerington) today. This compares with over 100 seen through September 1981 and about 70 seen December 16, 1981. About a foot of snow fell January 6, 1982 and temperatures were below zero every night. Hazel Ames said she saw several dead ones at her house when the weather was below zero. Also, I saw two dead at this location."

In the Fallon area, it was recorded 2303 times from January 1, 1960 to March 27, 1969 (Alcorn MS).

On August 19, 1967 an English Sparrow appearing to be an albino was seen at the Kelly Johnson Ranch west of Fallon.

Biale (MS) recorded this sparrow in the Eureka area 499 times between March 3, 1965 and December 31, 1967. They were present in all months. He (PC, 1986) considers them an abundant, permanent resident in the Eureka area.

Austin and Bradley (1971) reported for Clark County, "Occurs all year; a permanent resident in riparian areas."

Nest. Nest building begins in the Fallon area in March. On June 5, 1942 a pair of these sparrows had a nest in a gourd nailed to a tree. The nest, made of grass, was lined with feathers and contained three young that were over half–grown (Alcorn MS).

On April 27, 1959 a pair of House Sparrows were carrying food to young in a nest situated in an air vent hole under the eaves of a Fallon building. The young appeared large enough to fly and would stick their heads of the hole in anticipation of being fed as the adults arrived with food (Alcorn MS).

At Fallon on May 27, 1966 a female House Sparrow was seen feeding young in a bulky nest in a Chinese elm tree in front of the author's office. The nest was in the fork of the tree (Alcorn MS).

Food. On October 3, 1957 in Fallon, a flock of about 10 English Sparrows were seen feeding on Russian olive tree berries. They were eating the fruit or meat of the berries without picking them. Although a few of the berries fell to the ground, the majority stayed attached to the tree (Alcorn MS).

On May 27, 1966 one adult female House Sparrow was seen eating Chinese elm seeds. This sparrow picked up a seed from the ground and worked it around in its beak to extract the seed from the seed wings, then ate the seed before picking up another elm seed (Alcorn MS).

These birds are a nuisance to ranches with poultry because they eat the poultry feed. They especially like wheat and milo and they will not eat barley if other food is available. At feed stations they appear fond of bread and will fly away with a quarter of a slice at a time (Alcorn MS).

RESOURCES

Alcorn, A. A.

Albert A. Alcorn worked as a wildlife biologist for the Nevada Fish and Game
Department in many areas of the state. Bird records from his unpublished notes
and personal communications with the author are included in the book.

Alcorn, J. R.

1940.   "New and Noteworthy Records of Birds for the State of Nevada." **Condor**,
42:169–170.

1941a.   "New and Additional Records for Nevada." **Condor**,43:118–119.

1941b.   "Two New Records for Nevada." **Condor**, 43:294.

1942a.   "Birds Affected by Botulism at Soda Lake, Nevada."**Condor**, 44:80–81.

1942b.   "Food of the Barn Owl at Soda Lake, Nevada." **Condor**,44:128–129.

1942c.   "Notes on the Food of the Horned Owl Near Fallon, Nevada." **Condor**,
44:284–285.

1943a.   "Observations on the White Pelican in Western Nevada." **Condor**, 45:34–36.

1943b.   "Additions to the List of Nevada Birds." **Condor**,45:40.

1943c.   "Flight Feeding of the Ring–billed Gull." **Condor**, 45:199–200.

1944.   "Botulism in the Carson Sink, Nevada." **Condor**, 46:300.

1946.   "The Birds of Lahontan Valley, Nevada." **Condor**, 48:129–138.

1953.   "Food of the Common Merganser in Churchill County, Nevada." **Condor**,
55:151–152.

1967.   Bird problems in feedlots. Paper written for the Fourth Annual Pesticide
Conference. February 1–2, 1967.

1969.   Vertebrate Pest Control in Nevada. Paper written for the Sixth Annual Agricultural Chemicals Conference. University of Nevada, Reno. January 28–29, 1969.

1971.   Bird and Mammal Control in Nevada. Paper written for the Eighth Annual
Agricultural Chemicals Conference. University of Nevada, Reno. March 17,
1971.

Alcorn, J. R., and Frank Richardson

1951.   "The Chukar Partridge in Nevada." **Journal of Wildlife Managment**, 15:265–
275.

Aldrich, J. W.

1942.   "Specific Relationship of the Golden and Yellow Warblers." **Auk**, 59: 447–449.

1943.   "A New Fox Sparrow from the Northwestern United States." **Proc. Biol. Soc.
Wash.**, 50:163–166.

1946.   "New Subspecies of Birds from Western North America." **Proc. Biol. Soc.
Wash.**, 25:129–136.

1950.   Editor. Christmas Bird Count. **Audubon Field Notes.**

1951a.   Editor. Christmas Bird Count. **Audubon Field Notes.**

1951b.   "A Review of the Races of Traill's Flycatcher." **The Wilson Bulletin**,
63:192–197.

1967.   "Taxonomy, Distribution and Present Status: The Wild Turkey and its Management." Oliver K. Hewitt, editor. The Wildlife Society: Washington, D. C., pp.
17–44.

Aldrich, J. W. and Allen J. Duvall
  1955.   "Distribution of American Gallinaceous Game Birds." Circular 34, U.S. Fish
            and Wildlife Service, Washington, D. C.
  1958.   "Distribution and Migration of Races of the Mourning Dove." **Condor**,
            60:108–128.

Allen, A. S.
  1945.   **Chasing Wrens.** Berkeley, CA: Gillick Press. pp. 69.

Amadon, D.
  1966.   "Confused Nocturnal Behavior of (?California) Gulls." **Condor**, 68:397–398.

American Ornithologist's Union
  1931.   **Checklist of North American Birds.** 4th ed. Lancaster, PA: American Orni-
            thologist's Union. 526 pp.
  1957.   **Checklist of North American Birds.** 5th ed. Baltimore, MD: American
            Ornithologist's Union. 691 pp.
  1983.   **Checklist of North American Birds.** 6th ed. Lawrence, KS: American Orni-
            thologist's Union. 877 pp.

Arbib, Robert S.
  1975.   "The Blue List for 1976." **American Birds**, 29:1967–1968.

Austin, George T.
  1967.   "The Avifauna of the Spring Mountains, Nevada." Unpublished Master of
            Science Thesis, University of Nevada at Las Vegas.
  1968a.   "The Occurrence of Certain Nonpasserine Birds in Southern Nevada." **Condor**,
            70:391.
  1968b.   "Additional Bird Records for Southern Nevada." **Auk**, 85:692.
  1969.   "New and Additional Records of Some Passerine Birds in Southern Nevada."
            Condor, 71:75–76.
  1970a.   "Interspecific Territoriality of Migrant Calliope and Resident Broad–tailed
            Hummingbirds." **Condor**, 72:234.
  1970b.   "Experimental Hypothermia in a Lesser Nighthawk." **Auk**, 87:372–374.
  1970c.   "The Occurrence and Status of Certain Anatids in Southern Nevada." **Condor**,
            72:474.
  1970d.   "Breeding Birds of Desert Riparian Habitat in Southern Nevada." **Condor**,
            72:431–436.
  1970e.   "Migration of Warblers in Southern Nevada." **The Southwestern Naturalist**,
            15:231–237.
  1971a.   "Roadside Distribution of the Common Raven in the Mohave Desert." **Califor-
            nia Birds**, 2:98.
  1971b.   "Body Composition and Organ Weights of the Verdin *(Auriparus flaviceps)*."
            Great Basin Naturalist, 31:66–68.

Austin, George T. and W. Glen Bradley
  1965.   "Bird Records from Southern Nevada." **Condor**, 67:445–446.
  1966.   "Additional Records for Uncommon Birds in Southern Nevada." **Great Basin
            Naturalist**, 26:41–42.
  1968.   "Bird Records for Clark County Nevada." **Great Basin Naturalist**, 28:61–62.
  1969.   "Additional Responses of the Poorwill to Low Temperatures." **Auk**, 86:717–
            725.

1971. "The Avifauna of Clark County, Nevada." **Journal of the Arizona Academy of Science.** Vol. 6:283–303.

Austin, George T. and Amadeo M. Rea
1976. "Recent Southern Nevada Bird Records." **Condor,** 78:405–408.

Bailey, A. M.
1928. "A Study of the Snowy Herons of the United States." **Auk,** 45:430–440.

Baldwin, G. C.
1944a. "Uncommon Birds from the Boulder Dam Area." **Condor,** 46:35.
1944b. "Unusual records of birds from the Boulder Dam Area." **Condor,** 46:206–207.
1947. "New Records for the Boulder Dam Area, Nevada." **Condor,** 49:85.

Bangs, O., and T. E. Penard
1921. "Descriptions of Six New Subspecies of American Birds." **Proc. Biol. Soc. Wash.,** 34:89–92.

Banks, Richard C.
1964. "Geographic Variations in the White–crowned Sparrow *Zonotrichia leucophrys.*" **Univ. Calif. Publ. Zool.** 70. 123 pp.
1968. "Annotated Bibliography of Nevada Ornithology Since 1951." **Great Basin Naturalist,** 28:49–60.

Banks, Richard C. and Charles G. Hansen
1970. "Bird Records from Southern Nevada." **Condor,** 72:109–110.

Banks, Richard C. and Donald E. Lewis
1969. "Bird Records from Nevada." **Condor,** 71:439–441.

Banks, Richard C. and R. Guy McCaskie
1964. "Distribution and Status of the Wied–crested Flycatcher in the Lower Colorado River Valley." **Condor,** 66:250–251.

Banks, Richard C., Roy W. McDiarmid and Alfred L. Gardner
1987. "Checklist of Vertebrates of the United States, the U.S. Territories and Canada." U.S. Department of the Interior, Fish and Wildlife Service. Resource Publication 166. Washington, D. C., pp. 79.

Behle, William H.
1942. "Distribution and Variation of the Horned Larks *(Otocoris alpestris)* of Western North America." **Univ. Calif. Publ. Zool.,** 46:205–316.
1956. "A Systematic Review of the Mountain Chickadee." **Condor,** 58:51–70.
1976. "Mohave Desert Avifauna in the Virgin River Valley of Utah, Nevada and Arizona." **Condor,** 78:40–48.

Behle, William H. and Robert K. Selander
1951. "A New Race of Dusky Grouse *(Dendragapus obscurus)* from the Great Basin." **Proc. Biol. Soc. Wash.** 64:125–128.

Bent, A. C.
1926. "Life Histories of North American Marsh Birds, Orders Odontoglossae, Herodiones and Paludicolae." **U.S. Nat. Mus., Bull.** 135:xii + 490.
1927. "Life Histories of North American Shore Birds, Order Limicolae (Part I)." **U.S. Nat. Mus., Bull.** 142:ix + 420.

Biale, Art
> Art Biale has been watching birds in the Eureka area since 1955 and recording them since 1964. From 1960–68, he banded numerous birds and he participated in the breeding bird survey from 1969–78. From 1953 to 1961 he served on the Nevada Fish and Game Commission. Information from his many pages of unpublished notes are cited in this book.

Bond, Gorman M.
    1963.   "Geographic Variations in the Thrush *Hylocichla ustulata.*" **Proc. U.S. National Museum**, 114:373–387.

Bond, Richard M.
    1940a.   "A Goshawk Nest in the Upper Sonoran Lifezone." **Condor**, 42:100–103.
    1940b.   "New Bird Records for Lincoln County, Nevada." **Condor**, 42:220–222.
    1942.   "Food of the Burrowing Owl in Western Nevada." **Condor**, 44:183.

Borrell, A. E. and Ellis, R.
    1934.   "Mammals of the Ruby Mountains Region of Northeastern Nevada." **Journal of Mammalogy**, 15:12–44.

Bradley, W. Glen
    1964.   "The Vegetation of the Desert Game Range with Special Reference to the Desert Bighorn." Trans. Desert Bighorn Council, 1964:43–67.
    1967.   "A Geographical Analysis of the Flora of Clark County, Nevada." **Jour. Ariz. Acad. Sci.**, 4:151–162.

Bradley, W. Glen and James E. Deacon
    1965.   "The Biotic Communities of Southern Nevada." Desert Research Institute, No. 9, Reprint Series, University of Nevada, Las Vegas, Nevada.
    1967.   "The Biotic Communities of Southern Nevada." Nevada State Museum Anthropological Papers, 14:201–295.

Broadbooks, H. E.
    1946.   "The American Redstart in Southern Nevada." **Condor**, 48:141–142.

Bump, Cardiner and Wayne H. Bohl
    1964.   "Summary of Foreign Game Bird Propagation and Liberations 1960 to 1963." Spec. Sci. Report Wildlife no. 80, Bureau of Sport Fisheries and Wildlife.

Bureau of Fisheries and Wildlife
    1956.   Birds of the Ruby Lake National Wildlife Refuge. Refuge Leaflet 156. Mimeo. 3 pp.
    1967a.   Ruby Lake National Wildlife Refuge. Refuge Leaflet. 98–R Mimeo. 6 pp.
    1967.   Birds of the Desert National Wildlife Range. Refuge Leaflet 132–R–3. Mimeographed, 4 pp.

Burleigh, Thomas D.
    1960a.   "Geographic Variation in the Western Wood–Pewee *(Contopus sordidulus).*" **Proc. Biol. Soc. Wash.**, 73:141–146.
    1960b.   "Three New Subspecies of Birds from Western North America." **Auk**, 77:210–215.
    1968.   "The Indian Tree–Pipit *(Anthus hodgsoni)* Recorded for the First Time in North America." **Auk**, 85:323.

— Thomas Burleigh collected numerous specimens in western Nevada for the United States National Museum. Records of these specimens have been included in the book.

Burleigh, Thomas D., and H. S. Peters
1948. "Geographic Variation in Newfoundland Birds." **Proc. Biol. Soc. Wash.** 61:111–126.

Castetter, Richard C. and Herbert O. Hill
1979. "Additions to the Birds of Nevada Test Site." **Western Birds**, 10:4

Chambers, Glenn D.
1965. Summary of Foreign Game Bird Propagation; 1964, and Liberations; 1960–1964. Supplement to Special Scientific Report–Wildlife No. 80, Bureau of Sport Fisheries and Wildlife, Washington, D. C.
1966. Summary of Foreign Game Bird Propagation; 1965, and Liberations; 1960–1965. Supplement to Special Scientific Report–Wildlife No. 80, Bureau of Sport Fisheries and Wildlife, Washington, D. C.

Christensen, Glen C.
1954. **The Chukar Partridge in Nevada.** Nevada Fish and Game Commission, Bio. Bull.7, pp. 77.
1958. The Effects of Drought and Hunting on the Chukar Partridge. Trans., 23rd North American Wildlife Conference. pp. 329–341.
1963a. Exotic Game Bird Introductions into Nevada. Nevada Fish and Game Commission, Biol. Bull. no. 3.
1963b. "Sandgrouse Released in Nevada Found in Mexico." **Condor**, 65:67–68.
1966. "Nevada's Experiences with the Himilayan Snow Partridge." Trans., Calif–Nev. Section of the Wildlife Society.

Christensen, Glen C. and Wayne H. Bohl
1964. A Study and Review of the Common Indian Sandgrouse. Spec. Sci. Report Wildlife No. 84, Bureau of Sport Fisheries and Wildlife.

Christensen, Glen C. and Tom T. Trelease
1941. "A New Record of the Semipalmated Plover in Nevada." **Condor**, 43:156.

Clark, William H., Keith I. Giezentanner and James L. Hainline
1974. "First Record of Sabine's Gull in Nevada." **The Wilson Bulletin**, 2:169.

Cooke, W. W.
1913. "Distribution and Migration of North American Herons and Their Allies." U.S. Dept. Agric., Biol. Surv., Bull, 45:1–70.

Cottam, Clarence
1929. "A Shower of Grebes." **Condor**, 31:80–81.
1936. "Notes on the Birds of Nevada." **Condor**, 38:122–123.
1941a. "California Cuckoo in Southeastern Nevada." **Condor**, 43:160.
1941b. "European Starling in Nevada." **Condor**, 43:293–294.
1942a. "Slate–colored Junco in Nevada." **Condor**, 44:185.
1942b. "Records from Extreme Northeastern Nevada." **Condor**, 44:127–128.
1947. "Some Bird Records for Southern Nevada." **Condor**, 49:244.
1954. "Bird Records for Nevada." **Condor**, 56:223–224.

Cruickshank, Allan D.
   1955–71. Christmas Bird Count. **Audubon Field Notes** and **American Birds**. Numerous
           records were taken from Christmas Bird Counts. The issues will not be listed
           separately.

Curran, Jim
           Jim Curran has worked for the Nevada Department of Wildlife since 1966. He is
           currently the Regional Fisheries Supervisor stationed at Fallon, Nevada. His
           observations, which cover many areas in the state, are contained in the book.

Delino, Thomas J. et al.
   1966.   "Toxicity of DRC–1330 to Starlings." **Journal of Wildlife Management**,
           30:249–153.

Dill, H. H.
   1946.   "Unusual Visitors at the Ruby Lake NWR, Nevada." **Condor**, 48:96–97.

Dillon, Richard
   1966.   **The Legend of Grizzly Adams: California's Greatest Mountain Man.**
           Coward–McCann, Inc., New York.

Drennan, Susan Roney and Gary Rosenburg
   1985.   Christmas Bird Count. **American Birds**, Vol. 39.

Drewien, Roderick C., Robert J. Oakleaf and William H. Mullens
   1975.   "The Sandhill Crane in Nevada." (in prep.)

Earhart, Caroline M. and Ned K. Johnson
   1970.   "Size Dimorphism and Food Habits of North American Owls." **Condor**,
           72:251–264.

Ellis, Elizabeth and Joanna Nelson
   1952.   **The Birds of the Carson River Basin.** Record–Courier Press, pp. 15.

Ellis, Ralph
   1935.   "Bird Records from Northeastern Nevada." **Condor**, 37:86–87.

Evans, Raymond N.
   1967.   "Nest Site Movements of a Poorwill." **The Wilson Bulletin**, 79:453.

Evenden, Fred G., Jr.
   1952.   "Additional Bird Records for Nevada." **Condor**, 54:174.

Evermann, B. W.
   1923.   "The Pelicans of Pyramid Lake." **Overland Monthly Magazine**, pp. 16–18.

Fisher, A. K.
   1893.   Report on the ornithology of the Death Valley Expedition of 1891, comprising
           notes on the birds observed in southern California, southern Nevada and parts of
           Arizona and Utah. **North American Fauna,** no. 7:7–158.

Ford, Homer S. and J. R. Alcorn
   1964.   "Observations of Golden Eagle Attacks on Coyotes." **Condor**, 66:76–77.

Fremont, J. C.
   1842.   The Exploring Expedition to the Rocky Mountains in the Year 1842, and to
           Oregon and North California in the years 1843–1844. Under the orders of Col.

J. J. Abert, Chief of the Topographical Bureau. Printed by order of the Senate of the United States. Gales and Seaton Printers, 1845, Washington.

Funk, Fred
1969.   **The Sparrows of Nevada.** Nevada State Museum, 47pp. illus., biblio. Carson City.

Gabrielson, I. N.
1935.   "A Nevada Record of the Harris' Sparrow." **Murrelet**, 16:41.
1949.   "Bird Notes from Nevada." **Condor**, 5:179–187.

Gabrielson, Ira N. and Stanley G. Jewett
1970.   **Birds of the Pacific Northwest.** Dover Publications.

Gaines, David
1974.   "Review of the Status of the Yellow–billed Cuckoo in California: Sacramento Valley Populations." **Condor**, 76:204–209.

Giles, LeRoy W., and David B. Marshall
1954.   "A Large Heron and Egret Colony on the Stillwater Wildlife Management Area, Nevada." **Auk**, 71:322–325.

Gladding, Hope M.
1941.   "Yellow–headed Blackbird Nests near Minden, Nevada." **Condor**, 43:202.

Grater, R. K.
1939a.   Preliminary Bird Checklist of the Boulder Dam Recreational Area. National Park Service, Boulder City, Nevada. mimeo.
1939b.   "New Bird Records for Nevada." **Condor**, 41:30.
1939c.   "New Bird Records for Nevada." **Condor**, 41:121.
1939d.   "Rose–breasted Grosbeak in Nevada." **Auk**, 56:191.
1939e.   "New Bird Records for Clark County, Nevada." **Condor**, 41:220–221.

Gromme, O. J.
1930.   A Soujourn Among the Wild Fowl of Pyramid Lake, Nevada. Public Museum of Milwaukee, Yearbook, 10:268–303.

Gullion, G. W.
1951.   "The Fulvous–tree Duck in Northeastern California." **Condor**, 53:158.
1952a.   "The Hudsonian Curlew in Nevada." **Condor**, 54:62.
1952b.   "Recent Bird Records from Southern Nevada." **Condor**, 54:204.
1952c.   Nevada's Experience with the Chukar Partridge. Proc. 32nd West Assoc. State Game and Fish Comm., pp. 149–153.
1953.   "Additional Bird Records from Southern Nevada." **Condor**, 55:160.
1954.   Sage Hen Population Shows Steady Decline in Northeastern Nevada. Nevada Fish and Game Bulletin, 1:5.
1956a.   "An Ancient Murrelet in Northeastern Nevada." **Condor**, 58:163.
1956b.   "Evidence of Double–brooding in Gambel Quail." **Condor**, 58:232–234.
1956c.   "The Current Status of the Starling in Nevada." **Condor**, 58:446.
1956d.   Let's Go Desert Quail Hunting. Nevada Fish and Game Commission, Biol. Bull. 2. 76 pp.
1957a.   "Miscellaneous Bird Records from Northeastern Nevada." **Condor**, 59:70–71.
1957b.   "Perecocial Strutting in Sage Grouse." **Condor**, 59:269.

1957c.   "Gambel Quail Disease and Parasite Investigations in Nevada." **American Midland Naturalist**, 57:414–420.

1960a.   "The Migratory Status of Some Western Desert Birds." **Auk**, 77:94–95.

1960b.   "The Ecology of Gambel's Quail in Nevada and the Arid Southwest." **Ecology**, 41:518–536.

1962.   "Organization and Movements of Coveys of Gambel Quail Population." **Condor**, 64:401–415.

1964.   Wildlife Uses of Nevada Plants. U.S. National Arboretum Contributions toward a flora of Nevada, no. 49, 170 pp. mimeo.

1965.   "A Critique Concerning Foreign Game Bird Introductions." **The Wilson Bulletin**, 77:409–414.

Gullion, G. W. and Glen C. Christensen

1957.   "A Review of the Distribution of Gallinaceous Game Birds in Nevada." **Condor**, 59:128–138.

Gullion, G. W. and Aredelle M. Gullion

1961.   "Weight Variations of Captive Gambel Quail in the Breeding Season." **Condor**, 63:95–97.

1964.   "Water Economy of the Gambel Quail." **Condor**, 66:32–40.

Gullion, G. W. and Leonard W. Hoskins

1956.   "Noteworthy Bird Records from Northeastern Nevada." **Condor**, 58:295.

Gullion, G. W., Warren M. Pulich and Fred G. Evenden

1959.   "Notes on the Occurrence of Birds in Southern Nevada." **Condor**, 61:278–297.

Gustafson, John Roger

1974.   Vertebrate Structure of a Sagebrush Community, Washoe County, Nevada. (April, 1974). A dissertation submitted in partial fulfillment of the requirements for the degree of Doctor of Philosophy in Biology, University of Nevada, Reno.

Hainline, James L.

1974.   The Distribution, Migration and Breeding of Shorebirds in Western Nevada. Master's Thesis. University of Nevada, Reno.

Hall, E. Raymond

1925.   "Pelicans Versus Fishes in Pyramid Lake." **Condor**, 17:147–160.

1926.   "Notes on Water Birds Nesting at Pyramid Lake, Nevada." **Condor**, 28:87–91.

1938.   "Inyo Screech Owl at Fallon, Nevada." **Condor**, 40:259.

Hanford, F. S.

1903.   "The Summer Birds of Washoe Lake, Nevada." **Condor**, 5:50–52.

Hanna, W. C.

1904.   "Nevada Notes." **Condor**, 6:47–48;76–77.

Hardy, Ross

1949.   "Ground Dove and Black–chinned Sparrow in Southern Nevada." **Condor**, 51:272–273.

Hayward, C. Lynn, Merlin L. Killpack and Gerald L. Richards

1963.   "Birds of the Nevada Test Site." Brigham Young University Science Bulletin., Biological Series 3 (1):1–27.

Heilbrun, Lois H. and Kenn Kaufman
  1976–79. Christmas Bird Count. **American Birds**, Vol. 30–33.

Heilbrun, Lois H. and Gary Rosenburg
  1981–83. Christmas Bird Count. **American Birds**, Vol. 35–37.

Heilbrun, Lois H. and Douglas Stotz
  1980.   Christmas Bird Count. **American Birds**, Vol. 34.

Henshaw, H. W.
  1877.   Report on the Ornithology of Portions of Nevada and California. Ann. Rep.
            Geog. Surv. West 100th Mer. by George M. Wheeler. App. N. N. of the Ann.
            Rep. Chief of Engineers for 1877; pp. 1303–1322.
  1880.   Ornithological Report from Observations and Collections made in Portions of
            California, Nevada and Oregon. Ann. Rep. Geog. Survey West 100th Mer. by
            George M. Wheeler. App. L. of the Ann. Rep. Chief Engineers for 1879 (Feb.,
            1880): pp. 282–335.

Herman, Steven G., John B. Bulger and Joseph B. Buchanan
  1980.   The Distribution and Abundance of Snowy Plovers in Southeastern Oregon and
            Western Nevada. Progress Report, 3 pps.

Herron, Gary
            Gary Herron has worked for the Nevada Department of Wildlife since 1970. His
            records, especially those about hawks and owls, were gained through personal
            interviews, and have been incorporated into the book.

Herron, Gary, Craig A. Mortimore and Marcus S. Rawlings
  1985.   **Nevada Raptors.** Biological Bulletin No. 8. Nevada Department of Wildlfe, 114
            pps.

Hoffman, W. J.
  1881.   Annotated List of the Birds of Nevada. Bull. U.S. Geol. and Geog. Surv. Terr.
            VI, no. 2:203–256.

Hollister, N.
  1908.   "Birds of Region about Needles, California." **Auk**, 25:455–462.

Howard, Hildegarde
  1939.   "A Prehistoric Record of Holboell Grebe in Nevada." **Condor**, 41:32.
  1958.   "An Ancient Cormorant from Nevada." **Condor**, 60:411–413.

Hubbard, John P.
  1973–75. Christmas Bird Count. **American Birds**, Vol. 27–29.

Jaeger, L. C.
  1927.   "Birds of the Charleston Mountains of Nevada." **Occasional Papers**, Riverside
            Junior College, 2:1–8.

Jewett, S. G.
  1940a.   "Bendire's Thrasher in Lincoln County, Nevada." **Condor**, 42:126.
  1940b.   "The Lark Bunting in Southern Nevada." **Condor**, 42:307.

Johnson, Archibald
  1935.   "The Common Loon in Nevada." **Condor**, 37:286.
  1936.   "The Sparrow Hawk a Pet Among the Paiutes." **Auk**, 53:210.

Johnson, Ned K.
  1952.   "Additional Records of the Rough–legged Hawk in Nevada." **Condor**, 54:65.
  1953.   "Dipper Eaten by Brook Trout." **Condor**, 55:158.
  1954a.   "Food of the Long–eared Owl in Southern Washoe County, Nevada." **Condor**,
            56:52.
  1954b.   "Notes on Some Nevada Birds." **Great Basin Naturalist**, 14:15–18.
  1956a.   "Birds of the Pinon Association of the Kawich Mountains, Nevada." **Great
            Basin Naturalist**, 16:32–33.
  1956b.   "Recent Bird Records for Nevada." **Condor**, 58:449–452.
  1963.   "Biosystematics of Sibling Species of Flycatcher in the *Empidonax hammondii–
            oberholseri–wrightii* Complex." Univ. Calif. Publ. Zool. 66:79–238.
  1965.   "The Breeding Avifaunas of the Sheep and Spring Ranges in Southern Nevada."
            Condor, 67:93–124.
  1972.   "Breeding Distribution and Habitat Preference of the Gray Vireo in Nevada."
            California Birds, 3:73–78.
  1973.   "The Distribution of Boreal Avifaunas in Southern Nevada." **Occasional
            Papers. Biol. Society of Nevada, no. 36.**
  1974.   "Montane Avifaunas of Southern Nevada: Historical Change in Species Com-
            position." **Condor**, 76:334–337.
  1975.   "Controls of Number of Bird Species on Montane Islands." **Great Basin
            Evolution**, 29:545–567.
  1976.   "Breeding Distribution of Nashville and Virginia Warblers." **Auk**, 93:219–
            230.

Johnson, Ned K. and Richard C. Banks
  1959.   "Pine Grosbeak and Lawrence Goldfinch in Nevada." **Condor**, 61:303.

Johnson, Ned K. and Frank Richardson
  1952.   "Supplementary Bird Records for Nevada." **Condor**, 54:358–359.

Johnson, Ned K. and Ward C. Russell
  1962.   "Distributional Data on Certain Owls in the Western Great Basin." **Condor**,
            64:513–514.

Johnson, Richard E.
  1975.   "New Breeding Localities for Leucosticte in the Contiguous Western United
            States." **Auk**, 92:586–589.

Kelleher, James V.
  1970.   The Ecological Distribution of the Summer Birds of the Carson Range (eastern
            Sierra) in the Area of the Lake Tahoe, Nevada State Park. Masters Thesis,
            University of Nevada, Reno.

Kelleher, James V. and William F. O'Malia
  1971.   "Golden Eagle Attacks a Mallard." **Auk**, 88:186.

Kingery, Hugh E.
  1972–87. From 1972 to present, Kingery has been editor of the "Mountain West" section
            of **American Birds** (formerly **Audubon Field Notes**), a publication of bird
            records and sightings in the Americas. The "Mountain West" section contains
            records supplied by many contributors for Nevada and neighboring states.
            Because of the many volumes and issues cited, they will not be listed separately.

Kingery, Hugh E. and Charles S. Lawson
  1985.   Mountain West. **American Birds**, 39:330–335.

Knopf Fritz L. and Joseph L. Kennedy
  1980.   "Foraging Sites of White Pelicans Nesting at Pyramid Lake, Nevada." **Western Birds**, 11:175–180.

Lanner, Ronald M.
  1981.   **The Pinon Pine: A Natural and Cultural History.** UNR Press, pp. 45–48.

LaRivers, I.
  1941.   "The Mormon Cricket as Food for Birds." **Condor**, 43:65–69. 1962.
          Fishes and Fisheries in Nevada. Carson City, NV, Nevada State Fish and Game Commission. pp. 1–782.

LaRochelle, O. E. (Larry)
          Larry LaRochelle worked as a wildlife biologist at Ruby Lake National Wildlife Refuge for the U.S. Fish and Wildlife Service from June, 1962 to January, 1967. His unpublished notes and personal communications about Nevada birds are included in the text.

Lawson, Charles S.
  1973.   "Notes on Pelecaniformes in Nevada." **Western Birds**, 4:23–30.
  1973.   "Charadriiformes New to Nevada." **Western Birds**, 4:77–82.
  1974.   "First Nevada Record of Chestnut–collared Longspur." **Auk**, 91:432.
  1975.   "Fish Catching by a Black Phoebe." **Western Birds**, 6:107–109.
  1976.   In 1976, Chuck Lawson taped his comments on the birds of southern Nevada. The notes were transcribed and incorporated into the book.
  1977.   "Nonpasserine species new cr unusual to Nevada." **Western Birds**, 8:73–90.
  1977.   "Stilt Sandpiper and Hudsonian Godwit in Nevada." **Great Basin Naturalist**, 37:532.
  1979.   "Nevada Records of Roseate Spoonbill." **Western Birds**, 10:166.

Leopold, Starker, A.
  1977.   **The California Quail.** University of California Press, 281 pps.

Leukering, Tony and Gary H. Rosenburg
  1987.   Christmas Bird Count. **American Birds**, 41:4.

Lewis, Donald E.
          Donald Lewis collected many specimens from the Ruby Lake National Wildlife Refuge. These specimens are reported in this book, to the author's knowledge, for the first time.

Linsdale, Jean M.
  1936.   "The Birds of Nevada." **Pacific Coast Avifauna**, 23:1–145.
  1937.   "The Natural History of Magpies." **Pacific Coast Avifauna**, 25:1–234.
  1938a.   "Environmental Responses of Vertebrates in the Great Basin." **American Midland Naturalist**, 19:1–216.
  1938b.   "Geographic Variation in Some Birds in Nevada." **Condor**, 40:36–38.
  1938c.   "Bird Life in Nevada with Reference to Modification in Structure and Behavior." **Condor**, 40:173–180.
  1951.   "A List of the Birds of Nevada." **Condor**, 53:228–249.

Long, Pauline and Florence E. Poyser
   1965.   "A Record of the Groove–billed Ani in Southern Nevada. **Condor**, 67:357–358.

Lugaski, Thomas, David Woreley and Edgar Kleiner
   1972.   "The First Confirmed Occurrence of White–tailed Kites." **Occasional Papers.**
            Biological Society of Nevada, No. 28.

MacDonald, Duncan and Robert A. Jantzen
   1967.   "Management of Merriam's Turkey." **The Wild Turkey and its Management**,
            Oliver H. Hewitt, ed. The Wildlife Society, Washington, D. C.

Manville, Richard H.
   1963.   "Altitude Record for Mallard." **The Wilson Bulletin**, 75:92.

Marshall, David B.
   1951.   "New Bird Records for Western Nevada." **Condor**, 53:157–158.

Marshall, David B. and J. R. Alcorn
   1952.   "Additional Nevada Bird Records." **Condor**, 54:320–321.

Marshall, David B and LeRoy W. Giles
   1953.   "Recent Observations of Birds of Anaho Island, Pyramid Lake, Nevada."
            Condor, 55:105–116.

Marshall, Joe T., Jr.
   1942.   "*Melospiza melodia virginis* a Synonym of *Melospiza melodia fallax.*" **Condor**,
            44:233.
   1967.   Parallel Variation in North and Middle American Screech Owls. Monog. Western
            Foundations of Vertebrate Zoologyy, no 1. 72 pps.

Marshall, Joe T., Jr. and W. H. Behle
   1942.   "The Song Sparrows of the Virgin River Valley, Utah." **Condor**, 44:122–124.

McCaskie, R. G. and R. C. Banks
   1964.   "Occurrence and Migration of Certain Birds in Southeastern California." **Auk**,
            81:353–361.

Medin, Dean E. 1987.
            "Breeding Birds of an Alpine Habitat in the Southern Snake Range, Nevada."
            Western Birds, 18:163–168.

Mewaldt, Dr. L. Richard
   1979.   List of Birds for Northeastern California and Western Nevada. (In preparation).

Miller, Alden H.
   1935.   "Some Breeding Birds of the Pine Forest Mountains, Nevada." **Auk**, 52:467–
            468.
   1941a.   "Racial Determination of Bewick Wrens in the Western Great Basin Region."
            Condor, 43:250–251.
   1941b.   "A Review of Centers of Differentiation for Birds in the Western Great Basin
            Region." **Condor**, 43:257–267.
   1945.   "Birds of the Yellow Pine Association of Potosi Mountain, Southern Nevada."
            Condor, 43:130–131.
   1946.   "Vertebrate Inhabitants of the Pinon Association in the Death Valley Region."
            Ecology, 27:54–60.

1951.   "An Analysis of the Distribution of the Birds of California." **University of California Publ. Zool.** 50:531–644.

1955.   "The Breeding Range of the Black–rosy Finch." **Condor**, 57:306–307.

Miller, Alden H. and Loye Miller
1951.   "Geographic Variation of the Screech Owls of the Deserts of Western North America." **Condor**, 53:161–177.

Miller, Alden H. and Ward C. Russell
1956.   "Distributional Data on the Birds of the White Mountains of California and Nevada." **Condor**, 58:75–77.

Miller, Loye
1931.   "The California Condor in Nevada." **Condor**, 33:32.

Mills, Vernon
        At one time Nevada's only state game warden. Vernon Mills has spent his lifetime observing birds in the Fallon area. Over the past 50 years, the author has spent many hours in the field with Mills. His many observations and communications are included in the book.

Monson, Gale
1949–73. From 1949–54, 1959–60, 1964 and 1972–73, Gale Monson edited the "Southwest Region" section for **Audubon Field Notes** (now **American Birds**). The "Southwest Region" section contained records from many contributors for the southern tip of Nevada. Because a large number of issues are cited, they will not be listed separately.

Mowbray, Vince
1986.   Nevada State Museum Checklist of the Birds of Nevada. In preparation.
—   Vince Mowbray has spent many years in Nevada in the field birdwatching. Since 1962, he has lived in Las Vegas where he spends much of his time observing birds. His efforts have contributed much to knowledge of birds in southern Nevada.

Napier, Larry D.
        Larry Napier worked as a wildlife biologist for the U.S. Fish and Wildlife Service at Stillwater Wildlife Management Area near Fallon, Nevada from 1967 to 1974. His unpublished notes and frequent personal conversations add considerably to the information about birds in the Lahontan Valley.

Nappe, Leontine and Don A. Klebenow
1973.   Rare and Endangered Birds of Nevada. Agricultural Experiment Station. Max C. Fleischmann College of Agriculture, University of Nevada, Reno and Foresta Institute of Ocean and Mountain Studies, Carson City, Nevada.

Narrative Reports
1969–72. Ruby Lake National Wildlife Refuge. The reports from 1969 to 1972, submitted by Lynn C. Howard and a section by Lowell L. Napier.

Neal, Larry
        Larry Neal is a non–game biologist with the Nevada Department of Wildlife. Information from personal conversations and notes has been incorporated into the book.

Nelson, E. W.
  1875.   "Notes on Birds Observed in Portions of Utah, Nevada and California." **Proc.,**
           Bost. Soc. Nat. Hist., 17:338–365.

Oakleaf, Robert J.
  1974a.   Species Distribution and Key Habitats of Selected Non–game Species. Nevada
           Department of Fish and Game, Reno, Nevada. September job performance
           report.
  1974b.   Non–game Population Surveys. Nevada Department of Fish and Game, Reno,
           Nevada. October job performance report.

Oakleaf, Robert J. and Donald A. Klebenow
  1975.   Changes of Avifauna Populations on the Lower Truckee River. Special Report,
           Federal Aid in Wildlife Restoration, pp. 19.

Oates, E. W.
  1902.   Catalogue of the Collection of Birds' Eggs in the British Museum (Natural
           History). Vol. II. Carinatae (Charadriiformes–Strigiformes). London, published
           by the British Museum, 400 pp.

Oberholser, H. C.
  1911.   "A Revision of the Forms of the Hairy Woodpecker *(Dryobates vilosus*
           linnaeus)." **Proc. USNM,** 40:595–621.
  1914.   "A Monograph of the Genus *Chordeiles swainson*, Type of a New Family of
           Goatsuckers." **USNM Bulletin,** 86:1–120.
  1918a.   "Notes on the Subspecies of *Numenius americanus bechstein*." **Auk,** 35:188–
           195.
  1918b.   "The Northernmost Record of *Icterus parisorum*." **Auk,** 35:481.
  1918c.   "The Common Ravens of North America." **Ohio State Journal of Science,**
           18:213–225.
  1918d.   "New Light on the Status of *Empidonax traillii auduboni*." **Ohio Journal of**
           Science, 18:85–98.

Orr, Robert T., and James Moffitt
  1971.   **Birds of the Lake Tahoe Region.** California Academy of Science: San
           Francisco, California, pp. 150.

Osugi, Cathy T.
  1973.   Monitoring Program of Wildlife Habitat and Associated Use in the Truckee–
           Carson Irrigation District, Nevada. Progress Report no. 1. Report of Wildlife
           Management Study.
  1974.   Monitoring Program of Wildlife Habitat and Associated Use in the Truckee–
           Carson Irrigation District, Nevada. Progress Report no. 2. Report of Wildlife
           Management Study.

Panik, Howard Ronald
  1976.   The Vertebrate Structure of a Pinon–juniper Commmunity in Northwestern
           Nevada. PhD Dissertation, University of Nevada, Reno.

Page, Jerry and Donald Seibert
  1973.   "Inventory of Golden Eagle nests in Elko County, Nevada." Cal–Nev. Wildlife
           Transactions Western Sections of the Wildlife Society, p. 1–8.

Parmenter, H. E.
1924. "The Gray Jay at Lake Tahoe." **Condor**, 26:72.

Pelican, The
1964–74. The Lahontan Audubon Society, Dr. Fred Ryser, ed., 1:2–11:2.

Phillips, A. R.
1947. "The Races of MacGillivray's Warbler." **Auk**, 64:296–300.
1948. "Geographic Variation of *Empidonax traillii*." **Auk**, 65:507–514.

Phillips, Allan, Joe Marshall and Gale Monson
1964. **The Birds of Arizona.** University of Arizona Press, Tuscon, AZ. 212 pp.

Pulich, Warren M.
1952. "The Arizona Crested Flycatcher in Nevada." **Condor**, 54:169–170.

Pulich, Warren M. and Gordon W. Gullion
1953. "Black–and–White Warbler, Dickcissel and Tree Sparrow in Nevada." **Condor**, 55:215.

Pulich, Warren M. and Allan R. Phillips
1951. "Autumn Bird Notes from the Charleston Mountains, Nevada." **Condor**, 53:205–206.
1953. "A Possible Flight Line of the American Redstart." **Condor**, 55:99–100.

Ray, M. S.
1910. "From Tahoe to Washoe." **Condor**, 12:85–89.

Remsen, J. R., Jr. and Laurence C. Binford
1975. "Status of the Yellow–billed Loon *(Gavia adamsii)* in the Western United States and Mexico." **Western Birds**, 6:19.

Richards, Gerald L.
1962. "Wintering Habits of Some Birds at the Nevada Atomic Test Site." **Great Basin Naturalist**, 22:30–31.
1965. "Prairie Falcon Imitates Flight Pattern of the Loggerhead Shrike." **Great Basin Naturalist**, 25:48.
1971. "The Common Crow, *Corvus brachyrhynchos*, in the Great Basin." **Condor**, 73:116–118.

Richardson, Frank
1952. "A Second Record of the Indigo Bunting in Nevada." **Condor**, 54:63.

Rickard, W. H.
1960. "An Occurrence of the Rose–breasted Grosbeak in Southern Nevada." **Condor**, 62:140.
1961. "Notes on Bird Nests Found in a Desert Shrub Community Following Nuclear Detonations." **Condor**, 63:265–266.

Ridgway, R.
1874. "Breeding Ground of White Pelicans at Pyramid Lake, Nevada." **American Sportsman**, 4:19.
1877. United States Geological Exploration of the Fortieth Parallel. Clarence King, Geologist in charge. Part III: Ornithology, p. 303–669.

1884.   Catalogue of the Aquatic and Fish–eating Birds Exhibited by the United States National Museum. Bull. 27., **U.S. National Museum**, pp. 139–184.

1901–19. "The Birds of North and Middle America." **U.S. National Museum Bull.**, 50.

Rogers, Glenn E.
1963.   "Blue Grouse Census and Harvest in the United States and Canada." **Journal of Wildlife Management**, 27:579–585.

Rubega, Margaret A. and David Stejskal
1984.   Christmas Bird Count. **American Birds**, Vol. 38.

Ryder, R. A.
1967.   Ibises–Bird Banding. A Journal of Ornithological Investigation. vol. xxxviii, October, 1967, no.4 pp. 257–277.

Ryser, Fred A.
1963.   "Prothonotary Warbler and Yellow–shafted Flicker in Nevada." **Condor**, 65:334.

1964–75. **The Pelican. The Pelican** was a quarterly publication put out by the Lahontan Audubon Society from 1964–1975. Dr. Fred Ryser served as editor. Many new and significant bird records were published in these papers.

1970.   Checklist of the Birds of Nevada. University of Nevada, February 22, 1970.

1985.   **Birds of the Great Basin: A Natural History.** University of Nevada Press, 604 pps.

Saake, Norm
          Norm Saake is currently working with the Nevada Department of Wildlife as the state waterfowl biologist, a position he has held since 1967. Stationed at Fallon, Nevada, he makes frequent statewide waterfowl surveys by air and land. Some of his observations, gained through personal interviews, are included in the book.

Salvadori, T.
1895.   Catalogue of the Chenomorphane (Palamedeae, Phoenicopteri, Anseres), Crypturi and Ratitae in the Collection of the British Museum. Cat. Birds, 17:i–xv, 1–636.

Schultz, Vincent
1966.   "References on Nevada Test Site Ecological Literature." **Great Basin Naturalist**, 26:79–86.

Schwabenland, Peter
          Peter Schwabenland worked as a wildlife biologist for the U.S. Fish and Wildlife Service at Stillwater Wildlife Management Area near Fallon, Nevada from 1963–1966. His unpublished notes and personal communications appear in the book.

Scott, Oliver K., Dr.
1959–71. From 1959 to 1971, Dr. Scott edited the "Great Basin, Central Rocky Mountain Region" of **Audubon Field Notes** (now **American Birds**). This region covered all of Nevada except for the extreme southern tip. Because of the many records cited, the issues will not be listed separately.

Selander, Robert K.
1954.   "A Systematic Review of the Booming Nighthawks of Western North America." **Condor**, 56:57–82.

Short, L. L. Jr.
  1965.   "Hybridization in the flickers *(Colaptes)* of North America." **Bull. American Museum of Natural History**, 129:309–428.

Slipp, J. W.
  1942.   "Notes on the Stilt Sandpiper in Washington and Nevada." Murrelet, 22:61–62.

Smiley, D. C.
  1937.   "Water Birds of the Boulder Dam Region." **Condor**, 39:115–119.

Smith, Bill
  1966.   "A Second Record of Ancient Murrelet from Nevada." **Condor**, 68:511–512.

Snider, Patricia R.
  1965–71. From 1965–1971, Patricia Snider edited the "Southwest Region" of **Audubon Field Notes** (now **American Birds**). The "Southwest Region" covered the extreme southern tip of Nevada. Because of the many issues used for records, they will not be listed separately.

Stiver, San Juan
          San Stiver has served as a staff biologist for the Nevada Department of Wildlife since 1970. Some of his unpublished notes and personal conversations with the author have been incorporated into the book.

Taylor, W. P.
  1912.   "Field Notes on Amphibians, Reptiles and Birds of Northern Humboldt County, Nevada with Discussion of Some of the Faunal Features of the Region." **Univ. Calif. Publ. Zool.**, 7:39–319–436.

Thompson, Steven P.
          Steven Thompson has worked as a wildlife biologist for the U.S. Fish and Wildlife Service at Stillwater Wildlife Management Area near Fallon, Nevada since August, 1985. He has provided many records from his unpublished notes and his conversations with the author.

Tsukamoto, George K.
  1966.   "Some Notes on Birds of Elko County, Nevada." **Condor**, 68:103–104.

Unitt, Philip
  1987.   *"Empidonax traillii extimus*: An Endangered Subspecies." **Western Birds**, 18:137–162.

U.S. Fish and Wildlife Service
  1981.   Birds: Ruby Lake National Wildlife Refuge, RF14570–2. U.S. Fish and Wildlife Service, U.S. Department of Interior.
  1982.   Wildlife at Sheldon National Wildlife Refuge, Nevada (pamphlet).

van Rossem, A. J.
  1930.   "The Races of *Auriparus flaviceps* (Sundevall)." **Trans. San Diego Soc. Nat. Hist.**, 6:199–202.
  1931.   "Descriptions of New Birds from the Mountains of Southern Nevada." **Trans. San Diego Soc. Nat. Hist.**, 6:325–332.
  1936a. "Birds of the Charleston Mountains, Nevada." **Pacific Coast Avifauna**, 24:1–65.

1936b.   "The Bushtit of the Southern Great Basin." **Great Basin Naturalist**, 153:85–86.
1937.   "A Review of the Races of Mountain Quail." **Condor**, 39:20–24.
1942.   "Four New Woodpeckers from Western United States and Mexico." **Condor**, 44:22–26.
1946.   "Two New Races of Birds from the Lower Colorado River Valley." **Condor**, 48:80–82.

Voget, Kenneth W.
          Kenneth Voget is the assistant complex manager at Sheldon–Hart National Wildlife Refuge. He provided bird records for the wildlife refuge from his personal notes and refuge records.

Walkinshaw, L. H.
1973.   **The Cranes of the World.** Winchester Press: New York.

Watkins, J. R.
1968.   Avian Ecology of Ponderosa Pine forests Sheep Range Mountains, Clark County, Nevada. Unpubl. Masters Thesis, Northern Arizona University.

Wauer, Roland H.
1969.   "Recent Bird Records from the Virgin River Valley of Utah, Arizona and Nevada." **Condor**, 71:331–335.

Wauer, Roland H. and Richard C. Russell
1967.   "New and Additional Records of Birds in the Virgin River Valley." **Condor**, 69:420–423.

Weaver, Harold R. and William L. Haskell
1967.   "Some Fall Foods of Nevada Chukar Partridge." **Journal of Wildlife Management**, 31:582–584.

Wheat, Margaret M.
1967.   **Survival Arts of the Primitive Paiutes.** University of Nevada Press, Reno, 1967. p. 9–12.

Wick, William Q.
1955.   "A Recent Record of the Sharp–tailed Grouse in Nevada." **Condor**, 57:243.

Wilson, V. T. and R. H. Norr
1949–   From 1949–1951, Wilson and Norr edited the "Great Basin–Central Rocky
1951.   Mountain Region" of **Audubon Field Notes** (now **American Birds**). Because many issues are cited, they will not be listed separately.

Wolfe, L. R.
1937.   "The Duck Hawk Breeding in Nevada." **Condor**, 39:25.

Woodbury, W. Verne
1966.   The History and Present Status of Biota of Anaho Island, Pyramid Lake, Nevada. University of Nevada, Reno. A thesis submitted in partial fulfillment of the requirements for the degree of Master of Science in Biology.

Wooten, Michael and David B. Marshall
1965.   "Heermann Gull in Nevada." **Condor**, 67:83–84.

Zimmerman, D. A.
1962.   Spring Migration, Southwest Region. **Audubon Field Notes**, 16:436–439.

Index

409